Günter P. Merker · Christian Schwarz · Gunnar Stiesch · Frank Otto

Simulating Combustion

Günter P. Merker · Christian Schwarz
Gunnar Stiesch · Frank Otto

Simulating Combustion

Simulation of combustion and pollutant formation
for engine-development

with 242 figures

Prof. Dr.-Ing. habil Günter P. Merker
Universität Hannover
Institut Technische Verbrennung
Welfengarten 1 A
30167 Hannover
merker@itv.-uni-hannover.de

Prof. Dr.-Ing. Christian Schwarz
BMW-Group, EA 31
Hufelandstr. 8a
80788 München
christian.schwarz@bmw.de

Dr.-Ing. habil Gunnar Stiesch
MTU Friedrichshafen GmbH, Abtl. TKV
Maybachplatz 1
88045 Friedrichshafen
gunnar.stiesch@mtu-online.com

Dr. rer. nat. Frank Otto
DaimlerChrysler AG, Abtl. HPC G252
70546 Stuttgart
frank.otto@daimlerchrysler.com

Library of Congress Control Number: 2005933399

ISBN 10 3-540-25161-8 Berlin Heidelberg New York
ISBN 13 978-3-540-25161-3 Berlin Heidelberg New York

Originally published as
Günter P. Merker/Christian Schwarz/Gunnar Stiesch/Frank Otto:
Verbrennungsmotoren. Simulation der Verbrennung und Schadstoffbildung

© B. G. Teubner Verlag / GWV Fachverlage GmbH, Wiesbaden, 2004

This work is subject to copyright. All rights are reserved, whether the whole or part of the material is concerned, specifically the rights of translation, reprinting, re-use of illustrations, recitation, broadcasting, reproduction on microfilms or in any other way, and storage in data banks. Duplication of this publication or parts thereof is permitted only under the provisions of the German Copyright Law of September 9, 1965, in its current version, and permission for use must always be obtained from Springer. Violations are liable to Prosecution under the German Copyright Law.
Springer is a part of Springer Science+Business Media
springeronline.com

© Springer-Verlag Berlin Heidelberg 2006
Printed in Germany

The use of general descriptive names, registered names, etc. in this publication does not imply, even in the absence of a specific statement, that such names are exempt from the relevant protective laws and regulations and free for general use.
Cover design: medionet AG, Berlin
Typesetting: Digital data supplied by editors

Printed on acid-free paper   68/3020/m-5 4 3 2 1 0

Preface

The numerical simulation of combustion processes in internal combustion engines, including also the formation of pollutants, has become increasingly important in the recent years, and today the simulation of those processes has already become an indispensable tool when developing new combustion concepts. While pure thermodynamic models are well-established tools that are in use for the simulation of the transient behavior of complex systems for a long time, the phenomenological models have become more important in the recent years and have also been implemented in these simulation programs. In contrast to this, the three-dimensional simulation of in-cylinder combustion, i.e. the detailed, integrated and continuous simulation of the process chain injection, mixture formation, ignition, heat release due to combustion and formation of pollutants, has been significantly improved, but there is still a number of challenging problems to solve, regarding for example the exact description of sub-processes like the structure of turbulence during combustion as well as the appropriate choice of the numerical grid.

While chapter 2 includes a short introduction of functionality and operating modes of internal combustion engines, the basics of kinetic reactions are presented in chapter 3. In chapter 4 the physical and chemical processes taking place in the combustion chamber are described. Chapter 5 is about phenomenological multi-zone models, and in chapter 6 the formation of pollutants is described. In chapter 7 and chapter 8 simple thermodynamic models and more complex models for transient systems analyses are presented. Chapter 9 is about the three-dimensional simulation of combustion processes in engines.

We would like to thank Dr. B. Settmacher for reviewing and formatting the text, for preparing the layout, and for preparing the printable manuscript. Only due to her unremitting dedication and her excellent time management the preparation of this book has been possible in the given timeframe. Further on, we would also like to thank Mrs. C. Brauer for preparing all the figures and diagrams contained in this book. The BMW group and the DaimlerChrysler AG contributed to this book by releasing the figures they provided. Last but not least, we would like to thank the Springer-Verlag for the always excellent collaboration.

This book is largely a translation of the second German edition, which has been published in 2004 by the B.G. Teubner-Verlag, whereas the text has been updated if necessary. We would like to thank Mr. Aaron Kuchle for translating the text into English.

Hannover/München/Friedrichshafen/Stuttgart, July 2005

Günter P. Merker
Christian Schwarz
Gunnar Stiesch
Frank Otto

Table of contents

Abbreviations and symbols   XII

1 Introduction   1
    1.1 Preface   1
    1.2 Model-building   1
    1.3 Simulation   2

2 Introduction into the functioning of internal combustion engines   5
    2.1 Energy conversion   5
    2.2 Reciprocating engines   6
        2.2.1 The crankshaft drive   7
        2.2.2 Gas and inertia forces   9
        2.2.3 Procedure   11
    2.3 Thermodynamics of the internal combustion engine   12
        2.3.1 Foundations   12
        2.3.2 Closed cycles   17
        2.3.3 Open comparative processes   25
    2.4 Characteristic qualities and characteristic values   28
    2.5 Engine maps   31
        2.5.1 Spark ignition engines   31
        2.5.2 Diesel engines   33
    2.6 Charging   35
        2.6.1 Charging methods   35
        2.6.2 Supercharging   37
        2.6.3 Constant-pressure turbocharging   38
        2.6.4 Pulse turbocharging   41

3 Foundations of reaction kinetics   44
    3.1 Chemical equilibrium   44
    3.2 Reaction kinetics   47
    3.3 Partial equilibrium and quasi-steady-state   48
    3.4 Fuels   50
        3.4.1 Chemical structure   50
        3.4.2 Physical and chemical properties   53
    3.5 Oxidation of hydrocarbons   56

4 Engine combustion — 60

- 4.1 Spark ignition engines — 60
  - 4.1.1 Mixture formation — 60
  - 4.1.2 Ignition — 63
  - 4.1.3 The combustion process — 65
  - 4.1.4 Abnormal combustion — 69
  - 4.15 Controlled autoignition — 70
- 4.2 Diesel engines — 72
  - 4.2.1 Injection methods and systems — 73
  - 4.2.2 Mixture formation — 80
  - 4.2.3 Autoignition — 81
  - 4.2.4 Combustion — 83
  - 4.2.5 Homogeneous combustion — 86
- 4.3 Pressure trace analysis — 88
  - 4.3.1 Determination of the heat release rate — 88
  - 4.3.2 Loss distribution — 92
  - 4.3.3 Comparison of various combustion processes — 95

5 Phenomenological combustion models — 98

- 5.1 Diesel engine combustion — 98
  - 5.1.1 Zero-dimensional heat release function — 98
  - 5.1.2 Stationary gas jet — 99
  - 5.1.3 Packet models — 104
  - 5.1.4 Time scale models — 111
- 5.2 SI engine combustion — 113

6 Pollutant formation — 116

- 6.1 Exhaust gas composition — 116
- 6.2 Carbon monoxide (CO) — 117
- 6.3 Unburned hydrocarbons (HC) — 118
  - 6.3.1 Limited pollutant components — 118
  - 6.3.2 Non-limited pollutant components — 122
- 6.4 Particulate matter emission in the diesel engine — 127
  - 6.4.1 Introduction — 127
  - 6.4.2 Polycyclic aromatic hydrocarbons (PAH) — 128
  - 6.4.3 Soot development — 129
  - 6.4.4 Particle emission modeling — 131
- 6.5 Nitrogen oxides — 132
  - 6.5.1 Thermal NO — 133
  - 6.5.2 Prompt NO — 138
  - 6.5.3 NO formed via N_2O — 140
  - 6.5.4 Fuel nitrogen — 140

7 Calculation of the real working process — 141

7.1 Single-zone cylinder model — 142
- 7.1.1 Fundamentals — 142
- 7.1.2 Mechanical work — 144
- 7.1.3 Determination of the mass flow through the valves / valve lift curves — 144
- 7.1.4 Heat transfer in the cylinder — 147
- 7.1.5 Heat transfer in the exhaust manifold — 156
- 7.1.6 Wall temperature models — 157
- 7.1.7 The heat release rate — 160
- 7.1.8 Knocking combustion — 174
- 7.1.9 Internal energy — 178

7.2 The two-zone cylinder model — 187
- 7.2.1 Modeling the high pressure range according to Hohlbaum — 187
- 7.2.2 Modeling the high pressure phase according to Heider — 190
- 7.2.3 Results of NOx calculation with two-zone models — 193
- 7.2.4 Modeling the charge changing for a two-stroke engine — 195

7.3 Modeling the gas path — 197
- 7.3.1 Modeling peripheral components — 197
- 7.3.2 Model building — 199
- 7.3.3 Integration methods — 200

7.4 Gas dynamics — 201
- 7.4.1 Basic equations of one-dimensional gas dynamics — 201
- 7.4.2 Numerical solution methods — 205
- 7.4.3 Boundary conditions — 208

7.5 Charging — 214
- 7.5.1 Flow compressor — 214
- 7.5.2 The positive displacement charger — 224
- 7.5.3 The flow turbine — 225
- 7.5.4 Turbochargers — 236
- 7.5.5 Charge air cooling — 239

8 Total process analysis — 245

8.1 General introduction — 245

8.2 Thermal engine behavior — 245
- 8.2.1 Basics — 245
- 8.2.2 Modeling the pipeline system — 246
- 8.2.3 The cooling cycle — 248
- 8.2.4 The oil cycle — 251
- 8.2.5 Physical properties of oil and coolant — 256

8.3 Engine friction — 257
- 8.3.1 Friction method for the warm engine — 257
- 8.3.2 Friction method for the warm-up — 258

8.4	Engine control	261
	8.4.1 PID controller	261
	8.4.2 Load control	261
	8.4.3 Combustion control	262
	8.4.4 Control of exhaust gas recirculation	262
	8.4.5 Charger aggregate control	264
	8.4.6 The driver controller	266
8.5	Representing the engine as a characteristic map	267
	8.5.1 Procedure and boundary conditions	267
	8.5.2 Reconstruction of the torque band	269
8.6	Stationary simulation results (parameter variations)	272
	8.6.1 Load variation in the throttled SI engine	273
	8.6.2 Influence of ignition and combustion duration	274
	8.6.3 Variation of the compression ratio, load, and peak pressure in the large diesel engine	276
	8.6.4 Investigations of fully variable valve trains	277
	8.6.5 Variation of the intake pipe length and the valve control times (SI engine, full load)	279
	8.6.6 Exhaust gas recirculation in the turbocharged diesel engine of a passenger car	279
	8.6.7 Engine bypass in the large diesel engine	283
8.7	Transient simulation results	285
	8.7.1 Power switching in the generator engine	285
	8.7.2 Acceleration of a commercial vehicle from 0 to 80 km/h	287
	8.7.3 Turbocharger intervention possibilities	289
	8.7.4 Part load in the ECE test cycle	290
	8.7.5 The warm-up phase in the ECE test cycle	292
	8.7.6 Full load acceleration in the turbocharged SI engine	293

9 Fluid mechanical simulation 297

9.1	Three-dimensional flow fields	297
	9.1.1 Basic fluid mechanical equations	297
	9.1.2 Turbulence and turbulence models	303
	9.1.3 Numerics	313
	9.1.4 Computational meshes	320
	9.1.5 Examples	321
9.2	Simulation of injection processes	326
	9.2.1 Single-droplet processes	327
	9.2.2 Spray statistics	331
	9.2.3 Problems in the standard spray model	343
	9.2.4 Solution approaches	347
9.3	Simulation of combustion	354
	9.3.1 General procedure	354
	9.3.2 Diesel combustion	357

9.3.3 The homogeneous SI engine (premixture combustion) 365
9.3.4 The SI engine with stratified charge (partially premixed flames) 380

Literature 382

Index 391

Abbreviations

50 mfb	50 % mass fraction burned
bb	blow-by
BDC	bottom dead center
BTDC	before top dead center
CA	crank angle
CAC	charge air cooler
CAI	controlled auto-ignition
cc	combustion chamber
CD	combustion duration
CCBDC	charge change bottom dead center
CCTDC	charge change top dead center
CFD	computational fluid dynamics
DI	direct injection
DISI	direct injection spark ignition
DS	delivery start
dv	dump valve
eb	engine block
EGR	exhaust gas recirculation
EV	exhaust valve
EVC	exhaust valve close
EVO	exhaust valve open
EIVC	early intake valve close
FEM	finite element method
HCCI	homogeneous charge compression ignition
hrr	heat release rate
ID	injection duration
IGD	ignition delay
IND	injection delay
ip	injection pump
IP	injection process
IT	ignition time
ITDC	ignition top dead center
IV	intake valve
IVC	intake valve close
IVO	intake valve open
LES	large-eddy-simulation
LIVC	late intake valve close
mcp	mass conversion point
mfb	mass fraction burned
nn	neuronal network
oc	oil cooler
OHC-equ.	oxygen-hydrogen-carbon-equilibrium
op	oil pan

Abbreviations and symbols

PAH	polycyclic aromatic hydrocarbons
PDF	probability density function
rg	residual gas
SOC	start of combustion
SOI	start of injection
TC	turbocharging, turbocharger
TDC	top dead center
tv	throttle valve
VTG	variable turbine geometry

Symbols

A	surface [m^2]
	kinematics of the Bolzmann equation variable α
	parameter Zacharias
	temperature difference Heider [K]
A^*	temperature difference Heider [K]
A_{id}	ignition model parameter
A_{prem}	combustion model parameter
a	Vibe heat release rate constant
	sonic speed [m / s]
	thermal diffusivity [m^2 / s]
	gradient „crooked coordinates"
	parameter knocking criterion
	reference opening path thermostat
B	function Heider
B_0, B_1	breakup model constants
b	breadth [m]
	parameter knocking criterion
b_e	specific fuel consumption [g / kWh]
C	function Lax Wendroff
	constant
	heat transfer Woschni constant
C_1	Woschni constant
C_2	Woschni constant [m / (s K)]
C_3	Vogel constant
	constant of the particle path
C_4	constant of the particle path
C_A	contraction coefficient
C_{gl}	Heider constant
C_v	velocity coefficient
C_w	drag coefficient
CD	combustion duration [Grad]
Cou	Courant number

c	carbon component [kg / kg fuel]
	spring rate [N / m]
	progress variable
	velocity [m / s]
	constant
	length [m]
	parameter knocking criterion
	specific heat [J / (kg K)]
$c_{(i)}$	species mass fraction of the species no. i
ci	stock concentration
c_f	constant friction method fan
c_m	medium piston velocity [m / s]
c_p	piston velocity [m / s]
	specific heat at constant pressure [J / (kg K)]
c_u / c_m	swirl number
c_x	mixture fraction variance transport equation model constants
$c_{\varepsilon_1}, c_{\varepsilon_2}, c_{\varepsilon_3}$	ε-equation model constants
c_μ	turbulence model constant
c_v	specific heat at constant volume [J / (kg K)]
D	diffusion constant
	diameter [m]
	parameter Zacharias
	cylinder diameter [m]
D_R	inverse relaxation time scale of a drop in turbulent flow [s^{-1}]
$\dfrac{\partial}{\partial t}$	partial differential
d	wall thickness [m]
	diameter [m]
	damping factor [kg / s]
d_f	fan diameter [m]
d_m	medium turbine diameter [m]
E	energy [J]
$\dot{E}$	energy flow [J / s]
E_A	activation energy
E_{id}	ignition energy [K]
E_{kin}	kinetic spray energy [J]
EB	energy balance
EGR	exhaust gas recirculation [%]
e	eccentricity, crossing [m]
F	Lax Wendroff function
	flexibility of the engine [Nm s]
	force [N]
	function
F_g	gas force [N]
FA	parameter Zacharias

f	general function
	force density [N/m^3]
	distribution function
f_{rg}	mass fraction of the residual gas
fmep	mean friction pressure [bar]
G	formal field variable, which zeros localize the flame front position
	free enthalpy [J]
	function Lax Wendroff
	Gibbs function [J]
g	specific free enthalpy [J / kg]
H	enthalpy [J]
	heating value [J / kg]
h	hydrogen component [kg / kg Kst]
	specific enthalpy [J / kg]
	stroke [m]
h_1	parameter polygon hyperbolic heat release rate
h_2	parameter polygon hyperbolic heat release rate
h_3	parameter polygon hyperbolic heat release rate
I	impulse [(kg m) / s]
	current [A]
I_K	knocking initiating critical pre reaction level
ID	injection duration [Grad]
ifa	fan ratio
imep	indicated mean effective pressure [N / m^2]
iz	number of line sections
L	angular momentum [N m s]
	length scale [m]
K	combustion chamber dependent constant (Franzke)
K_d	differential coefficient
K_i	integral coefficient
K_p	proportional coefficient
	equilibrium constant
K_b	bearing friction constant
K_η	constant [m^3]
K_ρ	factor gap thickness
k	constant
	turbulent kinetic energy [m^2 / s^2]
	heat transfer coefficient [W / (m^2 K)]
	index
k_c	container stiffness [N / m^5]
k_f	velocity coefficient for the forward reaction
	pipe friction coefficient [m / s^2]
k_r	velocity coefficient for the reverse reaction
kp	knocking probability
L	swirl length [m]

L_{min}	minimal air requirement (stoichiometric combustion)
l	connecting rod length [m]
	length [m]
l_F	thickness of the turbulent flame front [m]
l_I	integral length scale
l_t	turbulent length scale [m]
lhv	lower heating value [J / kg]
M	mass [kg]
	molar mass [kg / kmol]
Ma	Mach number
m	mass [kg]
	Vibe parameter
$\dot{m}$	mass flow
mep	mean effective pressure [N / m^2]
N	normalization constant
Nu	Nußelt number
n	number of moles
	polytrope exponent
	speed [rpm]
n_i	quantity of substance [mol]
n_{wc}	number of working cycles per time
Oh	Ohnesorge number
P	power [W]
	term of production in k-equation [W]
Pe	Peclet number
Pr	Prandtl number
Pr$_k$	turbulent Prandtl number for k transport
Pr$_\varepsilon$	turbulent Prandtl number for ε transport
p	partial pressure [N / m^2]
	pressure [N / m^2]
	probability density, distribution function
p_0	pressure of the motored engine [N / m^2]
p_{Gauss}	distribution function with Gaussian distribution
p_{inj}	injection pressure [N / m^2]
p_β	distribution function with β-function distribution
Q	source term of a scalar transport equation
	amount of heat [J]
$\dot{Q}$	heat flow [W]
Q_f, Q_{chem}	heat release [kJ / KW]
q	specific heat [J / m^3]
	heat source [W]
R	electrical resistance [Ohm; Ω]
	gas constant [J / (kg K)]
	drop radius [m]
$\tilde{R}$	universal gas constant

R_0	universal gas constant [J / (mol K)]
$\dot{R}_{dis}$	drop radius change because of disintegration [m / s]
$\dot{R}_{evap}$	drop radius change because of evaporation [m / s]
R_m	molar gas constant [J / (mol K)]
R_{th}	thermal substitute conduction coefficient [W / (m^2 K)]
Re	Reynolds number
r	crankshaft radius [m]
	air content
	radius [m]
S	entropy [J / K]
	spray penetration [m]
S_{ij}	shear tensor [s^{-1}]
Sc	Schmidt number
SF	scavenging factor
Sh	Sherwood number
SMD	Sauter mean diameter [m]
s	flame speed [m / s]
	piston path, stroke [m]
	specific entropy [J / (kg K)]
s_b	bowl depth [m]
s_L	laminar flame speed [m / s]
s_t	turbulent flame speed [m / s]
T	Taylor number
	temperature [K]
	Torque [Nm]
$\dot{T}_{heat}$	drop temperature change because of heating [K / s]
t	time [s]
U	internal energy [J]
u	specific internal energy [J / kg]
	velocity component [m / s]
u'	turbulent velocity fluctuation [m / s]
u/c_0	type number
V	length scale [m / s]
	volume [m^3]
V_d	displacement (cubic capacity) [m^3]
v	velocity [m / s]
	specific volume [m^3 / kg]
v^+	standardized velocity (turbulent wall law)
v_{inj}	injection velocity [m / s]
v_τ	shear stress velocity [m / s]
W	work [J]
$\dot{W}$	power [W]
We	Weber number
w	velocity [m / s]
	specific work [J / kg]

w_i	indicated work [kJ / l]
X	control variable
Y	correcting variable
X_d	control deviation
x	component
	coordinate
	distance [m]
	random number
x_{rg}	amount of residual gas
y	coordinate
	component
y^+	standardized wall distance (turbulent wall law)
y_2^*	parameter polygon hyperbole heat release rate
y_4	parameter polygon hyperbole heat release rate
y_6	parameter polygon hyperbole heat release rate
z	component
	coordinate
	mixture fraction
	number of cylinders
	random number

Greek symbols

α	generic parameter
	flow coefficient
	coefficient Lax Wendroff
	variable set of the spray adapted Boltzmann equation
	heat transfer coefficient [W / (m^2 K)]
α_F	model parameter of the flame surface combustion model
β	generic parameter
	coefficient Lax Wendroff
	reduced variable set of the spray adapted Boltzmann equation
	angle [°]
β_F	model parameter of the flame surface combustion model
γ	angle [°]
Δ	difference
	combustion term
Δ_m	Vibe parameter
Δt	time increment [s]
Δx	length increment [m]
$\Delta \eta$	efficiency difference
$\Delta \phi$	combustion duration [Grad]
δ_0	thickness of the laminar flame front [m]
ε	dissipation rate [m^2 / s^2]

Abbreviations and symbols

	cooling coefficient
	compression ratio
Γ	Gamma function
	ITNFS function (premix combustion model)
η_{th}	thermal efficiency
η	dynamic viscosity $[(N\,s)/m^2]$
η_{conv}	degree of conversion
Θ	polar mass moment of inertia $[kg/m^2]$
ϑ	temperature $[K]$
κ	isentropic exponent
	von Karman constant (turbulence model)
Λ	wavelength in droplet breakup model $[m]$
λ	air-fuel ratio
	heat conductivity $[W/(m\,K)]$
λ^*	mixture stoichiometry
λ_0	air-fuel ratio Heider
λ_L	volumetric efficiency
λ_a	air expenditure
λ_e	eccentric rod relation
λ_f	pipe friction coefficient
μ	chemical potential
	flow coefficient
	1st viscosity coefficient (without index: laminar) $[(N\,s)/m^2]$
ν	kinematic viscosity $[m^2/s]$
	amount of substance $[mol]$
ν_i	stoichiometric coefficient
Π_{br}	branch pressure ratio
π	pressure ratio
	mathematical constant (3,14159)
π_t^*	reciprocal value turbine pressure ratio
π_c	compressor pressure ratio
ρ	density $[kg/m^3]$
σ	specific flame front $[m^2/kg]$
	transient function in Boltzmann equation
	variance
τ	time of flight $[s]$
	stress (also tensor) $[N/m^2]$
	time (ignition delay) $[s]$
τ_{corr}	correlation time of the velocity fluctuation affecting a drop $[s]$
τ_{id}	ignition delay $[s]$
τ_{trb}	turbulent time scale
Φ	generic transport variable
	ratio of equivalence
	specific cooling capacity $[W/K]$
ξ	part

	2nd laminar viscosity coefficient [$(N\,s)/m^2$]
φ	crankshaft angle [°KW]
Ψ	outflow function
ψ	relative clearance of a bearing [m]
Ω	growth rate of the wavelength Λ in droplet breakup model [s^{-1}]
ω	angular speed [s^{-1}]
	swirl number
ζ	pipe friction number

Operators

$\langle\,\rangle$	ensemble averaging
$\langle\,\rangle_F$	Favre averaging
$'$	fluctuation in ensemble average
$''$	fluctuation in Favre average

Indices

*	dimensionless quantity
$\cdot$	time differentiation
$-$	molar quantity
°	reference pressure 1 atm.
	standard status
~	molar quantities
0	idle condition
	drag
	index Runge Kutta
01	idle condition
1	supplied
	after throttling device
	before flow machine
	zone 1
	index Runge Kutta
	when inlet closes
	constant friction
1'	base
15	at 15°C
2	removed
	after flow machine
	zone 2
	index Runge Kutta
	constant friction

Abbreviations and symbols

2'	base
3	index Runge Kutta
	constant friction
4	constant friction
5	constant friction
6	constant friction
50 *mfb*	50 % mass fraction burned
75	at 75% conversion rate
$(i), (j), (k)$	number of species
A	starting point
	branch A
a	axial
a. c.	after compressor
a. t.	after turbine
act	actual
add	added
B	branch B
BDC	bottom dead center
b	bearing
	burned
b. c.	before compressor
b. comb.	before combustion
b. t.	before turbine
bb	blow-by
C	carbon
	crankshaft
C_3H_8	propane
CAC	charge air cooler
CD	combustion duration
CO	carbon monoxide
CO_2	carbon dioxide
c	Carnot process
	circumference
	compression, compressor
	cooler
c, conv	conversion
c, cyl.	cylinder
c.p.	crank pin
c.r.	connecting rod
ch, chem	chemical
cm	cooling mass
comp	compression
corr	correction, corrected
cs	calculation start

cyl	cylinder	
cylw	cylinder wall	
DS	delivery start	
diff.	diffusion	
dr	drop, droplet	
dx	length increment	
E	exhaust gas	
	end gas, characteristic crank angle knocking criterion	
EOC	end of combustion	
evap	evaporated	
EVO	exhaust valve open	
e	effective	
eg	exhaust gas	
env	environment	
F	flame, flame front	
f	foot (point)	
	formation	
	friction	
	fuel	
fg	gaseous fuel	
	fresh gas	
fl	liquid fuel	
Gl	glysantin	
g	gas	
	gas phase	
gl	global	
H_2	hydrogen	
H_2O	water	
ID	injection duration	
IND	injection delay	
IGD	ignition delay	
IT	ignition time	
IV	intake valve	
IVC	intake valve close	
IK	initiating knocking	
i	index	
	inlet	
	inner, inside	
id	inflammation duration	
imp	imperfect	
ind	indicated	
inj	injected, injection	
is	isentropic	
j	index	
K	knocking, characteristic crank angle knocking criterion	

Abbreviations and symbols

kpr	knocking probability region
krit.	critical
L	line
l	lower
l, lam	laminar
m	mass
	mechanic
	medium
	molar
max	maximal
min	minimal
N_2	nitrogen
n	speed
	component C
noz	nozzle
o	outer
	out, outlet
	standard state
osc	oscillated
p	constant pressure process
	(constant) pressure
p	packet
	piston
pre	premixed
r	radial
ref	reference
rem, remov.	removed
rg	residual gas
rot	rotating
SOC	start of combustion
SOI	start of injection
s	isentropic
	stroke
sc	short circuit
spray	spray adapted
sq	squeeze
suppl	supplied
sys	system
T	tangential
	temperature
	turbine
TC	turbocharger
TDC	top dead center
TG	gas tangential pressure
TM	mass tangential pressure

t	technical
	turbulent
tc	to be cooled
th	thermostat
	thermal
$th., theo.$	theoretic
tot	total
$turb$	turbulent
u	upper
u, ub	unburned
v	constant-volume process
	specific volume
	competitive process
v_p	Seiliger process
vol	volume
W, w	wall
x	point x
	starting point
x_{rg}	residual gas
y	component H
z	component O
	number of cycles
α	convective
ε	radiation

1 Introduction

1.1 Preface

One of the central tasks of engineering sciences is the most possibly exact description of technical processes with the goal of understanding the dynamic behavior of complex systems, of recognizing regularities, and thereby of making possible reliable statements about the future behavior of these systems. With regard to combustion engines as propelling systems for land, water, and air vehicles, for permanent and emergency generating sets, as well as for air conditioning and refrigeration, the analysis of the entire process thus acquires particular importance.

In the case of model-based parameter-optimization, engine behavior is described with a mathematical model. The optimization does not occur in the real engine, but rather in a model, which takes into account all effects relevant for the concrete task of optimization. The advantages of this plan are a drastic reduction of the experimental cost and thus a clear saving of time in developmental tasks, see Kuder and Kruse (2000).

The prerequisite for simulation are mechanical, thermodynamic, and chemical models for the description of technical processes, whereby the understanding of thermodynamics and of chemical reaction kinetics are an essential requirement for the modeling of motor processes.

1.2 Model-building

The first step in numeric simulation consists in the construction of the model describing the technical process. Model-building is understood as a goal-oriented simplification of reality through abstraction. The prerequisite for this is that the real process can be divided into single processual sections and thereby broken down into partial problems. These partial problems must then be physically describable and mathematically formulatable.

A number of demands must be placed upon the resulting model:

- The model must be formally correct, i.e. free of inconsistencies. As regards the question of "true or false", it should be noted that models can indeed be formally correct but still not describe the process to be investigated or not be applicable to it. There are also cases in which the model is physically incorrect but nevertheless describes the process with sufficient exactness, e.g. the Ptolemaic model for the simulation of the dynamics of the solar system, i.e. the calculation of planetary and lunar movement.
- The model must describe reality as exactly as possible, and, furthermore, it must also be mathematically solvable. One should always be aware that every model is an approximation to reality and can therefore never perfectly conform with it.
- The cost necessary for the solution of the model with respect to the calculation time must be justifiable in the context of the setting of the task.
- With regard to model-depth, this demand is applicable: as simple as possible and as complex as necessary. So-called universal models are to be regarded with care.

It is only by means of the concept of model that we are in the position truly to comprehend physical processes.

In the following, we will take a somewhat closer look into the types of models with regard to the combustion engine. It must in the first place be noted that both the actual thermodynamic cycle process (particularly combustion) and the change of load of the engine are unsteady processes. Even if the engine is operated in a particular operating condition (i.e. load and rotational speed are constant), the thermodynamic cycle process runs unsteadily. With this, it becomes obvious that there are two categories of engine models, namely, such that describe the operating condition of the engine (total-process models) and such that describe the actual working process (combustion models).

With respect to types of models, one distinguishes between:
- *linguistic models*, i.e. a rule-based method built upon empirically grounded rules, which cannot be grasped by mathematical equations, and
- *mathematical models*, i.e. a method resting on mathematical formalism.

Linguistic models have become known in recent times under the concepts "expert systems" and "fuzzy-logic models". Yet it should thereby be noted that rule-based methods can only interpolate and not extrapolate. We will not further go into this type of model.

Mathematical models can be subdivided into:
- parametric, and
- non-parametric

models. Parametric models are compact mathematical formalisms for the description of system behavior, which rests upon physical and chemical laws and show only relatively few parameters that are to be experimentally determined. These models are typically described by means of a set of partial or normal differential equations.

Non-parametric models are represented by tables that record the system behavior at specific test input signals. Typical representatives of this type of model are step responses or frequency responses. With the help of suitable mathematical methods, e.g. the Fourier transformation, the behavior of the system can be calculated at any input signal.

Like linguistic models, non-parametric models can only interpolate. Only mathematical models are utilized for the simulation of the motor process. But because the model parameters must be adjusted to experimental values in the case of these models as well, they are fundamentally error-prone. These errors are to be critically evaluated in the analysis of simulation results. Here too, it becomes again clear that every model represents but an approximation of the real system under observation.

1.3 Simulation

For the construction of parametric mathematical models for the simulation of temporally and spatially variable fluid, temperature, and concentration fields with chemical reactions, the knowledge of thermodynamics, fluid dynamics, and of combustion technology is an essential prerequisite, see Fig. 1.1.

1.3 Simulation

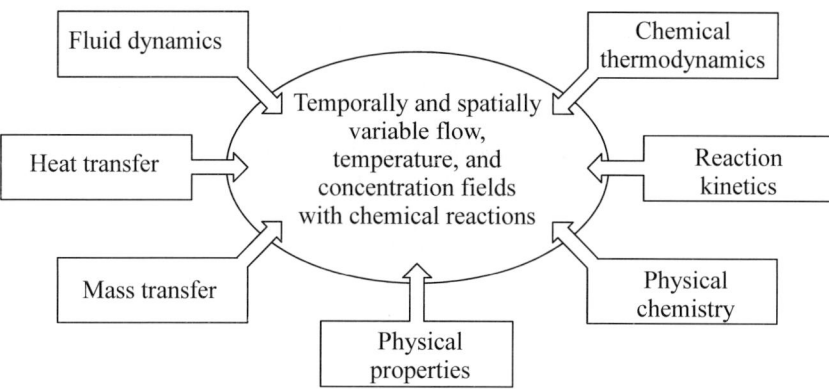

Fig. 1.1: Area of knowledge important for process simulation

With respect to the simulation of fluid fields with chemical reactions, it should be noted that physical and chemical processes can progress at very different temporal and linear scales. The description of these process progressions is usually simpler when the time scales are much different, because then simplifying assumptions can be made for the chemical or physical process, and it is principally very complex when the time scales are of the same order of magnitude. This is made clear by means of the examples in Fig. 1.2.

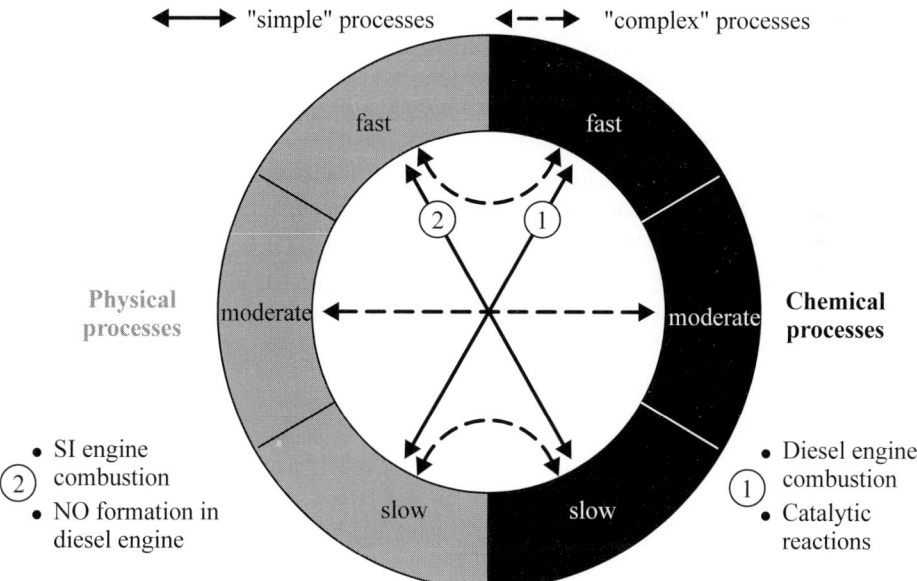

Fig. 1.2: Time scales of physical and chemical processes influencing process simulation

Yet in addition, knowledge of modeling methods is also necessary. Although some universally valid rules can be given for this, this step allows a lot of free room for the creativity and imagination of the modeler. Essentially, the modeling procedure can be subdivided into the following steps:

1st step: define the system and boundaries from the environment, determine the relevant reservoirs as well as the mass and energy flow between them.

2nd step: draw up balance sheets according to the unified scheme: temporal change of the reservoir is equal to the inflow minus the outflow.

3rd step: with the help of physical laws, describe the mass and energy flows.

4th step: simplify the resulting model, if necessary by neglecting secondary influences.

5th step: integrate the model numerically, i.e. execute the simulation.

6th step: validate the model, compare the calculated data with experimentally obtained data.

In the utilization of an existing simulation program for the solution of new tasks, the prerequisites which were met in the creation of the model must always be examined. It should thereby be clarified whether and to what extent the existing program is actually suitable for the solution of the new problem. One should in such cases always be aware of the fact that "pretty, colorful pictures" exert an enormous power of suggestion upon the "uncritical" observer.

The prerequisite for the acceptance of what we nowadays designate as computer simulation was a gradual alteration in philosophical thought and in the conceptualization and understanding of the world in which we live. In the past, humanity perceived the world and its processes predominately as linear and causal, and we are gradually comprehending the decisive processes flow in a non-linear and chaotic fashion. Only with the rise of the sciences and with the development of their methodological foundations could the basis for computer simulation be created.

Numeric simulation opens up unimagined possibilities. We can get an idea of what is to be expected in this field if we bear in mind the rapid development in the information sector and compare the present condition of "email" and the "internet" with that of ten years past.

With respect to technological progress and the ecological perspectives related to it, the reader is referred to Jischa (1993). Also, Kaufmann and Smarr (1994) have provided interesting insight into the topic of simulation.

2 Introduction into the functioning of internal combustion engines

2.1 Energy conversion

In energy conversion, we can distinguish hierarchically between general, thermal, and motor energy conversion.

Under *general energy* conversion is understood the transformation of primary into secondary energy through a technical process in an energy conversion plant, see Fig. 2.1.

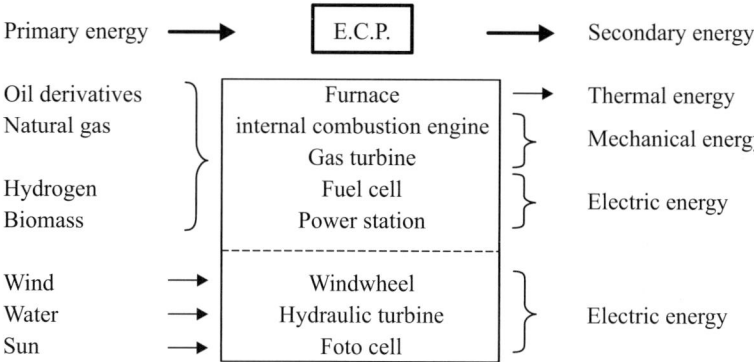

Fig. 2.1: Diagram of general energy conversion

Thermal energy conversion is subject to the laws of thermodynamics and can be described formally, as is shown in Fig. 2.2.

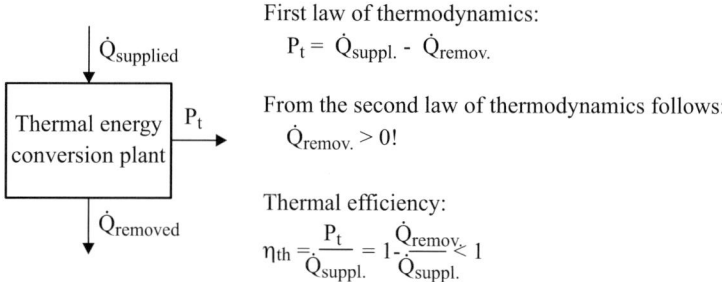

First law of thermodynamics:
$$P_t = \dot{Q}_{suppl.} - \dot{Q}_{remov.}$$

From the second law of thermodynamics follows:
$$\dot{Q}_{remov.} > 0!$$

Thermal efficiency:
$$\eta_{th} = \frac{P_t}{\dot{Q}_{suppl.}} = 1 - \frac{\dot{Q}_{remov.}}{\dot{Q}_{suppl.}} \leq 1$$

Fig. 2.2: Diagram of thermal energy conversion

The *internal combustion engine* and the *gas turbine* are specialized energy conversion plants, in which the chemical energy bound in the fuel is at first transformed into thermal energy in the combustion space or chamber, this being then transformed into mechanical energy by the motor. In the case of the stationary gas turbine plant, the mechanical energy is then converted into electrical energy by the secondary generator.

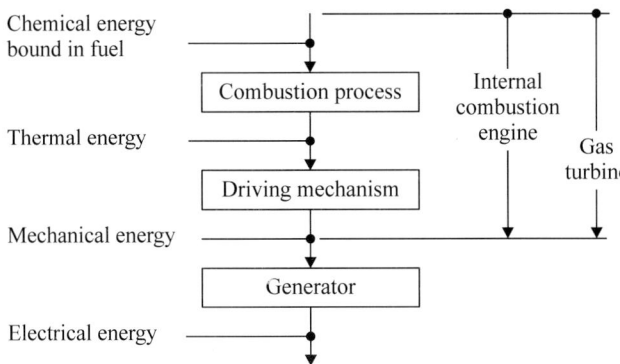

Fig. 2.3: Diagram of energy conversion in an internal combustion engine or gas turbine

2.2 Reciprocating engines

Internal combustion engines are piston machines, whereby one distinguishes, according to the design of the combustion space or the pistons, between reciprocating engines and rotary engines with a rotating piston movement. Fig. 2.4 shows principle sketches of possible structural shapes of reciprocating engines, whereby today only variants 1, 2, and 4 are, practically speaking, still being built.

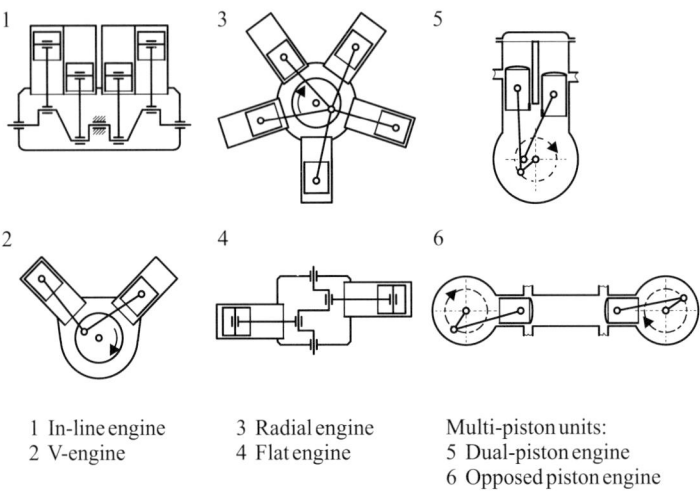

1 In-line engine
2 V-engine
3 Radial engine
4 Flat engine
Multi-piston units:
5 Dual-piston engine
6 Opposed piston engine

Fig. 2.4: Types of reciprocating engines

2.2.1 The crankshaft drive

For an extensive description of other models of the combustion engine, see Basshuysen and Schäfer (2003) and Maas (1979).

2.2.1 The crankshaft drive

The motor transforms the oscillating movement of the piston into the rotating movement of the crankshaft, see Fig. 2.5. The piston reverses its movement at the top dead center (TDC) and at the bottom dead center (BDC). At both of these dead point positions, the speed of the piston is equal to zero, whilst the acceleration is at the maximum. Between the top dead center and the underside of the cylinder head, the compression volume V_c remains (also the so-called dead space in the case of reciprocating compressors).

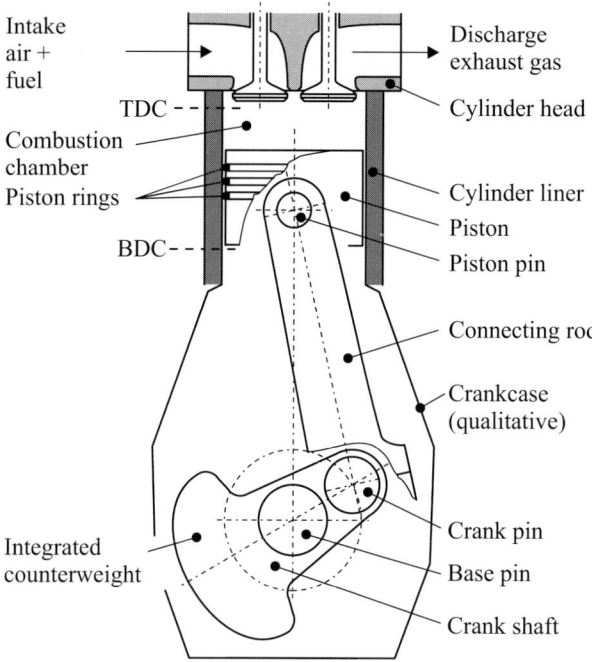

Fig. 2.5: Assembly of the reciprocating engine

Fig. 2.6 shows the kinematics of a crankshaft drive with crossing, in which the longitudinal crankshaft axle does not intersect with the longitudinal cylinder axle, but rather is displaced by the length e.

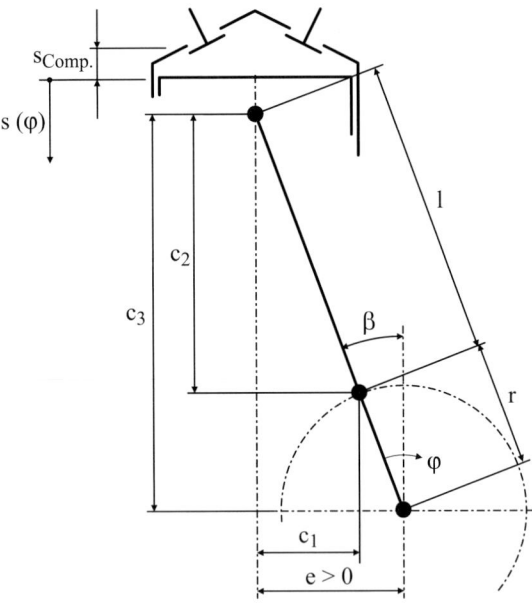

Fig. 2.6: Kinematics of the crankshaft drive

For the piston path $s(\varphi)$, it follows from Fig. 2.6:

$$s(\varphi) = c_3 - c_2 - r\cos(\varphi - \beta) \qquad (2.1)$$

from which with

$$\sin\beta = \frac{e}{r+l} \quad \text{and} \quad \beta = \arcsin\left(\frac{e}{r+l}\right), \text{ respectively}$$

$$c_1 = e - r\sin(\beta - \varphi),$$

$$c_2 = \sqrt{l^2 - c_1^2} \quad \text{and}$$

$$c_3 = \sqrt{(r+l)^2 - e^2}$$

finally

$$s(\varphi) = \sqrt{(r+l)^2 - e^2} - \sqrt{l^2 - [e + r\sin(\varphi - \beta)]^2} - r\cos(\varphi - \beta) \qquad (2.2)$$

results. The derivative provides for the piston speed the relation

$$\frac{ds}{d\varphi} = r\sin(\varphi - \beta) + \frac{r[e + r\sin(\varphi - \beta)]\cos(\varphi - \beta)}{\sqrt{l^2 - [e + r\sin(\varphi - \beta)^2]}}. \qquad (2.3)$$

With the definition of the cylinder volume

2.2 Reciprocating engines

$$V(\varphi) = V_c + D^2 \frac{\pi}{4} s(\varphi) \tag{2.4}$$

follows for the alteration of cylinder volume

$$\frac{dV}{d\varphi} = D^2 \frac{\pi}{4} \frac{ds}{d\varphi} . \tag{2.5}$$

With the eccentric rod relation $\lambda_e = r/l$, it follows finally for the limiting case $e = 0$

$$s(\varphi) = r \left\{ [1 - \cos(\varphi)] + \frac{1}{\lambda_e} \left[1 - \sqrt{1 - \lambda_e^2 \sin^2(\varphi)} \right] \right\} \tag{2.6}$$

and

$$\frac{ds}{d\varphi} = r \left[\sin(\varphi) + \frac{\lambda_e}{2} \frac{\sin(2\varphi)}{\sqrt{1 - \lambda_e^2 \sin^2(\varphi)}} \right] . \tag{2.7}$$

2.2.2 Gas and inertia forces

The motor is driven by the gas pressure $p(\varphi)$ present in the combustion space. With the piston area $A_P = D^2 \pi/4$, one then obtains for the gas force

$$F_g = D^2 \frac{\pi}{4} p(\varphi) . \tag{2.8}$$

Because of the masses in motion in the driving mechanism, additional and temporally variable inertia forces arise, which lead to rotating and oscillating unbalances and must at least partially be counterbalanced in order to guarantee the required driving mechanism running smoothness. The single components of the motor execute rotating (crank pin, $m_{c.p.}$), oscillating (piston block, m_P), or mixed (connecting rod) movements. If one distributes the mass of the connecting rod into a rotating ($m_{c.r.,rot}$) and an oscillating ($m_{c.r.,osc}$) portion, one then obtains for the rotating and oscillating masses of the driving mechanism

$$m_{rot} = m_{c.r.,rot} + m_{c.p.} ,$$
$$m_{osc} = m_{c.r.,osc} + m_p .$$

For small λ_e, the expression under the root in (2.6) corresponding to

$$\sqrt{1 - \lambda_e^2 \sin^2(\varphi)} = 1 - \frac{\lambda_e^2}{2} \sin^2(\varphi) - \frac{\lambda_e^4}{8} \sin^4(\varphi) - \ldots$$

can be developed into a Taylor's series, whereby the third term for $\lambda_e = 0.25$ already becomes smaller than 0.00048 and can thus be neglected as a rule. With the help of trigonometric transformations, one finally obtains for the *piston path*

$$\frac{s}{r} = 1 - \cos(\varphi) + \frac{\lambda_e}{4} (1 - \cos(2\varphi)) . \tag{2.9}$$

With the angular velocity ω

$$\frac{d\varphi}{dt} = \omega$$

one obtains for the *piston speed*

$$\frac{ds}{dt} = \frac{ds}{d\varphi}\frac{d\varphi}{dt} = \omega\frac{ds}{d\varphi}$$

the expression

$$\frac{ds}{dt} = r\omega\left[\sin(\varphi) + \frac{\lambda_e}{2}\sin(2\varphi)\right] \qquad (2.10)$$

and for the *piston acceleration*

$$\frac{d^2 s}{dt^2} = \frac{d^2 s}{d\varphi^2}\left(\frac{d\varphi}{dt}\right)^2 = \omega^2 \frac{d^2 s}{d\varphi^2}$$

finally

$$\frac{d^2 s}{dt^2} = r\omega^2\left[\cos(\varphi) + \lambda_e \cos(2\varphi)\right] . \qquad (2.11)$$

With that, one obtains the expression

$$F_{m,rot} = m_{rot}\, r\omega^2 \qquad (2.12)$$

for the *rotating inertia force*, which triggers an unbalance striking in the crankshaft axle and rotating with the speed of the crankshaft. For the *oscillating inertia force* one obtains the expression

$$F_{m,osc} = m_{osc}\, r\omega^2 \left[\cos(\varphi) + \lambda_e \cos(2\varphi)\right] . \qquad (2.13)$$

This consists of two parts, whereby the first rotates with simple crankshaft speed and the second with doubled crankshaft speed. One therefore distinguishes between inertia forces of first and second order,

$$F_1 = m_{osc}\, r\omega^2 \cos(\varphi) ,$$
$$F_2 = m_{osc}\, r\omega^2 \lambda_e \cos(2\omega) .$$

The inertia forces are proportional to ω^2 and are thus strongly contingent on speed. The resulting piston force consists of gas force and oscillating inertia force,

$$F_P = D^2 \frac{\pi}{4} p(\varphi) + m_{osc}\, r\omega^2 \left[\cos(\varphi) + \lambda_e \cos(2\varphi)\right] . \qquad (2.14)$$

Fig. 2.7 shows the progression of gas force and oscillating inertia force for a 4-stroke reciprocating engine over a full cycle. One recognizes that the peak load caused by the gas force is quickly diminished with rising speed because of the inertia force proportional to ω^2.

2.2 Reciprocating engines

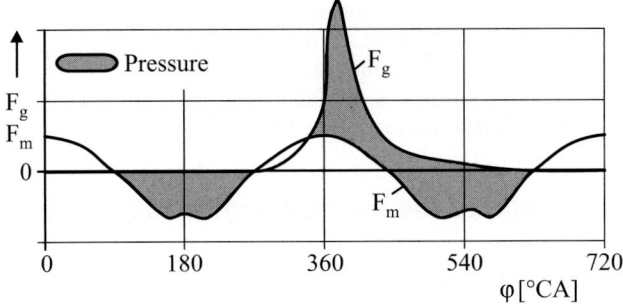

Fig. 2.7: Gas force and oscillating inertia force of a 4-stroke reciprocating engine

2.2.3 Procedure

With regard to charge changing in the reciprocating engine, one distinguishes between the 4-stroke and the 2-stroke methods and in reference to the combustion process between diesel and spark-ignition (SI) engines. In the case of the *4-stroke-procedure*, see also Fig. 2.8, the charge changing occurs in both strokes, expulsion and intake, which is governed by the displacement effect of the piston and by the valves. The intake and exhaust valves open before and close after the dead point positions, whereby an early opening of the exhaust valve indeed leads to losses during expansion, but also leads to a diminishment of expulsion work. With increasing valve intersection, the scavenging losses increase, and the operative efficiency decreases. Modern 4-stroke engines are equipped, as a rule, with two intake and two exhaust valves.

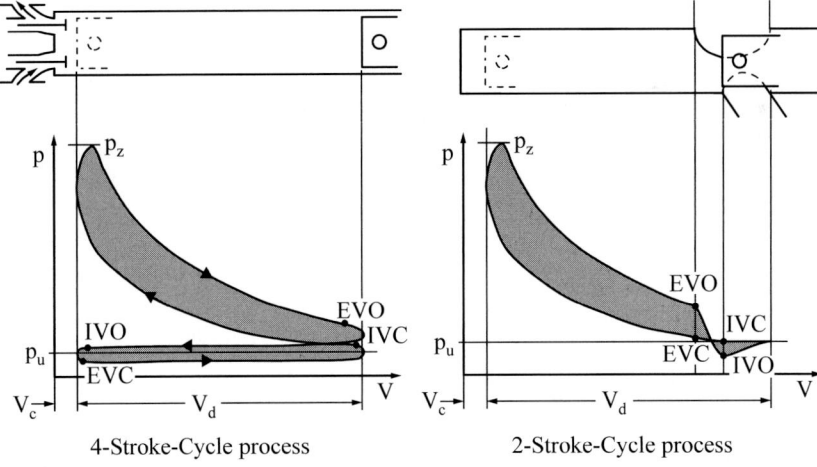

Fig. 2.8: p, V diagram for the 4-stroke and 2-stroke processes

In the case of the *2-stroke engine*, the charge changing occurs while the piston is near the BDC. With such so-called piston-valve engines, the exhaust gas is expelled out of the cylinder by the in-flowing fresh air, if the piston sweeps over the intake and exhaust sections arranged in the lower area of the cylinder. In the case of larger engines, exhaust valves are mostly used instead of exhaust ports, which are then housed in the cylinder head. Instead of so-called loop scavenging, one then has the fundamentally more effective uniflow scavenging. For more details, see Merker and Gerstle (1997).

2.3 Thermodynamics of the internal combustion engine

2.3.1 Foundations

Our goal in this chapter will be to explain the basic foundations of thermodynamics without going into excessive detail. Extensive presentations can be found in Baehr (2000), Hahne (2000), Lucas (2001), and Stephan and Mayinger (1998, 1999).

For the simulation of combustion-engine processes, the internal combustion engine is separated into single components or partial systems, which one can principally view either as closed or open thermodynamic systems. For the balancing of these systems, one uses the *mass balance* (equation of continuity)

$$\frac{dm}{dt} = \dot{m}_1 - \dot{m}_2 \qquad (2.15)$$

and the *energy balance* (1st law of thermodynamics)

$$\frac{dU}{dt} = \dot{Q} + \dot{W} + \dot{E}_1 + \dot{E}_2 \qquad (2.16)$$

with

$$\dot{E} = \dot{m}\left(h + \frac{c^2}{2}\right)$$

for the *open, stationary flooded system* shown in Fig. 2.9 (flow system), or

$$\frac{dU}{dt} = \dot{Q} + \dot{W} \qquad (2.17)$$

for the closed system shown in Fig. 2.10 (combustion chamber).

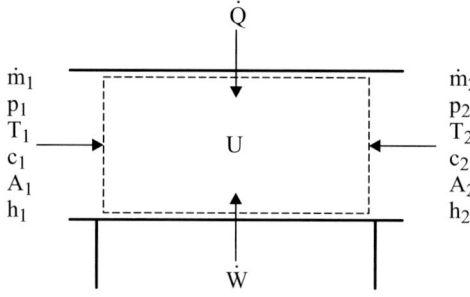

Fig. 2.9: Open thermodynamic system (----- system boundaries)

2.3 Thermodynamics of the internal combustion engine

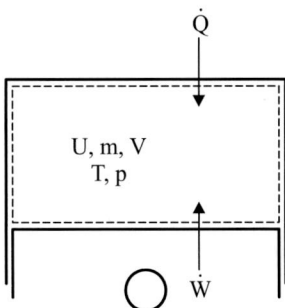

Fig. 2.10: Closed thermodynamic system (----- system boundaries)

In closed systems, no mass, and with that no enthalpy, flows over the system limits. Neglecting the blow-by losses, the *combustion chamber* (cylinder) can be viewed as a closed system during the so-called high pressure process (compression and expansion act). In contrast, in the case of an open system, e.g. a *reservoir* or a *line section*, masses can flow over the system boundaries.

Neglecting the friction or dissipation of mechanical work into heat, one obtains for the *volume work*

$$\dot{W} = -p\frac{dV}{dt} \ . \tag{2.18}$$

In the open system, one summarizes the thermal energy transferred to the system boundaries and the intake and expulsion work practically as enthalpy

$$h \equiv u + pv \ . \tag{2.19}$$

The *thermal state equation*

$$f(p,T,v) = 0 \tag{2.20}$$

ties together the three thermal condition magnitudes of pressure, temperature, and volume and the *caloric state equation*

$$u = u(T,v) \text{ and}$$
$$h = h(p,T), \text{ respectively} \tag{2.21}$$

describes the inner energy as a function of temperature and volume, or the enthalpy as a function of pressure and temperature. We will in the following view the materials under consideration first as *ideal gases*, for which the thermal state equation

$$pv = RT \tag{2.22}$$

is applicable. Because the inner energy of ideal gas is only dependent on temperature, follows from (2.19) with (2.22), that this is also valid for enthalpy. Thus for differential alteration of caloric magnitudes of the ideal gas we have:

$$du = c_v(T)dT \text{ and}$$
$$dh = c_p(T)dT \text{ , respectively.} \tag{2.23}$$

For ideal gas
$$R = c_p(T) - c_v(T) \tag{2.24}$$
and
$$\kappa = \frac{c_p}{c_v} \tag{2.25}$$
are applicable. For reversible condition alterations, the 2nd law of thermodynamics holds in the form
$$T\,ds = dq\,. \tag{2.26}$$
With that, it follows from (2.17) with (2.18)
$$du = -p\,dv + T\,ds\,. \tag{2.27}$$
With (2.23), it follows for the rise of the isochores of a ideal gas
$$\left(\frac{dT}{ds}\right)_s = \frac{T}{c_v}\,. \tag{2.28}$$
Analogous to this, it follows for the rise of isobars
$$\left(\frac{dT}{ds}\right)_s = \frac{T}{c_p}\,,$$
and for the isotherms and isentropes follows
$$\frac{dp}{dv} = -\frac{p}{v} \quad \text{or} \quad \frac{dp}{dv} = -\kappa\frac{p}{v}\,.$$
Fig. 2.11 shows the progression of simple state changes in the p,v and T,s diagram.

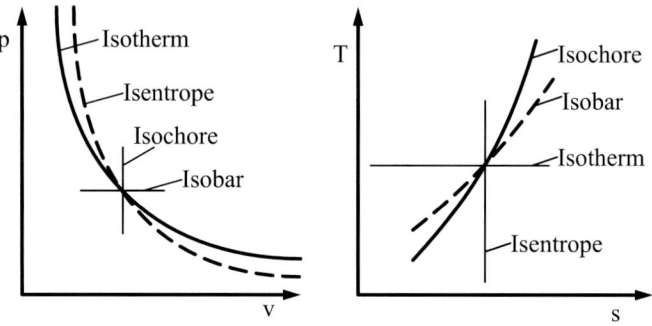

Fig. 2.11: Course of a simple change of state in the p, v- and in the T, s diagram

With the relations above, one finally obtains for the *energy balance of the closed system*

2.3 Thermodynamics of the internal combustion engine

$$m c_v \frac{dT}{dt} = \frac{dQ}{dt} - p \frac{dv}{dt} \; . \tag{2.29}$$

Under consideration of the enthalpy flows and the transferred kinetic energy to the system boundaries, one obtains for the *energy balance of the open system*

$$m c_v \frac{dT}{dt} + c_v T \frac{dm}{dt} = \frac{dQ}{dt} + \frac{dW}{dt} + \dot{m}_1 \left(h_1 + \frac{c_1^2}{2} \right) - \dot{m}_2 \left(h_2 + \frac{c_2^2}{2} \right) . \tag{2.30}$$

For stationarily flooded open systems, it follows for the case that no work is transferred

$$\dot{m} \left[(h_2 - h_1) + \left(\frac{c_2^2}{2} - \frac{c_1^2}{2} \right) \right] = \frac{dQ}{dt} \; . \tag{2.31}$$

With this relation, the flow or outflow equation for the calculation of the mass flows through throttle locations or valves can be derived. We consider an outflow process from an infinitely large reservoir and presume that the flow proceeds adiabatically. With the indices "0" for the interior of the reservoir and "1" for the outflow cross section, it follows with $c_0 = 0$ from (2.31)

$$\frac{c_1^2}{2} = h_0 - h_1 \; . \tag{2.32}$$

With the adiabatic relation

$$\frac{T_1}{T_0} = \left(\frac{p_1}{p_0} \right)^{\frac{\kappa - 1}{\kappa}} \tag{2.33}$$

it first follows

$$\frac{c_1^2}{2} = c_p T_0 \left(1 - \frac{T_1}{T_0} \right) = c_p T_0 \left[1 - \left(\frac{p_1}{p_0} \right)^{\frac{\kappa - 1}{\kappa}} \right] \tag{2.34}$$

and furthermore for the velocity c_1 in the outflow cross section

$$c_1 = \sqrt{\frac{2\kappa}{\kappa - 1} R T_0 \left[1 - \left(\frac{p_1}{p_0} \right)^{\frac{\kappa - 1}{\kappa}} \right]} \; . \tag{2.35}$$

With the equation for ideal gas, it follows for the density ratio from (2.33)

$$\frac{\rho_1}{\rho_0} = \left(\frac{p_1}{p_0} \right)^{\frac{1}{\kappa}} \; . \tag{2.36}$$

With this results for the mass flow

$$\dot{m} = A_1 \rho_1 c_1$$

in the outflow cross section the relation

$$\dot{m} = A_1 \sqrt{\rho_0 p_0}\ \Psi\!\left(\frac{p_1}{p_0}, \kappa\right), \qquad (2.37)$$

whereby

$$\Psi\!\left(\frac{p_1}{p_0}, \kappa\right) = \sqrt{\frac{2\kappa}{\kappa-1}\left[\left(\frac{p_1}{p_0}\right)^{\frac{2}{\kappa}} - \left(\frac{p_1}{p_0}\right)^{\frac{\kappa+1}{\kappa}}\right]} \qquad (2.38)$$

is the so-called outflow function, which is solely contingent upon the pressure ratio p_1/p_0 and from the isentrope exponent κ. Fig. 2.12 shows the progression of the outflow function for various isentrope exponents.

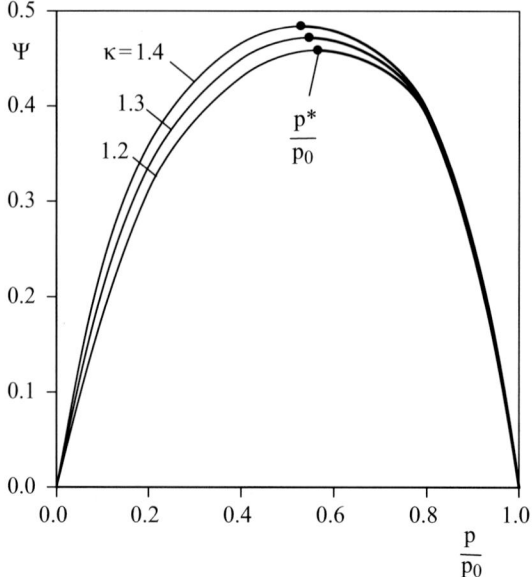

Fig. 2.12: Outflow function $\Psi\!\left(\dfrac{p_1}{p_0}, \kappa\right)$

The maximums of the outflow function result from the relation

$$\frac{\partial \Psi}{\partial \left(\dfrac{p_1}{p_0}\right)} = 0 \quad \text{for} \quad \Psi = \Psi_{max}. \qquad (2.39)$$

2.3 Thermodynamics of the internal combustion engine

With this, one obtains for the so-called critical pressure ratio the relation

$$\left(\frac{p_1}{p_0}\right)_{crit} = \left(\frac{2}{\kappa+1}\right)^{\frac{\kappa}{\kappa-1}} \quad \text{or} \quad \left(\frac{T_1}{T_0}\right)_{crit} = \frac{2}{\kappa+1}. \tag{2.40}$$

If we put this relation into (2.35) for the isentropic outflow velocity, then

$$c_{1,crit} = \sqrt{\kappa R T_1} \tag{2.41}$$

finally follows. From (2.36) follows

$$\frac{dp}{d\rho} = \kappa \frac{p}{\rho} = \kappa R T \tag{2.42}$$

for the isentropic flow. With the definition of sound speed

$$a \equiv \sqrt{\frac{dp}{d\rho}} \tag{2.43}$$

thus follows

$$a_1 = \sqrt{\kappa R T_1} \tag{2.44}$$

for the velocity of the outflow cross section. The flow velocity in the narrowest cross section of a throttle location or in the valve can thereby reach maximal sonic speed.

2.3.2 Closed cycles

The simplest models for the actual engine process are closed, internally reversible cycles with heat supply and removal, which are characterized by the following properties:

- the chemical transformation of fuel as a result of combustion are replaced by a corresponding heat supply,
- the charge changing process is replaced by a corresponding heat removal
- air, seen as a ideal gas, is chosen as a working medium.

- **The Carnot cycle**

The Carnot cycle, represented in Fig. 2.13, is the cycle with the highest thermal efficiency and thus the ideal process. Heat supply results from a heat bath of temperature T_3, heat removal to a heat bath with temperature T_1. The compression of $2 \to 3$ and $4 \to 1$ always takes place isentropically.

With the thermal efficiency

$$\eta_{th} = 1 - \frac{q_{supplied}}{q_{removed}}$$

we obtain the well-known relation

$$\eta_{th,c} = 1 - \frac{T_1}{T_3} = f\left(\frac{T_1}{T_3}\right) \qquad (2.45)$$

for the Carnot cycle.

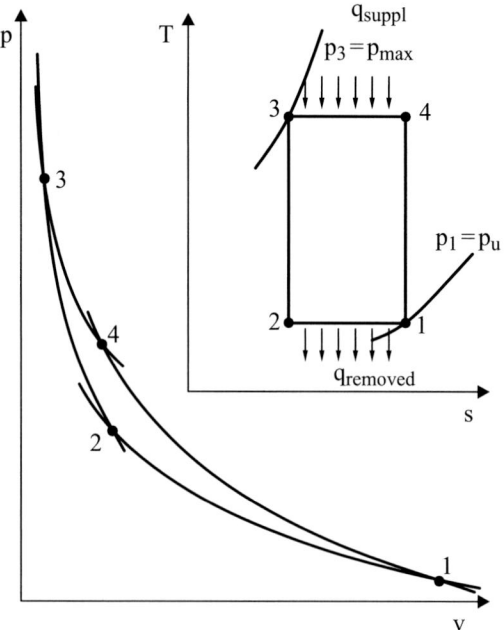

Fig. 2.13: Carnot cycle

The Carnot cycle cannot however be realized in internal combustion engines, because

- the isothermal expansion with $q_{supplied}$ at T_3 = const. and the isothermal compression with $q_{removed}$ at T_1 = const. are not practically feasible, and
- the surface in the p, v diagram and thus the internal work is extremely small even at high pressure ratios.

In accordance with the definition, for the medium pressure of the process is applicable

$$p_m = \frac{w}{v_1 - v_3}. \qquad (2.46)$$

For the supplied and removed heat amounts in isothermal compression and expansion

$$q_{supplied} = q_{34} = RT_3 \ln \frac{p_3}{p_4},$$

$$q_{removed} = q_{12} = RT_1 \ln \frac{p_2}{p_1}$$

2.3 Thermodynamics of the internal combustion engine

applies. With the thermally and calorically ideal gas, we obtain

$$\frac{p_3}{p_2} = \left(\frac{T_3}{T_2}\right)^{\frac{\kappa}{\kappa-1}} \quad \text{and} \quad \frac{p_4}{p_1} = \left(\frac{T_4}{T_1}\right)^{\frac{\kappa}{\kappa-1}}$$

for the isentrope, from which follows

$$\frac{p_3}{p_2} = \frac{p_4}{p_1} \quad \text{and} \quad \frac{p_3}{p_4} = \frac{p_2}{p_1}$$

because $T_1 = T_2$ and $T_3 = T_4$. At first we obtain for the medium pressure

$$p_m = \frac{R(T_3 - T_1)\ln\frac{p_3}{p_4}}{v_1 - v_3}$$

and finally by means of simple conversion

$$\frac{p_m}{p_1} = \frac{\frac{p_3}{p_1}\left(\frac{T_3}{T_1} - 1\right)\left(\ln\frac{p_3}{p_1} - \frac{\kappa}{\kappa-1}\ln\frac{T_3}{T_1}\right)}{\frac{p_3}{p_1} - \frac{T_3}{T_1}} \quad (2.47)$$

The relation

$$\frac{p_m}{p_1} = f\left(\frac{T_3}{T_1}, \frac{p_3}{p_1}, \kappa\right)$$

setting to zero and with extreme values is graphically presented in Fig. 2.14 for $\kappa = 1{,}4$.

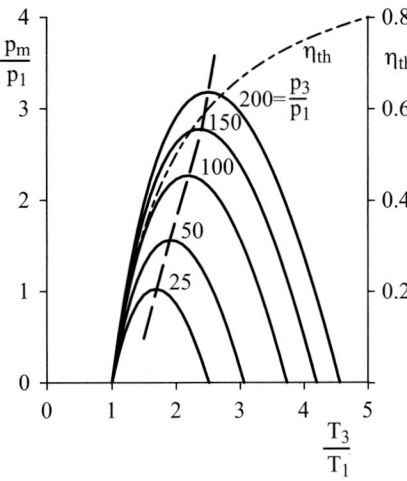

Fig. 2.14: Medium pressure of the Carnot cycle

While the thermal efficiency in an optimally run process at a pressure ratio of 200 with 0.6 achieves relatively high values, the reachable medium pressure still amounts only to $p_m = 3.18\, p_1$. The work to be gained is thus so small that an engine realizing the Carnot cycle could in the best scenario overcome internal friction and can therefore deliver practically no performance.

The Carnot cycle is thus only of interest as a theoretical comparative process. In this context, we can only point out to its fundamental importance in connection to energy considerations.

- **The constant-volume process**

The constant-volume process is thermodynamically efficient and, in principle, feasible cycle (see Fig. 2.15). In contrast to the Carnot process, it avoids isothermal expansion and compression and the unrealistically high pressure ratio. It consists of two isentropes and two isochores.

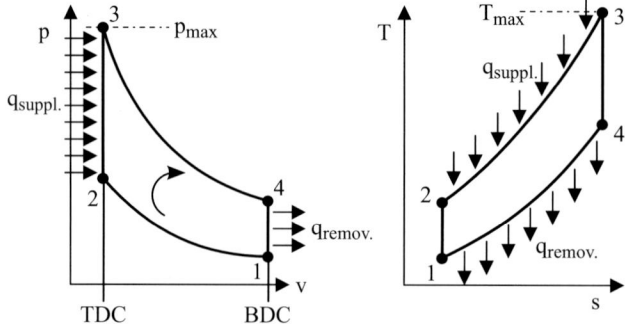

Fig. 2.15: Representation of the constant-volume cycle in the p, v and T, s diagram

It is called the constant-volume process because the heat supply (instead of combustion) ensues in constant space, i.e. under constant volume. Because the piston moves continuously, the heat supply would have to occur infinitely fast, i.e. abruptly. However, that is not realistically feasible. For the thermal efficiency of this process follows

$$\eta_{th,v} = 1 - \frac{q_{removed}}{q_{supplied}} = 1 - \frac{c_v\,(T_4 - T_1)}{c_v\,(T_3 - T_2)} = 1 - \frac{T_1}{T_2}\frac{\frac{T_4}{T_1} - 1}{\frac{T_3}{T_2} - 1}.$$

With the relations for the adiabatic

$$\left.\begin{array}{l}\dfrac{T_1}{T_2} = \left(\dfrac{v_1}{v_2}\right)^{\kappa-1} \\[4pt] \dfrac{T_3}{T_4} = \left(\dfrac{v_4}{v_3}\right)^{\kappa-1} = \left(\dfrac{v_1}{v_2}\right)^{\kappa-1}\end{array}\right\} = \dfrac{T_4}{T_1} = \dfrac{T_3}{T_2}$$

2.3 Thermodynamics of the internal combustion engine

and the compression ratio $\varepsilon = v_1/v_2$ follows finally for the thermal efficiency of the constant-volume process

$$\eta_{th,v} = 1 - \left(\frac{1}{\varepsilon}\right)^{\kappa-1}. \tag{2.48}$$

This relation, represented in Fig. 2.16, makes it clear that, after a certain compression ratio, no significant increase in the thermal efficiency is achievable.

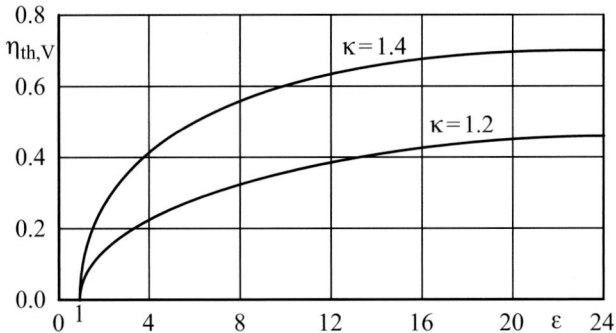

Fig. 2.16: Efficiency of the constant-volume cycle

- **The Constant-pressure process**

In the case of high-compressing engines, the compression pressure p_2 is already very high. In order not to let the pressure climb any higher, heat supply (instead of combustion) is carried out at constant pressure instead of constant volume. The process is thus composed of two isentropes, an isobar, and an isochore, see Fig. 2.17.

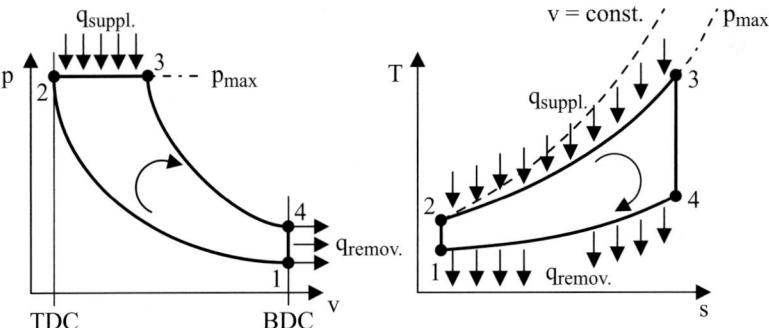

Fig. 2.17: The constant-pressure cycle in the p, v and T, s diagram

Again, for the thermal efficiency applies

$$\eta_{th,p} = 1 - \frac{q_{removed}}{q_{supplied}} = 1 - \frac{c_v(T_4 - T_1)}{q_{supplied}}.$$

As opposed to the constant-volume process, there now appear however three prominent volumes. Therefore, a further parameter for the determination of $\eta_{th,p}$ is necessary. Pragmatically, we select

$$q^* = \frac{q_{supplied}}{c_p T_1}$$

for this. With this, we first obtain

$$\eta_{th,p} = 1 - \frac{c_v(T_4 - T_1)}{c_p T_1 q^*} = 1 - \frac{1}{\kappa q^*}\left(\frac{T_4}{T_1} - 1\right).$$

And finally after a few conversions

$$\eta_{th,p} = 1 - \frac{1}{\kappa q^*}\left[\left(\frac{q^*}{\varepsilon^{\kappa-1}} + 1\right)^{\kappa} - 1\right]. \tag{2.49}$$

The thermal efficiency profile of the constant-pressure process in contingency on ε and q^* is represented in Fig. 2.18.

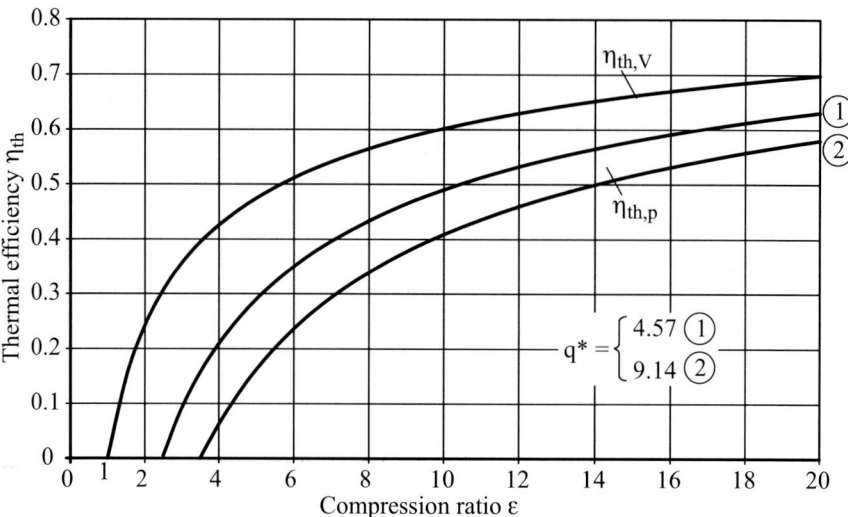

Fig. 2.18: Thermal efficiency of the constant-pressure cycle

- **The Seiliger cycle**

The Seiliger cycle, demonstrated in Fig. 2.19, represents a combination of the constant-volume and the constant-pressure processes.

2.3 Thermodynamics of the internal combustion engine

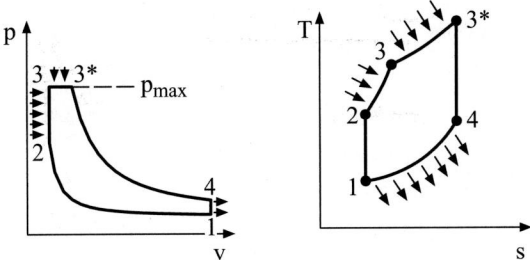

Fig. 2.19: The Seiliger cycle in the p, v and T, s diagram

One utilizes this comparative process when, at a given compression ratio, the highest pressure must additionally be limited. The heat supply (instead of combustion) succeeds isochorically and isobarically. With the pressure ratio $\pi = p_3/p_1$, we finally obtain the relation

$$\eta_{th,vp} = 1 - \frac{1}{\kappa q^*}\left\{\left[q^* - \frac{1}{\kappa \varepsilon}(\pi - \varepsilon^\kappa) + \frac{\pi}{\varepsilon}\right]^\kappa \left(\frac{1}{\pi}\right)^{\kappa-1} - 1\right\} \tag{2.50}$$

for the thermal efficiency, which is graphically represented in Fig. 2.20. From this it becomes clear that, at a constant given compression ratio ε, it is the constant-volume process, and at a constant given pressure ratio π, it is the constant-pressure process which has the highest efficiency.

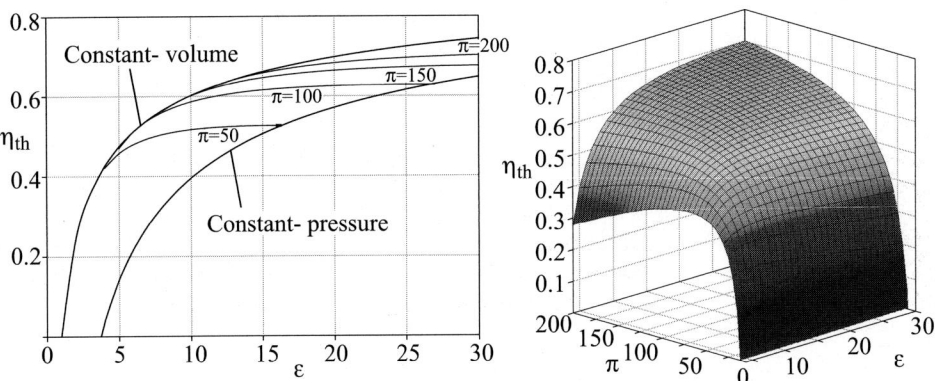

Fig. 2.20: Thermal efficiency of the Seiliger cycle

- **Comparison of the cycles**

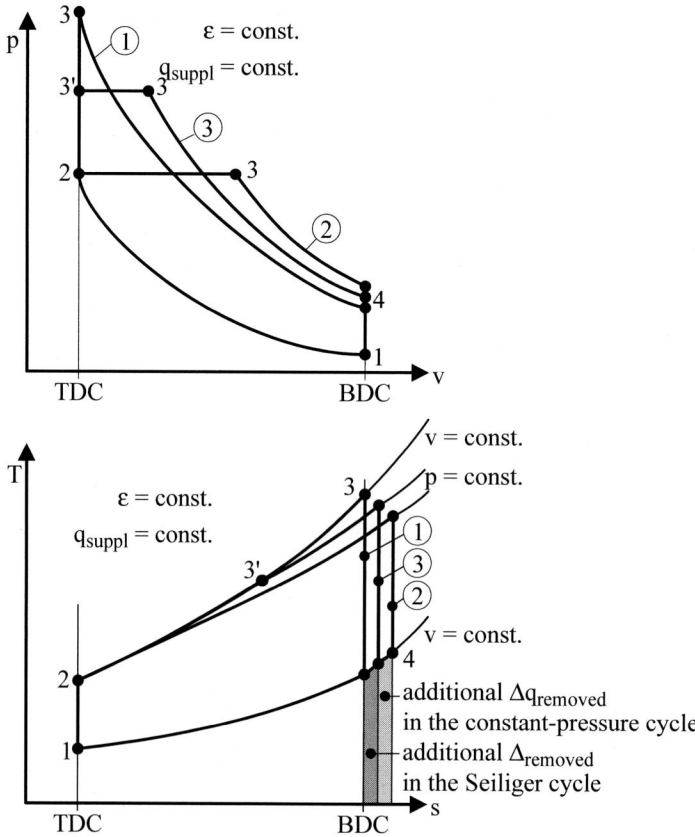

Fig. 2.21: Comparison of the closed cycles, ① = constant-volume, ② = constant-pressure, ③ = Seiliger cycle

For the efficiency of the particular comparative processes, the following contingencies result

$\eta_{th,c} = f\left(\dfrac{T_1}{T_3}\right)$ Carnot,

$\eta_{th,v} = f(\varepsilon)$ constant volume,

$\eta_{th,p} = f(\varepsilon, q^*)$ constant pressure,

$\eta_{th,vp} = f\left(\varepsilon, q^*, \dfrac{p_3}{p_1}\right)$ Seiliger.

In Fig. 2.21, the constant-volume, constant-pressure, and the Seiliger processes are represented together in a p, v and T, s diagram. The constant-volume process has the highest, while

2.3 Thermodynamics of the internal combustion engine

the constant-pressure process has the lowest efficiency. The efficiency of the Seiliger cycle lies between them. In this comparison, the compression ratio and the supplied heat amount are equal for all three cycles. This shows clearly that, in the case of the Seiliger cycle some, in the case of the constant-pressure process unmistakably more heat must be removed as in the constant-volume process and therefore that the thermal efficiency of these processes is lower.

2.3.3 Open comparative processes

- **The process of the perfect engine**

The simple cycles depart partially, yet significantly from the real engine process, such that no detailed statements are possible about the actual engine process. Therefore, for further investigations we also consider open comparative processes, which consider the chemical transformation involved in combustion instead of the heat supply and removal of the closed cycles.

As opposed to closed cycles, the open comparative cycles allow a charge changing, and they calculate the high pressure process in a stepwise fashion and are thus more or less realistically. However, the charge changing is, as a rule, also not considered more closely. The essential differences to the closed cycles are:

- compression and expansion are either considered isentropically, as previously, or they are described via polytropic state changes.
- energy release via combustion is calculated gradually, if also with certain idealizations with reference to the combustion itself.
- an approach to the consideration of energy losses resulting from heat transfer is made.

Fig. 2.22 shows an open comparative process in the T, s diagram. Exactly speaking, the fresh load "appears" at point 1 and the exhaust gas, which is viewed as mixture burned to an arbitrary extent, "disappears" at point 4.

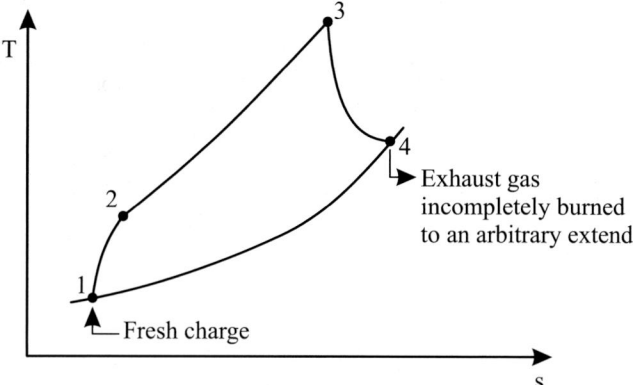

Fig. 2.22: Ideal process for internal combustion engines

The charge changing is not considered here any further. For additional details, refer to Pischinger et al. (2002).

- **Heat release through combustion**

A distinction is made between incomplete/complete and perfect/imperfect combustion.

For air ratios $\lambda \geq 1$, the fuel could in principle burn completely, i.e. the added energy $m_{fuel}\,lhv$ is completely converted into thermal energy

$$Q_{max} = Q_{th} = m_{fuel}\,lhv \ .$$

For complete combustion of hydrocarbons, the total reaction equations

$$H_2 + \tfrac{1}{2}O_2 = H_2O$$
$$C + O_2 = CO_2 \ .$$

are valid. Thus, only the products water and carbon dioxide are formed.

In actuality, however, combustion advances maximally until chemical equilibrium for air ratios $\lambda \geq 1$ as well, thus always incompletely.

For air ratios $\lambda < 1$, the fuel cannot completely burn as a result of a lack of O_2. In the case of such incomplete combustion, the combustion proceeds, in the best case, until chemical equilibrium.

Under all air ratios, combustion can further progress imperfectly, in the case that the oxygen present is not sufficiently optimally distributed (mixture formation), or in the case that single reactions progress slowly and thus that chemical equilibrium is never reached. In the exhaust, we therefore find not only CO_2 and H_2O, but also carbon monoxide, unburned hydrocarbons, soot particles, and nitrogen compounds.

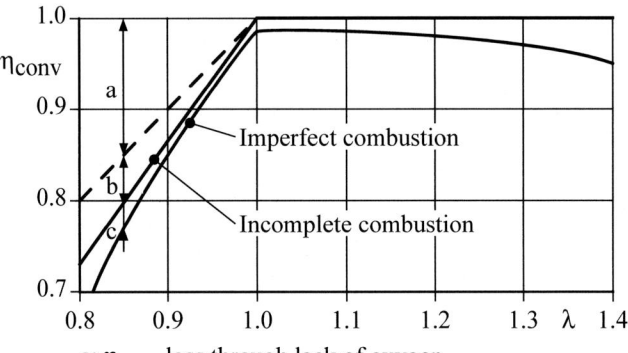

a: η_{conv}-loss through lack of oxygen
b: η_{conv}-loss through incomplete combustion
c: η_{conv}-loss through imperfect combustion

Fig. 2.23: Release of energy and degree of conversion

2.3 Thermodynamics of the internal combustion engine

The degree of conversion is defined as

$$\eta_{conv} = 1 - \frac{Q_{ub}}{m_{fuel} \, lhv} \, .$$

According to Pischinger et al. (1989), the total degree of conversion can be written thus:

$$\eta_{conv,total} = \eta_{conv,ch} \cdot \eta_{conv}$$

with ub = unburned and ch = equilibrium.

On the basis of reaction-kinetic estimations, Schmidt et al. (1996) provide the term

$$\eta_{conv,ch} = \begin{cases} 1 & \text{for} \quad \lambda \geq 1 \\ 1.3773\,\lambda - 0.3773 & \text{for} \quad \lambda \leq 1 \end{cases}$$

for the degree of conversion $\eta_{conv,ch}$. These conditions are clarified visually in Fig. 2.23.

- **The real engine process**

Proceeding from the perfect engine process, the actual efficiency of the real engine process can be established through gradual abandonment of particular idealizations. Practically, particular losses are thereby considered via corresponding reductions in efficiency, e.g.

Volumetric efficiency $\Delta\eta_{rch}$: Loss with respect to the entire engine because the actual cylinder filling at "intake closes" is smaller than that of the ideal engine

Combustion $\Delta\eta_{ic}$: Loss as a result of incomplete or imperfect combustion

Heat transfer p_{wh}: Heat losses through heat transfer to the combustion chamber walls

Charge changing $\Delta\eta_{cc}$: Charge changing losses

Blow-by $\Delta\eta_{bb}$: Leakage

Friction $\Delta\eta_m$: Mechanical losses due to engine friction (piston-piston rings-cylinder liner, bearing) and accessory drives (valve train, oil and water pump, injection pump).

Particular losses are visually clarified in Fig. 2.24. We will not go into any further detail, one reason being that these simple considerations are of increasingly less importance.

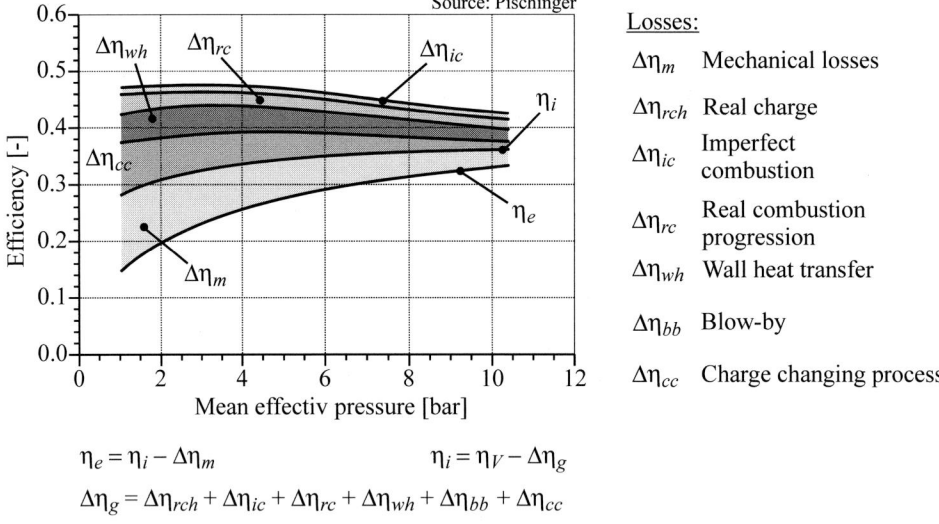

Fig. 2.24: Division of losses in the real internal combustion engine

2.4 Characteristic qualities and characteristic values

Characteristic quantities of internal combustion engines are important with reference to the interpretation and determination of engine measurements, the examination and establishment of the actual performance and the evaluation and comparison of various combustion machines.

The *medium pressure* is an important characteristic quantity for the evaluation of performance and of the technological standing of an internal combustion engine. From the definition of piston work

$$dW = p\, A_p\, dx = p\, dV$$

we obtain through integration over a working cycle for the indicated work per cycle

$$W_i = \oint p\, dV$$

and with the definition

$$W_i = imep\, V_d$$

for the indicated mean effective pressure

$$imep = \frac{1}{V_d} \oint p\, dV \ . \tag{2.51}$$

For the indicated or internal performance of a multi-cylinder engine follows

$$P_i = P_{i,cyl}\, z = z\, n_{wc}\, imep\, V_d \ .$$

2.4 Characteristic qualities and characteristic values

With the number of working cycles per time

$$n_{wc} = i \cdot n \quad \text{with} \quad i = \begin{cases} 0.5 & \text{for 4-stroke} \\ 1 & \text{for 2-stroke} \end{cases}$$

we finally obtain for the indicated total performance for

$$P_i = i\, z\, n\, imep\, V_d \,. \tag{2.52}$$

Analogous to this, we obtain, with the actual mean effective pressure *mep*

$$P_e = i\, z\, n\, mep\, V_d \tag{2.53}$$

for the actual total performance. The actual performance is the difference from indicated performance and friction performance

$$P_e = P_i - P_{fric}, \tag{2.54}$$

from which follows the relation

$$fmep = imep - mep \tag{2.55}$$

for the friction mean effective pressure.

The internal performance of an engine follows from the so-called indicator diagram, the actual performance follows from

$$P_e = T\, 2\pi\, n \tag{2.56}$$

whereby the torque T and the speed n are determined on an engine torque stand.

The *efficiency* of a thermal energy conversion machine is quite generally the relation of benefit to expenditure. In the case of the internal combustion engine, the benefit the indicated or actual engine performance and the expenditure the energy added with the fuel mass flow $\dot{m}_{fuel}\, lhv$. With this, it follows

$$\eta_{i,e} = \frac{P_{i,e}}{\dot{m}_{fuel}\, lhv} \,. \tag{2.57}$$

The ratio of actual to indicated performance is the mechanical efficiency

$$\eta_m = \frac{\eta_e}{\eta_i} = \frac{P_e}{P_i} = \frac{mep}{imep} \,. \tag{2.58}$$

The *specific fuel consumption* is the fuel consumption related to the engine performance

$$b_e = \frac{\dot{m}_{fuel}}{P_e} = \frac{1}{\eta_e\, lhv} \tag{2.59}$$

With a medium value for the lower heating value of gasoline and diesel oil of about

$$lhv \approx 42{,}000 \left[\frac{kJ}{kg}\right]$$

one obtains the simple rule-of-thumb between the specific fuel consumption and the actual efficiency

$$b_e \approx \frac{86}{\eta_e} \left[\frac{g}{kWh} \right].$$

An actual efficiency of $\eta_e = 40\%$, for example, leads to a specific fuel consumption of $b_e = 215\,g/kWh$.

We distinguish between upper and lower heating values. In the determination of the upper heating value, the combustion products and cooled back to intake temperature, the water contained in it is condensed out and is thus liquid. As opposed to this, in the determination of the lower heating value, the water is not condensed out and is thus in a vapor state. For internal combustion engines, the lower heating value must be used because of the relatively high exhaust gas temperature.

A so-called mixture heating value is occasionally utilized, under which is meant the added energy flow in reference to the fresh charge. For SI and diesel engines, we retain different expressions for this, because one takes in a gasoline-air mixture and the other pure air.

$$H_m = \frac{m_{fuel} lhv}{V_m} \quad \text{with} \quad V_m = \frac{m_{air} + m_{fuel}}{\rho_m} \qquad \text{SI engine}$$

$$H_m = \frac{m_{fuel} lhv}{V_{air}} \quad \text{with} \quad V_{air} = \frac{m_{air}}{\rho_{air}} \qquad \text{diesel engine}.$$

The ratio of added fresh charge to the theoretically possible charge mass m_{th} is designated as air expenditure,

$$\lambda_{air} = \frac{m_m}{m_{th}} = \frac{m_m}{V_d \rho_{th}}. \tag{2.60}$$

The theoretical charge density is the density of the intake valve.

As opposed to the air expenditure, the *volumetric efficiency* designates the ratio of charge mass actually found in the cylinder after charge changing in comparison to the theoretically possible charge mass.

$$\lambda_L = \frac{m_{cyl}}{m_{th}} = \frac{m_{cyl}}{V_d \rho_{th}}. \tag{2.61}$$

Volumetric efficiency λ_L is contingent above all on valve overlap in the charge changing-TDC. An optimization of the volumetric efficiency can, with constant control times, only ensue for one speed. With variable valve timing (e.g. by retarding the camshaft), volumetric efficiency can be optimized across the entire speed range. For 4-stroke engines with little valve overlap, $\lambda_L \approx \lambda_a$ is valid.

In addition to the above-cited quantities, a few other characteristic quantities are used. The medium piston speed is a characteristic speed for internal combustion engines,

$$c_m = 2\,s\,n. \tag{2.62}$$

The maximal piston speed is dependant on the eccentric rod ratio and lies in the range $c_{max} = (1.6 - 1.7)c_m$.

The compression ratio is the total cylinder volume in reference to the compression volume,

$$\varepsilon = 1 + \frac{V_d}{V_c} \tag{2.63}$$

The displacement (cubic capacity) is the difference between total volume and compression volume. With the piston path s results for this

$$V_d = \frac{\pi}{4} D^2 s \ . \tag{2.64}$$

As a further characteristic quantity, the bore/stroke ratio

$$\frac{s}{D} \tag{2.65}$$

is used.

2.5 Engine maps

2.5.1 Spark ignition engines

In the case of the conventional *spark ignition (SI) engine*, gasoline is sprayed in the intake port directly in front of the intake valve (multi point injection). Through this, a mixture of air and fuel is taken in and compressed after the closure of the intake valve. Before arrival at the top dead center, the compressed mixture is ignited by means of a sparkplug (spark ignition). Because two phases, the intake and compression strokes, are available for mixture formation, the mixture is nearly homogenous at the end of compression. In conventional engines, the amount of inducted air is regulated by means of a throttle valve positioned in the suction line. At light throttle conditions, this throttle valve is almost closed, while at full loads is it completely open. The amount of inducted air is measured and the fuel injected in proportion to the amount of air, normally such that the medium air ratio $\lambda = 1$ is maintained (see chapter 4). Because the amount of mixture is regulated in SI engines, we speak of a quantity regulation. So that no autoignition takes place in the compressed mixture, the compression ratio ε has to be restricted.

Fig. 2.25 shows the p, v diagram for a 4-stroke SI engine at partial load (left) and full load (right). The throttle valve, nearly closed at partial load, results in high pressure losses in the suction hose. It thus leads to a "large charge changing slide" and finally to inferior efficiency.

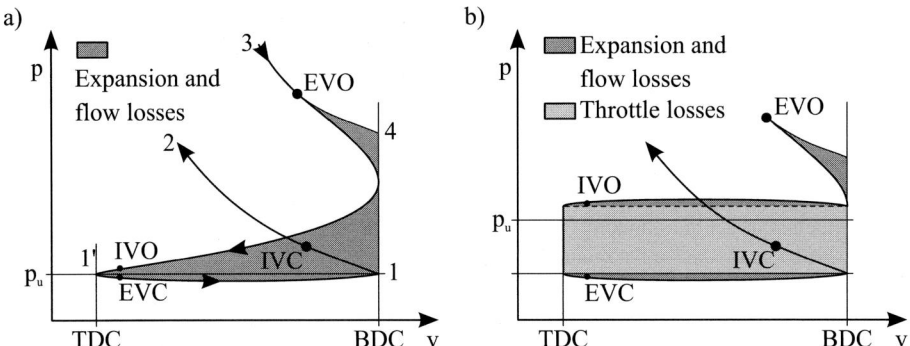

Fig. 2.25: p, v diagram for a 4-stroke SI engine under a) full load and b) partial load

Fig. 2.26 shows the engine map for a 4-stroke SI engine. The characteristic map is restricted by the idle and limit speeds as well as by the maximal torque line. Because $P \sim T \cdot n$, the lines of constant performance are hyperbolas in the engine map. The so-called chondiodal curves are lines of constant specific consumption.

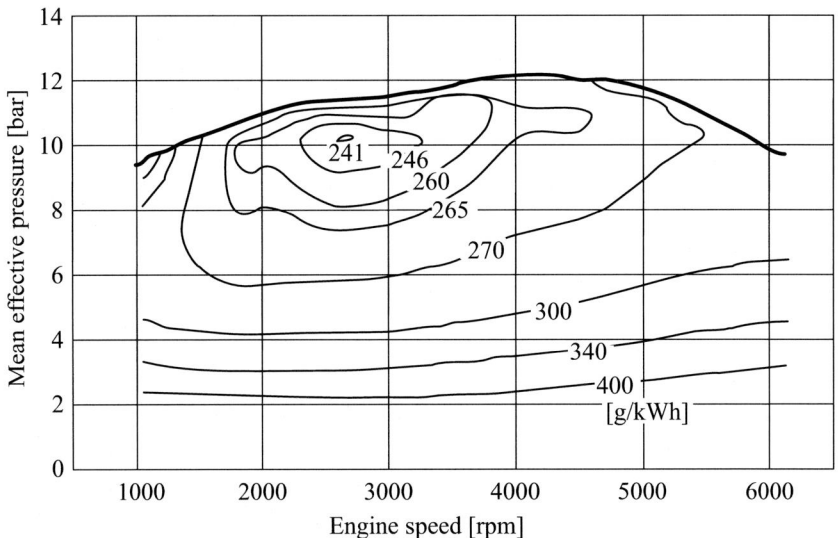

Fig. 2.26: Engine map for a 4-stroke SI engine with lines of constant consumption

The behavior and characteristics of the internal combustion engine can be read off from the engine map. From the relations for performance and for torque (load)

$$P_e = i\, n\, z\, mep\, V_d = T_e\, 2\pi n$$

we obtain the contingencies

2.5 Engine maps

power: $P_e \sim n \, mep$

load: $T_e \sim mep$.

The load thereby corresponds to the torque and not to performance!

For the ratio of actual to indicated fuel consumption follows

$$\frac{b_e}{b_i} = \frac{P_i}{P_e} = \frac{P_e + P_{fric}}{P_e} = 1 + \frac{P_{fric}}{P_e} = 1 + \frac{fmep}{mep}.$$

The friction mean effective pressure *fmep* is nearly proportional to the speed. Therefore at a constant speed, b_e must increase with a sinking actual medium pressure.

Flexibility F of the internal combustion engine is the slope of the torque at the rated load point

$$F \equiv -\left(\frac{dT}{dn}\right)_{n_{rated}}. \tag{2.66}$$

As an approach,

$$F^* \equiv \frac{T_{max}}{T_{rated}} \frac{n_{rated}}{n_{T_{max}}} \tag{2.67}$$

is also often used.

In a flexible engine, the speed lowers under increasing loads, but under constant performance less so as in an inflexible one.

2.5.2 Diesel engines

In the case of the conventional diesel engine, only air is taken in and compressed. The fuel (diesel oil) is injected just before the top dead center into the hot air. Because of the high air ratios, the temperature of the compressed air is clearly higher than the autoignition temperature of the fuel, and after the so-called ignition lag time (see chapter 4) autoignition begins. In contrast to the SI engine, no homogeneous mixture can form in the short time between injection start and autoignition: injection, mixture formation, and combustion therefore proceed in partial simultaneity. The regulation of the diesel engine ensues with the amount of injected fuel, one thus refers to a quality regulation. While in the conventional SI engine the air ratio is always $\lambda = 1$, it varies in the case of the diesel engine with the load and moves within the region $1.1 \leq \lambda \leq 10$.

Fig. 2.27 shows the engine map of a 4-stroke diesel engine. We recognize that the speed spread is clearly narrower and that the actual medium pressure is evidently higher than in the case of the SI engine.

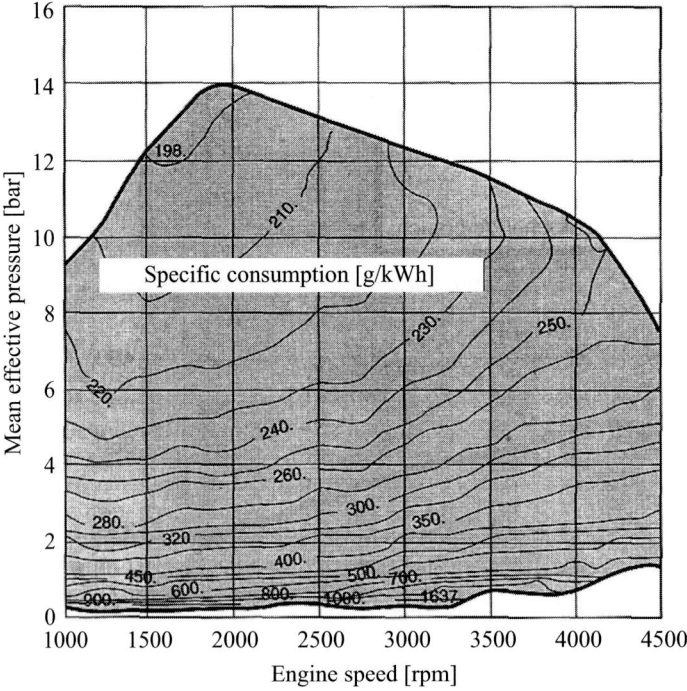

Fig. 2.27: Engine map of a 4-stroke diesel engine

With reference to their speeds, internal combustion engines are divided into high-speed engines, medium-speed engines, and slow-speed engines, with reference to their construction sizes, however, into vehicle, industrial, and large engines as well. On the other hand, engines built for racing have a special position, particularly because extremely light construction and high performance are principally stressed. With that we have the classification provided in Fig. 2.28.

Type	$\dfrac{n}{\text{rpm}}$	$\dfrac{\text{mep}}{\text{bar}}$	η_e	ε	$\dfrac{c_m}{\text{m/s}}$
Passenger cars - SI	< 7000	8 - 13	0.25 - 0.35	6 - 12	9 - 20
Passenger cars - diesel	< 5000	7 - 22	0.30 - 0.40	16 - 22	9 - 16
Trucks - diesel	< 3000	15 - 25	0.30 - 0.45	10 - 22	9 - 14
High-speed engines	1000 - 2500	10 - 30	0.30 - 0.45	11 - 20	7 - 12
Medium-speed engines	150 - 1000	15 - 25	< 0.5	11 - 15	5 - 10
Slow-speed engines	50 - 150	9 - 15	< 0.55	11 - 15	5 - 7
Racing engines		12 - 35	- 0.3	7 - 11	< 25

Fig. 2.28: Classification of internal combustion engines

This classification is certainly not compulsory and is not without a certain amount of arbitrariness, yet it is still practical and comprehensible. We could, in principle, expand it further in consideration of the categories mini-engines (model air planes), small engines (chain saws), and motorcycle engines.

2.6 Charging

Charging was originally considered a performance improvement method. It has, however, been playing an increasingly large role, whereby consumption and emission questions have stepped more into the foreground. For an exhaustive treatment, the reader is referred to Zinner (1985) and Pischinger et al. (2002). Jenni (1993) has provided an interesting portrait of the historical development of charging.

2.6.1 Charging methods

We distinguish between external and internal charging, see Fig. 2.29.

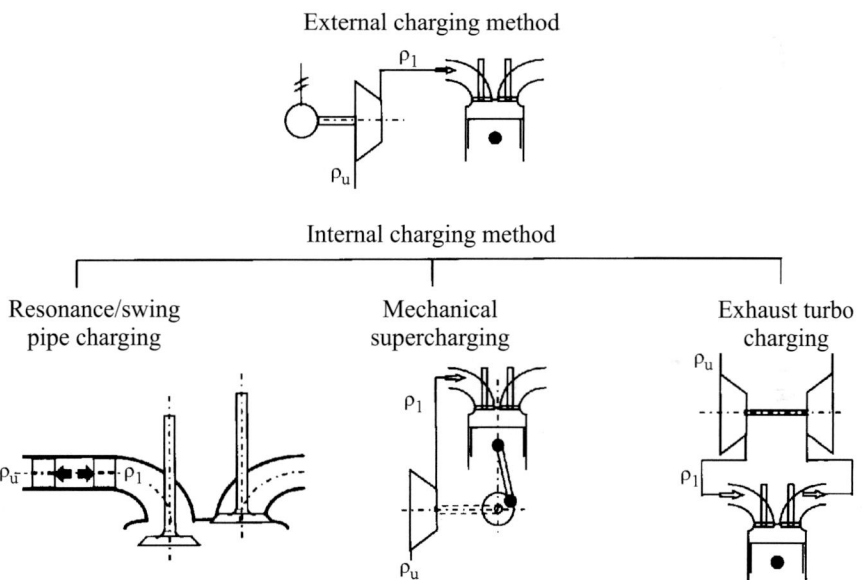

Fig. 2.29: Classification of charging methods

Under the concept of external charging we categorize the externally driven charge-air compressor in one-cylinder engines and the scavenging fan in large 2-stroke engines. Internal charging can be divided into

- *resonance or swing-pipe charging,*
  whereby pipe vibrations in the suction tube are used for charging through the adjustment of the length and diameter of this tube,

- *mechanical charging*,
  whereby a supercharger/compressor mounted on the engine is run,
- *turbocharging (TC)*,
  whereby a charge-air compressor, under the utilization of exhaust gas energy, is run by means of an exhaust turbine, i.e. the reciprocating engine is linked with the flow machine "turbocharger" only fluid-mechanically, see Fig. 2.30.

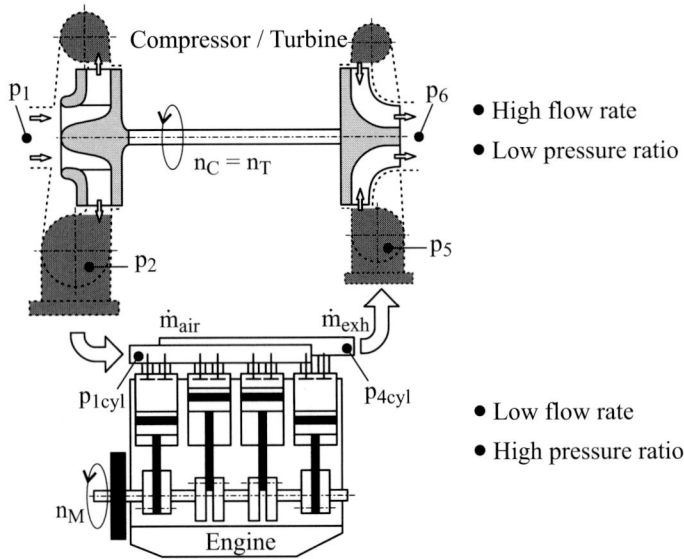

Fig. 2.30: Reciprocating engine and turbo-machine. Indices: 1: b.c., 2: a.c., 3: b.t., 4: a.t.

In the case of turbocharging, we distinguish between:
- constant-pressure turbocharging, whereby only the thermal energy of the exhaust gas is utilized; and under this we distinguish between
  - single-stage
  - and two-stage charging
- and pulse turbocharging, whereby both the thermal and the chemical energy of the exhaust gas is utilized.

Register charging is a one or two-stage constant-pressure turbocharging method, in which
- several equally large turbochargers, or
- several turbochargers of varying sizes

are connected one after the other, i.e. with ascending engine load and speed. It is employed in large medium-speed and high-speed high performance diesel engines, *two-stage register constant-pressure turbocharging*, however, is only used in high-speed high performance diesel engines.

2.6 Charging

We understand under the *concept of composite method* a combination of varying charging methods in one and the same engine, e.g.:

- mechanical charging for light load operation,
- TC for medium load operation, and
- TC and effective turbine for full load operation.

Fig. 2.31 shows the principle sketch of single- and two-stage turbocharging with intercooling. Under the presumption that the same boost pressure is reached respectively, as a result of intercooling (= low pressure-charge air cooler), the compression work in two-stage charging is smaller than in the single-stage method; correspondingly, the total amount of heat to be removed is also smaller. Under utilization of infinitely many intercoolers, an isothermal compression would theoretically be realized.

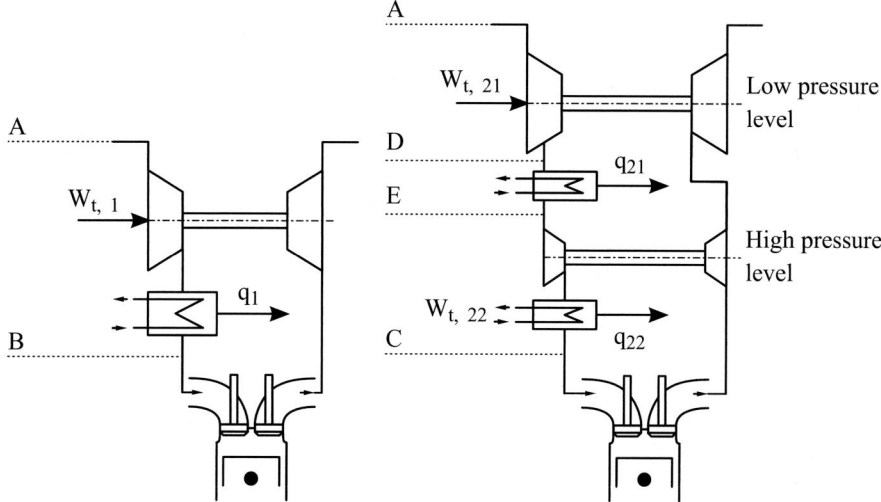

Fig. 2.31: Single- and two-stage exhaust turbocharging

2.6.2 Supercharging

In supercharging, the engine and compressor speeds are rigidly coupled. The actual performance of the engine while running a mechanically driven supercharger comes out to

$$P_e = P_i \, \eta_m - P_c \frac{1}{\eta_g} \qquad (2.68)$$

with a transmission efficiency η_g. For the compressor performance, we obtain

$$P_c = \dot{m}_c \, \Delta h_{is,c} \frac{1}{\eta_{is,c}} \frac{1}{\eta_{m,c}} \; . \qquad (2.69)$$

Should the flow velocities in front of and behind the compressor be about the same, it then follows from the 1st law of thermodynamics for the isentropic downhill grade

$$\Delta h_{is,c} = c_p \, T_{b.c.} \left[\left(\frac{p_{a.c.}}{p_{b.c.}} \right)^{\frac{\kappa-1}{\kappa}} - 1 \right]$$

and for the isentropic efficiency

$$\eta_{is,c} = \frac{\Delta h_{is,c}}{\Delta h_c} = \frac{T_{a.c.s} - T_{b.c.}}{T_{a.c.} - T_{b.c.}}.$$

With these relations, we obtain for the temperature ratio at the compressor

$$\frac{T_{a.c.}}{T_{b.c.}} = 1 + \frac{1}{\eta_{is,c}} \left[\left(\frac{p_{a.c.}}{p_{b.c.}} \right)^{\frac{\kappa-1}{\kappa}} - 1 \right]. \tag{2.70}$$

2.6.3 Constant-pressure turbocharging

In turbocharging, the reciprocating engine (displacement machine) is only connected to the TC (flow machine) fluid-mechanically, and therefore the speed of the TC is perfectly independent of the engine speed. In the case of the stationarily running TC, the performance provided by the turbine must be equal to that received by the compressor. For the performance balance at the TC is thus valid, see also Fig. 2.32,

$$P_c = P_t \tag{2.71}$$

with the performance of the compressor from (2.69)

$$P_c = \dot{m}_c \, \Delta h_{is,c} \, \frac{1}{\eta_{is,c} \, \eta_{m,c}}$$

and the performance of the turbine

$$P_t = \dot{m}_t \, \Delta h_{is,t} \, \eta_{is,t} \, \eta_{m,t}, \tag{2.72}$$

if the friction performance of the wheel assembly in the mechanical efficiency of the turbine and the compressor are taken into account.

In Fig. 2.33, the conditions before and after the compressor and turbine in the h, s diagram are illustrated.

2.6 Charging

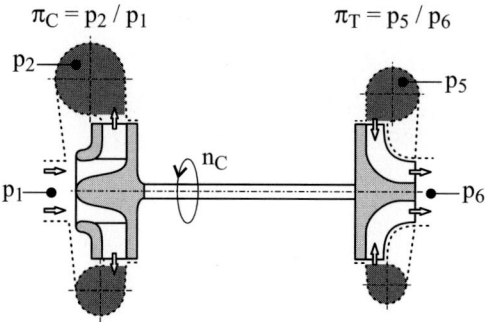

Fig. 2.32: Performance balance of a turbocharger, Indices: 1: b.c., 2: a.c., 3: b.t., 4: a.t.

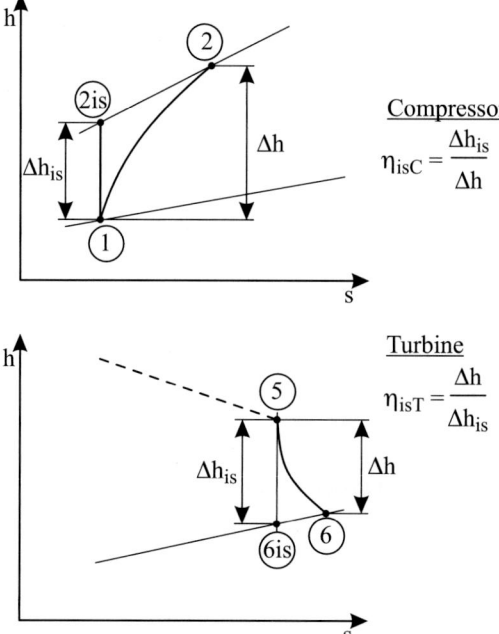

Fig. 2.33: h, s diagram for compressor and turbine

By equating the compressor and turbine performances, it follows from (2.71)

$$\underbrace{\underbrace{\frac{\dot{m}_t}{\dot{m}_c} \underbrace{\eta_{is,t}\, \eta_{is,c}}_{\eta_{TC}} \underbrace{\eta_{m,t}\, \eta_{m,c}}_{\eta_{n,TC}}}_{\eta^*_{TC}}} = \frac{\Delta h_{is,c}}{\Delta h_{is,t}} \;. \qquad (2.73)$$

With the isentropic downhill grade for the compressor

$$\Delta h_{is,c} = c_{pc} T_{b.c.} \left(\frac{T_{a.c.}}{T_{b.c.}} - 1 \right) = c_{pc} T_{b.c.} \left[\left(\frac{P_{a.c.}}{P_{b.c.}} \right)^{\frac{\kappa_c - 1}{\kappa_c}} - 1 \right] \quad (2.74)$$

and for the turbine

$$\Delta h_{is,t} = c_{pt} T_{b.t.} \left(1 - \frac{T_{a.t.}}{T_{b.t.}} \right) = c_{pt} T_{b.t.} \left[1 - \left(\frac{P_{a.t.}}{P_{b.t.}} \right)^{\frac{\kappa_t - 1}{\kappa_t}} \right] \quad (2.75)$$

one finally obtains the freewheel condition or also the so-called 1st fundamental equation of turbocharging

$$\frac{c_{pt}}{c_{pc}} \eta_{TC}^* \frac{T_{b.t.}}{T_{b.c.}} = \frac{\left(\frac{P_{a.c.}}{P_{b.c.}} \right)^{\frac{\kappa_c - 1}{\kappa_c}} - 1}{1 - \left(\frac{P_{a.t.}}{P_{b.t.}} \right)^{\frac{\kappa_t - 1}{\kappa_t}}} \quad (2.76)$$

For the pressure ratio at the compressor, it follows from this

$$\frac{P_{a.c.}}{P_{b.c.}} = f\left(\frac{P_{b.t.}}{P_{a.t.}}, \eta_{TC}^*, \frac{T_{a.t.}}{T_{b.c.}} \right). \quad (2.77)$$

Because the composition of the fresh air flowing through the compressor and the exhaust gas flowing through the turbine is different, a differentiation must be made between material values c_p and κ for both of these gas flows, which should be expressed by the additional indices c and t. Since both the mass flows $\dot{m}_c$ and $\dot{m}_t$ and the specific heat capacities c_{pc} and c_{pt} differ only slightly, an approach to considering these differences can be made in the modified efficiency η_{TC}^*.

The mass flow through the turbine, the so-called turbine absorption capacity, can be roughly determined with the help of the relation for flow through restriction (throttle). Thus, with the isentropic equivalent cross section of the turbine, we obtain

$$\dot{m}_t = A_{is,t} \rho_{is} v_{is}$$

with

$$\rho_{is} = \frac{P_{b.t.}}{R_t T_{b.t.}} \left(\frac{P_{a.t.}}{P_{b.t.}} \right)^{\frac{1}{\kappa_t}}$$

and

2.6 Charging

$$v_{is} = \sqrt{\frac{2\,\kappa_t}{\kappa_t - 1}\, R_t\, T_{b.t.} \left[1 - \left(\frac{p_{a.t.}}{p_{b.t.}}\right)^{\frac{\kappa_t - 1}{\kappa_t}}\right]}.$$

Substituting ρ_{is} and v_{is} in the relation for $\dot{m}_t$ and conversion finally provides the expression

$$\dot{m}_t \frac{\sqrt{T_{b.t.}}}{p_{b.t.}} = A_{is,t} \sqrt{\frac{2\,\kappa_t}{R_t\,(\kappa_t - 1)} \left[\left(\frac{p_{a.t.}}{p_{b.t.}}\right)^{\frac{2}{\kappa_t}} - \left(\frac{p_{a.t.}}{p_{b.t.}}\right)^{\frac{\kappa_t + 1}{\kappa_t}}\right]}. \qquad (2.78)$$

This is the so-called 2nd fundamental equation of turbocharging

$$\dot{m}_t \frac{\sqrt{T_{b.t.}}}{p_{b.t.}} = A_{is,t}\, f\!\left(\frac{p_{a.t.}}{p_{b.t.}},\, \kappa_t\right). \qquad (2.79)$$

With the help of this relation, reference or dimensionless quantities can be produced, with which we will deal further in chapter 7.

The isentropic turbine equivalent cross section is approximately constant and contingent only on the geometric turbine cross-section; dependence upon the speed of the turbocharger and the pressure ratio $p_{b.t.}/p_{a.t.}$ is thereby ignored.

We immediately see that the chosen turbine cross-section and thus the chosen turbine is only optimal for a certain operation point. For lower engine speeds as the selected one, a turbine with a smaller cross-section would be necessary, for higher speeds a larger turbine. At lower speeds, an excessively large turbine leads to a torque/performance drop. A modulation valid for a larger speed range can be achieved through

- charge-air release (boost pressure restriction)
- exhaust gas release (wastegate)
- a turbine with an adjustable geometry (VTG charger), i.e. with a variable turbine cross-section
- register charging.

2.6.4 Pulse turbocharging

In pulse turbocharging, we keep the volume of the exhaust pipe very small. In this way, a very fast filling of the pipe is achieved and a large part of the kinetic energy of the exhaust gas is thereby retained. As opposed to this, in the case of constant-pressure turbocharging, the conversion of kinetic energy into pressure energy is fraught with losses. In Fig. 2.34, pressure and pulse turbocharging are presented in an h, s diagram. The principle differences are clearly recognizable. The enthalpy downhill grade at the turbine is evidently larger in the case of pulse turbocharging. Yet the turbine is unsteadily admitted.

Because the pressure in the exhaust pipe is no longer constant, in the case of multicylinder engines, one can only amalgamate those cylinders in which the pressure impacts in the exhaust system do not reciprocally disturb the charge changing. Therefore, the ignition distance from the amalgamated cylinders should be larger than the opening duration of an exhaust valve, from which, according to experience, an ignition distance of at least 240° CA results for a disturbance-free region.

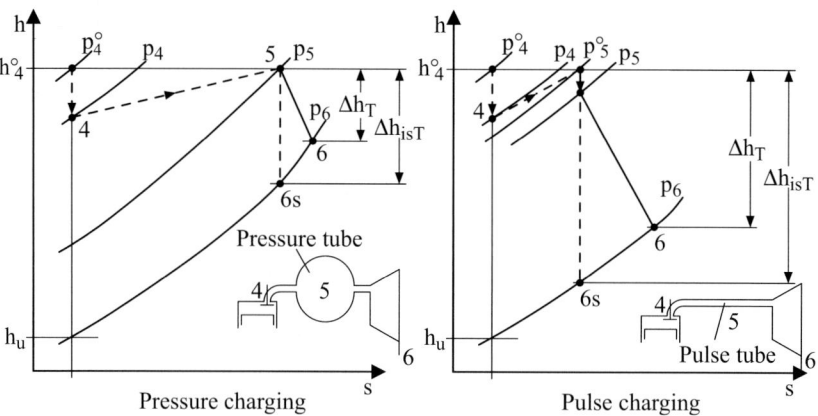

Fig. 2.34: Pressure and pulse turbocharging in the h, s diagram

Fig. 2.35 shows the diagram of a pulse turbocharged 6-cylinder 4-stroke diesel engine, whereby the exhaust pipes of cylinder 1, 2, and 3 as well as those of 4, 5, and 6 are respectively amalgamated. Both of these multiple vents are lead separately to the turbine. The turbine is equipped with a so-called twin spiral casing.

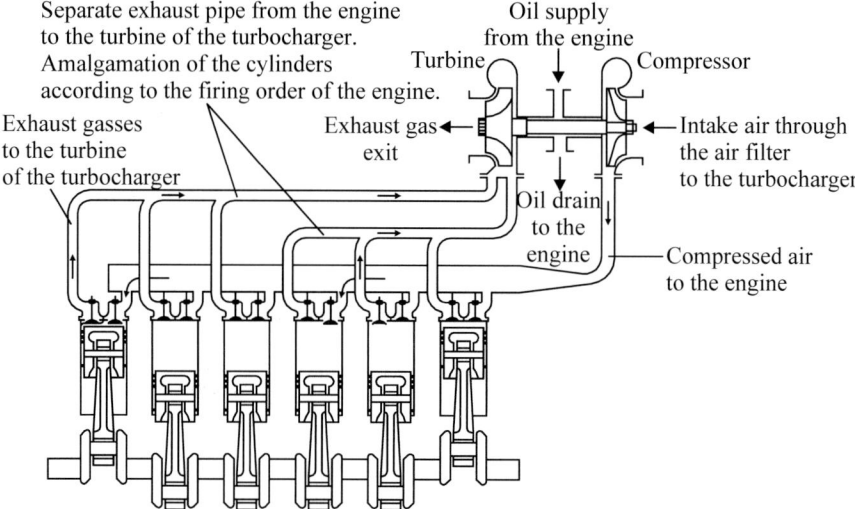

Fig. 2.35: Pulse turbocharged 6-cylinder, 4-stroke engine

2.6 Charging

In conclusion, Fig. 2.36 shows the engine map realizable with varying charging methods for a fast-running high performance diesel engine. With two-stage constant-pressure turbocharging in register mode, we see a very broad map, with single-stage pulse turbocharging, on the other hand, a relatively constant torque at high performance.

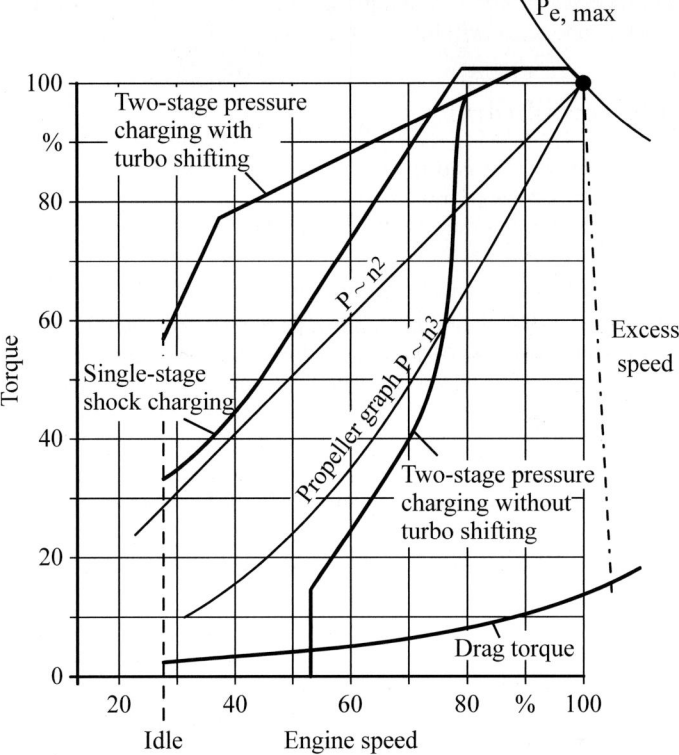

Fig. 2.36: Comparison of different charging methods

3 Foundations of reaction kinetics

In view of the processes of combustion and pollutant formation to be described later, the foundations of reaction kinetics will be introduced in the following.

3.1 Chemical equilibrium

A chemical reaction between the educts A_a, A_b, etc., which form the products A_c, A_d, etc. can be described in the following form

$$v_a A_a + v_b A_b + \cdots \rightarrow v_c A_c + v_d A_d + \cdots . \tag{3.1}$$

The v_i thereby designate the so-called stoichiometric coefficients of the reaction. Since every chemical reaction can in principle run both forwards as well as backwards, the reaction arrow in (3.1) can be replaced with an equal sign. We thereby obtain the general form of the reaction equation

$$\sum_i v_i A_i = 0 , \tag{3.2}$$

whereby the stoichiometric coefficients are conventionally negative for all educts and positive for all products.

Every chemical reaction strives towards its state of equilibrium, which is reached under the condition that sufficient time is available. This state of equilibrium can be interpreted as a situation, in which both the forward as well as the reverse reactions progresses with identical speed. The macroscopically visible reaction rate is thus reduced to zero, and the material composition no longer changes. This material composition in the state of equilibrium can, with the help of both laws of thermodynamics, be determined contingent upon the boundary values of temperature and pressure. This method will be demonstrated in the following.

For a closed, compressible system with constant temperature and constant pressure, the 1st and 2nd laws of thermodynamics read as follows

$$dU = dQ + dW = dQ - p\,dV , \tag{3.3}$$

$$dS = \frac{dQ}{T} + dS_{irr} . \tag{3.4}$$

dS_{irr} thereby designates the increase of entropy due to irreversibilities, which is always greater than or equal to zero. The combination of (3.3) and (3.4) produces therefore the inequality

$$T\,dS - dU - p\,dV \geq 0 . \tag{3.5}$$

If we introduce to this the Gibbs free enthalpy G,

$$G = H - TS = U + pV - TS , \tag{3.6}$$

we then obtain, after differentiation and corresponding conversion

3.1 Chemical equilibrium

$$dG - V\,dp + S\,dT \leq 0 \,. \tag{3.7}$$

For a closed system with constant temperature and constant pressure, the dissipation of Gibbs free energy is thus always less or equal to zero. That means that every alteration of the composition through chemical reactions reduces the value of G and that the condition

$$dG\big|_{T,p} = 0 \tag{3.8}$$

is satisfied in chemical equilibrium.

For a single-phase multicomponent-system, e.g. a combustible gas, the Gibbs free energy is a function of temperature, pressure and composition,

$$G = G(T, p, n_1, n_2, n_3, \ldots) \,, \tag{3.9}$$

whereby n_i represents the number of moles of the various species i. Now the chemical potential μ_i is introduced. This is defined as the partial derivative of the Gibbs free energy with respect to the number of moles of i,

$$\mu_i = \frac{\partial G}{\partial n_i}\bigg|_{T,p,n_j} \,, \quad j \neq i \,. \tag{3.10}$$

It can be shown that for an ideal gas – and this is an acceptable assumption for most combustible gasses – the chemical potential is identical to the molar Gibbs function (Moran and Shapiro, 1992).

$$\mu_i = \tilde{g}_i(T, p_i) = \tilde{g}_i^\circ + \tilde{R}\,T \ln \frac{p_i}{p^\circ} \,. \tag{3.11}$$

The index "$^\circ$" thereby represents the condition under a reference pressure of 1 atm. The first term on the right side of the equation (3.11) stands for

$$\tilde{g}_i^\circ = \tilde{h}_i^\circ(T) - T\,\tilde{s}_i^\circ(T) \tag{3.12}$$

and can thus be taken from tabulated property data. It should be pointed out that the molar enthalpy consists of the standard enthalpy of formation and a temperature dependant term

$$\tilde{h}_i(T) = \tilde{h}_{f,i}^\circ + \Delta \tilde{h}_i(T) \,. \tag{3.13}$$

If the chemical potential introduced in (3.10) is inserted into the condition of equilibrium (3.7), the form

$$dG\big|_{T,p} = \sum_i \mu_i \, dn_i = 0 \tag{3.14}$$

is obtained.

However, for a general chemical reaction corresponding to (3.2), changes in the material quantities dn_i are proportional to the corresponding stoichiometric coefficients, so that the equation

$$dn_i = \nu_i \, d\varepsilon \,, \tag{3.15}$$

with the proportionality factor $d\varepsilon$ is satisfied for all components i. The equation (3.14) can thus be simplified to the form

$$\sum_i \mu_i \nu_i = 0 , \tag{3.16}$$

which contains all the necessary information regarding the equilibrium composition. However, the solution of (3.16) according to the various species concentrations is only iteratively possible and thus very costly. In order to avoid this problem, the concept of equilibrium constant is introduced. Through the insertion of (3.11) into (3.16), we obtain the relation

$$\sum_i \nu_i \tilde{g}_i^\circ + RT \ln \prod_i \left(\frac{p_i}{p^\circ} \right)^{\nu_i} = 0 . \tag{3.17}$$

where the logarithmic term corresponds to the equilibrium constant

$$K_p = \prod_i \left(\frac{p_i}{p^\circ} \right)^{\nu_i} . \tag{3.18}$$

This equilibrium constant K_p now contains the information about the equilibrium material composition in terms of the partial pressures p_i of the various species i. Since the first term of (3.17) is exclusively dependant on the temperature (see (3.12)), it is evident that the equilibrium constant K_p is also only a function of T. It can easily be calculated with the help of thermodynamic property data and tabulated for any reaction. With (3.18), it follows from (3.17)

$$\ln K_p = \frac{-\sum_i \nu_i \tilde{g}_i^\circ}{\tilde{R}T} = \frac{-\Delta_R \tilde{g}^\circ}{\tilde{R}T} . \tag{3.19}$$

The numerator in (3.19) is customarily designated as free molar reaction enthalpy. With the help of (3.18), the equilibrium composition for a system in which a single chemical reaction elapses, e.g. $(CO + \frac{1}{2}O_2 = CO_2)$, can now be solved for specific temperature and pressure conditions. However, in addition to (3.18), two further conditions are required, since a total of three unknowns, namely the partial pressures of CO, O_2, and CO_2, must be calculated. Both of these conditions result from the fact that the absolute number of atoms of an element does not change during a chemical reaction. Because we usually work with the partial pressures of the components and not with their absolute atomic or molecular number, it is sensible to express the atomic balances as a ratio. This is possible because the ratio of two constants is always itself a constant. For the example reaction $(CO + \frac{1}{2}O_2 = CO_2)$, one obtains accordingly for the atomic number ratio $\xi_{C/O}$ of carbon to oxygen atoms before and after the reaction

$$\xi_{C/O}\big|_{educts} \equiv \xi_{C/O}\big|_{products} = \frac{p_{CO} + p_{CO_2}}{p_{CO} + 2p_{O_2} + 2p_{CO_2}} ,$$

whereby the atomic number ratio before the reaction is known from the material quantities of the educts which are to be reacted with each other

$$\xi_{C/O}\big|_{educts} = \frac{n_{CO}}{n_{CO} + 2n_{O_2}} .$$

Through the ratio formation of both atomic balances, one has however lost an independent equation. This can be substituted with Dalton's law, which states that the sum of all partial pressures corresponds to the system pressure

$$\sum_i p_i = p_{sys} .$$

3.2 Reaction kinetics

On the micro-scale, i.e. on the molecular level, a chemical reaction always progresses, as is given for example in (3.1), in both forward and reverse directions. The macroscopic reaction direction then results from the simple difference between forward and reverse reactions. Thus, the chemical equilibrium represents a special case, in which the forward and reverse reactions each run equally fast, so that no macroscopically visible material conversion occurs. On the molecular level however, reactions still progress in both directions. While the macroscopic reaction rate is always aimed in the direction of chemical equilibrium, the equilibrium analysis however does not provide any information regarding the absolute reaction rates, i.e. regarding the time necessary for the attainment of chemical equilibrium. This information is supplied by the *reaction kinetics*.

For the chemical reaction specified in (3.1), the temporal change of a species concentration, e.g. for $[A_c]$, can be given with the empirical formulation

$$\frac{d[A_c]}{dt} = v_c \left(\underbrace{k_f [A_a]^{v_a} [A_b]^{v_b}}_{\text{forward}} - \underbrace{k_r [A_c]^{v_c} [A_d]^{v_d}}_{\text{reverse}} \right), \qquad (3.20)$$

whereby the first term on the right side describes the reaction rate of the forward direction and the second term the rate of the reverse reaction. k_f and k_r are thereby the so-called rate coefficients of the forward and reverse reactions. They must be experimentally determined for every particular chemical reaction, e.g. in experiments in shock tubes. Since the speed coefficients of most reactions are strongly dependent on temperature, they are customarily represented with an Arrhenius formulation of the form

$$k = AT^b \exp\left(-\frac{E_A}{\tilde{R}T}\right). \qquad (3.21)$$

The constant A and the exponent b as well as the so-called activation energy E_A are summarized for many chemical reactions in extensive tables, for example Warnatz et al. (2001).

It is sufficient to be familiar with the speed coefficient of either the forward or the reverse reaction. The other respective coefficient can then be determined simply with the inclusion of the corresponding equilibrium constant. This becomes clear if we consider that, in the special case of chemical equilibrium, the integral conversion rate becomes zero, since the reaction in both directions progress equally fast. If we substitute these conditions in (3.20), we obtain

$$\frac{k_f}{k_r} = \frac{[A_c]^{\nu_c}[A_d]^{\nu_d}}{[A_a]^{\nu_a}[A_b]^{\nu_b}} \equiv K_c ,\qquad(3.22)$$

whereby K_c is the equilibrium constant defined in reliance on the species concentrations. Through the relation

$$K_c = K_p \left(\frac{p^\circ}{\tilde{R}T}\right)^{\sum \nu_i} \qquad(3.23)$$

it is clearly coupled with the equilibrium constant K_p, defined in reliance upon the partial pressures and introduced in (3.18) and (3.19). Since both the speed coefficients and the equilibrium constants depend exclusively on the temperature and not on the actual species concentrations, the relation

$$\frac{k_f}{k_r} = K_p \left(\frac{p^\circ}{\tilde{R}T}\right)^{\sum \nu_i} \qquad(3.24)$$

is valid not only for the equilibrium state but also in general.

3.3 Partial equilibrium and quasi-steady-state

In an extensive reaction system with a high number of reactions between participating species, we speak of a *partial equilibrium* if several reactions (not necessarily all) progress fast enough that the assumption of equilibrium between the species appearing in the reaction is justifiable at all times. This does not mean, however, that the absolute concentrations of the species found in partial equilibrium have to be temporally constant. Under altered boundary conditions, the species concentrations can certainly change in time. However, on the assumption of infinitely fast forward and reverse reactions, these concentration changes of all participating species are solidly coupled to each other, such that the determination of species concentrations can be very much simplified: in the case of partial equilibrium, the partial pressures of the corresponding species can be determined in analogy with the method for a single reaction. However, the number of unknowns (i.e. the partial pressures of the species) is now larger, so that additional equations must be set up in order to solve the system. These equations are obtained when we determine an equilibrium constant for every reaction found in the partial equilibrium and put this according to (3.18) in relation to the corresponding partial pressures.

One example for the appearance of partial equilibrium are the reactions between the species $CO, CO_2, H, H_2, H_2O, O, O_2$ and OH directly within the flame and also within the hot combustion products in internal combustion engines. The concentrations of these eight species, which contain a total of three different elements $(C, O$ and $H)$, can be calculated via five linearly independent reaction equations each found in partial equilibrium. These reactions are, for example,

3.3 Partial equilibrium and quasi-steady-state

$$\begin{aligned} 2H + M &= H_2 + M, \\ 2O + M &= O_2 + M, \\ \tfrac{1}{2}H + OH &= H_2O, \\ \tfrac{1}{2}O_2 + H_2 &= H_2O, \\ \tfrac{1}{2}O_2 + CO &= CO_2, \end{aligned} \tag{3.25}$$

whereby M is a collision partner which does not participate in the reaction. For a more extensive presentation, the reader is referred to Warnatz et al. (2001). The further three equations necessary for the solution of the eight unknown partial pressures is to be obtained from the atom balances of the three participating atoms.

A condition is generally characterized as *quasi-steady*, if in a two-step reaction

$$A \xrightarrow{k_{f,1}} B \xrightarrow{k_{f,2}} C \tag{3.26}$$

the second reaction step progresses much faster than the first, i.e. if $k_{f,1} \ll k_{f,2}$. In this case, it can be assumed that the total mass quantity of B, which is formed in the first partial reaction, is immediately decomposed through the much faster second partial reaction. With this, the temporal rate of change of the concentration of B is approximately equal to zero, i.e. B is quasi-steady

$$\frac{d[B]}{dt} \approx 0 \ . \tag{3.27}$$

Through this assumption, the time dependent determination of the concentrations of the participating species A, B and C can be extremely simplified, as will be shown in the following. In accordance with chapter 3.2, the concentration change rates

$$\frac{d[A]}{dt} = -k_{f,1}[A], \quad \frac{d[B]}{dt} = k_{f,1}[A] - k_{f,2}[B], \quad \frac{d[C]}{dt} = k_{f,2}[B] \tag{3.28}$$

result for the reaction (3.26). The integration of the equations (3.28) under consideration of initial conditions

$$[A]_{t=0} = A_0, \quad [B]_{t=0} = [C]_{t=0} = 0 \tag{3.29}$$

then produce the time dependent course of concentrations of A, B and C, which are schematically presented in Fig. 3.1 a) for the case $k_{f,1} \ll k_{f,2}$

$$[A] = A_0 \exp(-k_{f,1} t), \tag{3.30}$$

$$[B] = A_0 \frac{k_{f,1}}{k_{f,1} - k_{f,2}} \left[\exp(-k_{f,2} t) - \exp(-k_{f,1} t) \right], \tag{3.31}$$

$$[C] = A_0 \left[1 - \frac{k_{f,1}}{k_{f,1} - k_{f,2}} \exp(-k_{f,2} t) + \frac{k_{f,2}}{k_{f,1} - k_{f,2}} \exp(-k_{f,1} t) \right]. \tag{3.32}$$

If we substitute the simplification (3.27) for the rate of change of $[B]$ in (3.28), we obtain instead of (3.31)

$$k_{f,1}[A] = k_{f,2}[B] \quad \Leftrightarrow \quad [B] = \frac{k_{f,1}}{k_{f,2}} A_0 \exp[-k_{f,1} t]. \tag{3.33}$$

(3.32) thus finally simplifies itself to

$$[C] = A_0 \left[1 - \exp(-k_{f,1} t)\right]. \tag{3.34}$$

The concentration progressions of A, B and C for the assumption of quasi-steady-state of B are represented in Fig. 3.1 b). In comparison to the exact solution in Fig. 3.1 a), it becomes clear that only at the very start of the reaction is the concentration of B incorrectly illustrated. During the most part of the reaction time, the solution determined with the help of the quasi-steady-state assumption agrees however quite exactly with the exact solution. One typical example for a quasi-steady-state in engine combustion is the nitrogen atom N in thermal NO formation. We will go into this example further in chapter 6.5.

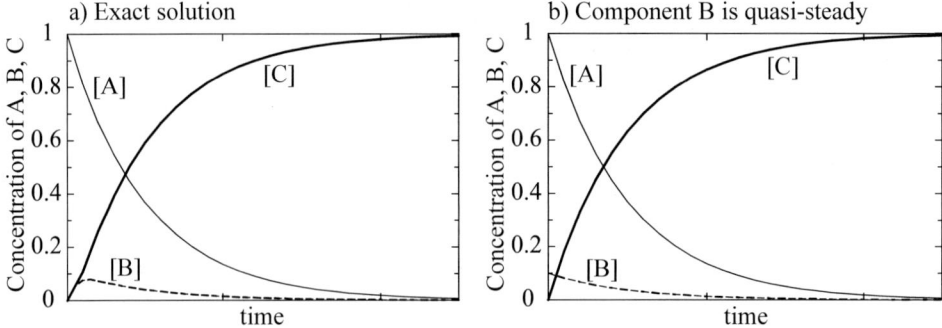

Fig. 3.1: Progression of concentrations of A, B and C under the condition $k_{f,1} \ll k_{f,2}$ in the sequential reaction ($A \xrightarrow{k_{f,1}} B \xrightarrow{k_{f,2}} C$) a) exact solution b) assuming that B is quasi-steady

3.4 Fuels

3.4.1 Chemical structure

SI and diesel engine fuels are each mixtures of several hundred different hydrocarbon components ($C_x H_y$). These components differ with reference to molecular size and structure and as a result have sometimes strongly varying physical and chemical properties. In the following, the chemical structure of the most important $C_x H_y$ compounds will be described. In addition, several oxygenous components ($C_x H_y O_z$) will be presented, as they are important during the combustion process.

3.4 Fuels

Alkanes $C_n H_{2n+2}$ (formerly: parrafins)
Chain hydrocarbons with only simple bonds

Normal-paraffins	Isoparaffin
straight-chain formed 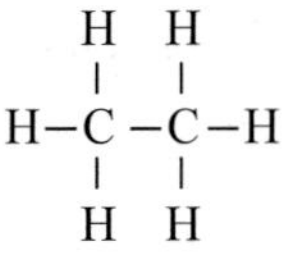	branched-chain-formed 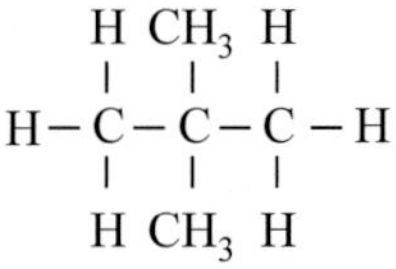

```
        H  H                H  CH₃ H
        |  |                |   |  |
      H-C -C-H            H-C - C- C - H
        |  |                |   |  |
        H  H                H  CH₃ H
      Ethane             2,2 Dimethylpropane
```

Alkenes C_nH_{2n} (formerly: olefins)
Chain hydrocarbons with double bonds (DB)

Alkenes (monoolefins)	Alkadienes (diolefins)
Chain-formed, <u>one</u> DBs	Chain-formed, <u>two</u> DBs

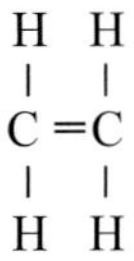

```
       H  H          C_nH_{2n-2}  H      H
       |  |                       |      |
       C =C                       C = C = C
       |  |                       |      |
       H  H                       H      H
      Ethene              Propadiene (allene)
```

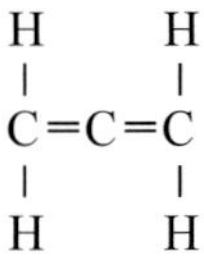

Alkines C_nH_{2n-2} (formerly: acetylenes)
Chain hydrocarbons with a triple bond

$$H-C \equiv C-H \qquad \text{Ethin}$$

Fig. 3.2: Aliphatic hydrocarbons

In the case of $C_x H_y$ compounds, we usually distinguish between alkanes (formerly: parraffins), alkenes (olefines), alkines (acetylenes), cycloalkanes (naphtenes) and aromatic compounds.

Alkanes are chain hydrocarbons with only simple bonds. One distinguishes between:
- n-alkanes, with straight chain structures and
- isoalkanes, with branched chain structures

Alkenes are chain hydrocarbons with one or two double bonds, whereby
- alkenes exhibit one double bond and
- alkadienes exhibit two double bonds.

Cykloalkanes are ring hydrocarbons with only single bonds.

Aromatic compounds are ring hydrocarbons with conjugated double bonds, the basic building block of which is the benzene ring.

The structural formulae for a few simple compounds are given for chain hydrocarbons in Fig. 3.2 and for ring hydrocarbons in Fig. 3.3.

Cycloalkane C_nH_{2n} (formerly naphtene)
Ring hydrocarbons with simple bonds

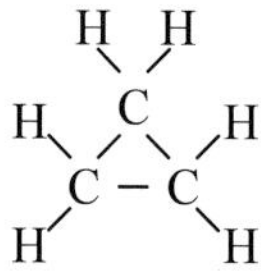

Cyclopropane 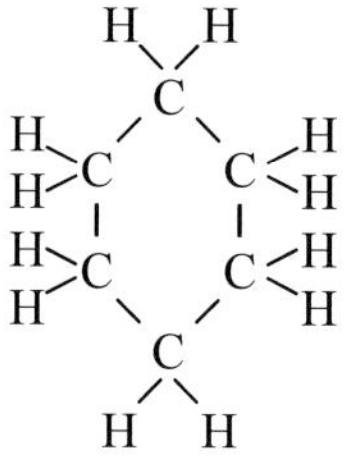   Cyclohexane

Aromatic compounds
Ring hydrocarbons with conjugated double bonds

Foundation is the benzene ring

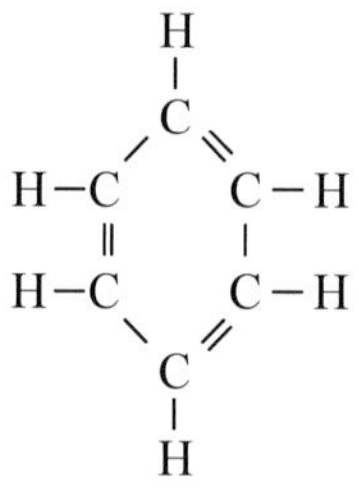

Benzene 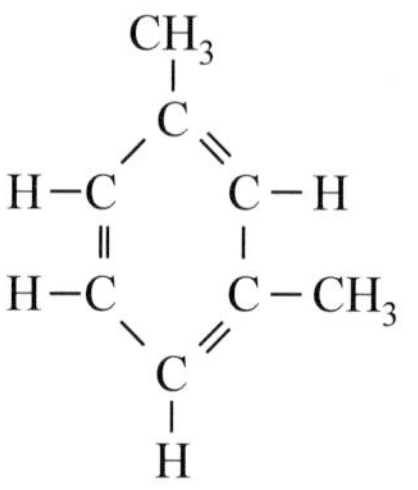   1,3 Dimethylbenzene

Fig. 3.3 Alicyclic and aromatic hydrocarbons

Oxygenous hydrocarbons are chain compounds, by which we distinguish between alcohols, ethers, ketones and aldehydes.

Alcohols contain a hydroxyl group ($R-OH$). The simplest alcohols are methyl alcohol (methanol, CH_3OH), and ethyl alcohol (ethanol, C_2H_5OH). *Ethers* are hydrocarbon chains connected together over an oxygen bridge (R_1-O-R_2). *Ketones* are hydrocarbon remains connected together over a carbonyl group (R_1-CO-R_2). *Aldehydes* contain a CHO group, e.g. formaldehyde $HCHO$. One should not confuse the CHO group with the OH group (-COH) connected to carbon. The differences are made clear by the two structural formulae

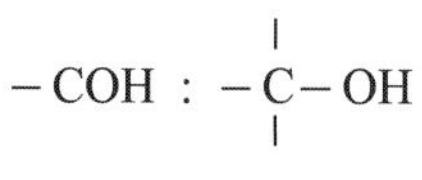

$$-COH : -\overset{|}{\underset{|}{C}}-OH \quad \text{and} \quad -CHO : -C\overset{H}{\underset{O\,..}{\diagup}}$$

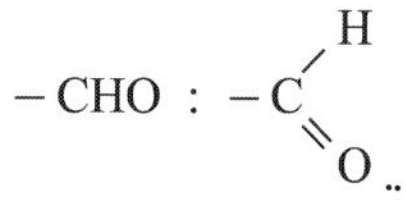

3.4 Fuels

Alcohols, R - OH

contain a hydroxyl group -OH

Methanol   Ethanol
(Methyl alcohol)   (Ethyl alcohol)

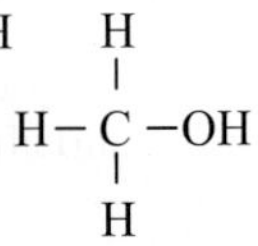

CH_3OH C_2H_5OH

$$H-\underset{\underset{H}{|}}{\overset{\overset{H}{|}}{C}}-OH \qquad H-\underset{\underset{H}{|}}{\overset{\overset{H}{|}}{C}}-\underset{\underset{H}{|}}{\overset{\overset{H}{|}}{C}}-OH$$

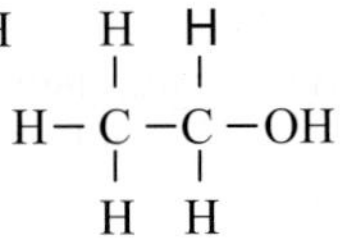

Ethers, R_1 - O - R_2

are hydrocarbon remains bound together over an O-bridge (R_1,R_2)

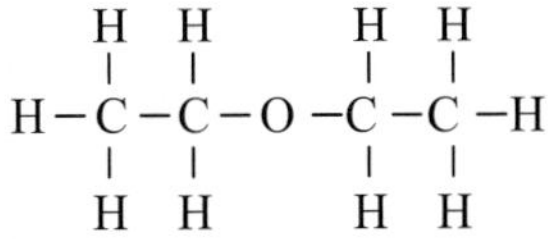

$$H-\underset{\underset{H}{|}}{\overset{\overset{H}{|}}{C}}-\underset{\underset{H}{|}}{\overset{\overset{H}{|}}{C}}-O-\underset{\underset{H}{|}}{\overset{\overset{H}{|}}{C}}-\underset{\underset{H}{|}}{\overset{\overset{H}{|}}{C}}-H$$

Diethyl ether
$C_2H_5-O-C_2H_5$

Ketones, R_1 - CO - R_2

are hydrocarbon remains bound together over a carbonyl group

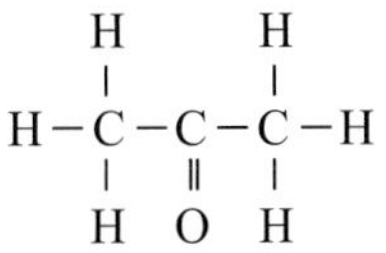

$$H-\underset{\underset{H}{|}}{\overset{\overset{H}{|}}{C}}-\underset{\underset{O}{\|}}{C}-\underset{\underset{H}{|}}{\overset{\overset{H}{|}}{C}}-H$$

Acetone
$CH_3-\underset{\underset{O}{\|}}{C}-CH_3$

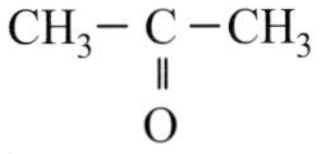

Aldehydes, R - CHO

contain a -CHO- group

Formaldehyde

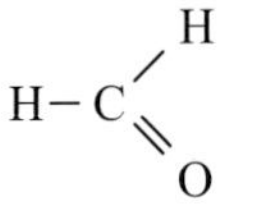

$$H-C\underset{\diagdown O}{\diagup H}$$

Fig. 3.4: Structural formulae for oxygenous hydrocarbons

In Fig. 3.4, the chemical structure of oxygenous hydrocarbons is depicted with the help of basic examples.

3.4.2 Physical and chemical properties

The *boiling temperature* of hydrocarbons rises with the number of carbon atoms, whereby the structure of the molecule itself is of lesser importance. The boiling graphs for SI and diesel fuels as well as for kerosine and water are shown in Fig. 3.5. Gasoline fuel is essentially less volatile than diesel fuel, while kerosine lies approximately in the middle of both. As a comparison, the boiling graph of water has also been drawn.

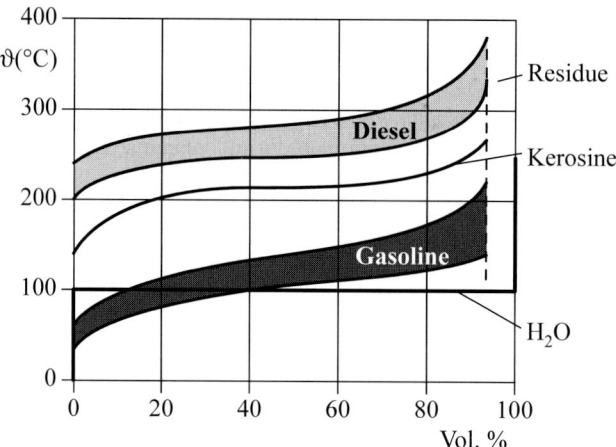

Fig. 3.5: Boiling graph for gasoline and diesel fuel, as well as for kerosine and water

The *lower heating value* of these compounds lies in the region

$$40.200 \, \text{kJ/kg} \, (\text{benzene}) < lhv < 45.400 \, \text{kJ/kg} \, (\text{n-pentane}) \,.$$

The *density* also increases with the number of carbon atoms and lies in the region

$$600 \, \text{kg/m}^3 < \rho < 900 \, \text{kg/m}^3 \,.$$

The *viscosity* also rises with the number of carbon atoms. The *ignition performance* of gasoline and diesel fuels is described by the octane and cetane numbers.

- **Definition of the octane number (ON) for gasoline fuels**

For the determination of ignition performance, we use a so-called comparison fuel, i.e. a two-component fuel consisting of

- n-heptane (C_7H_{16}) with ON = 0 and
- isooctane (C_8H_{18}) with ON = 100.

The octane number is defined as the isooctane fraction of the comparison fuel. It is determined with a standardized single-cylinder test engine described more closely in Tab. 3.1 with the following method:

1. The compression ratio ε of the test engine run with the gasoline fuel under investigation is raised until knocking occurs.
2. Then the equivalence ratio λ is adjusted so that maximal knocking intensity is reached.
3. Finally, the test engine is run with the comparison fuel, whereby the isooctane fraction is chosen such that the same knocking intensity occurs as during the operation with the gasoline fuel.

Tab. 3.1: Test engine for the determination of the engine and research octane number for gasoline fuels

	Engine ON (EON)	Research ON (RON)
Engine speed n	600	900
Compression ratio ε	4-10	4-10
Ignition start	13°CA before TDC	26°CA before TDC ($\varepsilon = 4$)
		14°CA before TDC ($\varepsilon = 10$)
Mixture preheating	none	149°C
Air preheating	52°C	25°C
Measurement precision	± 1%	± 1%

For the augmentation of the anti-knock index of gasoline fuels, so-called *additives* are utilized. These are metallo-organic compounds that decompose at high pressure and high temperature, whereby the metal vapor, due to its large surface, functions as a reaction-hindering catalyst during the ignition lag period.

- **Definition of the cetane number (CN) for diesel fuels**

In determining ignition performance, we use a comparison fuel, which is, in this case, a two-component fuel composed of:

- α-methyl naphthaline ($C_{11}H_{10}$) with $CN = 0$ and
- n-hexadecane (cetane) ($C_{16}H_{34}$) with $CN = 100$.

The components of the comparison fuels for the determination of the cetane and octane numbers are compiled in Fig. 3.6.

The cetane number is equal to the cetane fraction of the comparison fuel. The cetane number is determined in a single-cylinder swirl chamber test engine with $V_d = 0.613 \text{dm}^3$ at $n = 900 \text{min}^{-1}$ and start of injection $= 13° \text{CA BTDC}$ (whereby the compression ratio can be regulated in the region $7 < \varepsilon < 28$) according to the following method:

- Operation of the test engine with the diesel fuel under investigation. Adjustment of the injection start at TDC through alteration of the compression ratio.
- Operation of the test engine with the comparison fuel and alteration of the cetane number until the combustion also begins at TDC.

While for gasoline fuel a poorer ignition performance and thus a high anti-knock index is desired, this is exactly the opposite as far as diesel fuel is concerned. Between the cetane and the octane number the following correlation can be established

$$CN = 60 - 0.5 \, EON$$
$$CN = 100 - RON \quad \text{for} \quad RON > 80.$$

The RON descends with an ascending number of carbon atoms in the case of n-alkanes and alkenes and ascends with increasing branchings in the case of isoalkanes and with the number of components with double bonds.

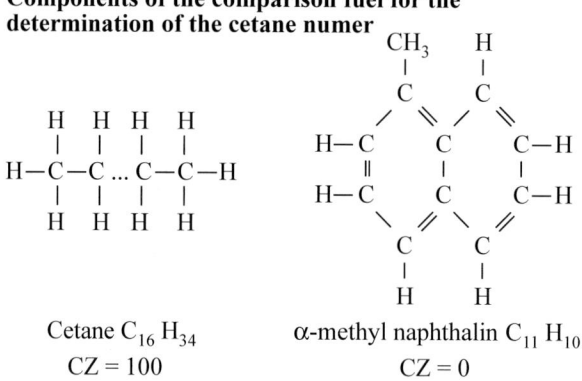

Fig. 3.6: Structural formulae of the components of substitute fuels for the determination of the cetane and octane numbers

3.5 Oxidation of hydrocarbons

The combustion of hydrocarbon components C_xH_y can be described overall via the gross reaction equation

(1) $\quad C_xH_y + \left(x + \dfrac{y}{4}\right) O_2 \rightarrow x\,CO_2 + \dfrac{y}{2} H_2O \qquad (3.35)$

However, chemical reactions proceed in general not according to the reaction diagram of this gross reaction, as one can immediately recognize in consideration of the example of methane oxidation

(2) $\quad CH_4 + 2\,O_2 \rightarrow CO_2 + 2\,H_2O \,. \qquad (3.36)$

3.5 Oxidation of hydrocarbons

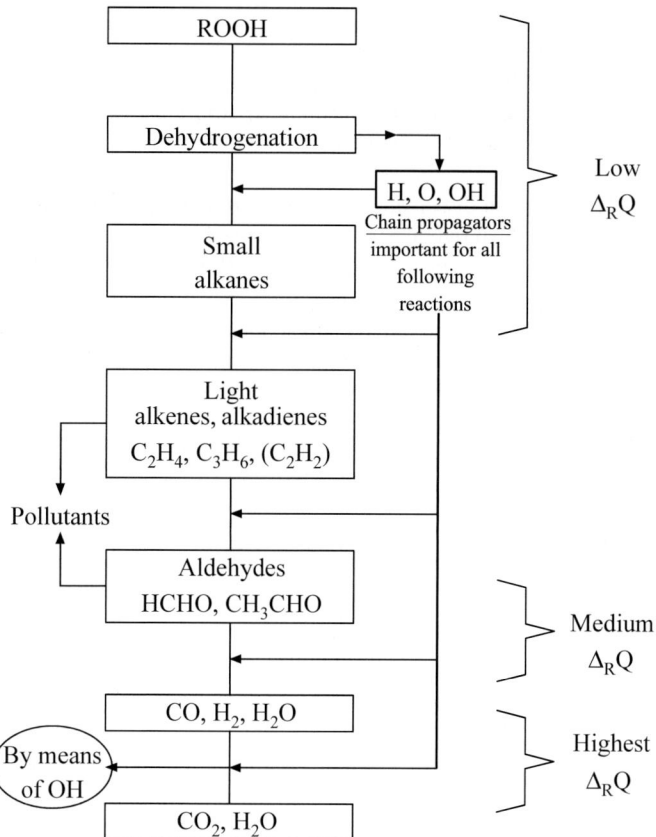

Fig. 3.7: Model of hydrocarbon oxidation

So that this reaction might occur, three molecules – two oxygen molecules and one methane molecules – should collide in a configuration appropriate for the reaction. Three-way collisions are however relatively rare. In the case of higher hydrocarbons, significantly more molecules had to collide. The probability that significantly more than three molecules simultaneously collide is however practically equal to zero. In actuality, oxidation of hydrocarbons progresses over a multitude of elementary reactions, in which a three-way collision is already relatively infrequent. We could take, as a simple example, the oxyhydrogen gas reaction with the gross reaction equation

(3) $\quad 2H_2 + O_2 \rightarrow 2H_2O$.

This reaction is relatively well described by the elementary reactions

(4) $\quad H_2 \qquad\qquad\qquad \to H^\bullet + H^\bullet$
(5) $\quad H^\bullet \quad + O_2 \quad \to OH^\bullet + O^\bullet$
(6) $\quad O^\bullet \quad + H_2 \quad \to OH^\bullet + H^\bullet$
(7) $\quad OH^\bullet \quad + H_2 \quad \to H^\bullet + H_2O$

While reaction (4) represents a dissociation process, reactions (5) and (6) are branching reactions, in which from one radical two new ones are formed. The final reaction (7) is a propagation reaction, in which water is formed as a final product.

Hydrocarbon combustion is extremely complex as has only nowadays been roughly understood. It proceeds approximately according to the oxidation diagram represented in Fig. 3.7.

In the first reaction phase, hydrocarbon peroxide (ROOH) is formed, which disintegrates through dehydrogenation into small alkanes. Through subsequent reactions with the radicals $H^\bullet, O^\bullet$ and $OH^\bullet$ (chain propagators) at first light alkenes and alkadienes and finally aldehydes, like formaldehyde HCHO and acetaldehyd CH_3CHO are developed. The formation of aldehyde, during which only about 10 % of the total energy release arises, is accompanied by the appearance of a cold flame. In the blue flame following this, CO, H_2, H_2O, and in the last place, in the hot flame finally CO_2 and H_2O are formed. In the oxidation of hydrocarbons to CO, about 30 % more and in oxidation from CO to CO_2 finally the remaining 60 % of the thermal energy in the fuel is released.

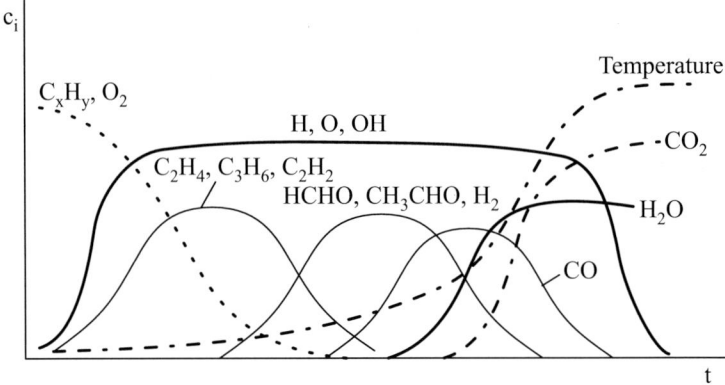

Fig. 3.8: Temporal progression of concentration in the combustion of hydrocarbons

The most essential heat release thus results only at the end of the reaction diagram in the oxidation of CO to CO_2. Fig. 3.8 qualitatively describes the course of the above-described groups and of temperature as a function of time. We thus recognize that only with the formation of CO_2 does the temperature noticeably rise. One should consider that the concentrations of the species are not represented to scale.

The reaction speed in engine combustion is essentially dependent on local conditions such as temperature, pressure, and composition and results from the reaction rates of the several hun-

3.5 Oxidation of hydrocarbons

dred to thousands of participating elementary reactions under these boundary conditions, see e.g. Warnatz et al. (2001). An illustration of all elementary reactions involved is however extremely costly, especially in the case of three-dimensional turbulent flames as they appear in internal combustion engines. Such an illustration is furthermore still full of uncertainties. Since, furthermore, the chemical reactions are often much faster than physical mechanisms such as turbulence and mixing, we are often satisfied with highly simplified formulations for a global reaction rate. Westbrook and Dryer (1981) have, for example, suggested a relation for the reaction rate of global reactions (3.35) for a selection of hydrocarbons

$$\frac{d[C_xH_y]}{dt} = -A\exp\left(-\frac{E_A}{RT}\right)[C_xH_y]^m[O_2]^n . \qquad (3.37)$$

The corresponding parameters A, m and n as well as the so-called activation temperature E_A/R are given in Tab. 3.2.

Tab. 3.2: Reaction rate parameters for the single-step reaction (3.36), acc. to Westbrook and Dryer (1981)

Fuel	A (mol, cm, s)	Activation temp. E_A/R (K)	m (-)	n (-)
CH_4	$8.3 \cdot 10^5$	15.098	-0.3	1.3
C_2H_6	$1.1 \cdot 10^{12}$	15.098	0.1	1.65
C_3H_8	$8.6 \cdot 10^{11}$	15.098	0.1	1.65
C_4H_{10}	$7.4 \cdot 10^{11}$	15.098	0.15	1.6
C_5H_{12}	$6.4 \cdot 10^{11}$	15.098	0.25	1.5
C_6H_{14}	$5.7 \cdot 10^{11}$	15.098	0.25	1.5
C_7H_{16}	$5.1 \cdot 10^{11}$	15.098	0.25	1.5
C_8H_{18}	$4.6 \cdot 10^{11}$	15.098	0.25	1.5
C_9H_{20}	$4.2 \cdot 10^{11}$	15.098	0.25	1.5
$C_{10}H_{22}$	$3.8 \cdot 10^{11}$	15.098	0.25	1.5
C_2H_4	$2.0 \cdot 10^{12}$	15.098	0.1	1.65
C_3H_6	$4.2 \cdot 10^{11}$	15.098	-0.1	1.85
C_2H_2	$6.5 \cdot 10^{12}$	15.098	0.5	1.25
CH_3OH	$3.2 \cdot 10^{12}$	15.098	0.25	1.5
C_2H_5OH	$1.5 \cdot 10^{12}$	15.098	0.15	1.6
C_6H_6	$2.0 \cdot 10^{11}$	15.098	-0.1	1.85
C_7H_8	$1.6 \cdot 10^{11}$	15.098	-0.1	1.85

4 Engine combustion

4.1 Spark ignition engines

4.1.1 Mixture formation

- **Fundamentals**

In a conventional spark ignition (SI) engine, the fuel is, as a rule, added to the air outside the combustion chamber, via carburetors in older engines and via injection into the intake manifold before the intake valve in newer engines. Load regulation of the engine occurs quantitatively, i.e. air and fuel always exist globally in the same (stoichiometric) ratio; the load is adjusted by adjusting the amount of mixture via the throttle. In direct fuel injection, the fuel is injected directly into the combustion chamber. Load regulation occurs in this case both quantitatively and qualitatively, i.e. via the mixture ratio between air and fuel. Engine performance is regulated, contingent on the load, through both fuel quantity as well as air mass.

The mixture formation process has the task of creating a mixture distribution optimal for the combustion process in question, e.g. a mixture which is as homogeneous as possible for the normal combustion process and a so-called stratified charge for the lean combustion process.

- **Multi point injection**

In the conventional method with external mixture formation, so-called single or multi point injection is used for the addition of fuel, the working method of which will be briefly described in the following. Fig. 4.1 shows the principle sketch of multi point injection, in which each cylinder intake is equipped with a separate injection valve for a consistent mixture formation. The fuel quantity specified by the electric control unit is injected with a constant pressure of up to 7 bar absolute through the injection valve with atomizing jets into the intake manifold directly in front of the intake valve. In order to achieve a fast as possible evaporation of the fuel and to restrain for the most part the development of a liquid wall-film, injection is aimed directly at the hot intake valve. The largest possible amount of time is thereby already utilized for the evaporation of the fuel before it flows into the combustion chamber. In induction pipe-synchronous injection, an attempt is made to spray directly through the open intake valve. A fast evaporation of the fuel and an even-as-possible blending of fuel and air at ignition timing in the combustion chamber are important requirements for the limitation of cyclical fluctuations characteristic for SI engines, which can, in the worst case, lead to engine knocking or misfiring (see chapter 4.1.3 and 4.1.4.). Besides the injection process, the intake flow from the intake manifold to the combustion chamber is also of essential importance for the quality of mixture formation. Mixture formation in the combustion chamber develops roughly according to the following pattern:

  - During the intake stroke, an extensive mixing between air and fuel takes place, whereby small drops with $d_{dr} \leq 20$ µm evaporate completely.
  - During the compression stroke, an intensive, small-scale mixing occurs, whereby the larger drops with diameters of $d_{dr} \approx 200$ µm also evaporate.

4.1 Spark ignition engines

- At the time of ignition, there are indeed no more drops detectable, yet inhomogenities still appear. The standard fluctuation of fuel concentration in the combustion chamber amounts to about 10-15 %.

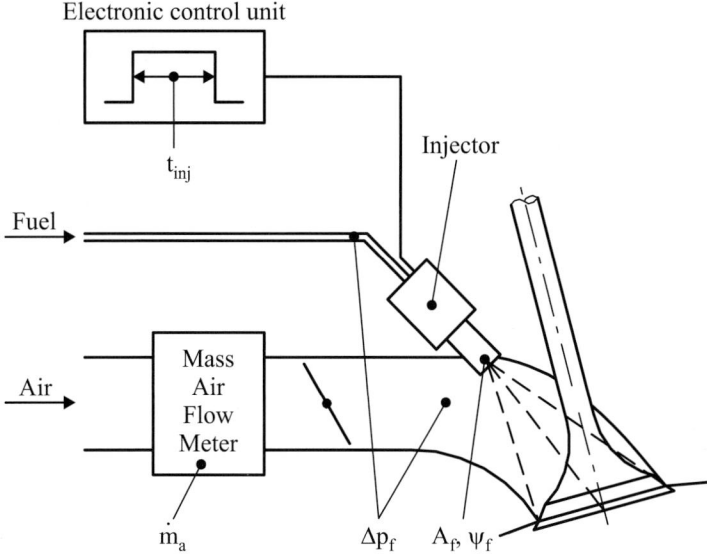

Fig. 4.1: Diagram of the principles of multi point injection

In principle, a high flow velocity and a high level of turbulence during the intake process lead to intense mixing and thus to small cyclical fluctuations. This behavior can be utilized, for example, through variabilities in intake valve control like variable camshaft adjustments or variable valve lifts, which at partial load only release a smaller flow cross section and thus increase the velocity of the air flow.

- **Direct injection**

The demand for a perceptible lowering of specific fuel consumption in the SI engine led to the development of combustion processes with internal mixture formation, i.e. with direct injection into the combustion chamber. The primary advantage that direct injection offers as opposed to conventional SI procedures with external mixture formation is the possibility to carry out a qualitative load regulation at part load instead of quantitative load regulation with the throttle valve. An engine can thereby take in air almost without throttling losses in the charge changing process. Because of the lean global air ratio ($\lambda > 1$) necessary for this combustion process, a so-called charge stratification must be realized in the combustion chamber, which secures that sufficient rich and thus ignitable mixture is in the region of the spark plug at ignition timing. These procedures are represented schematically in Fig. 4.2.

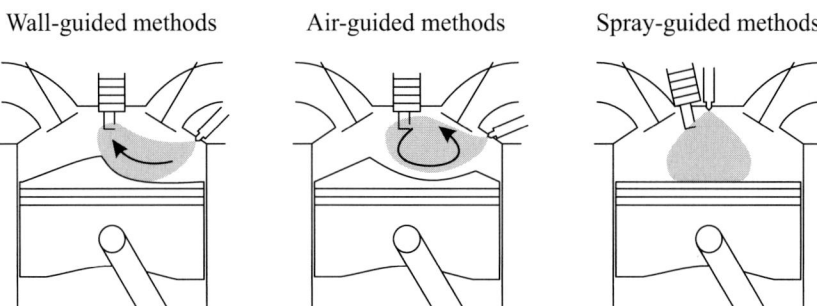

Fig. 4.2: Combustion process in direct injection spark ignition engines

In the wall-guided procedure, the mixture cloud resulting from the injection jet is led to the spark plug across a corresponding deformation ("nose") in the piston. In the air-guided process, this task is taken over by a tumble (an air swirl around a horizontal axis) directed toward the injection spray. This tumble must be generated across the geometry of the intake ports during the inflow process into the combustion chamber. Finally in the spray-guided method, the intake valve and the spark plug are located very close to each other, and the control of mixture formation and thus the obtainment of an ignitable mixture at the spark plug can result exclusively across the injection spray itself. It is obvious that this procedure represents the greatest challenge since flow condition in the combustion chamber vary significantly with engine speed and load. On the other hand, spray-guided injection offers the most potential in reference both to the lowering of fuel consumption and raw emissions and to the extension of the charge stratification area in the engine map.

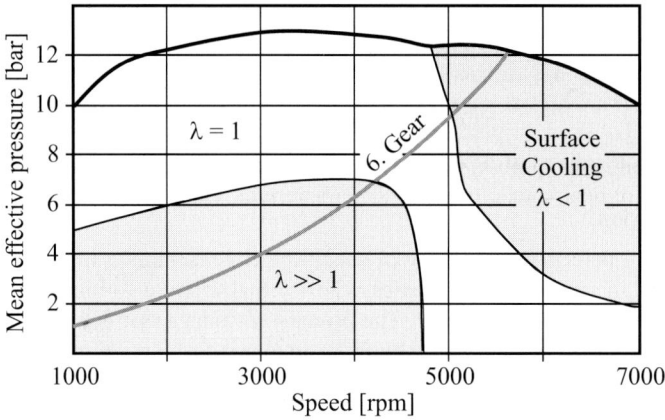

Fig. 4.3: Modes of operation in the characteristic map of a DISI engine acc. to Fröhlich et al. (2003)

Fig. 4.3 shows the extension of the charge stratification area for a spray-guided combustion process according to Fröhlich et al. (2003). As follows from this, a direct injection spark ignition (DISI) engine can not be operated in the entire speed/load map in stratified charge mode. At higher load and the large fuel quantity to be injected associated with it, the mixture cloud would be too rich and thus result in unacceptably high particulate emissions. On the other hand, at higher engine speeds, the absolute available time for mixture formation is too short to secure a sufficient mixture quality.

Besides the higher efficiency, the DISI engine can be operated at higher compression ratios (1-2 units), since the danger of knocking decreases as a result of the direct ignition (evaporation with heat absorption) of the fuel injected directly into the combustion chamber. Also resulting from the internal cooling is a greater filling and thus an approx. 3-5 % higher full load torque as opposed to an engine with multi point injection, since the latter already uses a part of the injection effect through the intake valve. As was described in chapter 2, the thermal efficiency of the engine in the entire operational range also climbs with the compression ratio.

The advantages of the DISI engine described here lead to the result that for the New European Driving Cycle, which is legally prescribed in Europe, a potential consumption advantage of approx. 15 % results compared with a customary SI engine with multi point injection.

One of the greatest challenges is still the subsequent treatment of exhaust gas for the reduction of pollutant emissions. Because of the globally lean equivalence ratio in stratified charge operation, the very effective and relatively inexpensive three-way catalytic converter known from the multi point injection engine cannot be utilized in the DISI engine. Instead, a storage-reduction catalytic converter is used for the reduction of nitric oxides, which must however be regenerated at short intervals through rich engine operation (rich air-fuel mixture), whereby fuel consumption is negatively influenced, see also Merker and Stiesch (1999). In this case, the sulfur content of the fuel in use also receives special importance, because even minute concentrations of sulfuric compounds strongly impair the catalyst function. On the one hand, these compounds block the NO_x storage locations in the catalytic converter, so that the consumption-intensive regeneration of the converter has to be carried out at shorter intervals. On the other hand, these compounds also attack the surface of the converter as a whole, so that it can become inoperative, and the service life of the converter is strongly reduced at high fuel sulfur contents. In this context, we speak of a poisoning of the converter. For this reason, the employment of SI engines with direct injection is a possible only on markets, in which fuel low in sulfur is abundantly available (e.g. Western Europe). In the USA, this is still today not the case, so that the utilization of DISI engines with a lean stratified charge concept is not on the forefront there at present.

4.1.2 Ignition

In the case of the conventional SI engine, ignition of the air-fuel mixture takes place shortly before the top dead center via spark discharge between the electrodes of the spark plug. Enough energy must be transmitted to the mixture through spark discharge that a so-called thermal explosion is introduced. The thermal explosion is explained by the theory of Semenov (1935), see below.

At the time of the beginning of the ignition, an ignitable mixture must be present at the spark plug. Mixtures that are too lean or too rich do not ignite. For gasoline fuel, the ignition boundaries lie in the area of $0.6 \leq \lambda \leq 1.6$, where λ is the inverse of equivalence ratio ϕ, i.e. $\lambda = 1/\phi$. In the area of the ignition spark, the ignition temperature of the mixture must be locally exceeded. From experiments we obtain a result of $3{,}000 \text{ K} \leq T \leq 6{,}000 \text{ K}$.

The *combustion delay* is the period of time between ignition timing and combustion start, whereby the latter is generally defined by a 5 % fuel mass conversion. The combustion delay amounts to about 1 millisecond. For a secure ignition start, the following requirements must be fulfilled: The ignition voltage is about 15 kV (normal) or 25 kV (cold start), the ignition energy thereby lies within the range of 30 – 150 mJ, and the *ignition duration* amounts to approx. 0.3 – 1 ms.

The ignition systems necessary for ignition can, in principle, be divided into two groups, namely into battery ignition and magneto ignition systems. Battery ignition is the standard today for SI engines. Magneto ignition systems are only found in small, cost sensitive engines, as well as when the maintenance of the battery cannot be guaranteed.

The ignition timing (or angle) is electronically controlled by the engine control unit contingent on engine speed, load, coolant temperature, etc., see Fig. 4.4.

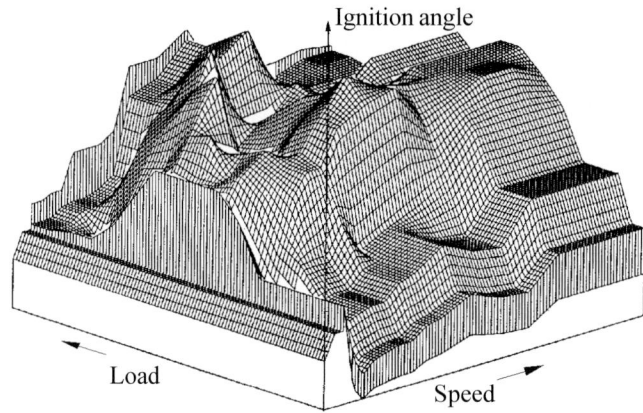

Fig. 4.4: Ignition angle map

These days, it is a matter of a static – thus electronically regulated – high-voltage distribution, in which no moving parts (i.e. distributors, contact breakers) are being used.

The *ignition timing* is not a constant value, but rather it is adjusted to the engine operation such that combustion progresses optimally, i.e. so that the combustion center is optimally adjusted. Where this is impossible, as for example under full load, the ignition is adjusted from late to early ignition start as long as necessary until a detonation sensor detects abnormal combustion (knocking). Then the ignition timing is adjusted later by a definite amount (approx. 6°CA) and the process of approaching the detonation boundary is commenced once again. Thus, via fully electronic ignition, consumption-optimal operation at varying boundary conditions (temperature, pressure of the intake air) and for varying fuels is made possible.

For a more extensive description of ignition phenomena, see Stiesch (2003).

4.1.3 The combustion process

In the homogeneously operated SI engine, the flame is initiated through the introduction of energy by means of the ignition spark, as described above. After this, a so-called premixed flame front begins to spread, which proceeds from the spark plug position through the entire combustion chamber. In Fig. 4.5, the positions of this flame front at different times is represented exemplarily for a modern 4-valve engine with a central spark plug. This illustration corresponds to a top view of the combustion chamber, whereby the small area in the middle of the cylinder describes the position of the spark plug, both large circles the projections of the intake valves, and the two intermediate circles the projections of the exhaust valves.

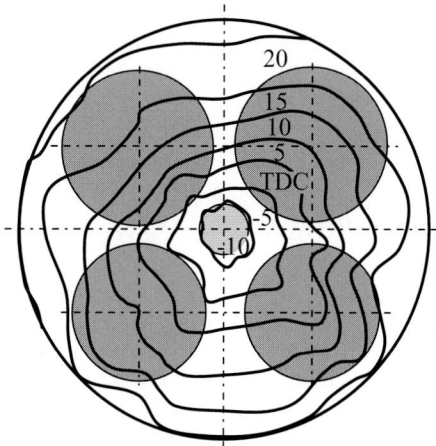

Fig. 4.5: Crank angle-dependent location of the flame front in a 4-valve engine with a central spark plug

- **Laminar and turbulent flame speed**

For the temporal energy release rate and thus also for the pressure increase in the cylinder that results from combustion, the propagation speed of the premixed flame front – the so-called flame speed – is of special importance. We hereby distinguish between laminar flame speed s_l and turbulent flame speed s_t. Laminar flame speed designates the propagation speed of a thin premixed flame front in a static (laminar) air-fuel mixture. Besides the reaction kinetics, it is dependent upon the heat conducting and diffusion processes inside the flame, and it can be assessed as a function of fuel, of the mixture ratio of fuel and air, of residual gas content, and of system pressure and the temperature of the educts. Often, the relation provided by Metghalchi and Keck (1980, 1982)

$$s_l = s_{l,0} \left(\frac{T_{env}}{T_0} \right)^\alpha \left(\frac{p}{p_0} \right)^\beta \left(1 - 2.1 f_{rg} \right) \tag{4.1}$$

is used, whereby f_{rg} designates the mass fraction of the residual gas and T_0 and p_0 the reference conditions at 298 K and 101.3 kPa. Both the exponents α and β as well as the flame speed under atmospheric conditions $s_{l,0}$ are fuel-contingent quantities. For propane, isooctane and methanol, the relations

$$\alpha = 2.18 - 0.8\left(\frac{1}{\lambda} - 1\right), \tag{4.2}$$

$$\beta = -0.16 + 0.22\left(\frac{1}{\lambda} - 1\right), \tag{4.3}$$

$$s_{l,0} = B_m + B_\lambda\left(\frac{1}{\lambda} - \frac{1}{\lambda_m}\right)^2 \tag{4.4}$$

are valid, whereby λ_m is the air-fuel equivalence ratio, at which $s_{l,0}$ reaches its maximal value of B_m. The parameters contained in (4.4) are summarized in Tab. 4.1.

Tab. 4.1: Parameters for (4.4)

fuel	λ_m	B_m [cm/s]	B_λ [cm/s]
methanol	0.9	36.9	-140.5
propane	0.93	34.2	-138.7
isooctane	0.88	26.3	-84.7
gasoline	0.83	30.5	-54.9

However, in real engine combustion chambers, the flow field is not laminar, but highly turbulent. Thus, we must additionally consider the influence of turbulence on the propagation speed of the flame front.

Through the interaction with turbulent swirls, the flame front, smooth in the laminar case, becomes wrinkled, so that its surface A_l becomes larger. But locally, the flame still burns with laminar speed. The enlargement of the surface via turbulent flame-wrinkling thus actually contributes to the increase of burning velocity. We for this reason introduce the turbulent flame speed s_t, which is calculated from s_l and the local turbulence level

$$u' = \sqrt{\frac{2k}{3}},$$

where u' is an average, local speed component due to turbulent fluctuations, see chapter 9.1.2. The Damköhler relation (Damköhler, 1940) represents a basic approach to this

$$s_t = \left(1 + C\frac{u'}{s_l}\right)s_l. \tag{4.5}$$

Equation (4.5) makes it clear that turbulent (actual) flame speed increases with higher levels of turbulence. This is the reason why SI engines can be run at much higher speeds than, for example, diesel engines. This is only possible because the level of turbulence in the combustion chamber caused by the inflow process climbs with increasing speeds, so that the flame

4.1 Spark ignition engines

speed increases and the mixture can still be fully converted despite the short amount of time available at high rpm's.

- **Medium pressure and fuel consumption**

As has been already shown, the air-fuel equivalence ratio λ considerably influences the flame speed and hence, across the combustion process, the reachable medium pressure and specific fuel consumption. For $\lambda > 1.1$, combustion begins to drag more and more due to the lower combustion temperature caused by the heating of surplus air and the resulting slower flame speed. The minimal fuel consumption is reached for air ratios of about $\lambda = 1.1$. Since the combustion velocity at $\lambda = 0.9 - 0.85$ is highest, the maximum medium pressure is engaged here. The optimal equivalence ratio lies thus in the region of $0.85 < \lambda < 1.1$. The "fish hook curve" shown in Fig. 4.6 represents the course of specific fuel consumption b_e contingent upon the mean effective pressure *mep*, whereby varying equivalence ratios λ are entered in for a SI engine at constant speed and constant throttle valve position. The effects of the air-fuel ratio upon pollutant formation will be explained in chapter 6.

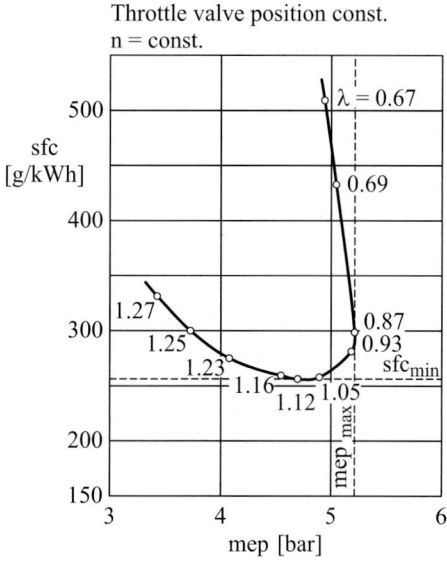

Fig. 4.6: Specific fuel consumption (b_e) and mean effective pressure (*mep*) contingent on the mean air-fuel equivalence ratio λ

- **Cyclical fluctuations**

Relatively large fluctuations in the pressure path from cycle to cycle, so-called cyclical fluctuations, are a typical feature of SI engine combustion. The causes for this are temporal and local fluctuations in the turbulent velocity field and in the mixture composition in the combustion chamber and thus also in the area of the ignition spark of the spark plug. Cyclical fluctuations in the ignition lag result from this. These can lead to fluctuations in flame propagation and therefore to more or less complete combustion as well.

Depicted at the top of Fig. 4.7 are the effects of cyclical fluctuations in the combustion of methanol and, on the bottom, the influence of the ignition angle on the pressure course. The comparison of both graphs shows that cyclical fluctuations have a similar effect as an adjustment of the ignition angle.

The reduction of cyclical fluctuations in SI engine combustion through the optimization of mixture formation, of ignition, and of flame spreading is a rewarding goal, also in consideration of the reduction of the specific fuel consumption and of HC emissions, see Bargende et al. (1997).

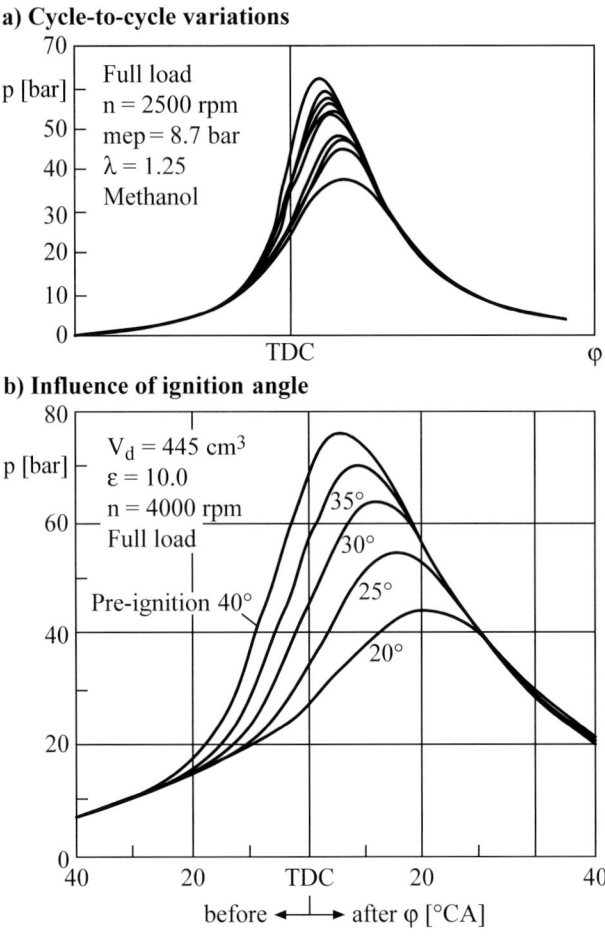

Fig. 4.7: Pressure trace in the combustion chamber a) cycle-to-cycle variations, b) influence of ignition angle

4.1.4 Abnormal combustion

Previously, "normal" combustion has been considered, in which the flame spreads relatively evenly across the combustion chamber with a speed of about 25 m/s. This leads to a so-called soft pressure path with a maximal pressure increase speed of about 2 bar/°CA. As opposed to this, in *knocking combustion*, considerable pressure fluctuations manifest themselves in the combustion chamber. In this case, the chemical pre-reaction started already in the unburned mixture during the compression stroke are very much accelerated. After the introduction of combustion through the ignition spark, the unburned mixture which has not been reached by the flame front is further compressed by the spreading flame. Thereby it is additionally heated such that the ignition boundary is exceeded, and finally a spontaneous autoignition begins in the unburned mixture, see Fig. 4.8. This, now almost isochorically proceeding combustion leads to steep pressure gradients, which expand in the form of pressure waves in the combustion chamber and cause the familiar knocking or ringing noise.

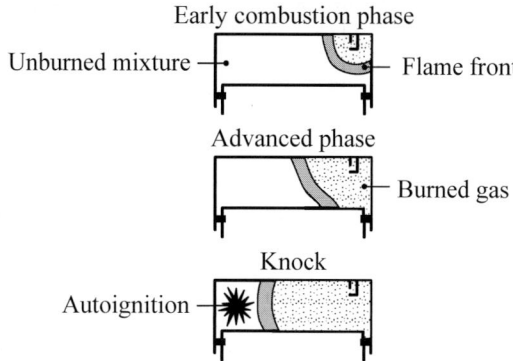

Fig. 4.8: Schematic representation of engine knock

Surface ignition is another undesirable combustion process. It is triggered by extremely hot zones on the combustion chamber walls, so-called "hot spots", with temperatures of about 1,200 K, which lie clearly over autoignition temperature. The most common "hot spots" are combustion residues that settle on the walls as hot flakes. The pressure progression in knocking combustion is qualitatively sketched in Fig. 4.9, whereby the start of the knocking combustion and of the surface ignition are drawn in. Knocking combustion can only appear after the beginning of combustion through the ignition spark; surface ignition, on the other hand, can begin earlier. Through the pressure waves, appearing both during knocking combustion as well as during surface ignition, mechanical material damages can occur, and through the elevated thermal load to melting of the piston and cylinder head.

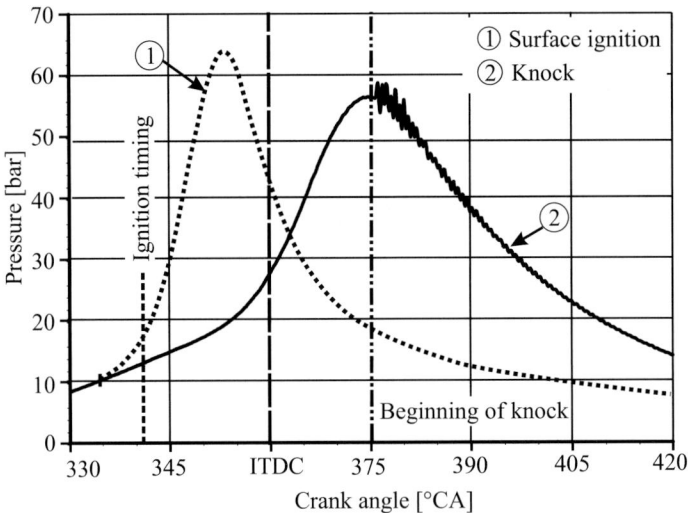

Fig. 4.9: Pressure trace in knocking combustion

For a more detailed description of the reaction-kinetic processes during knocking, also see Warnatz et al. (2001). The various forms of surface ignition as well as of the processes in the overlapping of surface ignition and knocking combustion are extensively described by Urlaub (1994).

The geometry of the combustion chamber has a crucial influence on the tendency towards knocking. Combustion chambers with little knocking tendency have

- short flame paths through compact construction and a centrally arranged spark plug,
- no hot spots at the end of the flame path through the positioning of the spark plug in the area of the exhaust valve,
- high flow velocities and hence favorable mixture formation as a result of swirl, tumble and squeeze flows, and a
- movement of charge in order to avoid a lag of the mixture at hot spots.

4.1.5 Controlled autoignition

In recent times, research activities have been intensified that have the goal of representing a considerably homogeneous combustion of gasoline fuel with autoignition. The boundaries between conventional SI and diesel engine combustion processes have thereby been increasingly blurred, and the systems are moving closer together. In diesel engines, we speak of so-called homogeneous charge compression ignition (HCCI), see chapter 4.2.5, while we often call the SI engine variant controlled autoignition (CAI). Both methods have the common goal of operating homogeneously and leanly especially at partial load, in order to realize both low consumption values and a minimal amount of raw emissions. The latter is essential so that lowest nitric oxide emissions can be achieved without expensive exhaust treatment, see chapter 6.5.

4.1 Spark ignition engines

In lean combustion with a spark plug, very low flame speeds can occur down to flame extinction. From this results a dragging combustion, which has an increase in consumption as a consequence. However, if the mixture – ideally at infinitely many locations in the combustion chamber – succeeds in igniting itself (we also speak of space ignition), the mixture can nevertheless be completely seized very quickly by the flame, and the global heat release rate integrated across the entire combustion space is sufficiently high to make possible a favorable fuel consumption. Yet the control of the autoignition timing under varying speed and load conditions represents a central problem. The boundary between a too premature and a too late ignition or even misfiring is very narrow. Not the least for this reason are homogenized combustion processes still a subject of research.

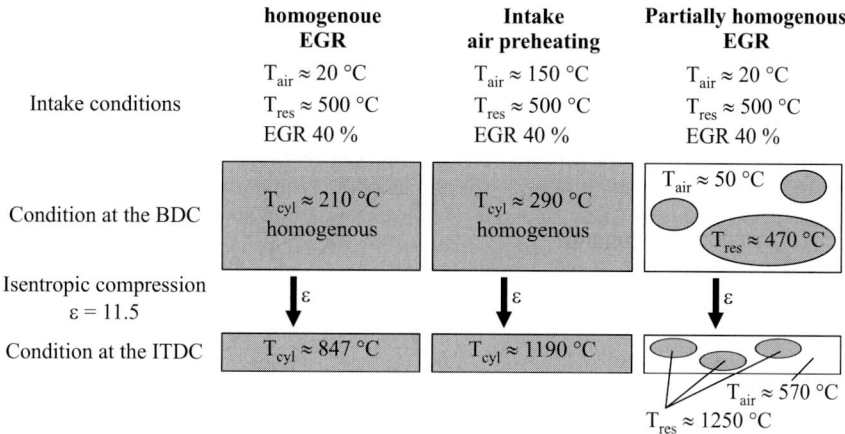

Fig. 4.10: Model of controlled auto-ignition (CAI), Wolters et al. (2003)

Gasoline fuel has the advantage in comparison to diesel fuel that it already evaporates at relatively low temperatures. This favors a largely homogeneous mixture formation. However, gasoline fuel is highly resistant to ignition, which is why autoignition of the mixture is only possible at a temperature between about 1,000 and 1,200 °C, which is far above the typical final compression temperature of a SI engine. Since the compression ratio of the engine can not be increased easily due to the danger of knocking at full load, the necessary boosting of the compression temperature can only be achieved by means of a preheating of the intake air or a recirculation of hot exhaust gas, e.g. by means of an negative valve overlap (combustion chamber recirculation) or of an external exhaust gas recirculation. Fig. 4.10 compares these possibilities each under constantly held boundary conditions: compression ratio ($\varepsilon = 11.5$), residual gas temperature ($T_{rg} = 500$ °C), and exhaust gas recirculation rate (EGR = 40 %). At an intake air temperature of 20 °C, the homogeneous mixture temperature of fresh air and recycled exhaust gas at bottom dead center amounts to 210 °C, and a final compression temperature of approx. 850 °C results, see Fig. 4.10 left. However, this is clearly too low for autoignition. In order to reach the final compression temperature necessary for autoignition, nearly 1,200 °C, with a complete mixing of fresh air and recycled exhaust gas, one would thus have to raise the mixture temperature at the bottom dead center to 290 °C. At a constantly held EGR rate and residual gas temperature, this means a preheating of the intake air temperature to 150 °C, see Fig. 4.10 middle. But similarly high temperatures at compression finish, and consequently autoignition, can be achieved without intake air preheating as well, if the recycled exhaust gas is not completely homogeneously mixed with the fresh air, but rather only partially mixed, and clusters of hot residual gas remain, see Fig. 4.10 right.

For the realization of the necessary internal exhaust gas recirculation, the variable valve control represents a key technology. Especially the necessity of controlling the combustion process both at changing loads and speeds as well as in transient engine operation places considerable demands on the variabilities in controlling the intake and exhaust valves. The realization of a combustion process with controlled autoignition is thus quite decisively bound to pending progress in the field of fully variable valve trains.

4.2 Diesel engines

Diesel engine combustion is characterized by the following features. Fuel is injected under high pressure towards the end of compression, as a rule before the top dead center, into the main combustion chamber (direct injection) or, in older engines, into a prechamber (indirect injection). The fuel spray which forms in this process evaporates, mixes with the compressed hot air, and ignites itself. As opposed to the SI engine, the diesel engine has only a very short time span for mixture formation available. A fast injection and an as good as possible atomization of the fuel are therefore the prerequisites for a fast and intensive blending of fuel and air.

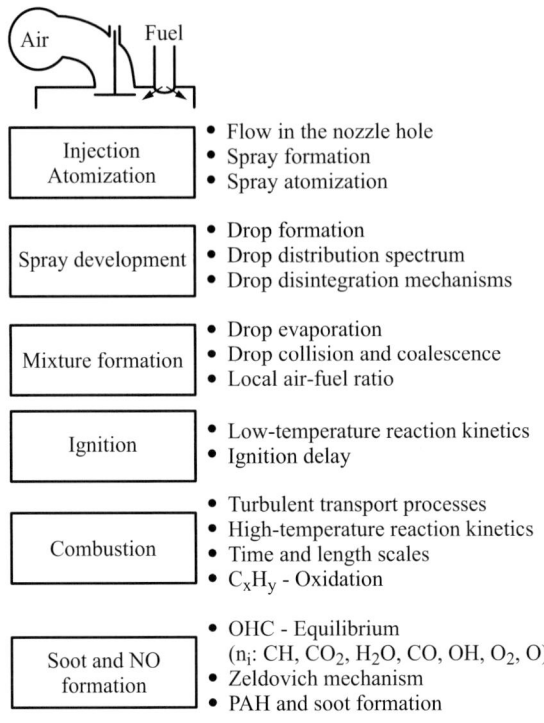

Fig. 4.11: Subprocesses of mixture formation and combustion in diesel engines

Fig. 4.11 qualitatively shows the subprocesses involved in diesel engine mixture formation and combustion. The particular subprocesses proceed largely simultaneously and interact with each other. The modeling of diesel engine combustion is thus extremely complex.

4.2.1 Injection methods and systems

- **Injection methods**

As opposed to formerly utilized injection into a prechamber or swirl chamber, direct injection into the combustion chamber is used today almost exclusively. In this case, the combustion chamber is accommodated as a bowl in the piston. The fuel is injected with a centrally arranged multi-hole nozzle. High injection pressures and many small holes in the injection nozzle see to a good mixture formation, which can be supported by a swirl flow in the combustion chamber gasses. The injected fuel should if possible not collide with the relatively cold piston wall, because in this way evaporation and then mixture formation are lagged and the formation of HC emissions is favored.

Direct injection methods in comparison to indirect ones have a clearly smaller specific fuel consumption. Because of high pressure increase speeds at the start of combustion, however, it also has a much higher noise level (so-called hard combustion). Beyond this, mixture formation is not supported by a fast charge movement in the channel between pre- and main chamber (200-500 m/s) as in pre-chamber engines. The total energy for the mixing of fuel and air is for the most part yielded by the jets injected into the combustion chamber, through which a considerably higher injection pressure is necessary. While in pre-chamber engines injection pressures of approx. 400 bar are sufficient, for direct injection they have to be between 1,200 and 2,000 bar. For vehicle diesel engines, higher injection pressures are also still being discussed.

- **Injection systems**

In the case of injection systems, we distinguish between conventional cam-operated systems like inline, single and distributor injection pumps, unit pumps and unit injectors as well as the common rail injection system developed recently. In *cam-operated injection systems*, pressure increase and fuel metering are coupled mechanically. The cam moves the tappet of the injection pump, which for its part "compresses" the fuel volume. The resulting climbing pressure opens a valve and thus releases the feeding pipe for the injection nozzle. The return line is opened via a trimming edge, and so the fuel pressure falls, the valve closes, and injection is over, see Fig. 4.12.

As opposed to this, pressure increase and fuel metering are completely separated in the common rail injection system. By means of a mechanically or electrically operated high pressure pump, fuel is continually delivered into a high pressure reservoir (common rail). With an electronically controlled injector, fuel is taken from the common rail and sprayed into the combustion chamber.

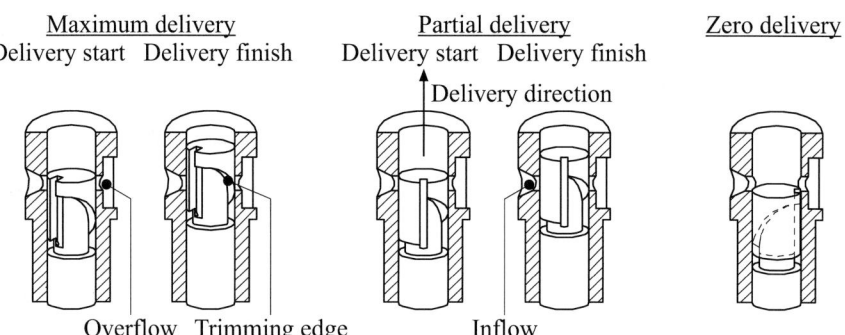

Fig. 4.12: Functional principle of delivery control

In the following, the distributor injection pump, the unit pump and the unit injector system as well as the common rail injection system will be briefly explained. For a more detailed account, refer to Basshuysen and Schäfer (2003).

- **Distributor injection pump (DIP)**

In the case of the distributor injection pump, only one pump unit exists for all cylinders. During one engine revolution, the DIP piston makes as many strokes (2-stroke engine) or half as many strokes (4-stroke engine) as there are cylinders, and via one rotation of the distributor head, the fuel is added to the single injection pipes.

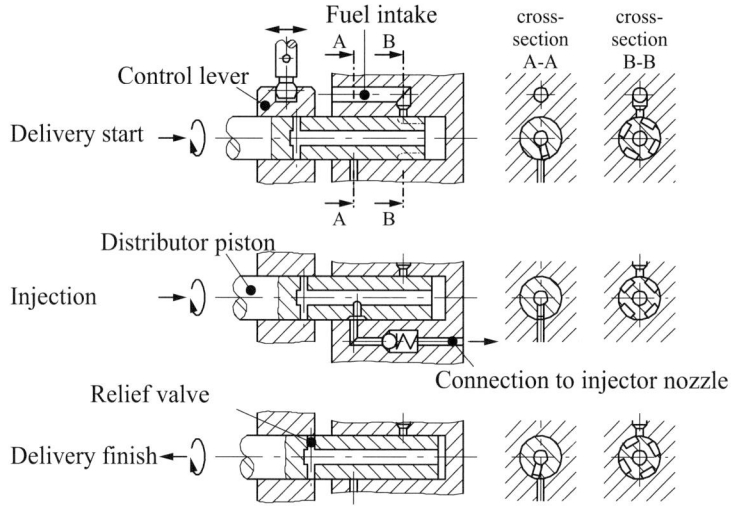

Fig. 4.13: Functional diagram of a distributor injection pump

Fig. 4.13 shows the functional diagram of a distributor injection pump as well as its most important structural groups. For smaller engines, DIP is less expensive than inline injection

4.2 Diesel engines

pumps or unit pump systems. It thus represents the current state of technology for passenger car engines with conventional injection systems.

Modern distributor injection pumps can create a maximum pump pressure of 800 – 1,000 bar. However, through purposeful exploitation of the pressure waves spreading in the injection pipe, a heightening of this maximum pressure to approx. 1,500 bar at the nozzle orifice is possible with existing systems.

- **Unit pump system (UPS)**

The UPS system is a modularly built high pressure injection system consisting of an injection pump, a short high pressure pipe and an injector-nozzle combination, see Fig. 4.14. Injection start and the injection amount are measured by means of a solenoid valve for every cylinder. With the help of the solenoid valve, access to a compensating volume is opened/closed. An opening of the valve causes a rapid decline in pressure in front of the injection nozzle and thus leads to a closing of the nozzle.

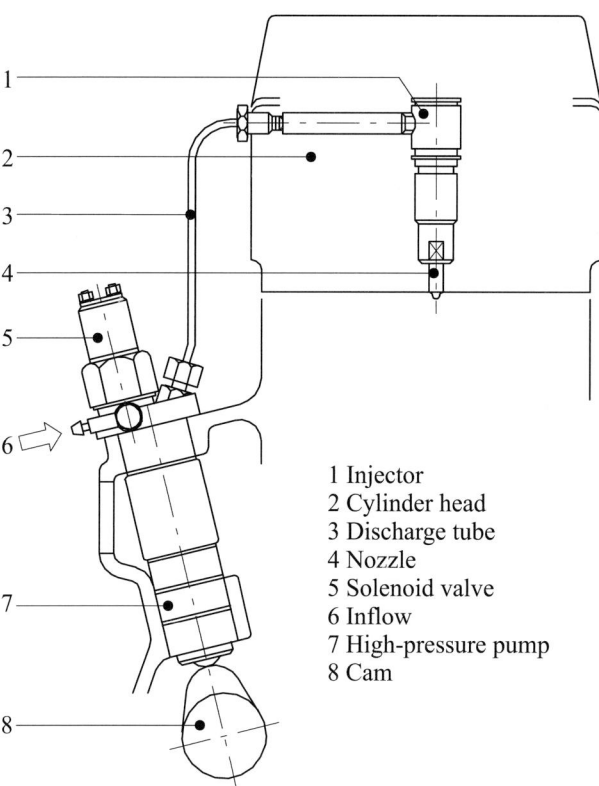

1 Injector
2 Cylinder head
3 Discharge tube
4 Nozzle
5 Solenoid valve
6 Inflow
7 High-pressure pump
8 Cam

Fig. 4.14: Construction of a unit pump system

- **Unit injector system (UIS)**

The injection pump and injection nozzle form a unit in this injection system, which is installed at every cylinder separately. A fast-switching solenoid valve controls the injection start and finish. It receives its shift signal from an electronic control unit, in the electronic module of which an injection map is stored.

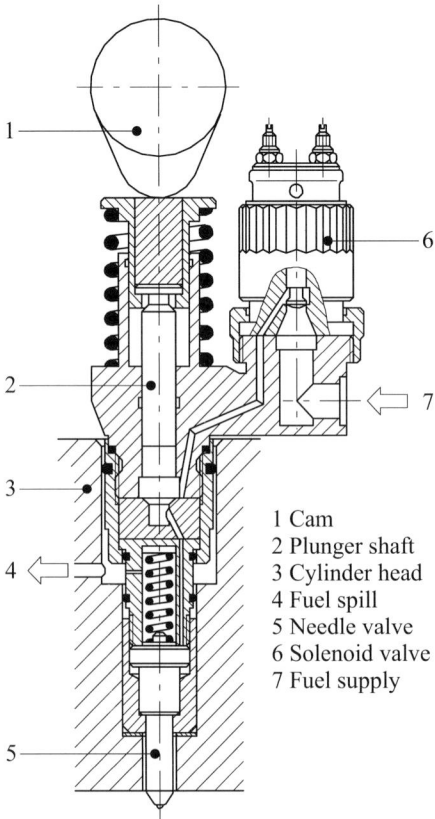

1 Cam
2 Plunger shaft
3 Cylinder head
4 Fuel spill
5 Needle valve
6 Solenoid valve
7 Fuel supply

Fig. 4.15: Functional diagram of a unit injector

In the UIS, injection pressures up to 2,000 bar can be represented. This makes low fuel consumption and emission levels possible. Fig. 4.15 shows the functional diagram of the UIS, which is also used in the TDI diesel engine of the Volkswagen AG for passenger cars.

- **The common rail system (CR)**

The injection systems described until this point are integrated systems for raising pressure and quantity regulation. Both of these functions are completely separated in the common rail system. In the electronically controlled common rail system, fuel is led to the common rail, a high-pressure reservoir built as a "pipe", with pressures in the area of $1,200 < p < 2,000$ bar.

4.2 Diesel engines

From there the fuel is regulated and led to each cylinders. Fig. 4.16 shows the functional principle of the CR injection system.

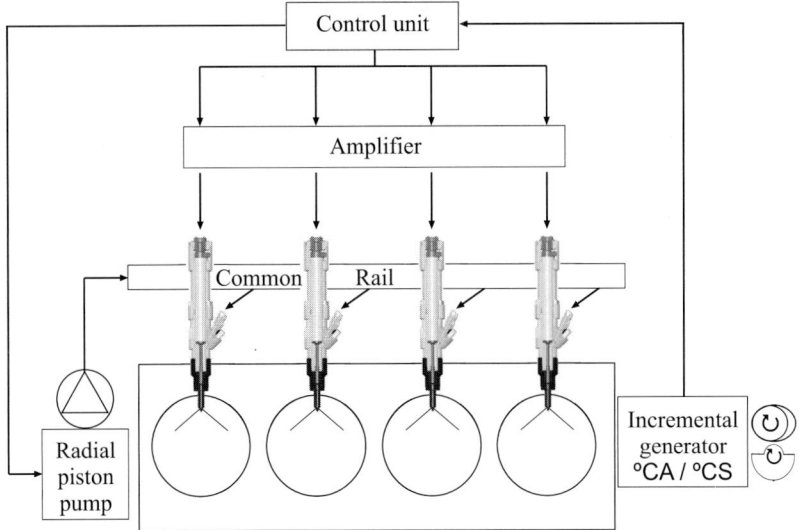

Fig. 4.16: Principle sketch of a common-rail system

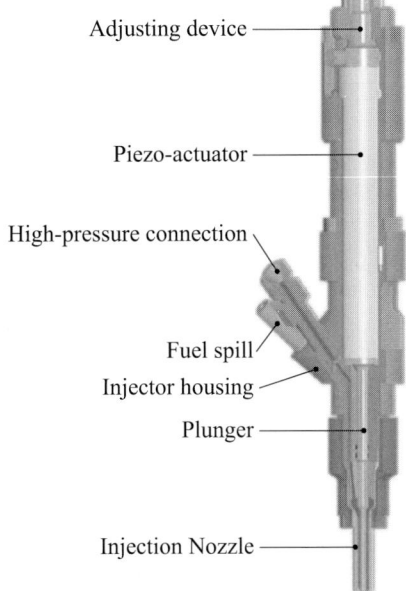

Fig. 4.17: Piezo-injector with direct control of the spray needle

With common rail injection system, almost any injection path can be represented, whereby the advantages of CR injection systems can only be realized when current solenoid valve/piezo valve controlled injectors are replaced with electronically controlled, directly activating piezo common rail injectors, see Meyer et al. (2002), whereby pressure-modulated piezo CR injectors represent a highly promising advance, see Stegemann (2003, 2004). Fig. 4.17 shows the cross section of a directly activating piezo common rail injector.

As opposed to cam-operated injection systems, the operating speed of the high pressure pump does not have to be rigidly coupled to the engine speed because of the common rail system's disassociation of pressure production and control functions. Through this, higher injection pressures can be realized even at smaller engine speeds, which causes better mixture formation and thus improved emission behavior. The common rail injection system should become universally successful in the near future due to its important advantages with regard to both pollutant reduction as well as constructive performance, whereby injection pressures well over 2,000 bar are under consideration.

- **Injection nozzles**

Fuel is injected into the combustion chamber through the bores in the injection nozzle. In the injection process, the fuel should be atomized to the highest possible degree in order to achieve a good air-fuel mixing.

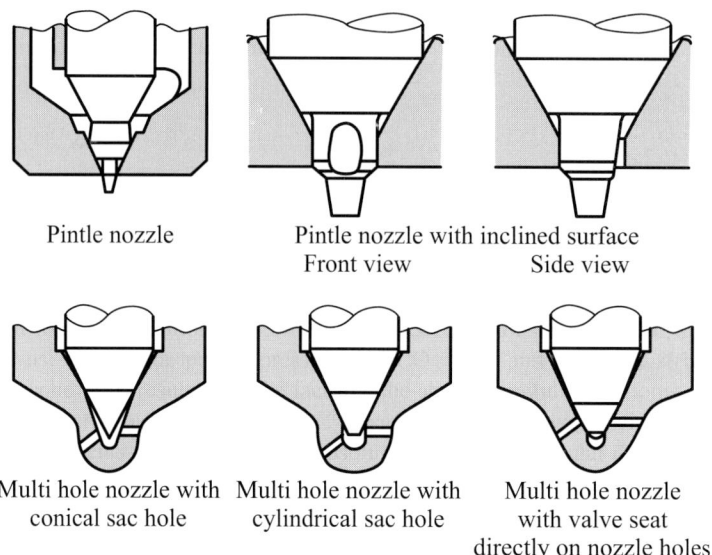

Fig. 4.18: Injection nozzle designs

For varying combustion processes and fuels, varying nozzle designs are utilized, see Fig. 4.18:

- Pintle nozzles are employed in pre- and swirl chamber engines. They have a stroke-dependent opening cross-section, are advantageous with respect to combustion noise, tend however towards carbonization (bung-hole nozzle).
- Multi-hole orifice nozzles are employed in direct injection diesel engines – sac hole nozzles typically for conventional injection systems and mini-sac hole nozzles as well as seat-type nozzles for common rail injection systems.

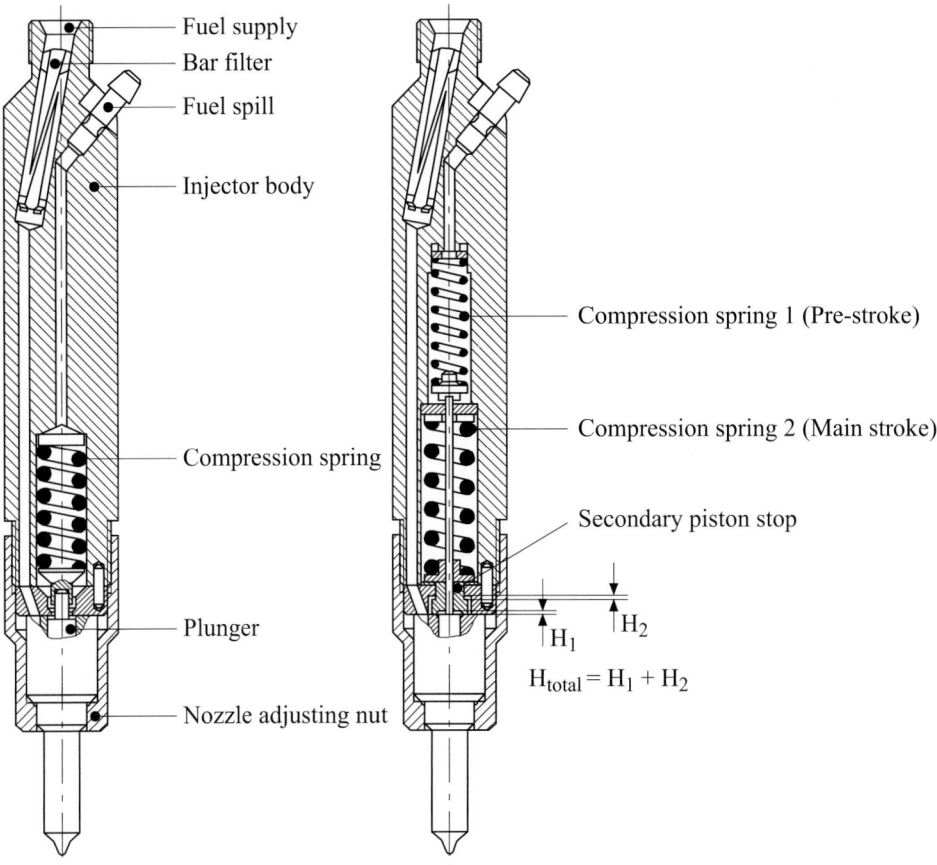

Fig. 4.19: Standard and two-spring injectors

The injection nozzle is integrated into an injector, which is screwed into the cylinder as a structural group. In Fig. 4.19, two injectors are illustrated, left a standard injector and right a two-spring injector for direct injection diesel engines with pre-injection. In the latter case, two springs with varying spring constants are employed. At injection start, the weaker spring only allows a restricted needle lift and thus a limited delivery rate. Only when the injection pressure exceeds the spring force of the second spring does full needle lift and the maximal injection rate become possible. Through the so-produced pre-injection of a smaller quantity of fuel, a softer pressure increase in the combustion chamber and thus a lower level of noise is achieved.

4.2.2 Mixture formation

- **Phenomenology**

Injection jets appear at high speed from the injection nozzle, and, as a result of the high relative speed of the surrounding highly compressed air and of high turbulence in the spray, disintegrate into small droplets. With progressing penetration into the combustion space, these droplets are then atomized. Fig. 4.20 shows a qualitative sketch of the fuel spray emerging from the injection nozzle.

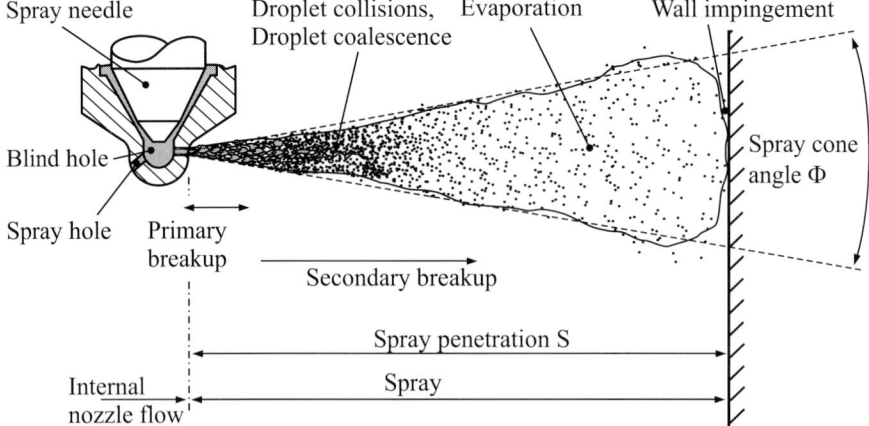

Fig. 4.20: Schematic representation of spray dispersion

Spray dispersion and the mixture formation related to it are determined by the injection parameters and the flow field (swirl, turbulence) in the combustion chamber. The turbulent kinetic energy of the spray is, however, at least one order of magnitude higher than that of the combustion air, so that the flow field in the cylinder only becomes significant towards the end of the injection, when the spray has already appreciably slowed down.

While the injection pressure in the case of conventional unit pump systems is strongly reduced towards the end of the injection duration, thereby causing an inferior atomization in this phase, the pressure remains at a constantly high level in the common rail system until the end, thus guaranteeing a continuously fine atomization.

At the fringe of the spray, fuel droplets mix with the hot air within the combustion chamber (*air entrainment*). In this way, the drops are heated up as a result of convective heat transfer and temperature radiation of the hot chamber walls, and the fuel finally begins to evaporate. Besides temperature, the rate of *drop evaporation* is determined by the diffusion of fuel from the drop surface (high vapor pressure) into the drop surroundings (lower vapor pressure of fuel).

In the diesel engine, mixture formation cannot be viewed independently from spray dispersion on the one hand and from combustion on the other. It is precisely the specificity of diesel engine combustion, that spray dispersion, mixture formation, and combustion run to some extent simultaneously. Only a small portion of the injected fuel mixes during the ignition delay with the air in the combustion chamber nearly homogeneously. At ignition, this quantity burns instantaneously (premixed-peak). Following this, mixture formation and combustion progress simultaneously, and combustion is controlled by the diffusion processes decisive for mixture formation.

Spray dispersion and mixture formation are nowadays, at least qualitatively, well understood and can be effectively described with semi-empirical models, see Baumgarten (2003), Ramos (1989), and Stiesch (2003).

4.2.3 Autoignition

The time span between the start of injection and that of combustion is designated as the *ignition delay*. The physical and chemical processes occurring in this interval are highly complex. The important *physical processes* are atomization of fuel, evaporation and mixing of fuel vapor and air until the formation of an ignitable mixture. The chemical processes are the pre-reactions described below in the mixture until autoignition, which ensue at a local air ratio of $0.5 < \lambda < 0.7$.

The oxidation of $C_x H_y$ fuels can be understood as a branched propagation process, in which the intermediate species formed in the course of the reaction act as chain propagators. The oxidation develops over hundreds of intermediate species. The reaction progression or path is highly dependent on the temperature and can be subdivided into three temperature ranges described in the following, see Warnatz et al. (2001).

At high temperatures above $T > 1,100$ K, the chain branching

$$H^\bullet + O_2 \rightarrow OH^\bullet + O^\bullet \tag{4.6}$$

is dominating. This reaction quickly loses importance at lower temperatures. In the medium temperature range $900 < T < 1,100$ K, the additional branchings

$$HO_2^\bullet + RH \rightarrow H_2O_2 + R^\bullet \quad \text{and} \tag{4.7}$$

$$H_2O_2 + M \rightarrow OH^\bullet + OH^\bullet + M \tag{4.8}$$

become more important, whereby the $OH^\bullet$ radicals are partially reformed into the originally employed $HO_2^\bullet$ radical.

In the low temperature range below $T < 900$ K, H_2O_2 disintegration is relatively slow, and degenerate branching processes become more important, which are characterized by the fact that precursors of the chain branching (e.g. RO_2) disintegrate again at higher temperatures. In this way, an inverse temperature dependence of the reaction rates is caused, which can be described as a 2-step reaction mechanism, see Warnatz et al. (2000). This 2-step reaction mechanism, which originally was developed to describe knocking combustion in the SI engine, leads to an extensive reaction diagram, since the residual molecules can have many isomeric structures. For oxidation of n-$C_{16}H_{34}$ alone, about 6,000 elementary reactions with 2,000 species have to be considered.

Various models have been developed for simulating the autoignition process. One model is that of Fieweger and Ciezki (1991), with whose help autoignition can be calculated in stoichiometric n-heptane air mixtures at varying pressures. Excellent agreement with experimental data proves the consistency of this model, see Fig. 4.21.

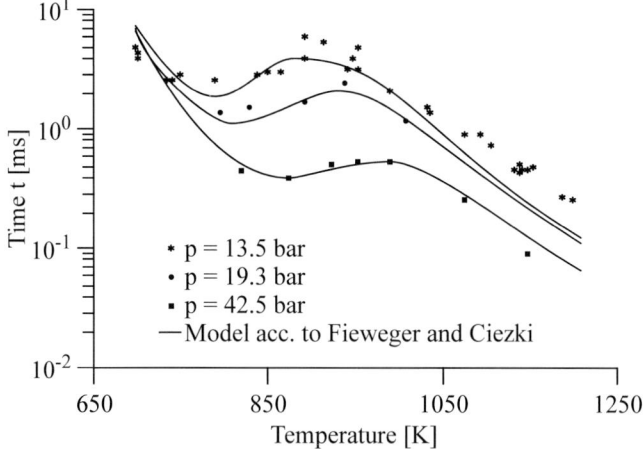

Fig. 4.21: Autoignition model, acc. to Fieweger and Ciezki (1991)

The Shell model of Halstead et al. (1977) includes a degenerate branching process with two reaction paths for the formation of branching products, as well as two chain break-off reactions. The reaction coefficients k_i are either of the standard Arrhenius type

$$k_i = A_i \exp\left(-\frac{E_i}{RT}\right) \quad (4.9)$$

or a combination of separate reaction coefficients (for three-stage propagation)

$$\frac{1}{k_i} = \frac{1}{k_{i1}[O_2]} + \frac{1}{k_{i2}} + \frac{1}{k_{i3}[RH]} \quad . \quad (4.10)$$

Reaction rates $\dot{x}_i$ result in the form

$$\frac{dx_i}{dt} = A_i \exp\left(-\frac{E_i}{RT}\right)[O_2]^{a_i}[RH]^{b_i} . \quad (4.11)$$

This autoignition model requires the adaptation of 26 reaction parameters. It is used, for example, in the FIRE code and is described in detail by Fuch et al. (1996). An extensive discussion and a comparison of different autoignition models can be found in Klaiß (2003).

For applications in diesel engines at typically higher temperatures, the above described complex models are often not really necessary, especially since these models are also not fundamentally valid and require the calibration of model constants. Therefore, a *single-equation*

4.2 Diesel engines

model is often used in praxis with fine results, which describes the ignition delay with the help of only one Arrhenius equation

$$\Delta t_{ID} = A \frac{\lambda}{p^2} \exp\left[\frac{-E}{RT}\right] \qquad (4.12)$$

contingent on pressure, temperature, and air-fuel ratio.

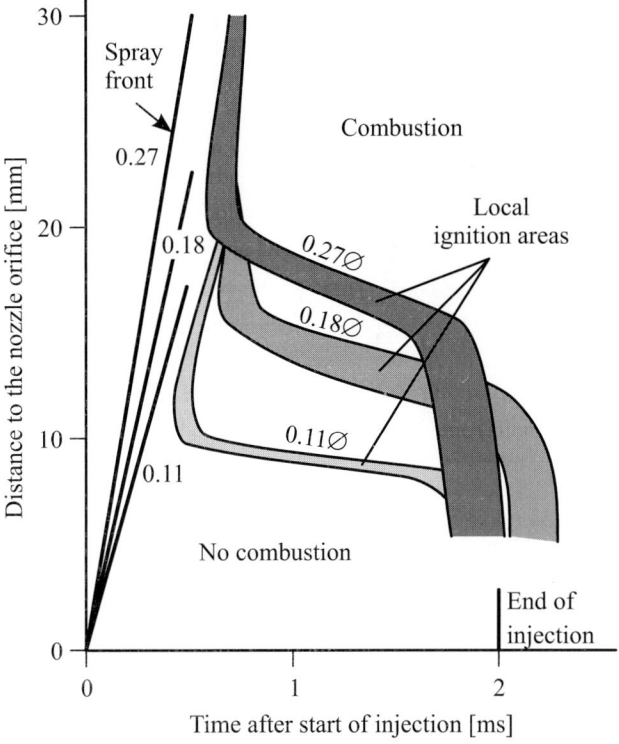

Fig. 4.22: Local ignition ranges contingent on the time after the start of injection and the distance from the nozzle orifice, acc. to Winkelhofer et al. (1991)

In summary, Fig. 4.22 shows local ignition regions contingent on time after injection and the distance from the nozzle for three different diameters of injection bore according to Winkelhofer et al. (1991). One recognizes that the mixture ignites after about one-half millisecond and that the ignition region lies closer to the nozzle orifice the smaller the diameter of the injection bore is.

4.2.4 Combustion

The course of diesel engine combustion can, crudely speaking, be divided into three phases, see Fig. 4.23.

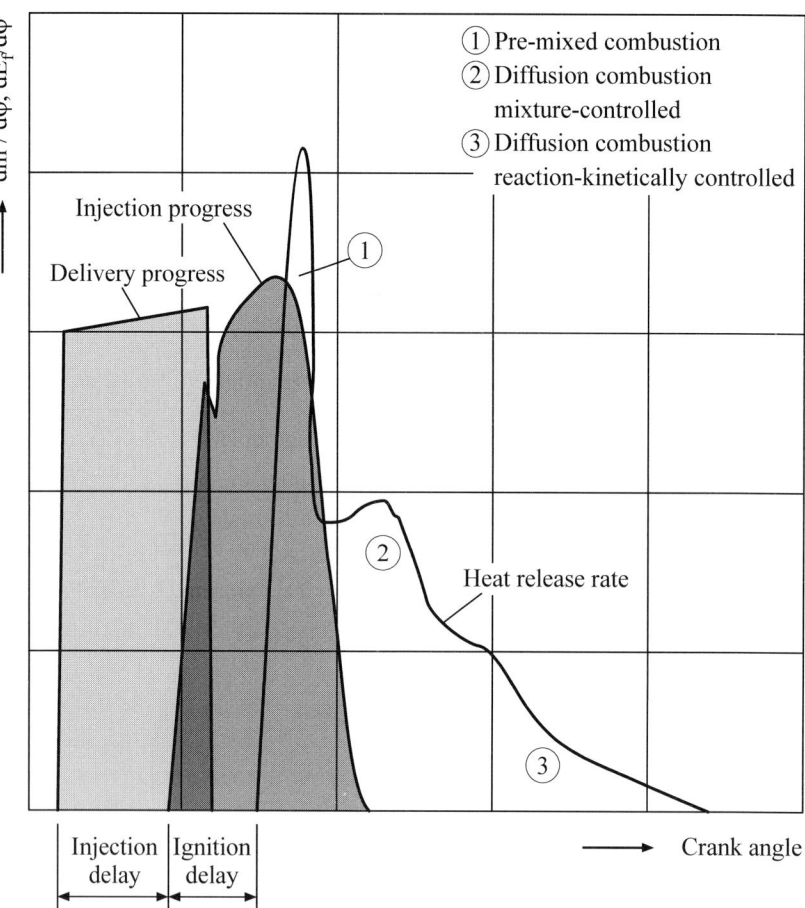

Fig. 4.23: Delivery progress, injection progress and heat release rate in the diesel engine

- **Phase I: premixed combustion**

The fuel injected during the ignition delay mixes with air in the combustion chamber and forms a nearly homogeneous and reaction-capable mixture. After the ignition delay period, which is physically and chemically controlled, this mixture burns very quickly (premixed peak). This premixed combustion is thus similar to SI engine combustion. The combustion noise typical for the diesel engine is caused by the high pressure increase speed $dp/d\varphi$ at combustion start. This pressure increase speed can be influenced by a change in the injection time, whereby is valid: an early injection start leads to a "hard" and a later to a "soft" combustion, see, Fig. 4.24. Beyond this, the combustion noise can be reduced by a pilot injection, where only a small fuel quantity of about 5 % is injected at first, and only after successful autoignition does the main injection begin.

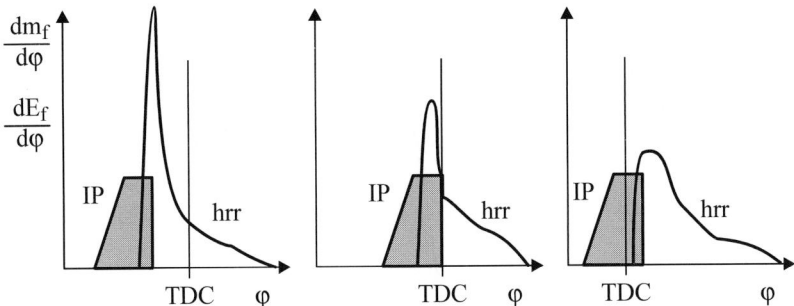

Fig. 4.24: Injection progression (IP) and heat release rate (hrr) for early (left) and late (right) combustion

- **Phase II: diffusion combustion**

Mixture formation processes continue during the main combustion phase and decisively influence both the combustion course itself as well as pollutant formation. The chemistry of this phase is very rapid; the combustion process is controlled by the mixing rate. We therefore also refer to it as mixing-controlled diffusion combustion. The end of the main combustion phase is characterized by the attainment of the maximum temperature in the combustion chamber.

- **Phase III: post-combustion**

Towards the end of combustion, pressure and temperature in the flame front have decreased so much that chemical reactions become slow in comparison with the simultaneously progressing mixture processes. Diffusion combustion is thus increasingly controlled by reaction kinetics. Besides the conversion of as yet unburned fuel, which decreases considerably towards the end, intermediate products are also further oxidized during post-combustion, which originated during the main phase as a result of local lack of oxygen. This last combustion phase is thus decisive especially for the oxidation of previously formed soot. Over 90 % of the total soot produced is broken down again in the combustion chamber during this phase.

For the thermodynamic quality of the total combustion process, the released thermal energy is decisive

$$\frac{dE_{fuel}}{d\varphi} = f(\varphi). \tag{4.13}$$

It leads to the warming up of the air-fuel mixture in the cylinder and thus to temperature and pressure increase. As an example, Fig. 4.25 shows the pressure progression and heat release rates at full and partial load in a high speed heavy-duty diesel engine with relatively late injection.

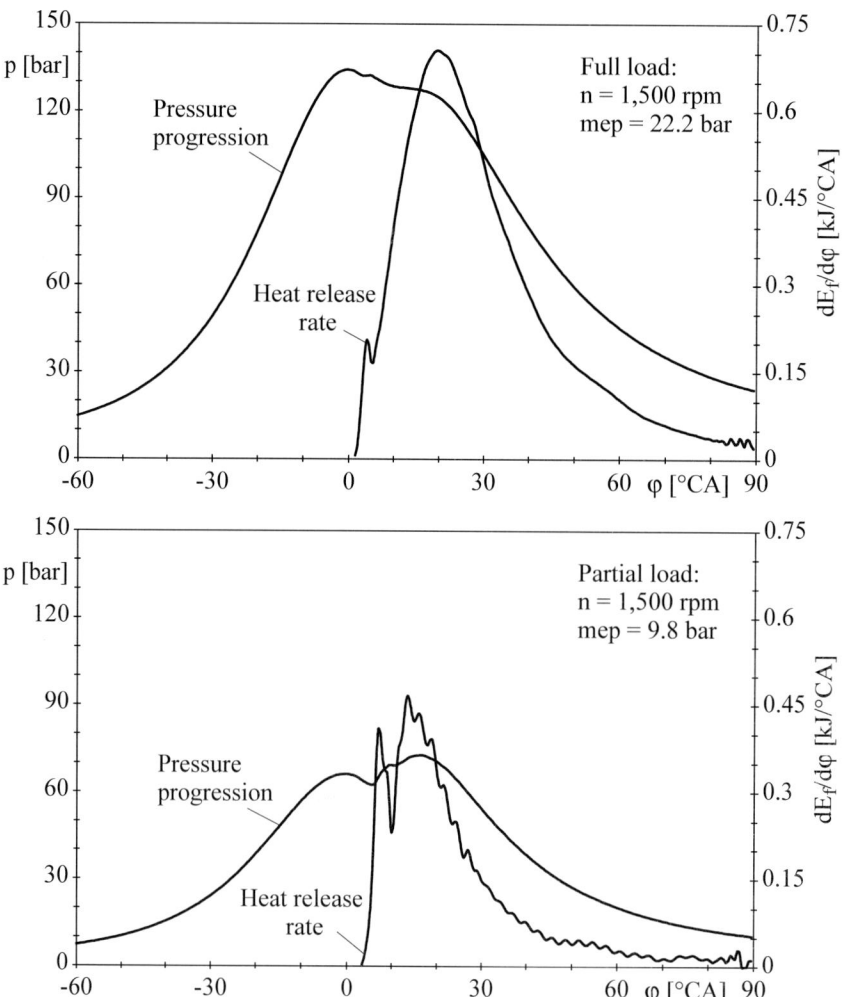

Fig. 4.25: Pressure progression and heat release rate in a high speed diesel engine at full and partial loads

4.2.5 Homogeneous combustion

In order to hinder as much as possible the production of pollutant components nitric oxide (NO_X) and soot, which are dominant in diesel engine combustion, a new combustion process, so-called homogeneous charge compression ignition (HCCI), is being developed. The NO_X-formation generally occurs at temperatures above 2,000 K, soot formation, on the other hand, in rich area of the mixture with $\lambda < 0.8$ and at temperatures above 1,400 K. In the HCCI method, these areas are avoided, see Fig. 4.26.

4.2 Diesel engines

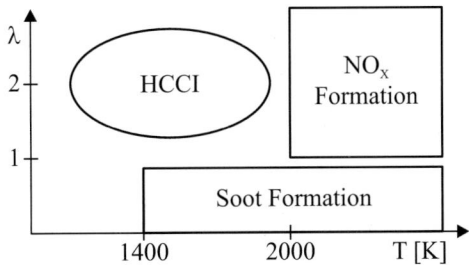

Fig. 4.26: The λ, T range of the HCCI method

In homogeneous diesel combustion, the fuel is already injected very early into the combustion chambers as opposed to the conventional diesel process, so that a relatively large amount of time is available for mixture formation. This very lean mixture ideally ignites simultaneously at many locations in the combustion chamber (space ignition) and therefore burns through very quickly. Because of the lean mixture, no local temperature peaks appear, and thus is the thermal formation of NO impeded; also, in largely homogeneous mixtures, no local rich regions are found and thus practically no soot is produced.

Fig. 4.27 shows the injection and heat release rates dependent upon the crank angle for the conventional and HCCI methods. In the HCCI method, the injection takes place via a sequence of single injection impulses (split injections), the heat release takes place very quickly in the TDC area. Split injections can be represented quite straightforwardly by a common rail injection system.

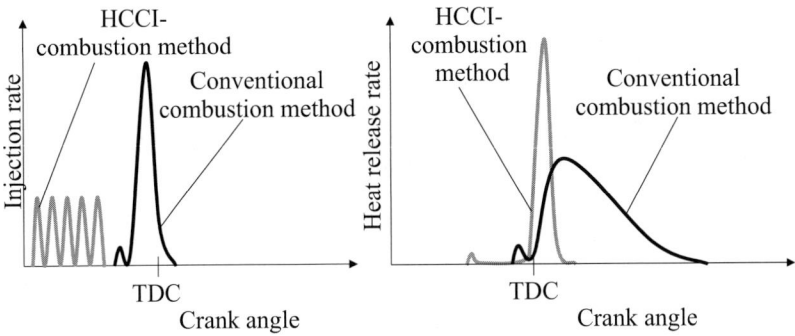

Fig. 4.27: Injection and heat release rates for the conventional and the HCCI method

For the control of the HCCI process, it is sensible to use the exhaust gas recirculation rate. As a result of the high level of charge dilution and very low combustion temperatures, there is a clear increase in CO and HC emissions. Both of these pollutant components must, if necessary, be reduced with the help of an exhaust gas after treatment by means of a catalytic converter.

4.3 Pressure trace analysis

Despite progress in optical measurement techniques, the analysis of the cylinder pressure trace cannot be ignored in the current development of internal combustion engines. Firstly, the cylinder pressure course is the most important quantity in recognizing knocking combustion online on the test stand. Secondly, important insight with respect to combustion (inflammation duration, ignition delay, heat paths and heat release rate) as well as to so-called loss distribution can be won from thermodynamic analysis. Furthermore, the pressure trace provides information about obeying the given peak pressure, the indicated work, the charge change behavior of the engine, and the residual gas in the combustion chamber.

4.3.1 Determination of the heat release rate

- **Pressure signal acquisition**

In determining the cylinder pressure, nowadays water-cooled pressure transducers are used, which function according to the piezo-electric measurement principle. These possess a high mechanical robustness despite small structural volume and a very high resolution. The pressure transducer is closed off from the combustion chamber with a membrane, which is acted upon by the combustion chamber pressure and leads to a force proportional to the pressure in the piezo elements. Proceeding from the cylinder pressure, the pressure receiver creates a charge that is transformed into an electric voltage proportional to the charge in the charge amplifier. This charge can be lead to an analog-digital converter; the cylinder pressure is thus available for thermodynamic evaluation.

The allocation of the pressure signal to the engine process (triggering) results by means of a so-called encoder, which is flange-mounted onto the engine crankshaft and supports itself with torsional vigor onto the crank housing. A resolution of approx. 1 °CA is sufficient for thermodynamic evaluations. Since the position of the encoder at TDC cannot be exactly determined, the distance between the sensor tip and the piston is usually measured with a capacitive sensor while the engine is motored; the geometric TDC of the engine is thereby determined for every speed. The precision requirement for this process amounts to ± 0.1 °CA.

In spark ignited engines (SI engines), considerable differences in combustion occur due to mixture inhomogeneities at the spark plug, since combustion starts a few degrees of crank angle earlier or later depending of the mixture condition at the spark plug. These so-called cyclical fluctuations are smoothed for thermodynamic evaluation by averaging over a large number of cycles. In the SI engine, an averaging of up to 250 cycles is desirable, see Fig. 4.9. Because of autoignition, these fluctuations are less marked in diesel engines, which is why an averaging over fewer then 50 cycles is usually sufficient.

- **Evaluation of the pressure signal**

If we first set the combustion chamber of the internal combustion engine as the balance volume, then the conditions of the enclosed gas – i.e. pressure, temperature, and internal energy – can be clearly described via the thermal equation of state and the mass and energy balance (see chapter 7.1).

4.3 Pressure trace analysis

The gas mass enclosed within the combustion chamber can be determined in the simplest case via the measurement of the fresh gas mass. The problem arises however that the degree of capture is usually insufficiently known. In addition, the allocation of fresh gas mass to the single cylinders in multi-cylinder engines proves problematic, since this quantity can usually only be integrally measured across all cylinders. In this case, a charge changing calculation, for example, can help as a back-up, for which a measurement of the pressure traces in the intake and exhaust ports is necessary. This measurement usually takes place by means of piezo-resistive pressure sensors. The pressure signals are impressed as boundary conditions onto a so-called mini-model, which, via the gas-dynamic relations described in chapter 7.4, describes the pipe system between the measurement location on the intake side and on the exhaust side.

In engines with high amounts of residual gas and external exhaust gas recirculation, a measurement of CO_2 concentration in the exhaust gas and in the gas mixture taken in by the cylinder – hence in the intake manifold after the mixing location – has proven practical for the determination of the residual gas mass within the cylinder. The internal exhaust gas recycling rate can, practically speaking, only be determined via the above-described charge changing calculation.

Since combustion normally only occurs during the high pressure phase (exception: later burning through, e.g., post-injection), we can consider the combustion chamber for the high-pressure phase as a closed system. With this, the enthalpy flows over the system boundaries are zero, and the blow-by losses and evaporation enthalpy in gasoline direct injection can also at first be roughly set to zero.

$$\frac{dQ_{fuel}}{dt} = \frac{dU}{dt} - \frac{dQ_w}{dt} + p\frac{dV}{dt} - \left[\frac{dm_{bb}}{dt}h_{bb} - \left(\frac{dm_{fuel,evap}}{dt}\Delta h_{evap}\right)\right]. \quad (4.14)$$

The internal energy in (4.14) can be described in relation to pressure, temperature, and gas composition. The wall heat losses of the piston, cylinder head, and liners can also be represented as functions of pressure and temperature. In describing heat flow in the combustion chamber walls, the wall temperatures are necessary, which can be determined through measurement or calculation. The cylinder volume is only contingent upon geometrical quantities anyhow. The affiliated physical regularities are extensively described in chapter 7.1.

The average gas temperature can be easily determined with knowledge of the momentary combustion chamber volume, pressure, and the total gas mass in the chamber via the thermal equation of state. The pressure in the cylinder remains the only unknown for the determination of the heat release rate. The question in establishing the heat release rate is thus aimed at the determination of the pressure in the cylinder, which has been described above.

Pressure transducers that function according to the piezo-electric principle cannot measure absolute pressures. Therefore the pressure level of the measured pressure signal must be adjusted according to certain thermodynamic criteria.

This adjustment can take place via polytropic compression in the compression phase, in which the gas temperature is of the same order of magnitude as the cylinder wall temperature. This is favored by the fact that the heat transfer coefficient is very low in this range. A range between approx. 100 °CA and approx. 65 °CA before TDC proves to be favorable in a number of engines, but this cannot be generalized. For diesel engines, a polytrope exponent of

1.37 should be chosen, while for SI engines with external mixture formation, due to the amount of fuel in the mixture, a polytrope exponent of 1.32 ought to be selected.

An adjustment by means of the 1st law of thermodynamics represents an additional possibility. Between the closing of the intake valve and the ignition timing, the heat release rate integrally and temporally by combustion has to be equal to zero. If one provides the measured pressure with an additive pressure correction link, one can then solve according to this and iteratively obtain a very exact solution for the pressure adjustment.

The highest precision is possible with the above described charge changing calculation under the provision of the measured intake and exhaust pressures. Upon completion of the charge changing calculation, the measured cylinder pressure is adjusted to the pressure of the charge changing calculation at "intake valve closes". Moreover, with this method, an exact determination of the residual gas portion and thus an exact determination of the cylinder mass becomes possible, which can be used as an additional advantage for the improvement of evaluation precision.

However, the so-called energy balance is an essential quantity in judging the quality of the heat release rate evaluation. It is produced from the quotient of the energy quantity determined by the heat release rate evaluation and the maximum energy quantity released by the fuel.

The heat release rate is thereby integrated over the entire combustion duration and represents the numerator of the quotient of the energy balance.

The maximum released energy quantity – the denominator – can be calculated from the product of the fuel injected per cycle and the lower heating value, whereby – especially in the case of the SI engine – the energy of unburned exhaust gas components must be subtracted.

$$EB = \frac{\int_{\varphi_{SOC}}^{\varphi_{EOC}} \frac{dQ_{fuel}}{d\varphi}}{m_{fuel} lhv - Q_{ub}} = \frac{\int_{\varphi_{SOC}}^{\varphi_{EOC}} \frac{dQ_{fuel}}{d\varphi}}{m_{fuel} lhv \, \eta_{conv}}, \quad (4.15)$$

$$\eta_{conv} = 1 - \frac{Q_{ub}}{m_{fuel} \, lhv}. \quad (4.16)$$

The following relation is thereby valid for unburned components in the exhaust gas like CO, H_2, HC, and soot:

$$Q_{ub} = m_{CO} \, lhv_{CO} + m_{H_2} \, lhv_{H_2} + m_{C_3H_8} \, lhv_{C_3H_8} + m_C \, lhv_C. \quad (4.17)$$

A fluctuation in the energy balance in the range of 95-105 %, thus in the range of ± 5 %, can be viewed as very good in the context of achievable precision in measurement and in the description of thermodynamic relations.

Besides the heat release rate, other quantities important for the characterization of the combustion path can also be determined. These are portrayed in Fig. 4.28.

4.3 Pressure trace analysis

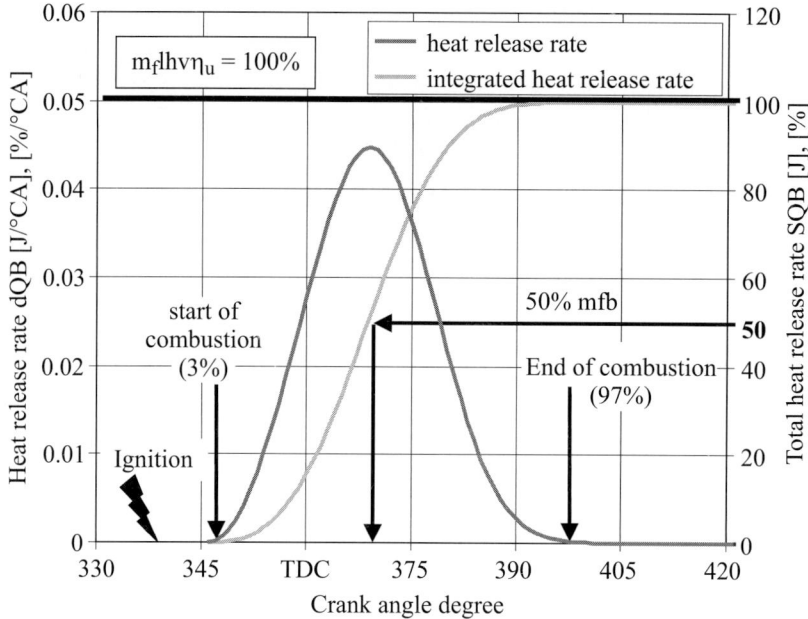

Fig. 4.28: Characteristic attributes of the combustion path

The time between ignition or injection in the diesel engine and start of combustion, which can be fixed at approx. 3 to 5 % of the total heat release rates, is called ignition delay. The time between start of combustion and end of combustion is designated as combustion duration. The center of the heat release rate (50 % mass fraction burned, 50 mfb) is defined as the degree of crankshaft, at which 50 % of the total heat quantity has been converted. Almost independently of the type of engine and the combustion process, consumption-optimal operating points result at 50 mfb position of approx. 8 °CA after the ignition TDC.

Fig. 4.29 shows the evaluation of a the heat release rate for a conventional SI engine at a speed of 1,000 rpm and a load of $p_i = 1$ bar. Represented are the heat release rate and the single components there of in accordance with the 1st law of thermodynamics, see (4.14). The exact procedure in determining the heat release rate is described by Witt et al. (1999) and others.

In summary, one can say that for a thermodynamically correct evaluation, a high precision in pressure indication and determination of all measured quantities is necessary. If all these prerequisites are fulfilled, then it is possible to determine not only the indicated mean effective pressure, but also the temporal release of heat release rate as a decisive requirement for an efficient simulation.

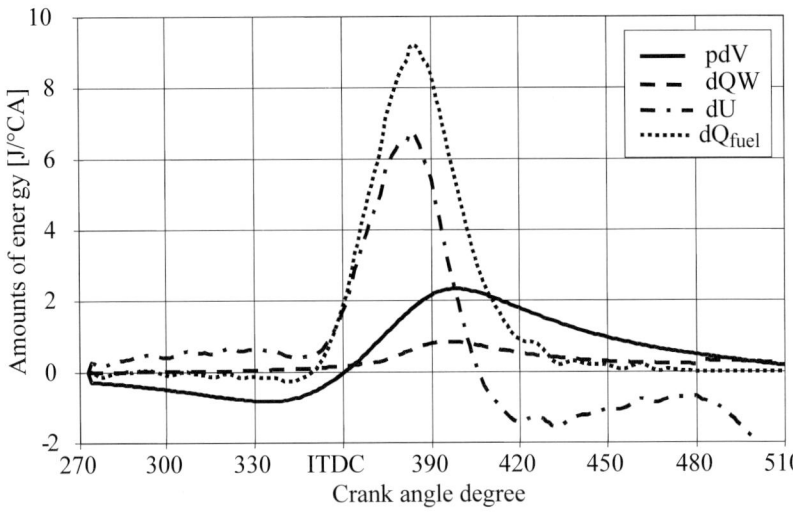

Fig. 4.29: Pressure progression and heat release rate for a SI engine at 1,000 rpm and $p_i = 1$ bar

4.3.2 Loss distribution

In order to be able to assess various combustion processes with respect to their potentials, we make use of so-called loss distribution. Individual loss fractions are systematically calculated proceeding from the perfect engine process and reconstructed for the real engine process.

The perfect engine process resembles the constant-volume process, as in the latter the total energy is added and removed at the top dead center as well. However, for the perfect engine, several deviating assumptions are made, which Witt (1999) summarizes as follows:

- calculation with ideal gas and real physical characteristics (c_v, c_p, $\kappa = f(T)$),
- an equal equivalence air ratio as in the real process,
- combustion progresses to the point of chemical equilibrium with consideration of dissociation,
- idealized heat release rate (heat supply at the TDC in the SI engine),
- no wall heat losses,
- no friction,
- no flow losses,
- the valve control times lie at the top and bottom dead centers (EVO in the BDC, EVC and IVO in the TDC and IVC in the BDC),
- pressure and temperature at start of compression are fixed such that the same comparison line between the perfect and the real process results,
- the charge-mass is the same as in the real process,
- equal amount of residual gas as in the real process.

4.3 Pressure trace analysis

Fig. 4.30 shows the efficiency of the perfect engine contingent upon the compression ratio and on the global air-fuel equivalence ratio according to Pischinger (1989).

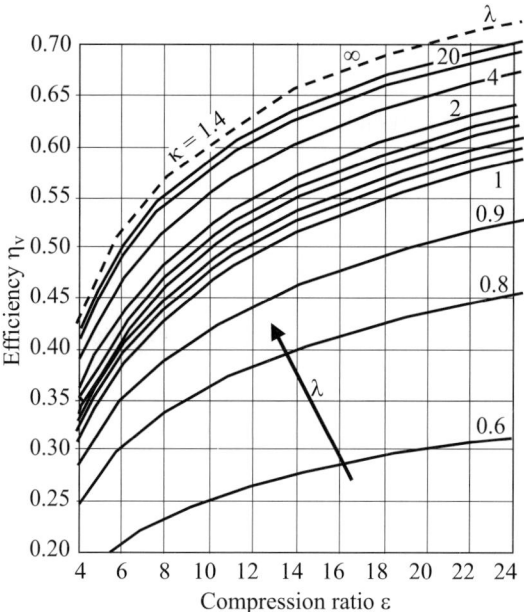

Fig. 4.30: Efficiency of the perfect engine contingent on the compression ratio and on the global air-fuel equivalence ratio, acc. to Pischinger (1989)

The real process distinguishes itself from the perfect process through losses from incomplete/imperfect combustion, through combustion losses, wall heat losses, charge changing losses, and friction losses. In order to quantify these losses, the cycle is calculated again in consideration of the respective sources of loss, and the difference to the previous cycle is evaluated.

- **Losses from incomplete/imperfect combustion**

We understand under losses from incomplete combustion those losses, which arise due to sub-stoichiometric combustion – i.e. due to lack of oxygen. These losses are already considered in the perfect engine, since in this case only the conversion of fuel until chemical equilibrium is taken into consideration anyhow. Losses from imperfect combustion arise when the fuel does not burn until chemical equilibrium. From this imperfect combustion, additional exhaust gas components result like CO, H_2, HC, and soot, which go beyond the level of combustion with a lack of oxygen. These components from incomplete and imperfect combustion are disclosed by exhaust gas analysis in its entirety. Losses from imperfect combustion reduce the heat quantity supplied isochorically to the process (related to 1 kg mixture mass) and are to be quantified as follows:

$$q_{ub, ic.} = q_{ub, tot} - q_{ub, chem} \cdot \quad (4.18)$$

Thereby, for the total losses from incomplete and imperfect combustion, the following is applicable from exhaust analysis

$$q_{ub,tot} = \left(v_{CO}\, lhv_{CO} + v_{H_2}\, lhv_{H_2} + v_{C_3H_8}\, lhv_{C_3H_8} + v_C\, lhv_C\right)\frac{1}{M_c} \qquad (4.19)$$

with

lhv_{CO} = 282,900 kJ/kmol,
lhv_{H_2} = 241,700 kJ/kmol,
$lhv_{C_3H_8}$ = 406,900 kJ/kmol,
lhv_C = 2,041,367 kJ/kmol,
M_c = 28.905 kg/kmol.

For incomplete combustion until chemical equilibrium, the following relation is valid according to Vogt (1975)

$$q_{ub,chem} = [1 - (1.3733\,\lambda - 0.3733)]\,lhv^* \qquad (4.20)$$

with

$$lhv^* = lhv\,\frac{1}{\lambda L_{min} + 1}\,.$$

- **Combustion losses**

Combustion losses arise in that the combustion heat is not supplied isochorically – i.e. in an infinitely short time span – in the real process, but rather in the form of the heat release rate (see 4.3.1), which covers several degrees of crank angle. The heat quantity supplied before the TDC is thereby acting against the compression while the heat quantity supplied after the TDC can no longer have an effect during the entire expansion. This loss can be determined through twice-repeated cyclical calculation – once with isochoric heat supply and once with the provision of the real combustion. One should hereby recognize that a reduction of the combustion losses as roughly isochoric combustion always goes along with an increase of wall heat losses in real engine operation, which is why the total optimum from combustion and wall heat losses does not lie in isochoric combustion.

- **Wall heat losses**

Two process calculations are necessary in determining wall heat losses as well. The wall heat flow is thereby calculated via known correlations, e.g. according to Woschni or Bargende (see chap. 7.1).

- **Charge changing losses**

The perfect engine possesses, according to its definition, no charge change losses, since the process control from BDC to BDC takes place with a heat removal. According to Witt (1999), in order to take the charge changing losses into exact consideration, a definition of the charge change losses according to the BDC-BDC method under additional consideration of the expansion and compression losses must be chosen. The reduction of the working surface in the p, V diagram through the sudden fall in pressure due to the opening of the exhaust valve

4.3 Pressure trace analysis

p, V diagram through the sudden fall in pressure due to the opening of the exhaust valve before the BDC is thereby taken into consideration. This is the case in the closing of the intake valve occurring after TDC as well. Corresponding compression losses must hereby be considered. These losses are, according to their respective cause added to the charge changing losses. The consideration of charge change losses leads to the indicated mean effective pressure and thus to the indicated efficiency. Losses such as leakages only marginally affect the result of the loss distribution.

4.3.3 Comparison of various combustion processes

In this chapter, the heat release rate and loss distributions for various example combustion processes will be presented. A throttled SI engine with multi point injection (MPI), a SI engine with fully variable mechanical valve lift control and multi point injection (VVH), a direct injecting SI engine with a spray-guided combustion process ($DISI_{spray}$), a SI engine with controlled autoignition (CAI), and a hydrogen engine with intake-pipe injection (H_2) are compared. A speed of 2,000 rpm and an indicated mean effective pressure of approx. 2 bar is common to all operating points.

- **Comparison of the heat release rate of various combustion processes**

Fig. 4.31 shows the heat release rates for the combustion processes described above. The difference in heat release rates between the throttled and unthrottled operation with fully variable valve lift control is only marginal in comparison with the heat release rates for controlled autoignition.

One clearly recognizes that the combustion duration in controlled autoignition is only approx. 10-16 °CA, which signifies a approximately three times smaller duration as in gasoline multi point injectors with stoichiometric combustion. This is caused by the many ignition points lying close to one another, whose surrounding mixture burns practically simultaneously.

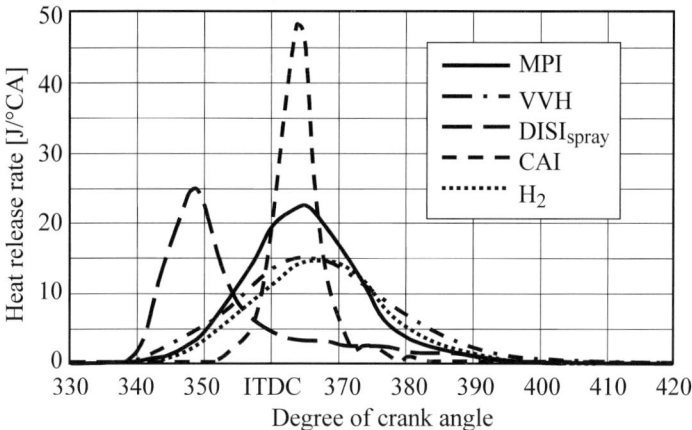

Fig. 4.31: Comparison of the characteristic combustion progressions of various combustion and load control processes

The DISI$_{spray}$ combustion process possesses a relatively early center position. One recognizes here the trade-off between an as good as possible mixture formation for securing a complete combustion with little emission (HC) and a late injection for the purpose of a consumption-optimal center position.

The hydrogen engine possesses a relatively similar burning duration as the multi point injecting SI engine, which results from the in principle high burning speed of hydrogen and the very lean mixture ($\lambda > 3$) which slows down the flame speed on the other hand.

- **Comparison of the loss distribution of various combustion systems**

The loss distributions for the operating conditions described above are represented in Fig. 4.32.

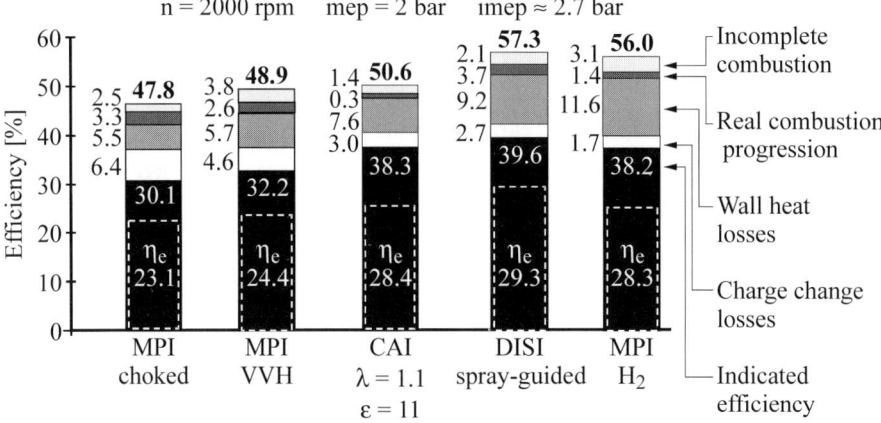

Fig. 4.32: Comparison of loss distribution in various combustion and load control processes

The combustion process with the fully variable intake valve lift control (WH) possesses greater potential for perfect engine efficiency than the throttled engine (MPI) because of higher residual gas capability. However, the losses via imperfect combustion are clearly higher and partially compensate for the significantly smaller charge changing losses.

The spray-guided DI engine (DISI$_{spray}$) possesses the highest potential of the perfect engine at over 57 %, since it has a very high global air-fuel equivalence ratio in the described operation point because of its ability to run in stratified-charge mode. The higher compression ratio of 12 as opposed to 10.5 in other combustion processes reinforces the higher basic potential. At this operation point, we recognize the small losses due to imperfect combustion. Through the early center point, the combustion losses are, however, higher as those of multi point injectors, which has a further effect on wall heat loss. These effects compensate for a part of the large potential, yet they show at the same time a large potential for improvements. The reduced charge changing losses, which cause in sum a approx. 7 % higher indicated efficiency, are clearly recognizable. This means a fuel consumption improvement of about 20 % in comparison with variable valve lift control at this operating conditions. Hereby not under consideration are, however, losses through exhaust gas treatment measures (purging).

4.3 Pressure trace analysis

Clearly recognizable is the great potential of controlled autoignition, in which, despite smaller basic potential from the standpoint of the perfect engine, extremely minute combustion losses, relatively small wall heat losses, and very minimal charge changing losses lead to a high indicated efficiency. This is obviously higher than that of variable valve lift control and only narrowly below that of direct injection, spray-guided combustion.

In the case of the hydrogen engine, the high compression ratio and above all the high air-fuel equivalence ratio ($\lambda > 3$) due to the extremely wide ignition boundaries of hydrogen have a very favorable effect on the basic potential of the perfect efficiency. The efficiency of the perfect engine amounts to about 56 %. Combustion with an optimal position does reduce combustion losses. However, clearly higher wall heat losses also result from the higher combustion temperatures in hydrogen combustion, which destroys a large part of its potential. Nevertheless, a comparatively high indicated efficiency results.

It can be recognized from the operation points that a short combustion duration and, related to that, a small amount of combustion loss cause higher wall heat losses. In this case, a compromise has to be found in order to be able to realize a low fuel consumption. The same is valid for the relation between a higher compression ratio for the sake of a higher efficiency of the perfect engine and wall heat losses.

5 Phenomenological combustion models

For the calculation of engine combustion processes, various model categories can be exploited, which are partially extremely diverse in their level of detail, but also in their calculation time requirements, see Stiesch (2003). Calculation models are customarily designated as phenomenological models that can calculate combustion and pollutant formation contingent upon important physical and chemical phenomena like spray dispersion, ignition, reaction kinetics, etc., see Fig. 4.11.

Because a spatial subdivision of the combustion space into zones of varying temperature and composition is often necessary, the models are also referred to as quasidimensional models. Phenomenological (or quasidimensional) models differ on the one hand from zero-dimensional (or thermodynamic) models, which simplify the combustion chamber as being ideally mixed at every point in time and are based on empirical approaches for the combustion rate. Examples of this are the VIBE and the polygon-hyperbola heat release rate, see chap. 7.1. On the other hand, phenomenological combustion models differ from the CRFD codes (see chap. 9), in that we consciously do without an explicit solution of the turbulent three-dimensional flow field in order to reduce the calculation time. In this way, the calculation time for one engine revolution lies in the region of seconds in phenomenological models, while in CRFD codes it takes hours.

In the following, a few of the most important phenomenological combustion models known in the literature will be introduced. The primary goal of each of these models is to calculate the heat release rate in advance contingent upon characteristic physical and chemical quantities without the need of pressure indication. Moreover, if statements must be made about pollutant formation, it is necessary to carry out a subdivision of the combustion space into zones of varying temperature and composition. This is because the reaction rates of the chemical reactions decisive for pollutant formation are in general exponentially contingent upon temperature, so that knowledge of the arithmetically averaged cylinder temperature does not suffice on its own, see chap. 6. Some of the phenomenological combustion models to be described in the following carry out such a subdivision of the combustion chamber into zones of varying temperature and composition automatically, so that the corresponding pollutant formation model can be directly coupled to it. To this belongs, for example, the packet models described in chapter 5.1.3. In the case of other phenomenological approaches, this zone subdivision is not yet implicitly contained, so that it has to be completed later in order to calculate not only the combustion rate, but also pollutant emissions. For this, we can, for example, utilize the two-zone cylinder model explicated in chapter 7.2.

5.1 Diesel engine combustion

5.1.1 Zero-dimensional heat release function

A relatively simple and thus calculation-time-efficient model for heat release in the diesel engine has been presented by Chmela et al. (1998). This model borders between zero-dimensional and phenomenological models, as it does not carry out a quasidimensional sub-

5.1 Diesel engine combustion

division of the combustion space into zones of varying composition and temperature. Yet nevertheless gives the heat release rate not empirically, e.g. with a VIBE function, but coupled rather to a few characteristic influence parameters of above-average importance. These parameters are the fuel mass available at every point in time, thus the difference between injected and burned fuel mass, as well as the specific turbulent kinetic energy, which is taken as representative for the mixing speed of air and fuel. Fig. 5.1 shows a typical temporal development of these two quantities as well as the combustion rate resulting from the product

$$\frac{dQ_{fuel}}{d\varphi} = C f_1(M_{fuel}) f_2(k) = \left(M_{fuel} - \frac{Q_{fuel}}{lhv} \right) \exp \frac{\sqrt{k}}{\sqrt[3]{V_{cyl}}} \ . \tag{5.1}$$

The temporal progression of the injection rate $dM_{fuel}/d\varphi$ is thereby given as a boundary condition, and the specific turbulent kinetic energy k can be calculated easily with the help of a few relations, see Chmela et al. (1998).

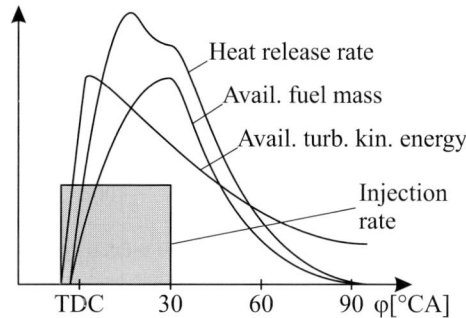

Fig. 5.1: Description of the combustion rate as a function of available fuel mass and turbulent kinetic energy, acc. to Chmela et al. (1998)

The extremely short calculation time and simple application are advantages of this model approach, as well as the fact that the effects of the injection system (e.g. injection pressure, nozzle orifice cross-section and number) on the heat release rate can, as a rule, be fairly accurately depicted. On the other hand, there is also the restriction that neither the ignition delay nor the typical premixed fraction of diesel engine combustion can be described with this model. Both phenomena are substantially influences by the evaporation speed of the fuel, the additional consideration of which in the model would cause a significantly higher requirement in calculation time.

5.1.2 Stationary gas jet

More extensive model approaches, which are each based on the gas jet theory of Abramovich (1963), have been chosen, e.g., by de Neef (1987) and Hohlbaum (1992), in order to calculate heat release in a DI diesel engine. Under the assumption that evaporation progresses quickly as opposed to mixture formation, injection is described as a quasi-stationary gas jet in an idealized solid body rotational flow, see Fig. 5.2. The combustion rate is then calculated as a direct function of the mixture formation rate, thus of the intermixing of fuel vapor and air.

The propagation speed of the spray front as well as its change in direction via the charge movement results analytically from mass and momentum balances of the spray which has been reduced to its central axis. In accordance with Fig. 5.2, the momentum balances in the radial, tangential, and vertical direction of the cylindrical coordinate system are:

$$\frac{d}{dt}(dm_{jet}\,\dot{r}) = dF_r \,, \tag{5.2}$$

$$\frac{1}{r}\frac{d}{dt}\left(dm_{jet}\,r^2\,\dot{\varphi}\right) = \frac{d}{dt}(dm_a)r\omega + dF_t \,, \tag{5.3}$$

$$\frac{d}{dt}(dm_{jet}\,\dot{z}) = 0 \tag{5.4}$$

whereby dm_{jet} designates the mass of a spray-disc with thickness dx. The magnitudes dF_r and dF_t are the radial and tangential forces, which affect the spray-disc, and the index a designates the unburned air surrounding the spray. The radial force is caused by the radial pressure gradient resulting from the rotational movement,

$$dF_r = -dV\frac{dp}{dr} = -\frac{dm_{jet}}{\rho}\rho_a\,r\omega^2 \tag{5.5}$$

and the tangential force is roughly

$$dF_t = 0{,}1\frac{1}{c}\frac{v_{inj}}{b}r(\omega - \dot{\varphi})dm_f \,, \tag{5.6}$$

with $b = b(x)$ as the position-contingent radius of the circular spray-disc. The over-bar designates the value mass-averaged over the entire spray cross-section.

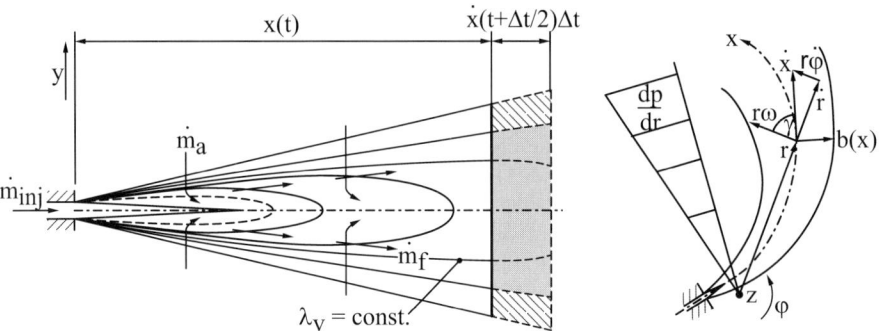

Fig. 5.2: Model of quasi-stationary gas jet in a solid body rotation flow-field

With the help of the above-mentioned relations, we receive the motion equations of the spray front in the three cylinder coordinates,

$$\ddot{r} + \overline{c}\frac{d}{dt}\left(\frac{1}{\overline{c}}\right)\dot{r} = r\left[\dot{\varphi}^2 - (1-\overline{c})\omega^2\right], \tag{5.7}$$

5.1 Diesel engine combustion

$$\ddot{\varphi} + 2\frac{\dot{r}}{r}\dot{\varphi} = \left[\bar{c}\frac{d}{dt}\left(\frac{1}{\bar{c}}\right) + 0{,}1\frac{v_{inj}}{b}\right](\omega - \dot{\varphi}),\tag{5.8}$$

$$\ddot{z} + \bar{c}\frac{d}{dt}\left(\frac{1}{\bar{c}}\right)\dot{z} = 0,\tag{5.9}$$

with spray speed $\dot{x}$ and penetration depth S,

$$\dot{x} = \sqrt{\dot{r}^2 + (r\dot{\varphi})^2 + \dot{z}^2},\qquad S = x = \int_0^1 \dot{x}\,dt.\tag{5.10}$$

The spray angle and thus the change in the spray radius along the spray axis has a considerable influence upon the rate of air entrainment to the fuel spray. For combustion processes with no or only a minor amount of air swirl, a standard value of

$$(db/dx)_{\omega=0} = 0{,}16 \tag{5.11}$$

is recommended. However, this value has to be adjusted if necessary in order to be able to illustrate the real spray angle, which is influenced, for example, by injection pressure, nozzle geometry, or physical characteristics of air and fuel. For combustion processes with marked swirl flow, de Neef (1987) provides the following correction for the spray angle

$$\frac{db}{dx} = \frac{1 - C(r\omega/v_{inj})}{1 + C(r\omega/v_{inj})}\left(\frac{db}{dx}\right)_{w=0},\tag{5.12}$$

with

$$C = \frac{r\dot{\varphi}}{x} - \frac{1}{2}\sqrt{2}\frac{\dot{r}}{\dot{x}} \tag{5.13}$$

and

$$v_{inj} = c_D \sqrt{\frac{2\Delta p_{inj}}{\rho_f}}.\tag{5.14}$$

In order to be able to determine the mixture distribution within the spray, the fuel mass fraction averaged over the spray cross-section $\bar{c}$ along the spray coordinate x is at first calculated with the help of mass conservation. Under the presumption that the fuel mass contained in a spray disc of thickness dx is constant ($dm_{jet}\,\bar{c}=\text{const.}$) and that the averaged spray density $\bar{\rho}$ within this disc is very small in comparison with the density of the liquid fuel ρ_f, the temporal change of the mean fuel mass fraction contingent upon the spray angle (db/dx) can be expressed as follows

$$\frac{d}{dt}\left(\frac{1}{\bar{c}}\right) = \frac{4}{d_{noz}^2\,v_{inj}}\frac{\rho_a}{\rho_f}\left[2\left(\frac{db}{dx}\right)b\dot{x}^2 + b^2\ddot{x}\right].\tag{5.15}$$

With the known fuel mass fraction $\bar{c}(x)$, averaged over the spray cross-section, the local fuel mass fraction $c(x,y)$ can be calculated in a further step. For this, an empirical dependence on the radial position in the spray is assumed

$$c = c_m \left[1 - \left(\frac{y}{b}\right)^{\frac{3}{2}}\right], \qquad (5.16)$$

whereby c_m corresponds to the fuel mass fraction on the central axis of the spray.

In the model of de Neef (1987), it is now assumed that the combustion rate is limited by the mass of fuel that is processed per unit of time in stoichiometric relation to air. This quantity is determined as follows. Since the fuel mass fraction is known at every position in the spray, the iso-contours of the air-fuel equivalence ratio λ within the spray represented in Fig. 5.2 can be determined. The dimensionless radius y/b of a definite air-fuel equivalence ratio λ_v contingent on the axial position in the spray is

$$\frac{y}{b}(\lambda_v, x) = \left[1 - \frac{c(\lambda_v)}{c_m(x)}\right]^{\frac{2}{3}}. \qquad (5.17)$$

As the injection spray is assumed to be stationary, the λ distribution within the spray does not change with time. In every numerical advance in time Δt, merely a new disc of thickness Δx is added to the spray, see Fig. 5.2. Due to mass conservation, the fuel mass contained in it is identical to the injected mass during this time increment ($\dot{m}_{inj}\,\Delta t$). Therefore, the fuel mass that goes beyond a certain boundary of λ_v = const. within a time step (hatched surface in Fig. 5.2) must be equal to the difference between the injected fuel mass and the fuel which is found within the λ_v boundary (gray surface), thus in the richer mixture.

$$\Delta m_{f,\lambda_v} = \dot{m}_{inj}\,\Delta t - \pi\, y^2(\lambda_v)\rho_a\, c_m\left[1 - \frac{4}{7}\left(\frac{y(\lambda_v)}{b}\right)^{\frac{3}{2}}\right]\dot{x}\,\Delta t. \qquad (5.18)$$

In order to determine the fuel mass being processed in the entire spray in stoichiometric relation with air, (5.18) has to be integrated between the rich ignition limit λ_R and $\lambda = 1$. Since only a fraction of $d\lambda_v$ of the fuel that crosses from ($\lambda = \lambda_v$) to ($\lambda = \lambda_v + d\lambda_v$) becomes newly prepared with air (the remaining fraction has already been prepared in previous time steps), we receive the relation

$$\Delta m_{f,stoic} = \lambda_{v,R}\,\dot{m}_{f,\lambda_{v,R}}\,\Delta t + \int_{\lambda_{v,R}}^{\lambda_v=1}\dot{m}_{f,\lambda_v}\,d\lambda\,\Delta t. \qquad (5.19)$$

Simplifying, it is assumed after injection finish that the area of spray close to the nozzle is no longer in existence, while the remaining part of the spray located further downstream is still behaving stationarily. This behavior is considered in so far as a second (virtual) spray is calculated that begins to spread out at injection finish and is subtracted from the original spray.

The combustion rate is described with a quasi-kinetic approach which expresses the combusted portion of the stoichiometrically prepared fuel mass

5.1 Diesel engine combustion

$$X = \frac{m_{f,b}}{m_{f,stoic}} \quad (5.20)$$

with the Arrhenius function

$$dX = A\rho_{jet} T_{jet}^\beta \frac{af_{stoic}(1-X)^2}{af_{stoic}-1} \exp\left[-\frac{E_A}{R_m T_{jet}}\right] dt . \quad (5.21)$$

In this, T_{jet} and ρ_{jet} are the values for temperature and density averaged over the entire spray. The Arrhenius constants A, β and E_A have to be adjusted empirically for a certain engine in order to be able to depict experimentally determined combustion rates.

Since, with the model of stationary gas spray, neither fuel atomization nor drop evaporation are explicitly described, it is hardly possible to model the ignition delay in a detailed fashion. It is instead assumed that combustion begins at the moment in which the air ratio on the spray axis goes beyond the lower ignition boundary λ_R for the first time. At this time however, a certain quantity of fuel found in the outer spray regions has already been mixed stoichiometrically with air. This can now be converted very quickly, so that the typical premix-peak of diesel engines results in the heat release rate, see Fig. 5.3.

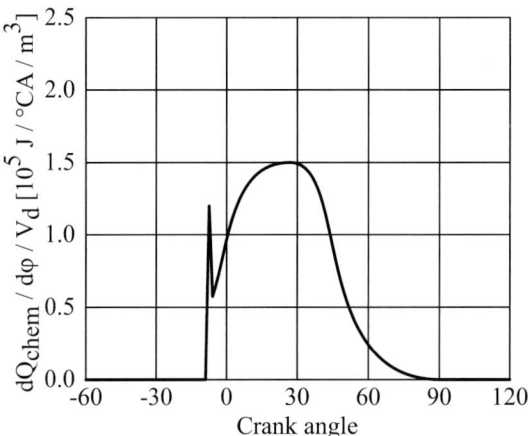

Fig. 5.3: Calculated heat release rate of a high speed high-performance diesel engine at rated power, acc. to Hohlbaum (1992)

It should be observed that the spray opening angle in this combustion model, which is to be empirically determined, is a parameter of decisive importance, since it substantially influences the mixing speed of fuel and air and thus also the combustion rate. Beyond this, it is apparent that the assumption of an undisturbed stationary gas spray no longer is applicable if the spray collides with a combustion chamber wall. For this reason, the model appears to be most appropriate for describing large engines with distinct air swirl.

5.1.3 Packet models

One frequently applied model approach for describing diesel engine combustion is the so-called packet model (Hiroyasu et al., 1983), depicted in Fig. 5.4. The injection jet is subdivided in this approach into many small zones, so-called packets, which illustrate in sum the contour of the entire spray. Each of these single spray packets is then considered as a separate thermodynamic control volume, for which the respective mass and energy balances are solved. And, within these limits, the most important subprocesses like drop evaporation or combustion and pollutant formation rates are calculated (see Fig. 4.11). From this results, for every packet, a distinctive composition and temperature history. Via simple addition of burning rates in each packet, we obtain finally the total heat release rate for the cylinder.

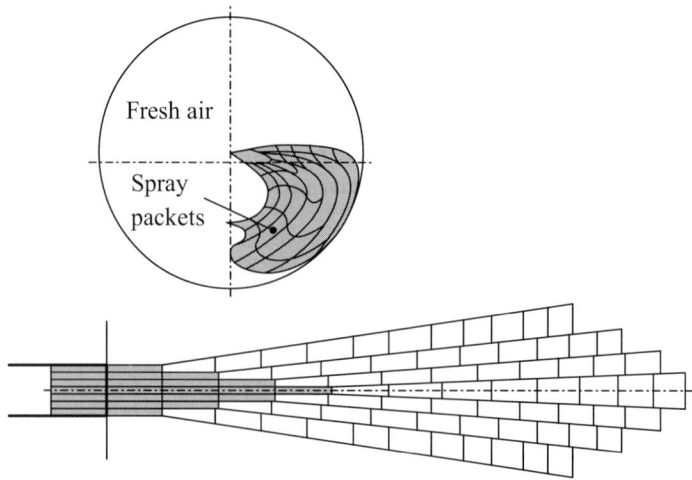

Fig. 5.4: Packet model, see Hiroyasu at al. (1983)

The phenomenological combustion model of Stiesch (1999), which will be described in more detail in the following, is based on the fundamental packet approach of Hiroyasu. During the compression stroke, only one zone exists, which extends itself over the entire combustion chamber and is viewed as ideally mixed. Fresh charge air and, in the case of exhaust gas recycling, combustion products are found in this zone. During the duration of injection, additional so-called spray packets are continuously generated, which reproduce the global form of the injection spray and subdivide it in both axial and radial directions. Independently of the number of nozzle bores, only one single fuel spray is viewed; an interaction of various sprays can thus not be considered. During the injection period, a new axial "disc" of packets is generated at each computational time step, whereby individual packets show a ring form because of their radial subdivision.

At the generation instant, only liquid fuel is found in the packet. After the progression of a characteristic time, the liquid fuel is atomized into small drops, and the entrainment of gasses from the surrounding zone of fresh air to the single spray packets begins. The fuel droplets are heated up by the hot gasses which have entered the packets and evaporate. After the end

5.1 Diesel engine combustion

of the ignition delay, the fuel-air-mixture begins to burn, by means of which the packet temperature increases again and pollutant formation (NO and soot) begins as well.

Both atomization and drop evaporation as well as ignition and combustion proceed within the packet boundaries and must therefore be calculated separately for each single packet. After combustion start, the packets can thus contain not only liquid fuel and fresh air, but also fuel vapor and combustion products, see Fig. 5.5. A mixing of various spray packets or an exchange of energy between them does not occur. With exception of air-entrainment into the spray (and thus to the packets) and wall heat transfer, all transport processes thus proceed within packet boundaries.

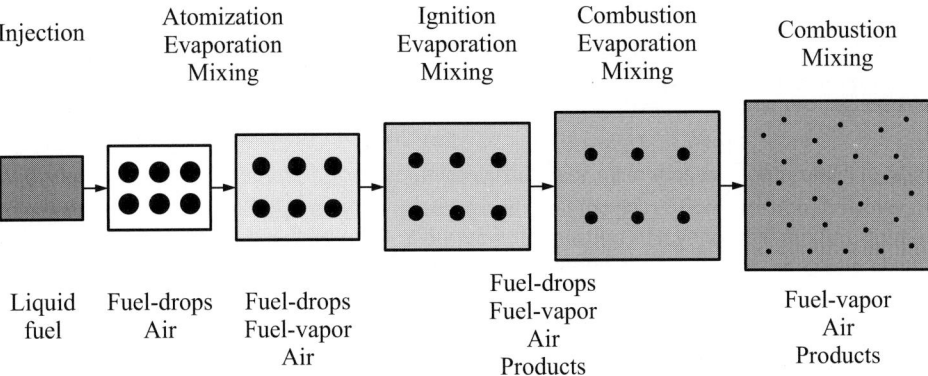

Fig. 5.5: Composition of spray packets

With the help of mass and energy balances as well as an equation of state, the change rates of composition, temperature, and volume can be calculated for each single packet as well as for the fresh air zone. On the other hand, pressure is viewed as independent of location and only as a function of time. This assumption is justified because of the high speed of sound at high pressures during the combustion phase.

- **Spray development and mixture formation**

Immediately after injection start, one spray packet is viewed as a continuous liquid phase, which moves into the combustion chamber with the constant speed

$$v_{inj} = 0.39 \sqrt{\frac{2 \Delta p_{noz}}{\rho_{f,fl}}} \qquad (5.22)$$

until atomization begins. The liquid fuel mass $m_{f,p}$ per packet with the momentary injection rate $\dot{m}_{inj}$, the number of packets in the radial direction k_{max} and the length of computational time step Δt amounts to

$$m_{f,p} = \frac{\dot{m}_{inj} \Delta t}{k_{max}} \ . \qquad (5.23)$$

After a characteristic time span, the liquid phase disintegrates into small drops. This so-called breakup time amounts on the spray axis to

$$t_{bu,c} = 28.65 \frac{\rho_{f,fl} \, D_{noz}}{\sqrt{\rho_a \, \Delta p_{noz}}} \,. \tag{5.24}$$

Because the interaction between fuel and air at the spray boundary is more marked than on the spray axis, spray breakup in the outer packets begins earlier corresponding to

$$t_{bu,k} = t_{bu,c} \left(1 - \frac{k-1}{k_{max}}\right), \tag{5.25}$$

if a linear decrease in breakup time is assumed over the spray radius. Through the entrainment of gasses from the fresh air zone into the spray packet, packet speed is reduced. For packets on the spray axis

$$v_{tip,c} = 1.48 \left(\frac{\Delta p_{noz} \, D_{noz}^2}{\rho_a}\right)^{1/4} \frac{1}{\sqrt{t}} \tag{5.26}$$

is valid, and for packets further outside, it is roughly assumed that a speed profile decreasing exponentially towards the spray border is engaged

$$v_{tip,k} = v_{tip,c} \exp\left(-C_{rad} \, (k-1)^2\right). \tag{5.27}$$

When five spray packets are viewed radially ($k_{max} = 5$), and it is further assumed that the speed of the outer packet amounts to approx. 55 % of the speed on the axis, a value of 0.374 for the constant C_{rad} results. The injection process itself also changes the flow pattern in the combustion chamber decisively. The kinetic energy of the injection jets is approximately two orders of magnitude above the kinetic energy of swirl and squeeze flows at injection start. As a result of this, the spray packets generated first are much more strongly slowed down by the surrounding gas phase than the ones created towards the end of injection, which move as it were in the "wind shadow". The packet speed after spray breakup is therefore corrected according to

$$v_{i,k} = C_1 \, v_{tip,k} \left[1 + \left(\frac{i-1}{i_{max}-1}\right)^{C_2} \frac{\Delta t_{inj}}{C_3}\right], \tag{5.28}$$

whereby $i = 1$ designates the packet generated first and $i = i_{max}$ the last. The constant C_1 can be slightly higher that 1, C_2 has the approximate value of 0.5 and C_3 describes the absolute speed difference between the first and the last packet.

The air entrainment rate is calculated with the principle of impulse conservation of the spray packets

$$v_{i,k} \left(m_{f,p} + m_{a,p}\right) = \text{const.} \,. \tag{5.29}$$

5.1 Diesel engine combustion

- **Drop distribution spectrum**

After the breakup time, the liquid fuel of the spray packet disintegrates into many small drops, the integral behavior of which can be described with the Sauter mean diameter. The Sauter mean diameter is thereby the diameter of a representative drop, which has the same volume to surface area ratio as all drops integrated over the entire spray. For this we find the relation

$$SMD = 6156 \cdot 10^{-6} \, v_{f,fl}^{0.385} \, \rho_{f,fl}^{0.737} \, \rho_L^{0.06} \, \Delta p_{noz}^{-0.54} \tag{5.30}$$

with SMD in [m], v in [m²/s], ρ in [kg/m³] and pressure difference Δp in [kPa]. The number of fuel drops in a packet under the assumption that all drops are equally large amounts to

$$N_{dr,p} = \frac{m_{f,p}}{\rho_{f,fl} \frac{\pi}{6} SMD^3} \tag{5.31}$$

For a more detailed description of atomization and thus also of the following evaporation process, the drop size distribution function

$$g(r) = \frac{r^3}{6\bar{r}^4} \exp\left(\frac{-r}{\bar{r}}\right) \tag{5.32}$$

can be utilized with the radius

$$\bar{r} = \frac{SMD}{6} \tag{5.33}$$

of the most frequently appearing drop.

- **Drop evaporation**

In describing evaporation, the mixing model is often used, in which the inside of the drop is always assumed to be isothermal. As a comparison fuel, pure tetradekan ($C_{14}H_{30}$) will be used in the following, which has similar physical properties as real diesel fuel. For investigations with two-component comparison fuels, e.g. a mixture of 70 vol.% n-dekan ($C_{10}H_{22}$) and 30 vol.% α-methyl napthaline ($C_{11}H_{20}$), refer to Stiesch (1999).

With this, we obtain for the convective heat transfer from the gas phase to the drop with the help of the Nußelt number

$$\frac{dQ_{Tr}}{dt} = \pi \, SMD \, \lambda_S \, (T_p - T_{Tr}) \frac{z}{e^z - 1} Nu, \tag{5.34}$$

whereby z represents a dimensionless correction factor, which diminishes the transferred heat flux under the simultaneous appearance of mass transfer via evaporation corresponding to

$$z = \frac{c_{p,f,g}\,\dfrac{dm_{dr}}{dt}}{\pi\,SMD\,\lambda_S\,\text{Nu}} \quad . \tag{5.35}$$

We calculate the evaporation rate of a drop with the help of the relation for mass transfer as

$$\frac{dm_{dr}}{dt} = -\pi\,SMD\,C_{diff}\,\rho_S\,\ln\!\left(\frac{p_{cyl}}{p_{cyl} - p_{f,g}}\right)\text{Sh} \quad . \tag{5.36}$$

For the Nußelt and Sherwood number is valid

$$\text{Nu} = 2 + 0.6\,\text{Re}^{1/2}\,\text{Pr}^{1/3} \quad , \tag{5.37}$$

$$\text{Sh} = 2 + 0.6\,\text{Re}^{1/2}\,\text{Sc}^{1/3} \quad , \tag{5.38}$$

whereby the Reynolds number is calculated with a relative speed between drop and gas phase, which is assumed to be up to 30 % of the momentary packet speed $v_{i,k}$. Temperature change of the liquid fuel drops results finally from an energy balance over a drop

$$\frac{dT_{dr}}{dt} = \frac{1}{m_{dr}\,c_{p,dr}}\left(\frac{dQ_{dr}}{dt} + \frac{dm_{dr}}{dt}\,\Delta h_v\right) , \tag{5.39}$$

with drop mass

$$m_{dr} = \frac{\pi}{6}\,\rho_{dr}\,SMD^3 \tag{5.40}$$

contingent on diameter and drop temperature.

- **Ignition delay**

Ignition delay is often described by means of a simple Arrhenius method

$$\tau_{ID} = C_1\,\frac{\lambda_p}{p_{cyl}^2}\,\exp\!\left(\frac{C_2}{T_p}\right) \tag{5.41}$$

with $C_1 = 18$ and $C_2 = 6{,}000$.

- **Heat release**

As a simplification, we assume that after reaching ignition delay, the fuel is completely converted to CO_2 and H_2O in correspondence with the gross reaction equation. For a detailed consideration of this, the reader is referred to Stiesch (1999).

The maximum combustion rate in the packet is limited by the strictest of the following three criteria. Firstly, only the vaporous fraction of fuel can be burned

$$\dot{m}_{f,Ox,p} \leq \frac{m_{f,g,p}}{\Delta t} - \dot{m}_{dr,p}\,N_{dr,p} \quad . \tag{5.42}$$

Secondly however, the quantity of air in the packet also limits the conversion rate according to

5.1 Diesel engine combustion

$$\dot{m}_{f,Ox,p} \leq \frac{\dot{m}_{a,p}}{L_{min} \Delta t} + \frac{\dot{m}_{tot.,p}}{L_{min}} . \quad (5.43)$$

Thirdly, a maximal chemical conversion rate for premixed flames must still be considered, which is described by the Arrhenius function

$$\dot{m}_{f,Ox,p} \leq 5 \cdot 10^5 \, \rho_{mix} \, x_{f,g,p} \, x_{O_2,p}^5 \, \exp\left(-\frac{12{,}000}{T_p}\right) V_p \quad (5.44)$$

and which is important in the late combustion phase, when the temperature in the cylinder has sunk considerably and the chemistry is therefore slow. The thermodynamic balance equations necessary for further calculation are introduced in chapter 2 and more exhaustively in chapter 7. For the determination of thermodynamic state quantities of single components represented in the packets, we refer again to Stiesch (1999).

- **Model validation**

Fig. 5.6 and Fig. 5.7 show a comparison of a measured and calculated combustion and pressure traces for two operating conditions of a high speed diesel engine with 3.96 liter displacement volume per cylinder, 165 mm piston diameter and a speed of 1,500 rpm. The operation point represented in Fig. 5.6 was chosen as a reference point for the adjustment of the model, whereby more attention was paid to a good agreement of pressure paths than on an exact adjustment of the heat release rate. In total, we can recognize a good agreement between simulation and measurement results. Detailed investigations show however that a more complex heat transfer model, which explicitly considers the influence of soot radiation, is advantageous for a further improvement in agreement.

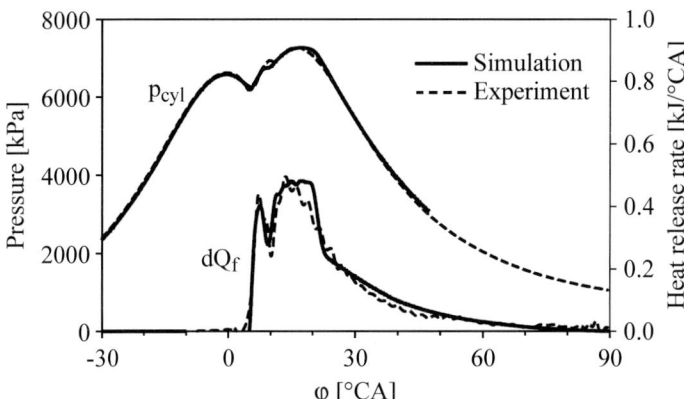

Fig. 5.6: Comparison of measured and calculated pressure progressions and heat release rate for a high speed diesel engine with 3,96 liter displacement volume per cylinder at n = 1,500 rpm and *mep* = 9.8 bar

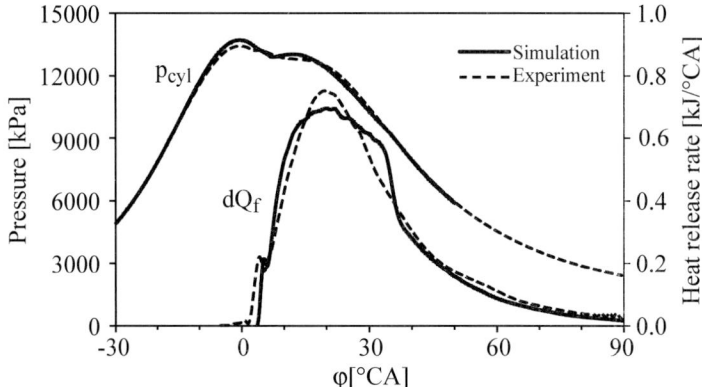

Fig. 5.7: Comparison of measured and calculated pressure progressions and heat release rate for a high speed diesel engine with 3,96 liter displacement volume per cylinder at n = 1,500 rpm and *mep* = 22.0 bar

- **Description of a pre-injection**

Thoma et al. (2002) expanded upon the packet method in order to describe diesel engine combustion processes with pre-injection as well. However, since the spray penetration curve (5.26) is only valid for continuously injected sprays and not for very small fuel quantities (Stegemann et al., 2002), Thoma et al. (2002) suggest a change of time contingency from $1/\sqrt{t}$ in $1/t$ for the pre-injection packets. Moreover: at the moment, in which the main injection phase begins, the pre-injection packets are compacted into a single so-called pre-injection zone, see Fig. 5.8. Because the pre-injection slows down so quickly, the packets of main combustion soon penetrate into the pre-injection zone, so that, instead of fresh air, gasses of the pre-injection zone are admixed into the main combustion packets. This entrainment of already hot gasses into the spray packets causes a reduction of the ignition delay of the main injection. From this results the well-known effect of pre-injection, namely, the clear reduction on the premix-peak in the heat release rate. Fig. 5.9 shows that this behavior of the model can be depicted very well.

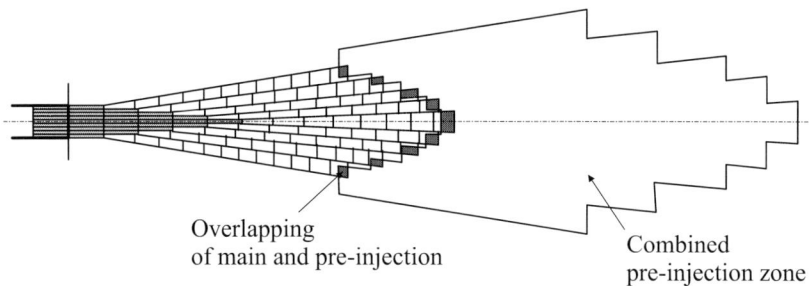

Fig. 5.8: Mixing of the pre- and main injection pulses, acc. to Thoma et al. (2002)

5.1 Diesel engine combustion

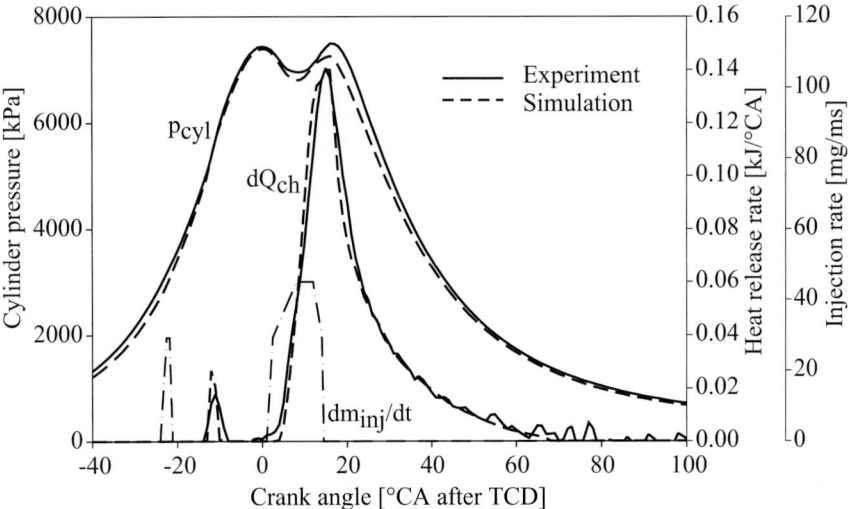

Fig. 5.9: Calculated and measured heat release rate and pressure progressions of a truck diesel engine with pre-injection, Thoma et al. (2002)

5.1.4 Time scale models

Weisser and Boulouchos (1995) have developed a phenomenological model for the heat release rate in the diesel engine, which is based on characteristic time scales, similar to the eddy breakup model which is often used within the CFD code (see chap. 9). In this case, two different time scales are considered for premixed and diffusion combustion, as it is assumed that premixed combustion is essentially influenced by fuel evaporation and reaction kinetics, while diffusion combustion is above all contingent on the speed of the turbulent mixing of fuel vapor and air.

The atomization and evaporation of fuel is very similarly modeled as in the packet method introduced above. However, the spray is here discretized only in the axial direction, and the spray penetration is calculated with the equation of Dent (1971):

$$S = 3.07 \left(\frac{\Delta p_{inj}}{\rho_g} \right)^{1/4} (d_{noz}\, t)^{1/2} \left(\frac{294}{T_g} \right)^{1/4}. \tag{5.45}$$

It is now assumed that the fuel fraction that already evaporated before the first ignition is converted as premixed combustion, while the remaining fuel is converted in diffusion combustion, which is controlled by the turbulent mixing. The ignition delay is again similarly determined as in the packet approach on the basis of an Arrhenius equation, see (5.41).

The time scale characteristic for reaction-kinetically controlled premixed combustion is assumed to be proportional to the ignition delay τ_{ID}, so that the conversion rate of the fuel can be given as

$$\frac{dm_{prem}}{dt} = C_{prem} \frac{1}{\tau_{ID}} f_{prep} \, m_{prem,av} \,, \tag{5.46}$$

whereby $m_{prem,av}$ is the total fuel mass added to the premixed combustion. The factor f_{prep} takes into account the fact that only a part of this mass has gone beyond the ignition delay and can actually be converted at the time under consideration.

The conversion rate of diffusion combustion is formulated in analogy to (5.46)

$$\frac{dm_{diff}}{dt} = C_{diff} \frac{1}{\tau_t} f_{A,turb} \, m_{diff,av} \,. \tag{5.47}$$

We have here the difficulty however that, within the phenomenological approach, the turbulent time scale τ_t cannot be determined as is customary in the CFD code (see chap. 8) directly from the knowledge of the turbulent flow field, but must rather be estimated with the help of a simplified method. For the sake of this, the turbulent mixing frequency – the reciprocal value of the turbulent time scale – is approached as the ratio of turbulent viscosity and of the square of a linear scale characteristic for the problem

$$\frac{1}{\tau_t} = \frac{u' l_I}{(X_{char})^2} \,. \tag{5.48}$$

In order to estimate the turbulent viscosity $u' l_I$ in the combustion chamber, a simplified method is chosen that assumes two sources of turbulence. The first source is the intake flow of charge-air, for which the turbulence intensity u' is set as proportional to the mean piston speed and the linear scale l_I as proportional to the clearance height. The second source of turbulence is the injection spray itself, for which the magnitudes u' and l_I can be solved with the help of conservation equations, see Heywood (1988). The initial values thereby result from the injection speed and the nozzle orifice diameter.

Correspondingly, for the sum of turbulent viscosity,

$$u' l_1 = (u' l_1)_{charge} + (u' l_1)_{inj} \tag{5.49}$$

is valid. The characteristic linear scale for the process of turbulent diffusion between fuel vapor and air is determined in contingency upon the momentary cylinder volume, the global air ratio, and the number of nozzle orifices

$$X_{char} = \left(\frac{V_{cyl}}{\lambda N_{noz}} \right) \,. \tag{5.50}$$

The factor $f_{A,turb}$ in (5.47) describes the increase of the actual surface of the flame front through turbulent folding (see chap. 4.1.3),

$$f_{A,turb} = \frac{u' l_I}{v} \,, \tag{5.51}$$

whereby v is the viscosity of the combustion gasses. With this, (5.47) can be written as

$$\frac{dm_{diff}}{dt} = C_{diff} \frac{u'l_I}{X_{char}^2} \frac{u'l_I}{V} m_{diff,av} . \qquad (5.52)$$

Finally, the fuel masses $m_{pre,av}$ and $m_{diff,av}$ available to both combustion types are determined through the integration of evaporation rate and combustion rate,

$$m_{i,av} \int_{t_{i,0}}^{t} \left(\frac{dm_{i,evap}}{dt} - \frac{dm_i}{dt} \right) dt , \qquad (5.53)$$

whereby the index i stands for both combustion types *pre* and *diff*.

5.2 SI engine combustion

In order to calculate heat release in the SI engine, the *entrainment model*, developed by Blizard and Keck (1976) and expanded by Tabaczinsky (1980), is often used and will be briefly explained in the following.

In this model, heat release or flame front propagation is analyzed into two partial steps. The first step describes the penetration of the flame because of the turbulent propagation mechanism without heat release into the as yet unburned mixture. The penetration speed is additively composed of the turbulent fluctuation speed (turbulence intensity) u' and the laminar flame speed s_l, and it amounts under consideration of the continuity condition for the charge-mass recorded per time to

$$\frac{dm_e}{dt} = \rho_u A_{ff} (u' + s_l) , \qquad (5.54)$$

whereby A_{ff} is the surface of the flame front and ρ_u is the density of unburned mixture.

The second partial step describes heat release through combustion, whereby the fresh gas swirl regions grasped by the flame are converted with laminar flame speed. The dominating swirl size is thereby the Taylor micro-length, which is defined with the integral linear length scale l_I by

$$l_t = \sqrt{\frac{15 l_I V}{u'}} . \qquad (5.55)$$

With that follows for the characteristic combustion time scale

$$\tau = \frac{l_t}{s_l} \qquad (5.56)$$

and with this for the conversion rate of the fuel mass found in the flame area

$$\frac{dm_x}{dt} = \frac{m_e - m_f}{\tau} . \qquad (5.57)$$

The turbulence intensity at ignition time is set proportionally to the mean piston speed

$$u'_{IT} = c_t c_m \qquad (5.58)$$

and changes in the course of combustion with the mixture density corresponding to

$$u' = u'_{IT} \left(\frac{\rho_M}{\rho_{M,IT}} \right)^{1/3} . \tag{5.59}$$

The integral length scale describes the extensive swirl structure and is assumed as proportional to the combustion chamber height

$$l_{I,IT} = c_L \, h_{cc} \tag{5.60}$$

and also changes in the course of combustion with mixture density in correspondence with

$$l_I = l_{I,IT} \left(\frac{\rho_{M,IT}}{\rho_M} \right)^{1/3} . \tag{5.61}$$

This entrainment model presumes a fully developed flame front. Therefore, an inflamed volume is given as a start value for the combustion, the mass of which corresponds to 1 % of the total charge-mass. In determining the ignition time, the time span between the ignition time and the 1 % mass conversion point, the so-called inflammation duration, must be calculated. We set for this

$$\Delta t_{id} = c_{id} \, \tau . \tag{5.62}$$

For the laminar flame speed, the method explained in chap. 4.1.3 ((4.1) to (4.4)) is utilized.

For a more extensive presentation of combustion modeling in SI engines, the reader is once again referred to Stiesch (2003).

- **Ignition model**

In the SI engine, combustion is introduced by a spark discharge at the spark plug. With the assumption that a constant adiabatic and isobaric temperature exist, which characterizes the ignition limit, the condition

$$h_{eg}\left(T_{ad,IT}, p_{IT}\right) = \frac{1 - \kappa_{rg}}{1 + \lambda \, L_{min}} \, lhv + h_{fg}\left(T_{u,IT}, p_{IT}\right) \tag{5.63}$$

can be derived by balancing a small volume element in the spark plug area.

The thus calculated ignition boundaries agree well in a wide temperature region with measured values, see Scheele (1999).

Fig. 5.10 shows a comparison of the characteristic mass conversions determined by pressure course analysis with those calculated by the entrainment model. The constants for the calculation of turbulence intensity is fixed at $c_t = 0.6$ and for the calculation of the integral length scale at $c_l = 0.35$, such that the ratio of the time periods between the mass conversion points agree with the values from the pressure path analysis. The assumption at combustion start of a turbulence intensity constant at various loads reproduces the change in burning duration with a sufficient degree of precision.

5.2 SI engine combustion

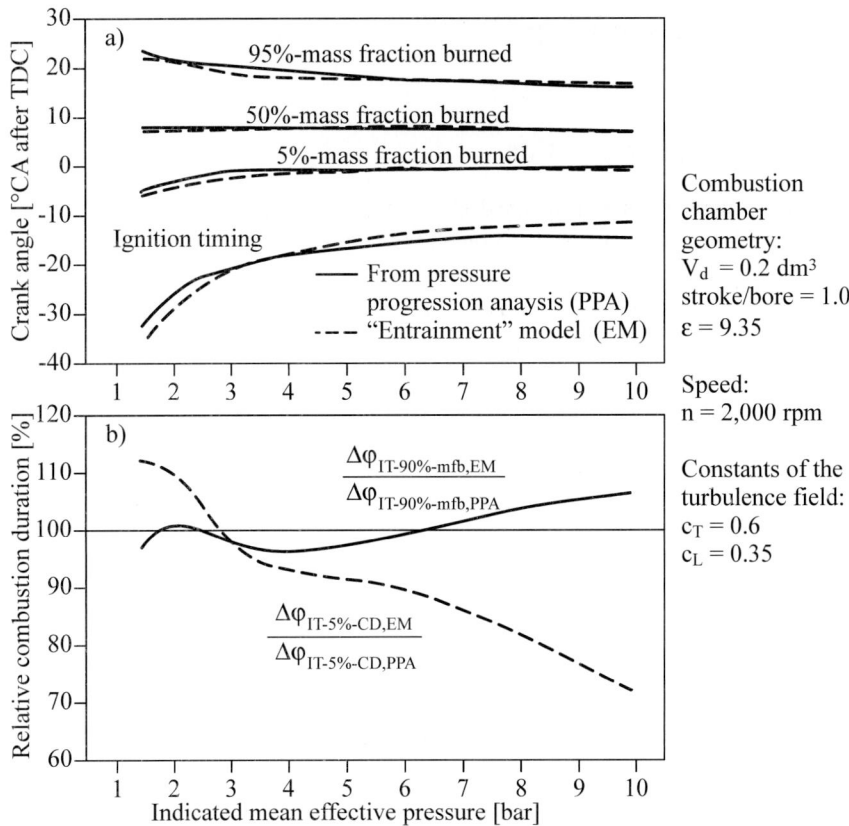

Fig. 5.10: Comparison of the mass conversions determined by the pressure progression analysis and calculated by the entrainment model (from: Scheele (1999))

6 Pollutant formation

6.1 Exhaust gas composition

In the complete combustion of a so-called C_xH_y fuel, consisting only of C and H atoms, the exhaust gas contains the components oxygen (O_2), nitrogen (N_2), carbon dioxide (CO_2), and steam (H_2O).

In real, incomplete combustion, carbon monoxide (CO), unburned hydrocarbons (HC), nitrogen oxide (NO_x), and particulates also appear in addition to the above components. As opposed to these substances, which are detrimental to human health, CO_2, which is partially responsible for the greenhouse effect, is not viewed as a pollutant, since it does not pose a direct health hazard and appears as the final product of every complete oxidation of a hydrocarbon. A reduction of CO_2 in the exhaust gas is thus only to be achieved through a reduction in consumption or through an altered fuel having a smaller amount of carbon with reference to its heating value.

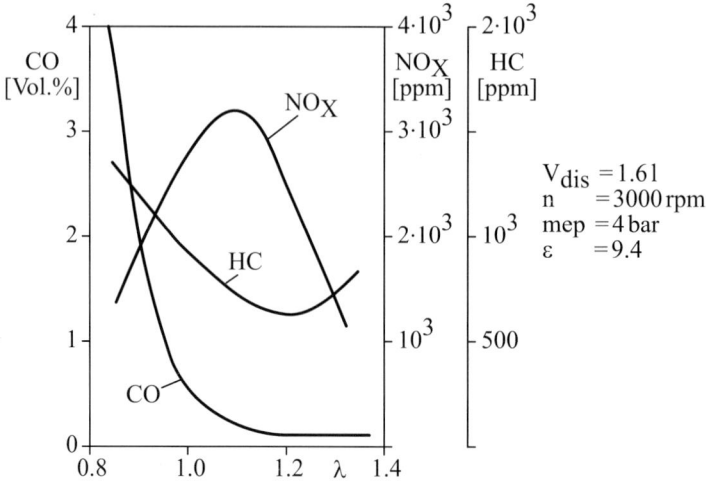

Fig. 6.1: Pollutant formation contingent upon the equivalence ratio

The formation of CO, HC and NO_x is primarily contingent on the air-fuel equivalence ratio λ and the combustion temperature coupled with it, see Fig. 6.1. While CO and HC rise as products of incomplete combustion in a rich mixture ($\lambda < 1.0$), NO_x formation is favored by a high temperature at sufficient levels of oxygen ($\lambda \approx 1.1$). With a lean mixture ($\lambda > 1.2$), the combustion temperature sinks, so that NO_x emissions fall off and HC emissions increase.

In Fig. 6.2, the compositions of the exhaust gasses (without a catalytic converter) of SI and diesel engines are shown. From this we see that the amount of pollutants has, from the point of view of energy, no significance in the engine process, but rather only from the point of view of its potential to jeopardize human health and the environment. Although the diesel engine only emits about a fifth the amount of pollutant that SI engines do, the absolute NO_x concentrations are not very different. While in the case of the diesel engine, particulate matter also represent a critical magnitude besides nitrogen oxides, CO is the dominate pollutant component in the SI engine.

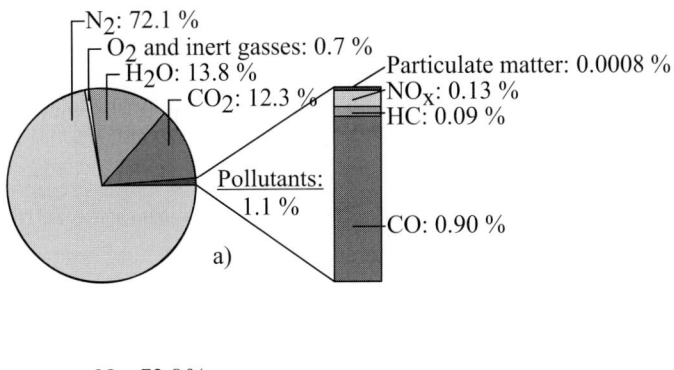

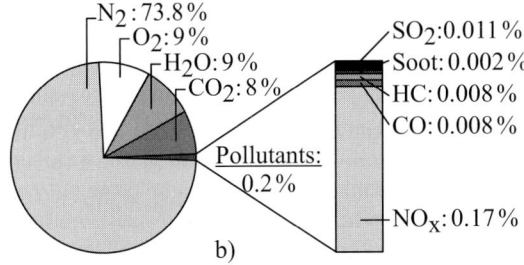

Fig. 6.2: Raw emissions (without catalyst) in percent by volume. a) SI engine and b) diesel engine

6.2 Carbon monoxide (CO)

Under a local lack of air ($\lambda < 1.0$), as a rule CO develops as a product of incomplete combustion. The oxidation of CO proceeds varyingly depending on the air-fuel equivalence ratio λ.

In the sub-stoichiometric range ($\lambda < 1.0$), CO oxidation progresses, due to a lack of O_2, in competition with H_2 oxidation

(1) $\quad CO + OH^{\bullet} \leftrightarrow CO_2 + H^{\bullet}$ and
(2) $\quad H_2 + OH^{\bullet} \leftrightarrow H_2O + H^{\bullet}$,

whereby the hydroxyl radical $OH^{\bullet}$ and atomic hydrogen $H^{\bullet}$ function as chain propagators. While reaction (2) is found in practical equilibrium, reaction (1) is kinetically controlled and thus advances much more slowly in the sub-stoichiometric range. With a climbing air ratio and temperature, the deviation of the kinetics of the OHC equilibrium becomes smaller

$$\frac{d[CO]}{dt} = f(\lambda, T) \tag{6.1}$$

and CO concentration thus decreases with an increasing air ratio λ.

In the stoichiometric range ($\lambda \approx 1.0$), reactions (1) and (2) can be described with a very good approximation as a gross reaction via the water gas reaction

(3) $CO + H_2O \leftrightarrow CO_2 + H_2$,

which in this case proceeds near equilibrium, because the surplus concentrations of the chain propagators $H^\bullet$ and $OH^\bullet$ are very large.

In the super-stoichiometric range ($\lambda > 1.0$), CO oxidation no longer progresses in competition with H_2 oxidation, but rather according to the following pattern

(1) $CO + OH^\bullet \leftrightarrow CO_2 + H^\bullet$,
(4) $O_2 + H^\bullet \leftrightarrow OH^\bullet + O^\bullet$.

In this range and during the act of expansion, relatively more $H^\bullet$ than $OH^\bullet$ exists because of the lack of equilibrium in reaction (1), and CO oxidation proceeds slowly.

In an extremely lean mixture ($\lambda > 1.0$), increased CO develops again due to the lower temperatures and incomplete combustion in the area near the wall of the combustion chamber. Generally, CO oxidation is highly contingent on temperature, so that reaction (1) becomes increasingly slow during expansion as well. The CO concentration in the exhaust gas thus corresponds approximately to the equilibrium concentration at 1,700 K.

6.3 Unburned hydrocarbons (HC)

In the combustion of C_xH_y fuels, no measurable HC concentrations appear "behind" the flame front assuming that $\lambda > 1$. HC thus originates in zones that are not completely or not at all involved in combustion. The unburned hydrocarbons are thereby composed of a number of different components, which are either completely unburned or already partially oxidized. Legislators today restrict only the sum of all HC components, which are usually determined with a flame ionization detector. In this way, no statement is made about the composition of these unburned hydrocarbons. The particular hazardous potential of certain components is thus not considered.

6.3.1 Limited pollutant components

In the SI engine, most of the unburned hydrocarbons are already emitted during the cold start and warm-up phases. HC development thereby has the following causes:
- frontal extinguishing of the flame while approaching a cold wall,
- flame extinguishment within a gap due to an excessive cooling of the flame front,

6.3 Unburned hydrocarbons (HC)

- flame extinguishment as a result of an insufficient flame speed during expansion (rapid temperature decline).

If the ratio of the thermal energy released in the flame to heat losses of the flame at the wall is lower than a certain value, the flame is extinguished. This can be roughly described by the Peclet number

$$\text{Pe}_{1,2} = \frac{\rho c_p w}{\lambda} x_{1,2} \,, \tag{6.2}$$

whereby the first two cases of wall and gap of flame extinguishment are differentiated with the indices 1 and 2. In this case, w stands for the flame speed and x_1 and x_2 for the wall distance and the gap width.

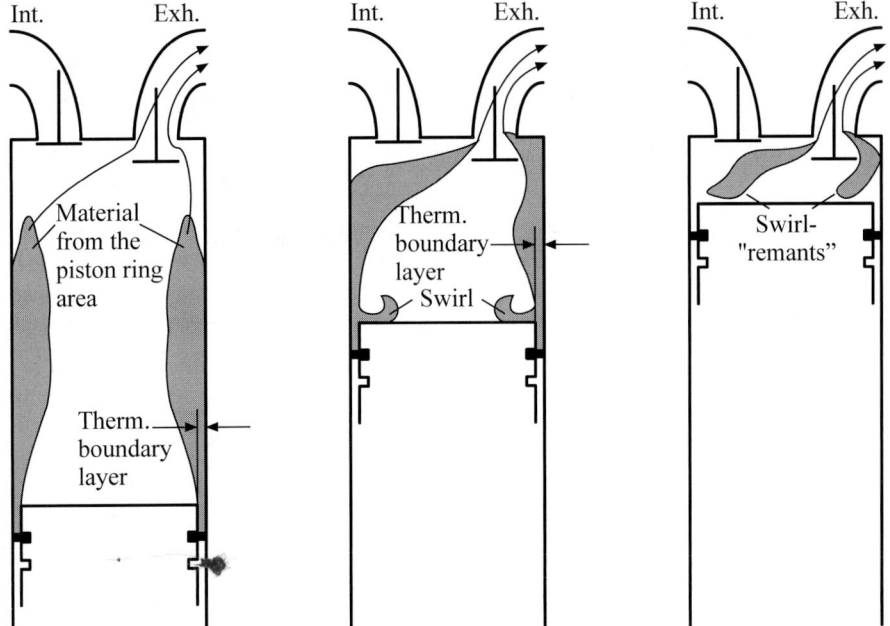

Fig. 6.3: Schematic representation of HC-formation

From numerous experiments, the values $\text{Pe}_1 \approx 8$ and $\text{Pe}_2 \approx 40$ result for the critical Peclet numbers and thus for the ratio $x_1/x_2 = 0.2$ and as orders of magnitude for the extinguishment distance

wall: $0.02 < x_1 < 0.2 \,\text{mm}$,

gap: $0.1 \;\; < x_2 < 1.0 \,\text{mm}$.

On the other hand, we obtain from considerations of the order of magnitude of the Nußelt and Reynolds numbers for the thickness of the thermal boundary layer at the combustion chamber wall $\delta_T \approx 1\text{mm}$. From this we see that the flame is extinguished at a temperature much closer to that of the wall than that of the gas.

At combustion start, up to 6 % of the mixture escapes into the piston ring gap, of which about 2 % flows back again towards the end of combustion. At this late time, the combustion chamber temperatures are often already so low that no further oxidation to complete combustion products is possible anymore. In the upward movement of the piston, the unburned hydrocarbons are then pushed out along with the remaining combustion exhaust gasses, see Fig. 6.3. For an extensive presentation of the influence of the combustion chamber form on HC emission, see in addition Borrmeister and Hübner (1997).

The different paths of HC formation mechanisms in SI engine combustion are represented in Fig. 6.4. This illustration summarizes recent research results and thus describes – if somewhat generalized – the present level of knowledge.

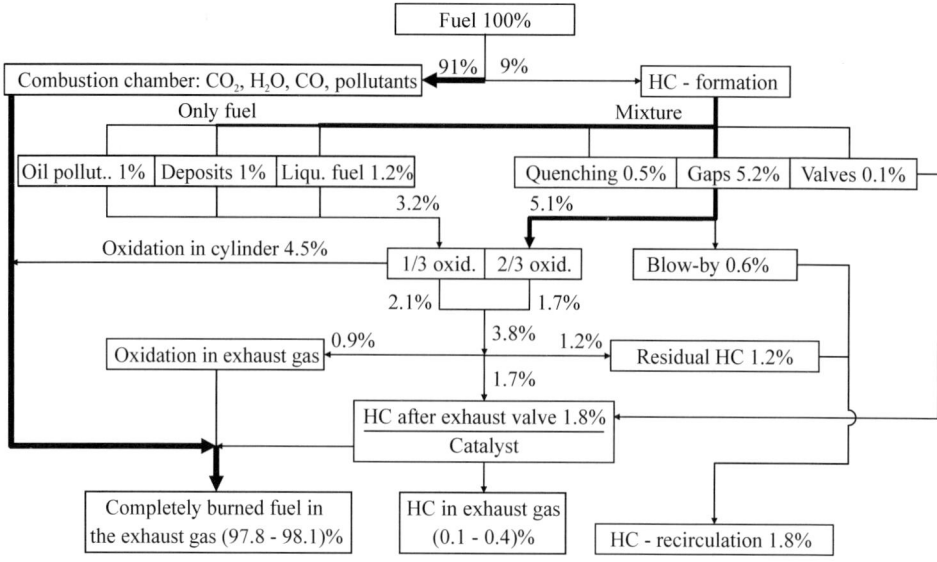

Fig. 6.4: HC-formation mechanisms in SI engine combustion

The so-called mechanisms of HC development are in total very complex, and a quantitative calculation of HC emission in the SI engine is thus not yet feasible. For the estimation of the oxidation speed of the amount of HC originating in the cylinder, the following global relation is often recommended

$$\frac{d[HC]}{dt} = - c_R A [HC][O_2] e^{-\frac{E}{RT}}, \tag{6.3}$$

with $E = 156 \text{ J/mol}$, $A = 6.7 \cdot 10^{21} \text{ m}^3/\text{mols}$.

6.3 Unburned hydrocarbons (HC)

The dimensionless factor c_R assists in the adjustment to experimental data and varies between approx. 0.1 and 1.0 as a result of the large differences previously described.

The unburned hydrocarbons can be generally divided into the following groups:

- aromatics, unsaturated compounds: 45 %
- alkanes, saturated compounds: 20 %
- alkenes: 30 %
- aldehydes: 5 %

Among the aromatic compounds, polycyclic aromatic hydrocarbons are also found, so-called PAHs, which are of importance in the formation of soot in the diesel engine, see chap. 6.4.2.

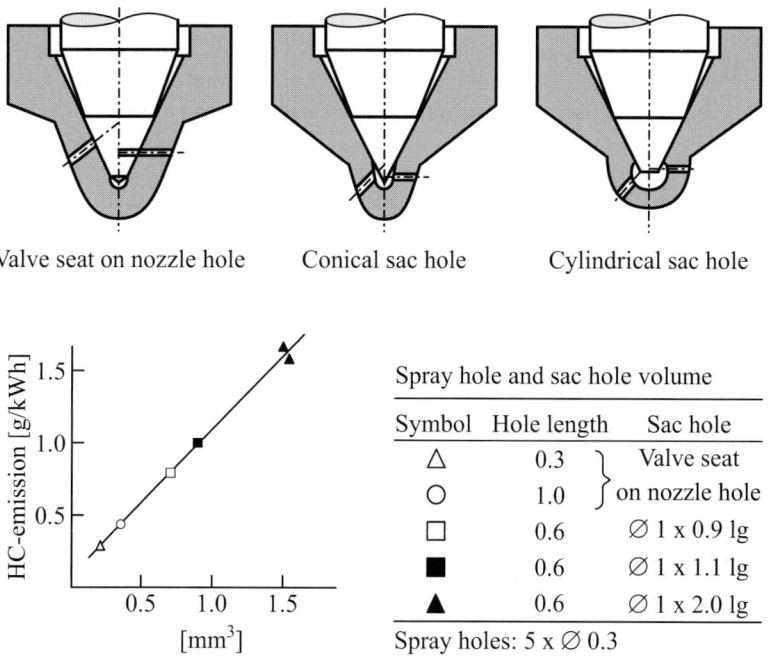

Fig. 6.5: HC-emission contingent on the blind hole volume

In the diesel engine, the process of HC formation is even more complex and thus even more problematic to calculate as in the SI engine. The most important sources of HC are:

- the periphery of the spray – the mixture composition lies outside the ignition area (too lean),
- the inner spray area – the mixture composition is too rich,
- extinguishment of the diffusion flame by rapid pressure and temperature decreases during expansion,
- fuel adhering to the wall is not completely oxidized due to insufficient temperatures,

- "after-injection" due to renewed opening of the nozzle needle after injection finish. From this results extremely large fuel drops, which can only evaporate and combust slowly.

With respect to the last point, the influence on the blind hole volume on HC emission in the diesel engine is represented in Fig. 6.5. One recognizes that HC emission climbs practically linearly with the volume of the blind hole.

6.3.2 Non-limited pollutant components

Among the total mass of unburned hydrocarbons are found several substances, the amount of which has still today not been explicitly limited yet, which are particularly important because of potential health hazards.

- **Carbonyl compounds**

Carbonyl compounds can harm the human organism by affecting it directly or via by-products formed in the atmosphere. They contribute, for example, towards the formation of ozone close to the ground (photochemical smog) in concert with nitrogen oxide.

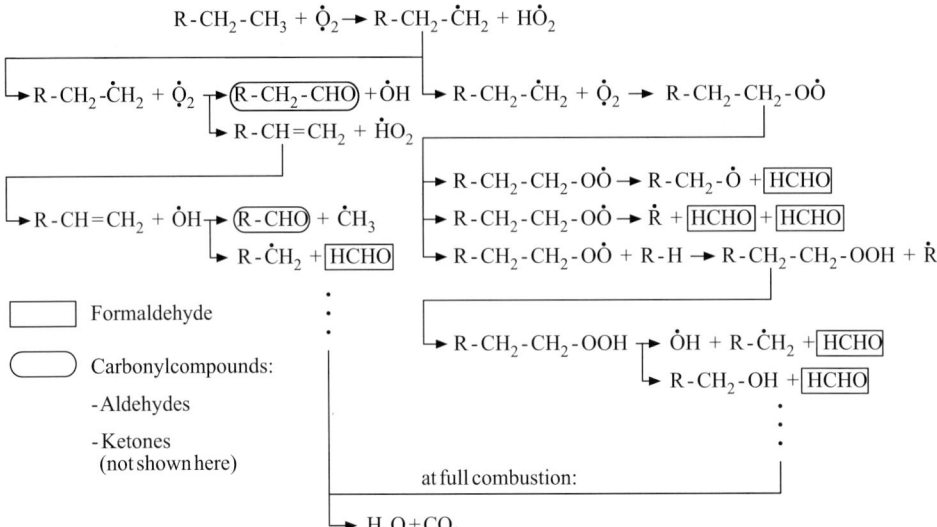

Fig. 6.6: Excerpt from the hydrocarbon oxidation diagram

With the carbonyl compounds rank aldehydes and ketones, which each have at least one characteristic carbonyl group at their disposal. They originate as partially burned combustion components, the complete oxidation of which had been prematurely aborted.

In Fig. 6.6, an extract from the hydrocarbon oxidation system is shown qualitatively with the aldehyde R-CHO, which appears in the final phase of oxidation, as well as formaldehyde HCHO. This presentation also conveys an idea of the complexity of the oxidation system underlying C_xH_y oxidation.

6.3 Unburned hydrocarbons (HC)

Supplementing this, in Fig. 6.7 are represented the carbonyl compounds detectable in the present day and in Fig. 6.8 the distribution of carbonyl compounds in the exhaust gas of the diesel engine of a commercial vehicle, see Lange (1996).

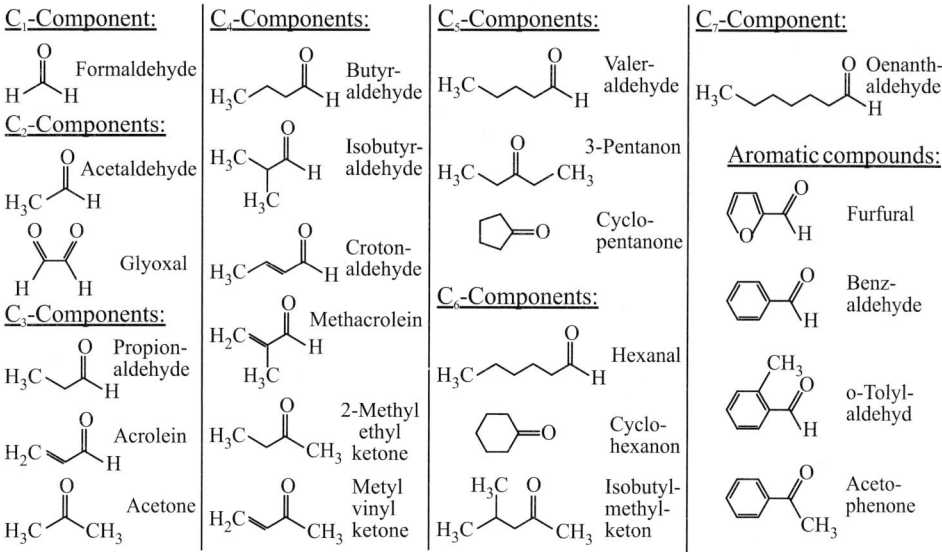

Fig. 6.7: Detectable carbonyl compounds

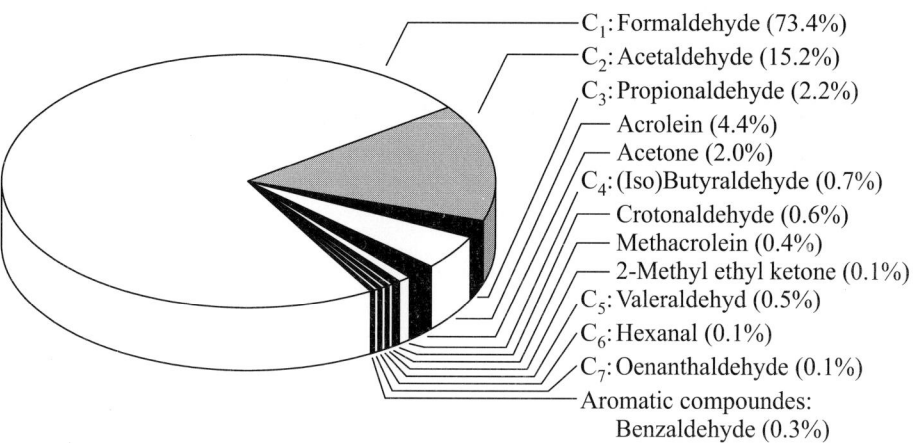

Fig. 6.8: Distribution of carbonyl compounds in the exhaust gas of a truck diesel engine

- **Dioxins and furans**

Dioxins are aromatic hydrocarbons with fully non-toxic to extremely toxic compounds. The definition frequently also includes the chemically and toxicologically related furan class. Since the chemical accident of Seveso in 1976 however, the extremely toxic 2,3,7,8 tetra-chlordibenzo-p-dioxin ("Seveso poison") released then is often representative for all dioxines. For the explanation of the chemical structure, the structural formulae of the benzene ring and of a few chlorinated as well as polycyclic aromatic hydrocarbons are shown in Fig. 6.9, and in Fig. 6.10 are shown the structural formulae of heterocyclic aromatic compounds pyridine, dioxine, and furane, as well as two substituted compounds.

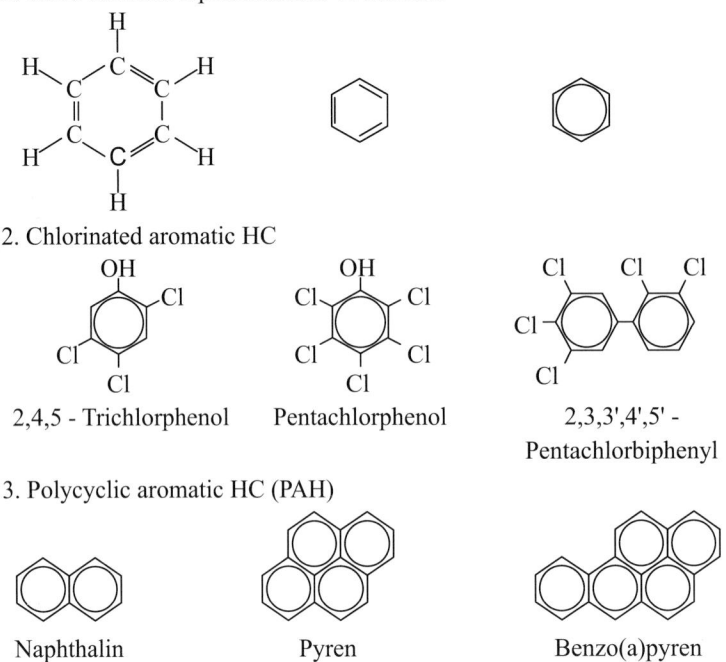

Fig. 6.9: Structure of various aromatic hydrocarbons

To be mentioned as precursors of dioxins and furans are the polycyclic aromatic hydrocarbons (PAH), and polycyclic biphenyls (BCB)

and polychlorinated terphenyls (PCT)

6.3 Unburned hydrocarbons (HC)

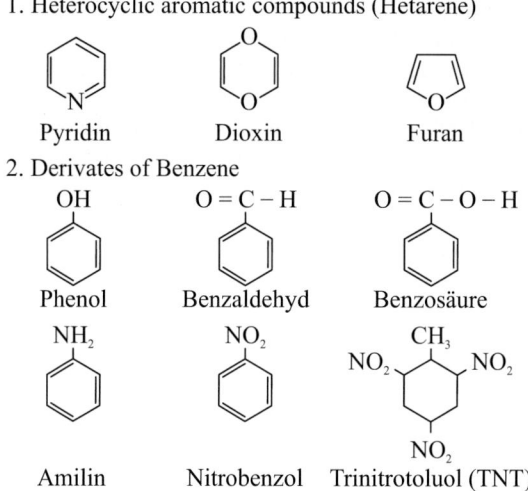

Fig. 6.10: Structure of heterocyclic aromatic hydrocarbons

In Fig. 6.11, the structural formulae of dibenzofuran, dibenzodioxin, so-called Seveso poison, as well as the number of possible derivatives are given.

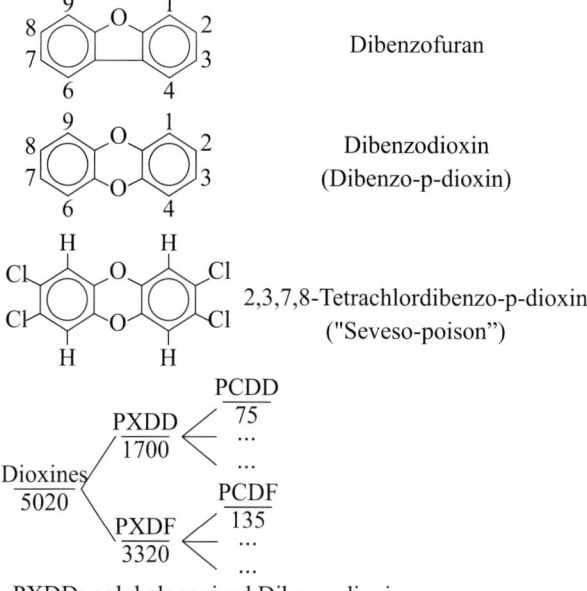

PXDD: polyhalogenized Dibenzodioxines
PCDD: polychlorized Dibenzodioxines

Fig. 6.11: Dioxine structures

In Fig. 6.12, the orders of magnitude of the concentrations of various pollutant components in the exhaust gas of an internal combustion engine is given. Different engine construction types as well as gasoline and diesel fuels do not differ as far as the order of magnitude is concerned. We recognize that the concentrations of all dioxines and furanes lie in the order of magnitude of 10^{-9} kg per kg in the exhaust gas and the concentrations of the infamous Seveso poison in the order of magnitude 10^{-14} kg per kg in the exhaust gas, i.e. far beneath the detection margin of present-day measurement techniques.

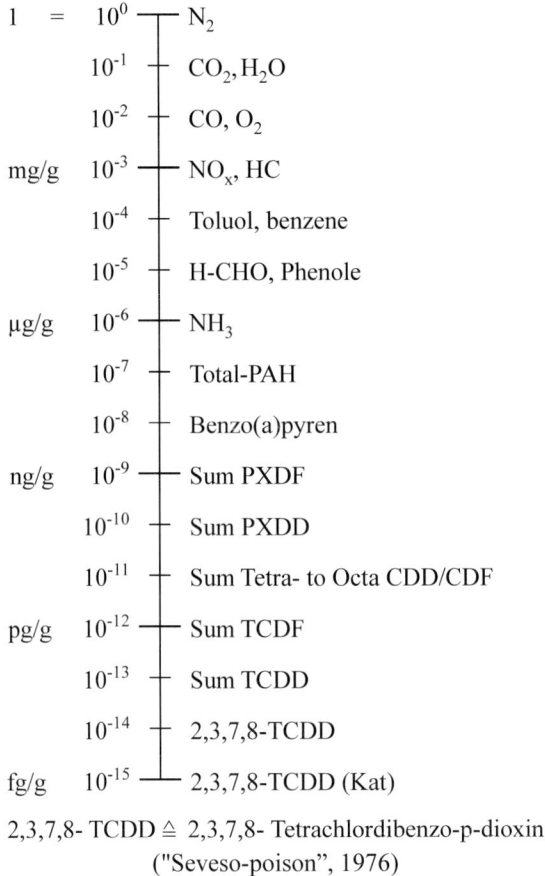

Fig. 6.12: Pollutant concentrations in the exhaust of combustion engines

For further explanation, see Bühler (1995) and Bühler et al. (1997).

6.4 Particulate matter emission in the diesel engine

6.4.1 Introduction

As the particulate matter content in the exhaust gas is designated the quantity of all substances that are captured by a certain filter after the exhaust gas has been diluted according to a defined method and cooled down to $\vartheta < 52\ °C$. The composition of all particulate matter in the diesel exhaust can be seen in Fig. 6.13. According to it, diesel particles consist up to 95 % of organic (PAH and soot) and up to 5 % of inorganic components.

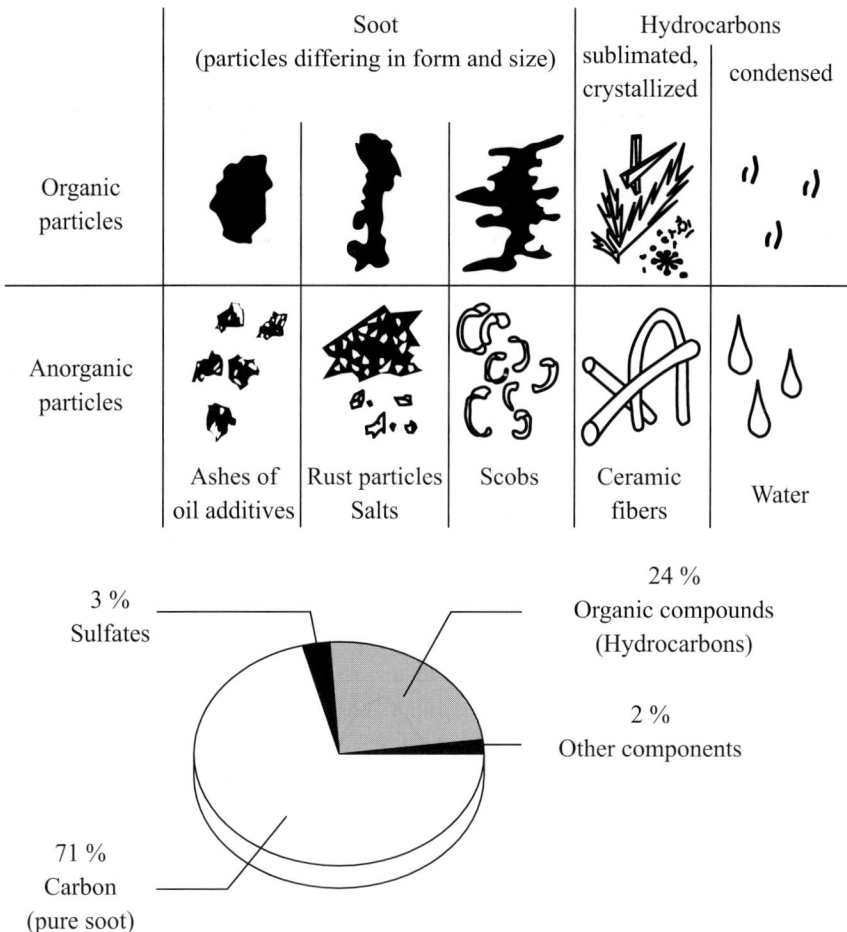

Fig. 6.13: Composition of diesel particles

The chemical and physical processes occurring in the origination of soot particles have been understood roughly, but still insufficiently in many details. The modeling of soot formation is thus quite problematic. Particle origination progresses according to present knowledge approximately according to the following scheme:

- chemical reduction of the fuel molecules to ethin (acetylene, C_2H_2) and to C_3H_3 ions, formation of the first benzene ring,
- formation of polycyclic aromatic hydrocarbons via polymerization of rings and advancing dehydration, also proportional increase of C atoms,
- condensation and formation of soot kernels (nucleation) with dimensions of approx. 1 to 2 nm,
- coalescence of soot kernels to form primary soot particles (surface growth) with diameters of about 20-30 nm and then addition of various substances,
- coalescence of primary soot particles to long, chain structures (agglomeration),
- breaking down of soot particles and intermediate species via oxidation with O_2 molecules and OH radicals.

6.4.2 Polycyclic aromatic hydrocarbons (PAH)

The formation of the first aromatic hydrocarbon (benzene) ring may be explained by two different hypotheses, the acetylene and the ion hypotheses.

The acetylene hypothesis assumes that several molecules of ethin (acetylene, C_2H_2), which originated in rich combustion, join together to a first benzene ring with the addition of $H^\bullet$ and the splitting off of H_2. In this case, two different reaction paths are possible according to the local temperature, as is shown in Fig. 6.14.

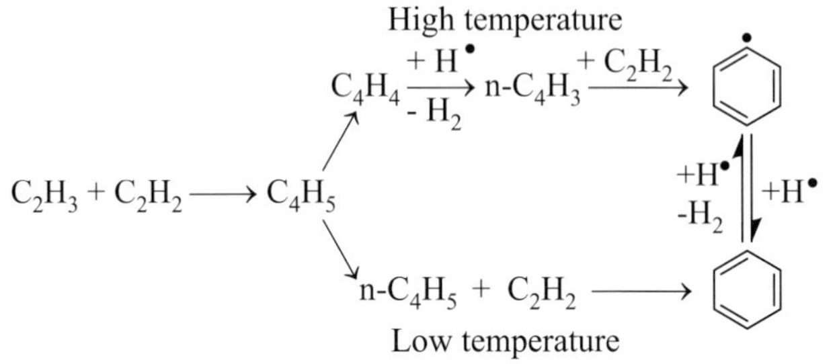

Fig. 6.14: Reaction path in the formation of benzene rings, acc. to Frenklach and Wang (1994)

The ion hypothesis states, on the contrary, that the ethin molecules combine themselves at first with CH- or CH_2 groups also found in the fuel-rich mixture to form C_3H_3 ions. Two such C_3H_3 ions can then unite with the rearrangement of two H atoms to form a ring, see Fig. 6.15.

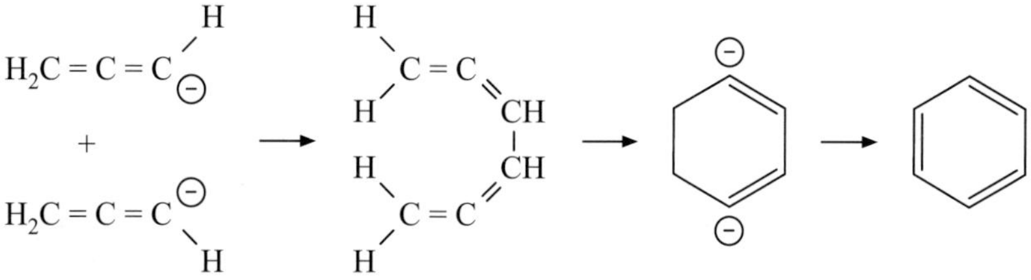

Fig. 6.15: Development of benzene rings, acc. to Warnatz et al. (1997)

Via progressing splitting off of H and C_2H_2 addition, the so-called HACA mechanism (H-abstraction, C_2H_2 addition), PAH rings originate, see Fig. 6.16 a). However, benzene rings can also directly unite themselves, thus constructing complex ring compounds, see Fig. 6.16 b).

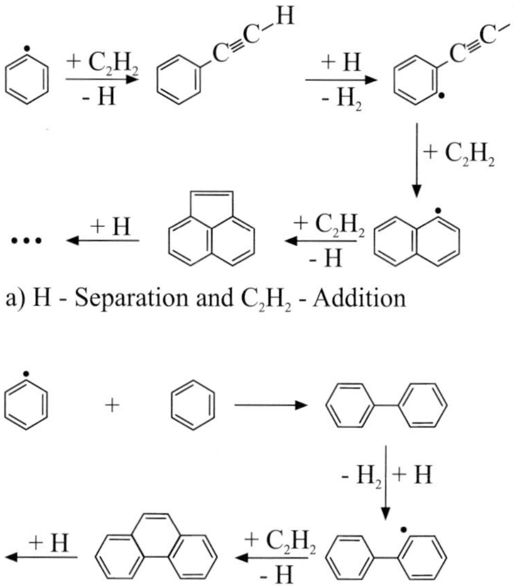

a) H - Separation and C_2H_2 - Addition

b) Ring-amalgamation

Fig. 6.16: PAH growth, acc. to Frenklach and Wang (1994)

6.4.3 Soot development

Polycyclic aromatic hydrocarbons develop into larger and larger formations. Customarily we then speak of soot particles, in which the PAHs are no longer arranged on one plane, but rather represent a three-dimensional object. Fig. 6.17 shows a principle sketch of soot formation in premixed flames, acc. to Bockhorn (1994).

The cumulated soot volume is given by

$$V_P = \frac{\pi}{6} N(d_P) d^3 = \frac{\pi}{6} N(d_P)\left[\sum \frac{1}{N} N_i d_{P,i}^3\right] \quad (6.4)$$

whereby $N(d_p)$ represents the soot particle magnitude distribution or the particle spectrum. Particle size covers a wide range of $10 < d_{P,i} < 150$ nm, whereby however particles with dimensions of up to 10 µm can also form; the distribution maximum lies however at about 100 nm. The density of the soot particles lies at about 2,000 kg/m³, the surface/mass ratio between 20 and 200 m²/g.

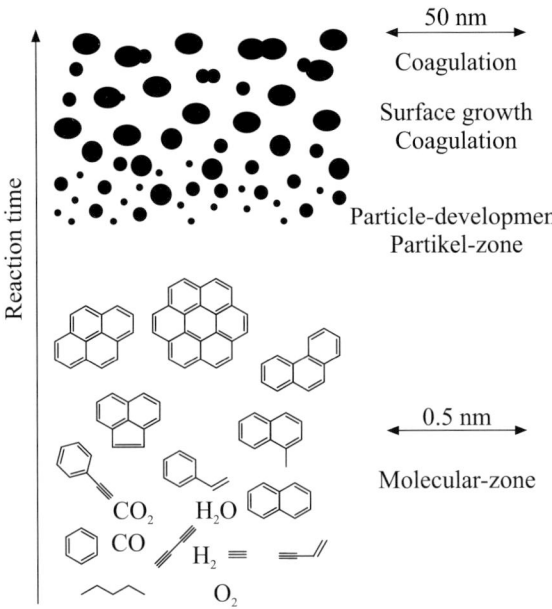

Fig. 6.17: Principle sketch of soot formation, acc. to Bockhorn (1994)

The influence of temperature on soot formation is still controversial, because a high temperature favors both formation (pyrolyse) and oxidation; the temperature window $1,500 < T < 1,900$ is critical for soot formation. This is made clear in Fig. 6.18, in which the percentage soot yield is represented as contingent upon the air-fuel equivalence ratio and temperature. We recognize the critical temperature range $1,500 < T < 1,900$ and an extreme increase in soot emission for equivalence ratios $\lambda < 0.6$.

The pyrolyse-oxidation problem is clarified in Fig. 6.19. Here we have illustrated the soot mass fraction as a function of the crank angle in a diesel engine. One can recognize that, at the beginning of combustion, a relatively large amount of soot is produced, which is however for the most part oxidized again during the main and post-combustion phases. The amount of particles measured in the exhaust gas is thus only a fraction (approx. 0.1 to 1 %) of the maximum formed particles. This very much problematizes the understanding and the coverage of the dominating chemical and physical processes in soot formation.

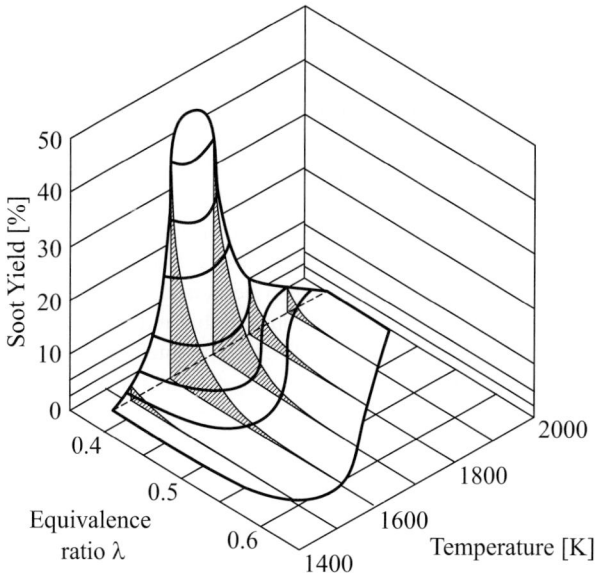

Fig. 6.18: Soot yield as function of equivalence ratio and temperature

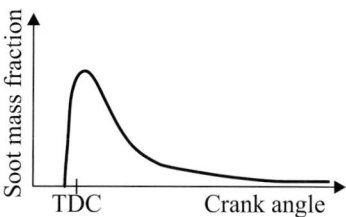

Fig. 6.19: Temporal progression of soot in the DI diesel engine

6.4.4 Particle emission modeling

The modeling of soot development and oxidation is still in its initial stages. It is however generally accepted that local temperature and local fuel vapor and oxygen concentrations are the controlling factors for the respective reaction speeds. Thus, we often utilize a simple 2-equation model, in which formation and oxidation are each described with one empirical equation. The net change in soot mass result from the difference of both of these quantities, see Nishida and Hiroyasu (1989)

$$\frac{dm_{P,f}}{dt} = A_f m_{f,g} p^{0.5} \exp\left[-\frac{6{,}313}{T}\right], \tag{6.5}$$

$$\frac{dm_{P,ox}}{dt} = A_{ox}\, m_P\, x_{O_2}\, p^{1.8} \exp\left[-\frac{7{,}070}{T}\right],\qquad (6.6)$$

$$\frac{dm_P}{dt} = \frac{dm_{P,f}}{dt} - \frac{dm_{P,ox}}{dt}.\qquad (6.7)$$

With this simple model, trend statements about soot formation can be achieved, but no quantitatively reliable results. Thus, it has been further developed, e.g. by Belardini et al. (1994) or Fusco et al. (1994), such that intermediate species such as ethin (acetylene C_2H_2) are also balanced. Both the intermediate species and the final soot particles can be broke down again through oxidation in the course of combustion, so that altogether one model with eight differential equations instead of the two above-named results. For a closer approach towards real conditions, an oxidation by OH radicals can also be considered in the model in addition to oxidation by O_2 molecules, which can gain considerable importance in diesel engine combustion, see Nagle and Strickland-Constable (1962). The interested reader is hereby referred to the cited literature, and in particular also to Bockhorn et al. (2003).

In addition to the local temperature and the time available for the chemical reactions, a sufficient concentration of oxidants ($O^{\bullet}, O_2, OH^{\bullet}$) at the reaction location is also decisive for the progression of soot oxidation. The assumption of a sufficiently fast mixing under all conditions between the oxidants on the one hand and the particles on the other is, however, a crude simplification. Mixture quality depends in reality upon the turbulent flow field. Recent considerations thus assume that, for the particle oxidation, mixing with the oxidants is necessary on the smallest observable dissipative quantity of turbulence elements in the flow field, so that the chemical reaction becomes possible on the molecular level. With this, the course of soot oxidation is also determined by the so-called Kolmogorov time, which is the characteristic revolution time of the smallest appearing eddies in the flow field

$$t_K = \left(\frac{v}{\varepsilon}\right)^{1/2} \qquad (6.8)$$

or the turbulent mixing frequency $1/t_K$. Numerous experiments, which have shown a decline in soot emission at intensified turbulence, confirm these considerations.

Despite recent progress, considerable effort is still necessary in order to understand fundamental aspects of particle formation and oxidation to an adequate extent.

6.5 Nitrogen oxides

In the troposphere, nitrogen oxides (NO_x) favor the formation of ozone close to the ground and photochemical smog. In engine combustion, mainly nitrogen monoxide (NO) develops, which however is converted after a longer period of time almost completely into nitrogen dioxide (NO_2) under atmospheric conditions. In combustion, NO can be formed in three different ways. In this case, we distinguish between so-called thermal NO, which is formed among the combustion products at high temperatures according to the Zeldovich mechanism from atmospheric nitrogen, so-called prompt NO, which develops already in the flame front

6.5 Nitrogen oxides

via the Fenimore mechanism from air nitrogen, and finally so-called fuel NO, which is produced by nitrogen portions in the fuel.

6.5.1 Thermal NO

Thermal NO formation proceeds "behind" the flame front in the burned gas region and was first described by Zeldovich (1946). The simple reaction mechanism provided by him was later expanded upon by Baulch et al. (1991). This extended Zeldovich mechanism consists of three elementary reactions

(1) $\quad O^\bullet + N_2 \xleftrightarrow{k_1} NO + N^\bullet$

(2) $\quad N^\bullet + O_2 \xleftrightarrow{k_2} NO + O^\bullet$

(3) $\quad N^\bullet + OH \xleftrightarrow{k_3} NO + H^\bullet$

with experimentally determined speed constants k_i. Although thermal NO formation according to the Zeldovich mechanism has been one of the most investigated of all reaction mechanisms, there is still uncertainty concerning the choice of speed constants. Deviating values are suggested for this in the literature, of which a few are summarized in Tab. 6.1.

In the exceptional case of chemical equilibrium, the forward (index r) and reverse reactions (index l) each progress equally fast, e.g.

$$k_{1,r}[O][N_2] = k_{1,l}[NO][N] \tag{6.9}$$

or according to the corresponding reformulation:

$$\frac{k_{1,r}}{k_{1,l}} = \frac{[NO][N]}{[O][N_2]} \equiv K_{C,1} . \tag{6.10}$$

Since both speed constants $k_{1,r}$ and $k_{1,l}$ as well as the equilibrium constant $K_{C,1}$ are contingent only on temperature, not however on the actually existing concentrations, the relation

$$K_C = \frac{k_r}{k_l} \tag{6.11}$$

is also generally valid, see chap 3.2. The equilibrium constant K_C is to be determined from thermodynamic data, e.g. NIST (1993). Thus, the speed constant k_l for the reverse reaction can be easily calculated with (6.11).

For the NO formation rate, we obtain with reaction equations (1) to (3)

$$\frac{d[NO]}{dt} = k_{1,r}[O][N_2] + k_{2,r}[N][O_2] + k_{3,r}[N][OH]$$
$$- k_{1,l}[NO][N] - k_{2,l}[NO][O] - k_{3,l}[NO][H], \tag{6.12}$$

and for the temporal change of nitrogen atom concentrations follows

$$\frac{d[N]}{dt} = k_{1,r}[O][N_2] - k_{2,r}[N][O_2] - k_{3,r}[N][OH]$$
$$- k_{1,l}[NO][N] + k_{2,l}[NO][O] + k_{3,l}[NO][H]. \tag{6.13}$$

If the momentary NO concentration is beneath the equilibrium concentration of the corresponding temperature, as is the case in extensive segments of engine combustion, the forward reaction has a decisive influence on the total conversion. Only when the momentary NO is above the equilibrium concentration of the corresponding temperature is the total conversion substantially determined by the reverse reaction. In the engine however, this situation appears at best towards the end of the expansion stroke, when temperature is already quite low.

Tab. 6.1: Speed coefficients for the forward reaction of the Zeldoch mechanism

Reaction i	$k_{i,r}$ [cm³/mol s]	Author
1	$1.8 \cdot 10^4 \exp\left[-\dfrac{38,400}{T}\right]$	Baulch et al. (1991)
	$0.544 \cdot 10^{14} T^{0.1} \exp\left[-\dfrac{38,020}{T}\right]$	GRI-MECH 3.0 (2000)
	$0.76 \cdot 10^{14} \exp\left[-\dfrac{38,000}{T}\right]$	Heywood (1988)
2	$6.4 \cdot 10^9 T \exp\left[-\dfrac{3,150}{T}\right]$	Baulch et al. (1969)
	$9.0 \cdot 10^9 T \exp\left[-\dfrac{3,280}{T}\right]$	GRI-MECH 3.0 (2000)
	$1.48 \cdot 10^8 T^{1.5} \exp\left[-\dfrac{2,860}{T}\right]$	Pattas (1973)
3	$3.0 \cdot 10^{13}$	Baulch et al. (1991)
	$3.36 \cdot 10^{13} \exp\left[-\dfrac{195}{T}\right]$	GRI-MECH 3.0 (2000)
	$4.1 \cdot 10^{13}$	Heywood (1988)

We know by the speed coefficients for the forward reactions that NO formation via reaction (1) progresses much more slowly than via reactions (2) and (3). For a temperature of $T = 1,800$ K, for example, we obtain

$$k_{1,r} \approx 1.0 \cdot 10^2, \; k_{2,r} \approx 2.0 \cdot 10^9, \; k_{3,r} \approx 2.8 \cdot 10^{10} \left[\dfrac{m^3}{\text{kmols}}\right].$$

6.5 Nitrogen oxides

The first reaction possesses a high activation energy because of the stable N_2 triple bond and thus proceeds sufficiently fast only at high temperatures; hence the designation "thermal". It is therefore the rate-limiting step. The above numerical values show that at 1,800 K, the first reaction progresses at around seven to eight decimal powers slower than the second and third. Fig. 6.20 shows the path of speed coefficients $k_{1,r}$ in [g/(mol s)] contingent on temperature T.

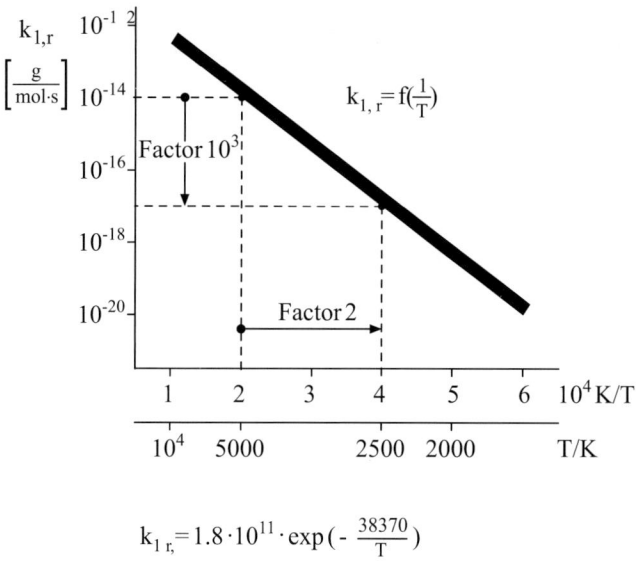

$$k_{1\,r} = 1.8 \cdot 10^{11} \cdot \exp\left(-\frac{38370}{T}\right)$$

Fig. 6.20: Reaction coefficient of the first Zeldovich reaction

We see that a doubling of the temperature raises thermal NO formation by the factor 10^3, and at a temperature elevation from 2,000 to 2,500 K the thermally formed NO climbs about 50-fold. Because of this considerable temperature-dependency, we refer to a kinetically controlled NO formation. That means that the chemical reaction kinetics at the temperatures in the combustion chamber is slow in comparison to the physical time scales of the flow field and thus that chemical equilibrium cannot be achieved. This should be clarified by Fig. 6.21, in which NO concentrations under assumption of equilibrium and under consideration of kinetics are qualitatively illustrated according to Zeldovich. The kinetically controlled process produces, according to Zeldovich, essentially less NO than under the assumption of equilibrium (Δ_1) at first, but in the late combustion phase, because of the extremely slow process at lower temperatures, produced NO does not regress over the reverse reactions (Δ_2). We speak of a reaction "freeze".

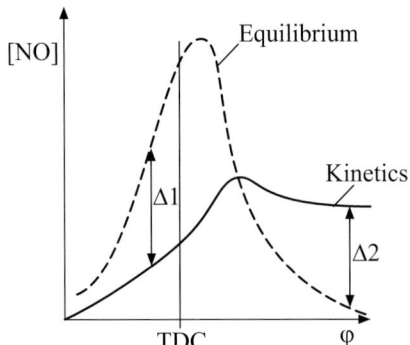

Fig. 6.21: NO-concentrations under condition of equilibrium / kinetically controlled NO-formation

Because the reaction speed of the forward reactions (2) and (3) is higher than that of reaction (1) by several decimal powers, the atomic nitrogen formed in the first reaction step is immediately further converted to NO in the second and third steps. The concentration of atomic nitrogen thus remains constant after a short initial phase. The concentration of [N] can thus be assumed to be quasi-steady, see chap. 3.3

$$\frac{d[N]}{dt} \approx 0 \, . \tag{6.14}$$

With this follows after adding the relations (6.12) and (6.13):

$$\frac{d[NO]}{dt} = 2k_{1,r}[O][N_2] - 2k_{1,l}[NO][N] \, . \tag{6.15}$$

For the unknown concentrations of nitrogen atoms [N], we obtain via the transformation of equation (6.13) under observance of equation (6.14)

$$[N] = \frac{k_{1,r}[O][N_2] + k_{2,l}[NO][O] + k_{3,l}[NO][H]}{k_{1,l}[NO] + k_{2,r}[O_2] + k_{3,r}[OH]} \, .$$

With this, equation (6.15) contains, besides the concentration of NO, only the easily determined concentration of N_2 and the concentrations O, O_2, OH, and H of the OHC system. The OHC system can be described by the following equation system

(4) H_2 =   2 H ,

(5) O_2 =   2 O ,

(6) H_2O =   1/2 H_2 + OH ,

(7) H_2O =   1/2 O_2 + H_2 ,

(8) CO_2 =   CO + 1/2 O_2 ,

6.5 Nitrogen oxides

Since the five reactions (4) to (8) proceed very fast compared with the reactions of the Zeldovich mechanism (1) to (3) at the temperatures in the combustion chamber, it can be assumed that the components are always in partial equilibrium, e.g.

$$\frac{[H]^2}{[H_2]} = K_{C,4}, \quad \frac{[O]^2}{[O_2]} = K_{C,5}, \quad \frac{[OH][H_2]^{1/2}}{[H_2O]} = K_{C,6}, \text{ etc.}$$

From these relations, together with the mass conservation of atoms O, H, and C and the condition, that the sum of partial pressures of all components must correspond to the total pressure, a non-linear equation system results, which is explicitly solvable due to its linear independence, for example, with the Newton-Kantorowitsch method (Bronstein and Semendjagew, 1997). An example concentration distribution of OHC components as a function of temperature is shown in Fig. 6.22.

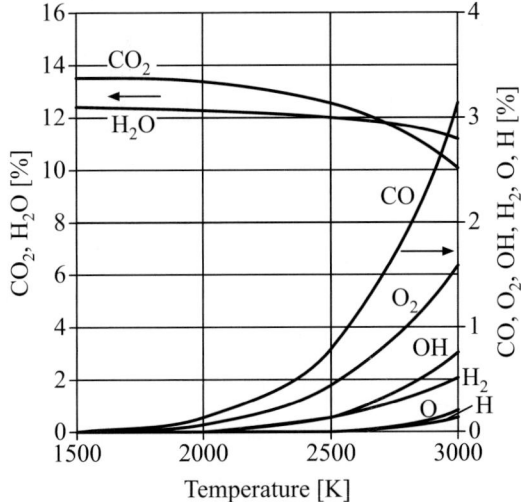

Fig. 6.22: Concentrations of OHC-components under condition of equilibrium

For a simplified estimation of nitrogen oxide formation, the reverse reactions can be neglected at first for the above mentioned reasons. Equation (6.15) is then simplified to

$$\frac{d[NO]}{dt} = 2k_{1,r}[O][N_2] \,. \tag{6.16}$$

Instead of solving for the partial equilibrium of the OHC system (4) to (8), the still unknown oxygen atom concentration in (6.16) can also be roughly approached according to Bockhorn (1997) by

$$[O] = 0.13[O_2]^{1/2} \exp\left\{-\frac{29{,}468}{T}\right\}, \tag{6.17}$$

whereby [O] and $[O_2]$ are to be inserted as $\frac{\text{kmol}}{\text{m}^3}$.

With the relation for the speed constant $k_{1,r}$ given above follows for the NO formation rate

$$\frac{d[NO]}{dt} = 4.7 \cdot 10^{13} [N_2][O_2]^{1/2} \exp\left\{-\frac{67{,}837}{T}\right\}. \tag{6.18}$$

This equation can, however, only be seen as a crude estimation, as it predicts an excessively high NO concentration. The reason for this is that the – no matter how minimal – reverse reactions of equations (1) to (3) slow down actual NO production and can even lead to a minor reduction of NO concentrations towards the end of the working cycle, see Fig. 6.21. Equation (6.18), on the other hand, cannot take on a negative value!

In Fig. 6.23, NO formation and disintegration in a thermal reactor at 60 bar and $\lambda = 1$ is represented for varying temperatures as a function of time. The illustration shows that equilibrium is reached more quickly the higher the temperature; for $T = 2{,}400$ K after about 20 ms and for $T = 2{,}800$ K already after about 3 ms.

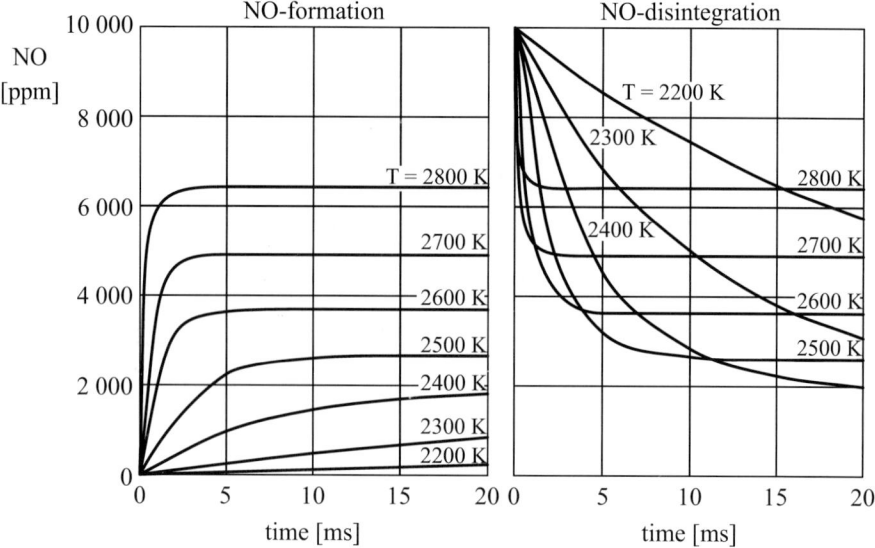

Fig. 6.23: NO-formation and disintegration in a thermal reactor, $p = 60$ bar; $l = 1.0$

Since a temperature of 2,800 K is reached at best during an extremely short time interval directly in the flame front and because the temperature in the burned mixture rapidly declines as a result of the admixture of unburned fresh air, equilibrium cannot be achieved within the time available. NO formation in engine combustion must therefore be calculated with the help of the reaction kinetics.

6.5.2 Prompt NO

The formation of prompt NO in the flame front itself is much more complicated than thermal NO formation, because this process is closely related to the formation of the CH radical,

6.5 Nitrogen oxides

which can react in many ways. So-called prompt NO formation was first described by Fenimore (1979). In the rate-limiting step, intermediately formed CH reacts with N_2 to form HCN (hydrocyanic acid) and then rapidly NO,

(9) $\quad CH + N_2 \xrightarrow{k_9} HCN + N \rightarrow ... \rightarrow NO$

with

$$k_{9,r} = 4.4 \cdot 10^9 \exp\left(-\frac{11{,}060}{T}\right) \frac{m^3}{kmol \, s} \, .$$

Ethin (acetylene, C_2H_2) as a precursor of the CH radical is only formed under fuel-rich conditions in the flame front, hence the concept "prompt NO". Because of the relatively low activation energy of the reaction prompt NO formation proceeds already at temperatures of about 1,000 K. For the oxidation of HCN and CH, the following reactions

(10) $\quad HCN + O \xrightarrow{k_{10}} NCO + H$,

(11) $\quad CN + O_2 \xrightarrow{k_{11}} NCO + O$,

(12) $\quad HCN + OH \xleftarrow{k_{12}} CN + H_2O$

with speed coefficients

$$k_{10,r} = 2.3 \cdot 10^4 \cdot T^{1.71} \exp\left(-\frac{3{,}521}{T}\right) \frac{m^3}{kmol \, s} \, ,$$

$$k_{11,r} = 8.7 \cdot 10^9 \exp\left(\frac{216}{T}\right) \frac{m^3}{kmol \, s} \, ,$$

$$k_{12,r} = 4.7 \cdot 10^9 \exp\left(-\frac{5{,}174}{T}\right) \frac{m^3}{kmol \, s} \, ,$$

$$k_{12,l} = 7.4 \cdot 10^9 \exp\left(-\frac{3{,}715}{T}\right) \frac{m^3}{kmol \, s}$$

are rate-limiting. Because the secondary reaction (conversion of NCO to NO) is relatively fast in comparison, for prompt NO formation is roughly valid

$$\frac{d[NO]}{dt} = k_{10,r}[HCN][O] + k_{11,r}[CN][O_2] \, . \tag{6.19}$$

[OH] and [O] concentrations are again determined via reactions (4) to (8) of the OHC system.

However, information available in the literature concerning reaction constants $k_{9,r}$ of the rate-limiting reaction (9) is basically still very inconsistent, see Warnatz et al. (1997). The calculation of prompt NO formation is thus also bound up with much larger uncertainties than the calculation of thermal NO. With the help of an estimation of reaction (9), it becomes clear however that prompt NO as opposed to thermal NO – see reaction (1) – is not as dependent on temperature (lower activation temperature) but is, on the other hand, highly dependent over CH and HCN on local fuel concentration.

In typical engine combustion, approx. 5-10 % of nitrogen originates via the Fenimore mechanism (prompt NO) and 90-95 % via the Zeldovich mechanism (thermal NO). For further information, see Teigeler et al. (1997).

6.5.3 NO formed via N_2O

This reaction mechanism is only significant when lean air-fuel mixtures repress the formation of CH, and thus little prompt NO is formed, and furthermore if low temperatures stifle the formation of thermal NO.

N_2O is formed analogously to the first, rate-limiting reaction of the Zeldovich mechanism,

(13) $N_2 + O + M \rightarrow N_2O + M$.

Stabilization occurs via a molecule M, so that N_2O develops instead of NO. NO formation then results trough oxidation of N_2O corresponding to

(14) $N_2O + O \rightarrow NO + NO$.

Because the N_2O is only formed in a three-way collision reaction, this reaction path progresses preferably at high pressures. Low temperatures hardly slow down this reaction. NO formed via N_2O is the essential NO source in lean premixed combustion in gas turbines. This mechanism is however also observable in SI engine lean combustion.

6.5.4 Fuel nitrogen

The conversion of nitrogen bound in the fuel to nitrogen oxide does not play a role in engine combustion, because fuels for internal combustion engines contain hardly any nitrogen. But it is of importance in the combustion of coal, since even "clean" coal contains about 1 % of nitrogen.

The rate-limiting steps are both of the reactions competing for nitrogen atoms known from the Zeldovich mechanism

(15) $N + OH \rightarrow NO + H$,

(16) $N + NO \rightarrow N_2 + O$,

for which sufficiently exact data exists, so that the formation mechanism of NO from nitrogen bound to fuel is well understood.

7 Calculation of the real working process

In the filling and emptying method, a *zero-dimensional* model, in which the process quantities depend only on time, not on location, particular *subsystems* of the engine, e.g.:

- combustion chamber,
- intake and exhaust pipes,
- valves and flaps, as well as the
- charge system

are physically and mathematically described either via *substitute systems*

- volume, pipes, or
- orifice plates

or with *characteristic maps*.

Fig. 7.1 shows an "engine model" for a turbocharged diesel engine.

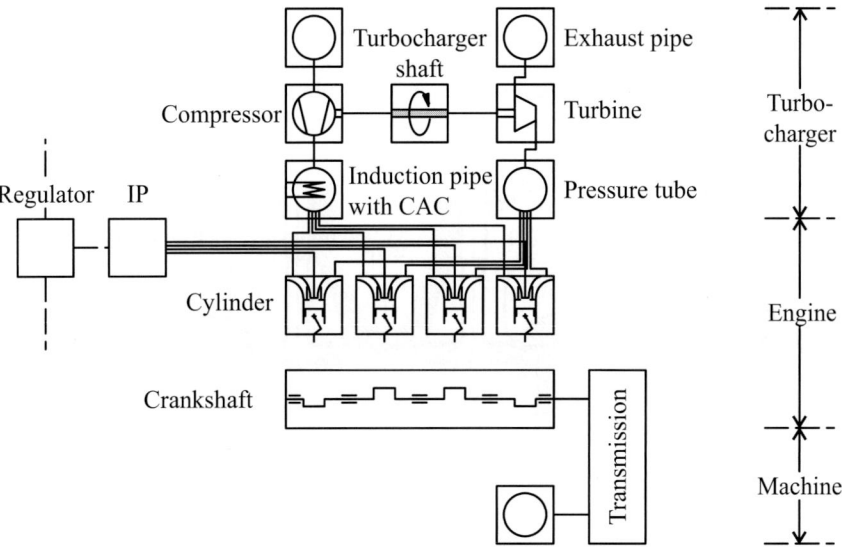

Fig. 7.1: Engine model

The centerpiece of the internal combustion engine, the combustion chamber, is thereby described with the thermodynamic model of the "ideally mixed volume", the turbocharger on the other hand with characteristic maps for the compressor and the turbine.

In the filling and emptying method, only the laws of conservation for *mass* and *energy*, but not the *law of conservation* for impulse is considered (see chap. 7.4). Because in this way no flow fields are considered, the resulting model is also designated as a zero-dimensional thermodynamic model. For a general examination, see Ramos (1989).

In describing processes in the fresh air and exhaust gas system of intake engines (today almost exclusively SI engines), the process quantities must be described with the help of one-dimensional gas dynamics as a function of time and location. Such complex systems live from the dynamics of pressure waves surging to and for and the interaction of these pressure waves with the valve control times for the purpose of as good a filling as possible. The intake and exhaust system can be represented with a number of pipe components connected over pipe branches, the volumes, orifice plates, the cylinders, and sometimes flow machines.

Chapter 7.5 goes into the particular requirements for the simulation of charging aggregates and air cooling.

7.1 Single-zone cylinder model

7.1.1 Fundamentals

Fig. 7.2 shows the combustion chamber of an internal combustion engine. This is limited by the combustion chamber walls, the piston, and the valves. The combustion chamber walls simultaneously represent the system boundaries. The entire combustion space is viewed as an ideally mixed volume, whereby heat release through combustion is described with the help of a substitute heat release rate. We will later take a closer look at *multi-zone models*, with which we can describe, for example, NOx formation in the combustion chamber.

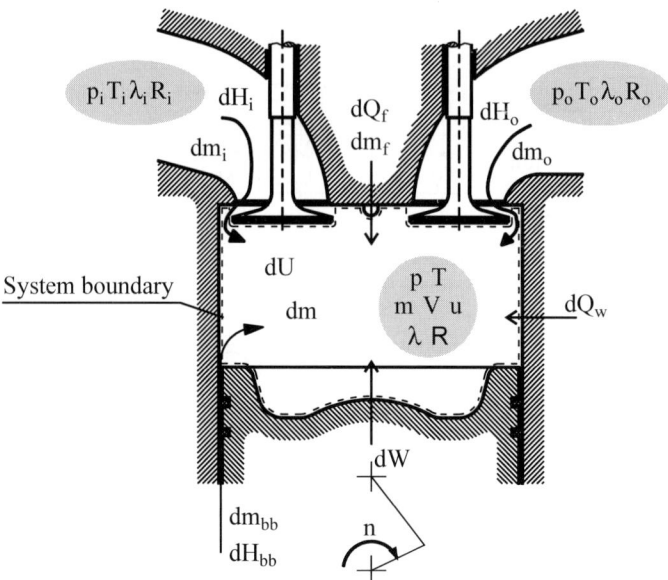

Fig. 7.2: Single-zone cylinder model

7.1 Single-zone cylinder model

It must thereby be considered that the volume, but not necessarily the mass of the combustion chamber continually changes with time (or crank angle) as a result of piston movement. In considering the balance equations in the combustion chamber, we must distinguish between different concepts of fuel introduction. While the fuel – as long as one does not calculate with a wall film model – is taken in proportionally to the fresh air in the mixture-intaking SI engine, in the direct injection SI engine the fuel is injected either during the opened intake valve (homogeneous operation, injection before TDC) or shortly before ignition (stratified operation). The fuel must then be prepared/evaporated in the combustion chamber. In the diesel engine, the fuel is directly injected. An evaporation of the fuel is generally not considered (exception: multi-zone models with fuel breakup models).

The *mass balance* for the cylinder provides for the description of all above mentioned possibilities for fuel insertion

$$\frac{dm_{sys}}{dt} = \frac{dm_i}{dt} + \frac{dm_o}{dt} + \frac{dm_{bb}}{dt} + \frac{dm_{fuel,evap}}{dt} \,. \tag{7.1}$$

The mass flow entering the engine can be, as already mentioned, pure air, an air-fuel mixture, an air-exhaust gas mixture, or a combination of air, fuel and exhaust gas.

The *energy balance* or the *1st law of thermodynamics* provides for the cylinder neglecting the kinetic energy

$$\frac{dE_{sys}}{dt} = \frac{dU}{dt} = \frac{dQ_{fuel}}{dt} + \frac{dQ_w}{dt} - p\frac{dV}{dt} + \frac{dm_i}{dt}h_i + \frac{dm_o}{dt}h_o + \\ + \frac{dm_{bb}}{dt}h_{bb} + \frac{dm_{fuel,evap}}{dt}h_{fuel,evap} + \frac{dQ_{evap.}}{dt} \,, \tag{7.2}$$

see also (2.30). As long as the fuel is injected into the combustion chamber but is not yet evaporated, it takes up such a small volume that it is irrelevant to the thermodynamic system. The introduced fuel only becomes "effective" for the mass and energy balance when it is evaporated and hence in the gaseous phase. Before evaporation, the fuel has to be heated up, for which a corresponding amount of heat is taken from the gas. The situation is the same for the heating up of fuel steam to the gas temperature. Normally, these effects in the lower heating value are considered such that the evaporation enthalpy of the fuel (350-420 kJ/kg) is added to the lower heating value. The same is valid for the re-condensation of fuel and for the water content in the air when the gas is greatly expanded (e.g. in load control via early "intake closes").

Either time t or the crank angle φ can be chosen as the *independent variable*. In more recent equations, time t is usually given precedence. Valid for the relation of time with the degree of crank angle is

$$\varphi = \omega t$$
$$d\varphi = \omega dt \,. \tag{7.3}$$

For the solution of the mass and energy balance, we need the thermal condition equation already mentioned

$$pV = mRT \,. \tag{7.4}$$

For the solution of this equation system, we still need relations for energy release via combustion, a so-called "combustion model", a relation for the heat transfer between the gas mixture and the combustion chamber walls, a so-called "heat transfer model", as well as a charge changing model (e.g. a two-zone model for 2-stroke engines) and, under certain circumstances, an evaporation model. The volume path is given by a crankshaft drive model. We will take a closer look at particular partial models in the following.

7.1.2 Mechanical work

The output at the piston dW/dt can be calculated from the cylinder pressure and the change in cylinder volume

$$\frac{dW}{dt} = -p\frac{dV}{dt} = -p\omega\frac{dV}{d\varphi} \quad . \tag{7.5}$$

In chap. 2.2.1, the geometrical relations at the crankshaft drive are represented. The crankshaft drive can be described by the geometrical quantities crankshaft radius r, connecting rod length l, eccentricity e, and cylinder diameter D, from which the volume change $dV/d\varphi$ can be determined.

7.1.3 Determination of the mass flow through the valves / valve lift curves

In the valve gap, the flow is constricted. This has the result that the actual cross-section surface is smaller than the geometrical. Because of friction in the ports, the actual mass flow is also smaller than the theoretical. This fact is taken into consideration by the introduction of a flow coefficient

$$\mu \equiv \frac{\dot{m}_{act.}}{\dot{m}_{theo.}} \quad . \tag{7.6}$$

For the determination of the mass flow through a valve, we utilize the flow function derived in chapter 2.3.1. For this, the actual mass flow contingent upon valve lift is determined for the valves on a so-called flow test bench and set in relation to the theoretical mass flow (see (2.37)).

$$\dot{m}_{theo} = A_{geo}\sqrt{p_1\rho_1}\,\Psi\!\left(\frac{p_2}{p_1},\kappa\right) \quad .$$

Fig. 7.3 shows the conditions at the flow test bench in principle. The flow coefficient is usually related to a circular surface at the port entrance. Since varying cycle calculation programs possess varying definitions of the valve opening surface, usually a corresponding recalculation is necessary. It must thereby be taken into consideration that the actual cross-section surface remains the same independent of the definition of the reference cross-section surface for the respective valve position.

7.1 Single-zone cylinder model

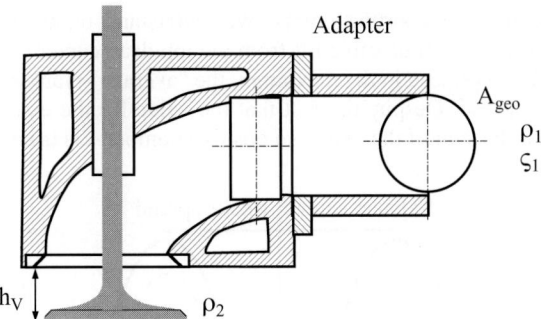

Fig. 7.3: Determination of the flow coefficients on the flow bench

Fig. 7.4 qualitatively shows the flow coefficients determined at the flow test bench contingent on the valve lift.

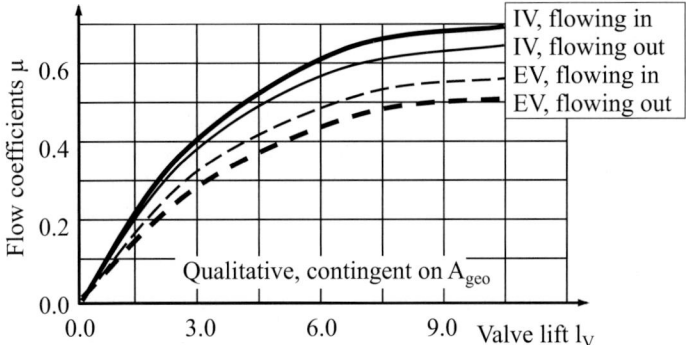

Fig. 7.4: Flow coefficients $\mu = f(l_v)$

We recognize thereby that varying flow coefficients are adjusted for varying flow directions at the intake and exhaust valves. The main flow direction for the intake valve is "inflow". However, an outflow from the intake valve is also possible. Due to the geometrical relations, this flow case is worse than that of inflow. The situation is the same for the exhaust valve. In this case however, the main flow case is "outflow", from which an, in principle, inferior flow behavior results.

The flow coefficients can be determined either in a stationary test or calculated by means 3D-CFD codes in order to make statements about the quality of ports without concrete hardware. In this case, the port geometry and the cylinder are duplicated and a corresponding pressure gradient applied at the margins. In discrete steps, the valve lift is changed and the "actual" mass flow calculated. This can then be compared with the theoretical flow in the same way as in measurement.

Fig. 7.5 shows the valve lift graphs for a conventional valve train. For the calculation of the working process, it is sufficient to provide rate of valve lift curves in steps of 1 to 5 °CA and to interpolate between the supporting points. We understand under the word "spread" the distance between the maximum of valve lift from the top dead center of the charge changing. If a valve lift curve has a plateau in the maximum, the "average" crank angle value is used for the definition of the spread. Despite the fact that the value for the exhaust spread should be calculated negatively, the sum of the value – a positive number – is usually given for this.

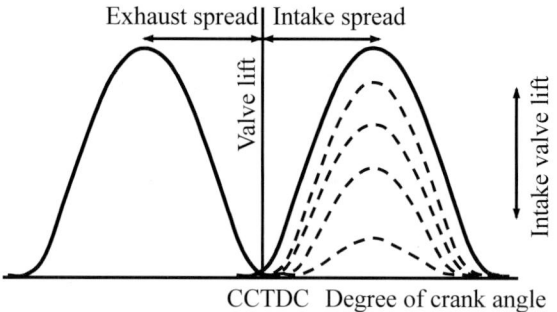

Fig. 7.5: Valve lift curve for a mechanical valve train

In Fig. 7.5, the rate of lift graphs for a fully variable mechanical valve train is also included, in which a continuous adjustment of valve lift is possible. With this variability for a quantity-regulated SI engine, a load control without a throttle valve is possible, since the flow and thus the fresh gas mass can be adjusted via the valve lift. For the low load range, the valve lift graduation has to be given in the range of tenths of a millimeter; after approx. 3 mm a range of half to entire millimeter steps is sufficient. Intermediate steps are thereby linearly interpolated. Nothing changes in curves for the flow coefficients for fully variable valve operation, since the flow coefficients only depend on the valve lift. Only in the range of small valve lifts is a finer grid recommendable here as well.

The behavior of so-called electromechanical valve trains should be seen completely differently for simulation. The electromechanical valve train concerns a one mass oscillator, which is usually attracted in a regulated manner and then held at the respective final positions by a magnet. In this case, ideally only the loss energy from swinging from one final position to the other via the magnet is added. In an ideal case, the valve lift curve in electromechanical valve trains is thus only contingent on time and not on the degree of crank angle. For varying speeds, the valve lift curves result as given in Fig. 7.6. In this example, the valve is move upwards again as soon as the lower final position is reached and is not held at this final position. The movement differential equations for the one mass oscillator is

$$m\ddot{x} + d\dot{x} + cx = F_{\text{fric}}(t) + F_{\text{magnet}}(t) + F_{\text{valve disc}}(t) + F_{\text{adh.}}(t) \ . \tag{7.7}$$

Not under consideration are furthermore the so-called adhesive forces at the actuator, which are contingent on its thermal state and on the presence, for example, of oil. For the values presented in Fig. 7.6, the attenuation is constant and all external forces have been set to zero. Thus results as an idealized solution for the differential equation a cosine function

7.1 Single-zone cylinder model

$$x(t) = x_{\max}\left[\frac{1}{2} - \frac{1}{2}\cos\left(\frac{2t}{\tau}\pi\right)\right] \quad \text{with } 0 < t < \tau . \tag{7.8}$$

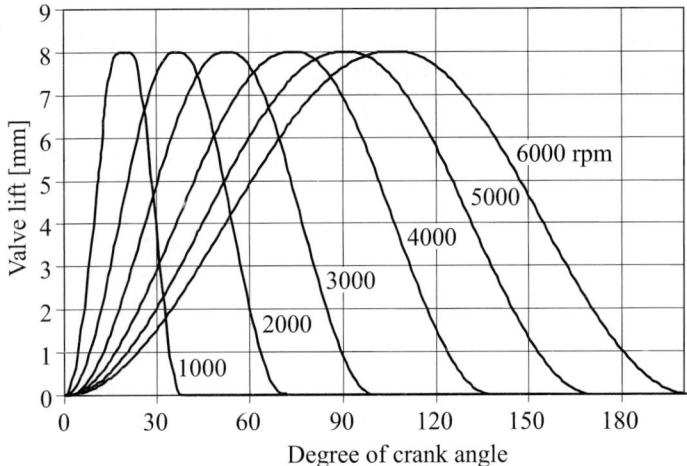

Fig. 7.6: Valve lift curve for an electromechanical valve train

The so-called flight time τ serves as a measure for the period duration. We will in this context do without a more extensive modeling.

7.1.4 Heat transfer in the cylinder

The description of heat transfer in the internal combustion engine places the highest demands on modeling and is usually based on a global inspection of very complex relations. Heat transfer consists of a convective and a radiation component

$$\frac{dQ_W}{dt} = \frac{dQ_\alpha}{dt} + \frac{dQ_\varepsilon}{dt} . \tag{7.9}$$

Usually the radiation component dQ_ε/dt is added to the convective heat transfer coefficient, although the maxima of dQ_α/dt and dQ_ε/dt actually appear phase-shifted with reference to the crank angle.

Proceeding from the Newtonian method, for the description of the wall heat flow

$$\frac{dQ_w}{dt} = \sum_i \alpha_i\, A_i\, (T_{w,i} - T_{gas}) \tag{7.10}$$

is valid.

We usually subdivide the combustion chamber thereby into three areas (see Fig. 7.2):

- the piston,

- the cylinder head, and
- the portion of the liner released by the piston including deck height and piston junk.

The valves are usually calculated with the cylinder head, or, in very detailed modeling, as their own area. The surfaces for the pistons and the cylinder head are usually larger than the cylinder cross-section surface, since these describe, for example, the roof form in a SI engine combustion process or the piston bowl shape in a diesel or SI engine combustion process. The part of the liner released by the piston amounts to

$$A_{liner} = A_{piston\ junk} + A_{deck\ height} + D\pi s(\varphi) \ . \tag{7.11}$$

The allocation between piston path $s(\varphi)$ and the position of the crankshaft has already been seen in chap. 2.2.1.

The calculation of heat transfer with the help of the Newtonian approach and the heat transfer coefficients requires an exact description of gas and all temperatures. The mean gas temperature results from the local averaging of gas temperature in the combustion chamber. Since the combustion chamber system is usually seen as an ideally mixed volume, the mean gas temperature is easy to determine from the condition equation for an ideal gas. At respective wall temperatures, we are dealing with the internal wall temperature averaged over one working cycle. For the piston and the cylinder, usually local constant temperatures are used. In the case of the liner, the wall temperature depends a lot on the engine type and on whether the liner is completely or only partially surrounded by the water jacket. In giving the temperature for the liner, we usually subdivide it into several areas, or we provide a temperature profile over the length of the liner. The temperatures can either be determined by measuring, or we can use a simple, iterative method for the calculation of internal wall temperature for stationary operating points. For this however, knowledge of the temperatures in at least one operating point is required. For non-stationary calculations, neither method is sufficient any more, which is why we use a more concrete heat conduction model, which takes the thermal inertias of the respective wall into consideration. All models are described in chap. 7.1.6.

For the calculation of heat transfer coefficients, semi-empirical methods are usually used, since many influence factors can only be determined experimentally. Therefore, exterior quantities are utilized as influence parameters that characterize the operating point. In this section, essentially two approaches will be introduced: The method of Woschni, which was constructed for diesel engines in 1969 and has continually been further developed, and that of Bargende, which was introduced in 1990 for SI engines. Besides these, a multitude of further approaches exist in the literature, e.g. from Hohenberg (1980) and Kleinschmidt (1993), which we will however not go into details here any further.

- **Heat transfer according to Woschni**

The model of Woschni (here 1970) assumes a stationary, fully turbulent pipe flow. For the dimensionless heat transfer coefficient, the Nußelt number, we obtain from a dimension analysis the semi-empirical power equation

$$\mathrm{Nu} = C\ \mathrm{Re}^{0.8}\ \mathrm{Pr}^{0.4}\ , \tag{7.12}$$

with the Nußelt number

7.1 Single-zone cylinder model

$$\text{Nu} = \frac{\alpha D}{\lambda}, \quad (7.13)$$

the Reynolds number

$$\text{Re} = \frac{\rho w D}{\eta} \quad (7.14)$$

and the Prandtl number

$$\text{Pr} = \frac{\eta}{\rho} a . \quad (7.15)$$

If we view the gas in the combustion chamber as an ideal gas,

$$\rho = \frac{p}{RT}, \quad (7.16)$$

thus follows at first

$$\frac{\alpha D}{\lambda} = \left(\frac{p}{RT} \frac{wD}{\eta}\right)^{0.8} \text{Pr}^{0.4} \quad (7.17)$$

and from that by means of conversion for the convective heat transfer coefficient

$$\alpha = C D^{-0.2} p^{0.8} w^{0.8} \frac{\text{Pr}^{0.4} \lambda}{(RT\eta)^{0.8}} . \quad (7.18)$$

With the physical characteristics

$$\text{Pr} = 0.74; \quad \frac{\lambda}{\lambda_0} = \left(\frac{T}{T_0}\right)^x ; \quad \frac{\eta}{\eta_0} = \left(\frac{T}{T_0}\right)^y \quad (7.19)$$

and with the assumption that the characteristic speed w is equal to the mean piston speed, i.e. $w \equiv c_m$, we further obtain

$$\alpha = C^* D^{-0.2} p^{0.8} c_m^{0.8} T^{-r} \quad \text{with} \quad r = 0.8(1 + y) - x . \quad (7.20)$$

Through a comparison with the measurement values, the exponent r for the temperature dependence is determined as $r = 0.53$ and the constant as $C^* = 127.93$. For fired engines, a modification of the characteristic speed is additionally introduced, which takes the change in heat transfer as a result of combustion into consideration. We thus obtain

$$\alpha = 127.93 \, D^{-0.2} \, p^{0.8} \, w^{0.8} \, T^{-0.53} \quad \left[\frac{W}{m^2 K}\right] \quad (7.21)$$

with

$$w = C_1 c_m + \underbrace{C_2 \frac{V_d T_1}{p_1 V_1}(p - p_0)}_{\text{Combustion term}} \tag{7.22}$$

and p_1, T_1, V_1 at compression start, i.e. at "intake closes". For the constants C_1 and C_2 we obtain via adjustment to measured values.

$$C_1 = \begin{cases} 6.18 + 0.417 \dfrac{c_c}{c_m} & : \quad \text{charge changing process} \\ 2.28 + 0.308 \dfrac{c_c}{c_m} & : \quad \text{compression / expansion} \end{cases} \tag{7.23}$$

$$C_2 = \begin{cases} 6.22 \cdot 10^{-3} \left[\dfrac{m}{s\,K}\right] : & \text{prechamber – engine} \\ 3.24 \cdot 10^{-3} \left[\dfrac{m}{s\,K}\right] : & \text{DI – engine} \end{cases} \tag{7.24}$$

For the intake swirl c_c/c_m, the range of validity is given as $0 \leq c_c/c_m \leq 3$. The swirl is determined in the stationary flow experiment on the flow test bench with the Tipelmann or the impeller method. In this case, an impeller of diameter d is arranged at a distance of 100 mm beneath the cylinder head in the cylinder liner. The flow through the intake valve is thereby adjusted such that this impeller receives a flow of mean piston speed c_m. With the speed of the impeller to be measured n_d we correspondingly obtain

$$c_c = D \pi n_d \tag{7.25}$$

the circumference speed and thus the swirl.

Contingent on the respective phases of a cycle, several terms or parameters in the heat transfer equation are altered. This leads, for example, in the transition between expansion and charge changing during the opening of the exhaust valve, to a leap in the constant C_1. In the same way, the term with the constant C_2 is only valid after the start of combustion. However, the transition between the compression phase and combustion is continuous via the term $(p - p_0)$.

With the term $(p - p_0)$, the difference is given between the cylinder pressure during combustion and the cylinder pressure during motored operation. Pressure p_0 can be calculated over a polytrope relation from the cylinder volume. The determination of the polytrope exponent $\bar{n}$ occurs shortly before combustion, as the polytrope exponents for the, for example, last 10 °CA before combustion are averaged. For p_0 then results

$$p_0(\varphi) = p_{v.comb.} \left(\frac{V_{v.comb.}}{V_{cyl.}(\varphi)} \right)^{\bar{n}} . \tag{7.26}$$

Fig. 7.7 demonstrate an example for wall heat flow in a turbocharged diesel engine at a speed of 2,000 rpm and an actual pressure of 2 bar.

7.1 Single-zone cylinder model

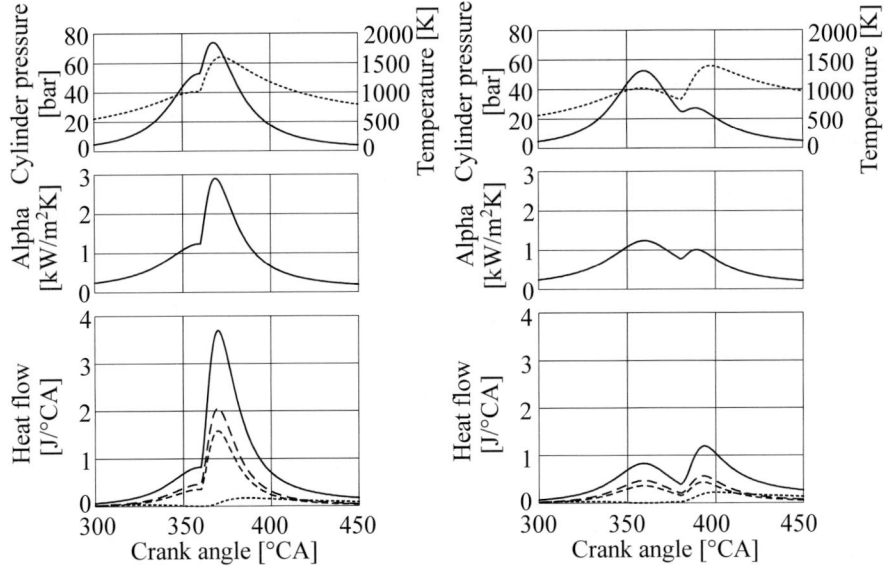

Fig. 7.7: Heat transfer in a turbocharged diesel engine, start of combustion 355 °CA (left) and 268 °CA (right)

In the lower part of the diagram, the heat flows of the piston, cylinder head, and the liner, as well as the total heat flow are represented. Above them the heat transfer coefficients according to Woschni are shown, as well as the mean mass temperature and the pressure in the cylinder. In the left side, a start of combustion of 355 °CA is recorded, in the right a start of combustion of 368 °CA. We recognize hereby that the heat transfer coefficient contingent on pressure and temperature is evidently smaller at a later start of combustion.

- **Modifications of the heat transfer equation according to Woschni**

Kolesa's investigations on heat transfer with insulated combustion chamber walls (1987) arrived at the result that the wall heat coefficient at wall temperatures over 600 K climbs significantly. For the constant C_2, Schwarz (1993) developed a constant function

$$C_2^* = C_2 + 23 \cdot 10^{-6} (T_W - 525) \quad \text{for } T_W \geq 525 \text{ K} . \tag{7.27}$$

The speed corrected by Woschni with the combustion term in the equation provides for motored engines and, in the lower load range however, excessively small values, as Huber (1990) has shown. Therefore, the heat transfer equation has been corrected for low loads. For

$$2 C_1 c_m \left[\frac{V_c}{V(\varphi)} \right]^2 imep_i^{-0.2} \geq C_2 \frac{V_d T_1}{p_1 V_1} (p - p_0) \tag{7.28}$$

is valid

$$w = C_1 c_m \left[1 + 2 \left(\frac{V_c}{V} \right)^2 imep_i^{-0.2} \right]. \qquad (7.29)$$

Further valid is: $imep = 1$ for $imep \leq 1$.

Additional investigations on the heat transfer coefficient – especially on the effect of insulation on combustion chamber wall deposits (soot, oil coke) – have been carried out by Vogel (1995). From these investigations have resulted further changes to the equation altered by Huber. For

$$2 C_1 c_m \left[\frac{V_c}{V(\varphi)} \right]^2 C_3 \geq C_2 \frac{V_h T_1}{p_1 V_1} (p - p_0) \qquad (7.30)$$

is valid

$$w = C_1 c_m \left[1 + 2 \left(\frac{V_c}{V} \right)^2 C_3 \right]. \qquad (7.31)$$

The constant C_2 for direct injection diesel engines is enlarged in its range of validity for SI engines.

$C_2 = 3.24 \cdot 10^{-3} \left[\frac{m}{s\,K} \right]$: DI engine, SI engine (gasoline).

Introduced as new constants are:

$C_2 = 4 \cdot 10^{-3} \left[\frac{m}{s\,K} \right]$: SI engine (methanol),

$C_3 = 0.8$: for gasoline,

$C_3 = 1.0$: for methanol and

$C_3 = 1 - 1.2 e^{-0.65 \lambda}$: for diesel.

In the case of large diesel engines running at medium speed, some deviations result in the calculation of the exhaust gas temperature in comparison with the measurement of about approx. 20 K. The exhaust gas temperature calculated too low leads to a low enthalpy at the turbine and thus to a slightly too low charge pressure. For the design of large diesel engines, this is however decisive, since they are mostly optimized at a stationary operating point. For this reason, Gerstle (1999) has modified the heat transfer according to Woschni for the charge changing. The constant C_1 is thereby valid beyond the point of exhaust opens, until the intake valve opens. Then the constant is raised by the factor 6.5 to 7.2.

$C_1 = 2.28 + 0.308 \frac{c_u}{c_m}$: compression / expansion / exhaust valve open.

7.1 Single-zone cylinder model

$$C_1 = k\left(2.28 + 0.308\frac{c_u}{c_m}\right) : \text{intake valve open and}$$

$k = 6.5$ *to* 7.2 .

Fig. 7.8 shows the modified path of the heat transfer coefficient.

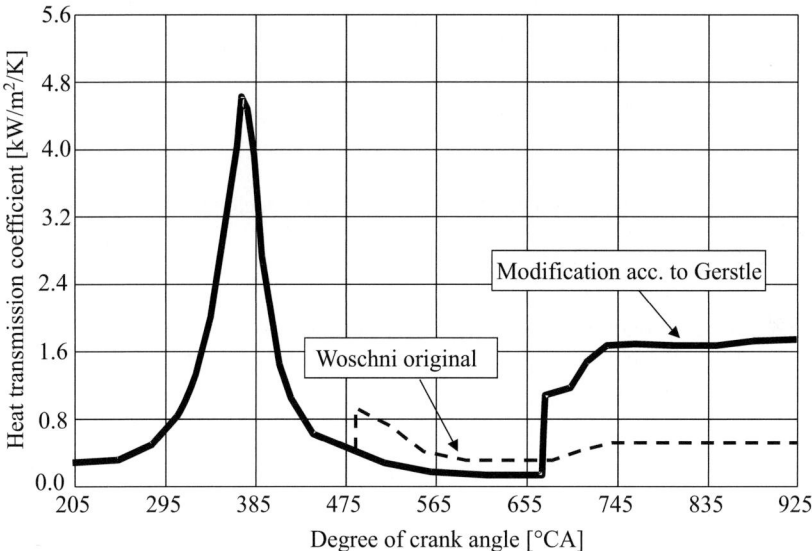

Fig. 7.8: Modified heat transfer coefficient for a medium speed diesel engine

- **Heat transfer according to Bargende**

Bargende (1990) also initially proceeds from Woschni's basic assumptions (see (7.12)) in describing the Nußelt number. The constants in the equation are determined for multi-point injection SI engines.

$$\text{Nu} = C\,\text{Re}^m\,\text{Pr}^n \quad \text{with } m = 0.78 \text{ and } n = 0.33\ .$$

Solved according to the heat transfer coefficient

$$\alpha = C\,\underbrace{D^{-0.22}}_{\text{Char. lenght}}\,\underbrace{\lambda\left(\frac{\rho}{\eta}\right)}_{\text{Material quantities}}\,\underbrace{w^{0.78}}_{\text{Gas velocity}}\,\underbrace{\Delta}_{\text{Combustion}} \qquad (7.32)$$

results. The characteristic length is expressed over the diameter of a sphere, which has the same volume as the momentary cylinder volume

$$D^{-0.22} \cong 1.11 V^{-0.073}\ . \qquad (7.33)$$

For the material quantities heat conductivity and dynamic viscosity is valid contingent on the air content r of the gas

$$\lambda = (1.15 r + 2.02) 10^{-4} T^{0.805} \quad \left[\frac{W}{m\,K}\right] \quad \text{and} \tag{7.34}$$

$$\eta = (2.57 r + 3.55) 10^{-4} T^{0.644} \quad \left[\frac{N\,s}{m^2}\right]. \tag{7.35}$$

The air content r of the gas is defined as

$$r = \frac{\lambda - 1}{\lambda + \dfrac{1}{L_{min}}} \quad \text{for } 0 \le r \le 1. \tag{7.36}$$

The density of the gas is expressed on the other hand via the ideal gas equation

$$\rho = \frac{p}{RT}.$$

In describing the gas conditions on the boundary layer, the average value from the gas temperature and the wall temperature is utilized, since the gas temperature falls off to the wall temperature on the boundary layer.

$$T_m = \frac{T_{gas} + T_{wall}}{2}. \tag{7.37}$$

The consideration of material quantities thus comes to

$$\lambda \left(\frac{\rho}{\eta}\right)^{0.78} \cong \frac{(1.15 r + 2.02)}{[R(2.57 r + 3.55)]^{0.78}} 10^{5.36} T_m^{-0.477} p^{0.78}. \tag{7.38}$$

The speed relevant to heat transfer is calculated by means of an approach from the specific turbulent energy k, which is determined by a simplified k,ε-model, see also chap. 9.1, and the piston speed c_p

$$w = \frac{\sqrt{\dfrac{8k}{3} + c_p^2}}{2}. \tag{7.39}$$

Valid for the change in specific turbulent energy is

$$\frac{dk}{dt} = \left[-\frac{2}{3}\frac{k}{V}\frac{dV}{dt} - \varepsilon\frac{k^{1.5}}{L} + \left(\varepsilon_s \frac{k_s^{1.5}}{L}\right)_{\varphi > TDC}\right]_{IVC \le \varphi \le EVO}, \tag{7.40}$$

with $\varepsilon = \varepsilon_s = 2.184$ and the characteristic swirl length $L = \sqrt[3]{6/(\pi V)}$. For the kinetic energy of the squeeze flow in a bowl-shaped combustion chamber with a bowl diameter of d_{bowl} is valid

7.1 Single-zone cylinder model

$$k_s = \frac{1}{18}\left[w_r\left(1+\frac{d_{bowl}}{d_{cyl.}}\right)+w_a\left(\frac{d_{bowl}}{d_{cyl.}}\right)^2\right]^2. \tag{7.41}$$

For the radial speed component w_r and the axial component w_a results

$$w_r = \frac{dV}{dt}\frac{1}{V}\frac{V_{bowl}}{V-V_{bowl}}\frac{d_{cyl}^2-d_{bowl}^2}{4d_{bowl}}, \tag{7.42}$$

with bowl volume V_{bowl} and

$$w_a = \frac{dV}{dt}\frac{1}{V}s_{bowl} \tag{7.43}$$

with bowl depth s_{bowl}. The specific kinetic start energy at "intake valve closes" results as

$$k_{IVC} = \frac{1}{16}\left[\frac{c_m d_{cyl}^2 \lambda_L}{d_{iv} h_{iv} \sin(45°)}\right]^2. \tag{7.44}$$

In this equation, the mean piston speed c_m, the volumetric efficiency λ_L as well as the intake valve diameter d_{iv} and the intake valve lift h_{iv} are introduced.

The combustion term Δ can be written with the help of the temperature of an intended zone for burned matter T_b and a zone for unburned matter T_{ub}

$$\Delta = \left[X\frac{T_b}{T_{gas}}\frac{T_b-T_{wall}}{T_{gas}-T_{wall}}+(1-X)\frac{T_{ub}}{T_{gas}}\frac{T_{ub}-T_{wall}}{T_{gas}-T_{wall}}\right]^2 \tag{7.45}$$

with

$$X = \frac{Q_f(\varphi)}{Q_f}.$$

The temperature for the unburned zone calculates over a polytrope compression to

$$T_{ub} = T_{gas,it}\left(\frac{p}{p_{it}}\right)^{(n-1)/n} \quad \text{with } 1.34 \leq n \leq 1.37. \tag{7.46}$$

With that results for the temperature of the burned gas

$$T_b = \frac{1}{X}T_{gas}+\frac{X-1}{X}T_W. \tag{7.47}$$

The heat transfer coefficient according to Bargende is only valid for the high pressure period. Fig. 7.9 shows an example for the heat transfer coefficient according to Bargende for a SI engine at 1,500 rpm and $imep = 7.35$ bar.

In the case of the equations for the description of the heat transfer in the cylinder described here, we are dealing with semi-empirical methods, in which the parameters have been determined by means of measured values. We do not thereby distinguish particularly between heat transfer via convection and via radiation. Despite diverse efforts and numerous heat transfer reactions which have been discovered to this day, there is still a need for a relatively simple to use, but physically better founded relation, which physically accurately reproduces especially the portions of heat transfer which result from convection and radiation.

Moreover, the validity of the heat transfer equations should be critically examined for SI engines with unconventional valve timing, as can appear, e.g. in the case of electromechanical valve trains, and for SI engines with direct injection and adjusted with suitable measurements.

The procedure in the determination of the heat transfer coefficient from measured values is described by Merker and Kessen (1999).

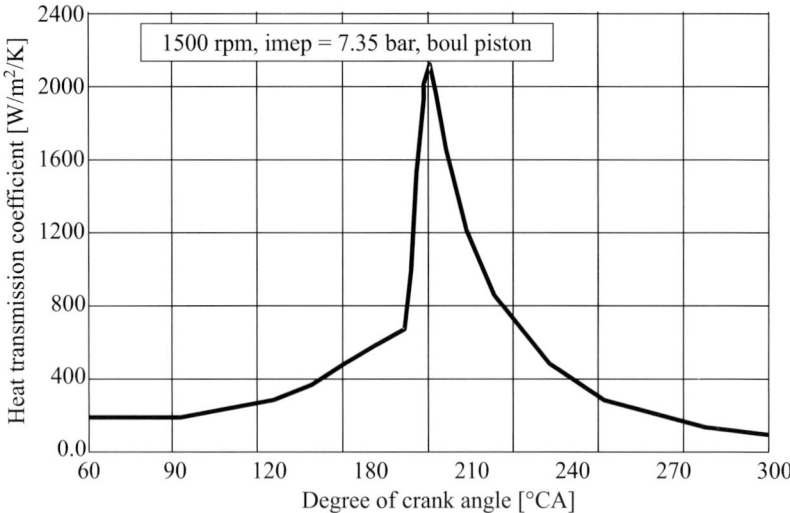

Fig. 7.9: Heat transfer coefficient α for a SI engine, see Bargende (1990)

7.1.5 Heat transfer in the exhaust manifold

Heat transfer in the exhaust manifold in charged engines plays a decisive role in the exact determination of the exhaust gas temperature and thus in the determination of exhaust gas enthalpy before the turbine. The wall heat losses do not belong to the 1st law for the cylinder, since its system boundaries end at the valves. Zapf suggested a relation in 1969 for manifold wall heat transfer which is still used today

$$\alpha_{manifold} = 0.001791\left(1 - 0.797\frac{h_{valve}}{d_i}\right)\sqrt{\dot{m}_{exhaust\ gas}}\ T^{-0.41}\ d_{port}^{-1.5}\ \left[\frac{W}{m^2\ K}\right]. \quad (7.48)$$

According to Zapf (1969), the exhaust manifold heat transfer is contingent on the exhaust mass flow $\dot{m}_{exhaust\ gas}$, the port diameter d_{port}, the gas temperature T – which described

7.1 Single-zone cylinder model

the material quantities in a similar way as in cylinder wall heat transfer – as well as on the valve lift h_{valve} and the internal valve diameter d_i.

For the calculation of the wall heat flow, the Newtonian relation should also be used here

$$\frac{dQ_{port}}{dt} = \alpha_{port} A_{port} \left(T_{w, port} - T\right). \tag{7.49}$$

7.1.6 Wall temperature models

- **Stationary operation**

For the experimental determination of internal wall temperatures, there are in principle two possibilities. Since the exact course on the surface is not required as the internal wall temperature (see Fig. 7.10), but rather only the value averaged over a working cycle, we can determine the gradient of the temperature path by means of a differential measurement. In the combustion chamber wall, a constant temperature gradient appears in the area of the combustion chamber outside wall because of heat conduction according to the adjusted operating point. If we fix two thermocouples offset by a known distance in this area, we can determine the gradient from the difference of the temperatures and the knowledge of the offset distance and with this arrive at a conclusion about the temperature in the inside of the combustion chamber. The other method is the direct measurement of the internal wall temperature, variable over the cycle, by means of a surface thermocouple. Through the use of solutions for the Fourier heat conduction equation, the mean internal wall temperature is easy to determine. This method is used at the same time for the determination of the local heat flows and thus for the determination of the local heat transfer coefficient in the combustion chamber, see Merker and Kessen (1999), Bargende (1990), or Hohenberg (1980).

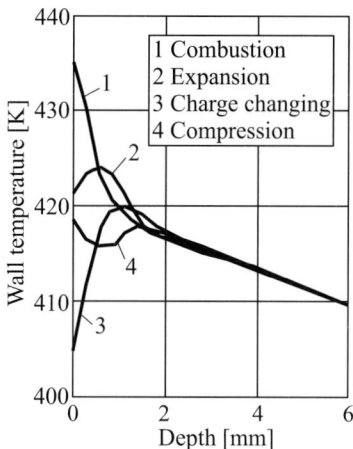

Fig. 7.10: Wall temperature progression during a working cycle

If the internal wall temperature is known for an operating point, the thermal substitute conduction coefficient can be determined from this quite easily for stationary operating points. The remaining quantities are known from process calculation (α, T_w) or as a boundary condition (T_c). The equilibrium of the heat flows is hereby valid, which result from the convective heat transfer (Newtonian equation), and the wall heat flow resulting from heat conduction through the combustion chamber wall (discretized Fourier equation).

$$\overline{\alpha} A \left(\overline{T}_{gas} - \overline{T}_w \right) = \frac{\lambda}{d} A \left(\overline{T}_w - \overline{T}_{w,c} \right) = \overline{\alpha}_c A \left(\overline{T}_{w,c} - \overline{T}_c \right) . \tag{7.50}$$

If we summarize heat conduction through the wall and the convective heat transfer – to the coolant, for example – we obtain the following relation

$$\overline{\alpha} \left(\overline{T}_{gas} - \overline{T}_w \right) = R_{th} \left(\overline{T}_w - \overline{T}_c \right) . \tag{7.51}$$

Solved with the substitute conduction coefficient, we obtain

$$R_{th} = \frac{\overline{\alpha} \left(\overline{T}_{gas} - \overline{T}_w \right)}{\left(\overline{T}_w - \overline{T}_c \right)} . \tag{7.52}$$

The values are averaged over one working cycle, which is indicated by the crossbar. Since the wall thickness in a real engine is not constant everywhere, this effect is also covered in the thermal substitute conduction coefficient. Thus, a local solution of wall temperature is not possible. Upon determining the substitute conduction coefficient, a calculation of the mean wall temperature from the data available at the end of a cycle is possible. On the other hand, since the wall heat flow is dependent on the wall temperature, an iteration loop over several working cycles is necessary. Here is valid

$$\overline{T}_w = \frac{\overline{\alpha} \, \overline{T}_{gas} + R_{th} \, \overline{T}_c}{\alpha + R_{th}} . \tag{7.53}$$

The stationary wall temperatures of the piston, cylinder head and liner can be calculated separately from each other according to this method. It is also practical for the determination of exhaust manifold-wall temperature.

- **Unsteady operation**

For the unsteady operation of an internal combustion engine, more expensive models must be considered, which take into consideration the storage capacity of the combustion chamber wall and the heat transfer to the coolant. One simple model for this can be found in Reulein (1998). A level plate as described in Fig. 7.10 serves as a basic model for this approach as well. From the wall, the physical properties heat capacity c, density ρ, and heat conductivity λ are known. The thickness d and the surface A of the wall must also be known. The mean wall temperature T_m can be calculated with the equation

$$\rho A d c \frac{dT_m}{dt} = \dot{Q}_i + \dot{Q}_o . \tag{7.54}$$

If the wall is divided, as indicated in Fig. 7.11, and a quasi-stationary heat conduction is assumed in both wall halves, the following equations also apply

7.1 Single-zone cylinder model

$$\dot{Q}_i = \frac{2\lambda A}{d}(T_{w,i} - T_m) \tag{7.55}$$

and

$$\dot{Q}_o = \frac{2\lambda A}{d}(T_m - T_{w,o}) . \tag{7.56}$$

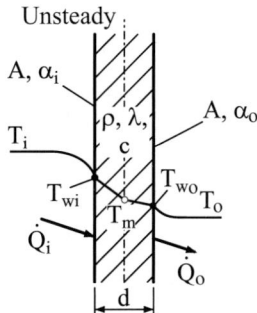

Fig. 7.11: Unsteady wall temperature model

The division of the wall into two planes takes place following FEM-models for the calculation of temperature distributions. Through the simulation of the wall with the help of at least two cells, actual temperature distribution within the wall during heating or cooling processes can be approached much more realistically.

In the case of the piston, we are dealing with a component that is not directly bathed by a coolant. For the stationary case, the substitute conduction coefficient covers the heat conduction over the piston rings and the oil to the liner and to the coolant. In the case of pistons under a full load, an oil injection cooling is employed. For modeling unsteady behavior of the piston, the model of Doll (1989) can be used. In this case, we assume that the heat flow falling in gaseous form into the piston distributes itself to the cylinder liner and the motor oil. The heat flow to the cylinder liner is thereby transferred partially over the piston rings, partially over the piston shaft. The particular heat flows can be determined with the following equations:

piston ring – cylinder wall

$$\dot{Q}_{ring} = A_{ring}\, \alpha_{ring}\, (T_p - T_{cylw}) , \tag{7.57}$$

piston shaft – cylinder wall

$$\dot{Q}_{shaft} = A_{shaft}\, k_{shaft}\, (T_p - T_{cylw}) , \tag{7.58}$$

piston underside – motor oil

$$\dot{Q}_{oil} = A_p\, k_{oil}\, (T_p - T_{oil}) . \tag{7.59}$$

The heat transfer coefficients k for the shaft amount to

$$k_{shaft} = \cfrac{1}{\cfrac{l_1 + \cfrac{l_2}{2}}{\lambda_p} + \cfrac{1}{\alpha_{shaft}}} \qquad (7.60)$$

and for the oil to

$$k_{oil} = \cfrac{1}{\cfrac{l_1 + l_2}{\lambda_p} + \cfrac{1}{\alpha_{oil}}} \; . \qquad (7.61)$$

l_1 signifies the free liner length and l_2 the covered liner length. For the heat transfer coefficients, the following assumptions can be made according to Pflaum and Mollenhauer (1977)

$$\begin{aligned} a_{ring} &= 2{,}500 \left[\frac{W}{m^2 \, K}\right] \\ a_{shaft} &= 1{,}000 \left[\frac{W}{m^2 \, K}\right] \\ a_{oil} &= 500 \left[\frac{W}{m^2 \, K}\right] \end{aligned} \qquad (7.62)$$

The calculation of the mean temperature of the piston then results by means of the differential equation

$$m_p \, c_p \, \frac{dT_p}{dt} = \dot{Q}_{gas} - \dot{Q}_{ring} - \dot{Q}_{shaft} - \dot{Q}_{oil} \; . \qquad (7.63)$$

7.1.7 The heat release rate

The heat release rate describes the temporal course of energy release in the combustion chamber. The integral of the heat release rate is designated as the total heat release rate or the burn-through function. In order to model combustion in the context of process calculation we use various approaches and mathematical modelings, which all have the goal of describing the actual heat release via combustion as exactly as possible as so-called substitute heat release rates. Another possibility for describing heat release are so-called phenomenological models, that pre-calculate the heat release rate, starting from the injection path of the diesel engine for example. In recent times, purely numerical methods are being used because of the increasing complexity of combustion processes (e.g. direct injection with stratified charge in the SI engine, common-rail injection in the diesel engine with multiple injection). In this case we are dealing with so-called neural networks, which have to be trained with results from measured operating points and can be contingent on a number of parameters.

- **The Vibe substitute heat release rate**

Proceeding from the "triangular combustion", Vibe (1970) provided the following relation for the total heat release rate with the help of reaction kinetic considerations

7.1 Single-zone cylinder model

$$\frac{Q_f(\varphi)}{Q_{f,total}} = 1 - e^{-a\left(\frac{\varphi - \varphi_{SOC}}{\Delta\varphi_{CD}}\right)^{m+1}} \quad \text{with } \varphi_{SOC} \leq \varphi \leq \varphi_{SOC} + \Delta\varphi_{CD} \,. \tag{7.64}$$

The entire amount of released energy is thereby to be calculated from the product of the fuel mass brought into the combustion chamber and the lower heating value

$$Q_{f,total} = m_{fuel}\, lhv \,. \tag{7.65}$$

Furthermore, φ_{SOC} signifies the start of combustion and $\Delta\varphi_{CD}$ the combustion duration. The so-called Vibe form parameter is designated with m.

At the end of combustion, i.e. at $\varphi = \varphi_{EOC}$, a certain percentage $\eta_{conv,total}$ of the total energy added with the fuel must be converted, for which is valid

$$\left.\frac{Q_f(\varphi)}{Q_{f,total}}\right|_{\varphi = \varphi_{EOC}} \equiv \eta_{conv,total} = 1 - e^{-a} \,. \tag{7.66}$$

From this follows for the factor a the relation

$$a = -\ln(1 - \eta_{conv,total}) \,,$$

from which we obtain the following numerical values

$\eta_{conv,total}$	0.999	0.990	0.980	0.950
a	6.908	4.605	3.912	2.995

If we derive the total heat release rate from the degree of crank angle, we obtain

$$\frac{dQ_f}{d\varphi} = Q_{f,total}\, a(m+1)\left(\frac{\varphi - \varphi_{SOC}}{\Delta\varphi_{CD}}\right)^m e^{-a\left(\frac{\varphi - \varphi_{SOC}}{\Delta\varphi_{CD}}\right)^{m+1}} \tag{7.67}$$

for the heat release rate.

Fig. 7.12 shows heat release rates for various Vibe form factors. One should note here that Vibe form factors smaller than 0 are also possible.

In order to adjust a real combustion by means of a Vibe substitute heat release rate, there are various methods for determining the three Vibe parameters start of combustion, combustion duration, and form parameter. The parameters can either be adjusted visually or can be determined using mathematical methods (e.g. that of the least square). It is however important that important process data such as peak pressure, indicated mean pressure and the exhaust gas temperature are calculated with the substitute heat release rate in agreement with reality. The determination of the Vibe parameters is solidly integrated into most pressure analysis programs.

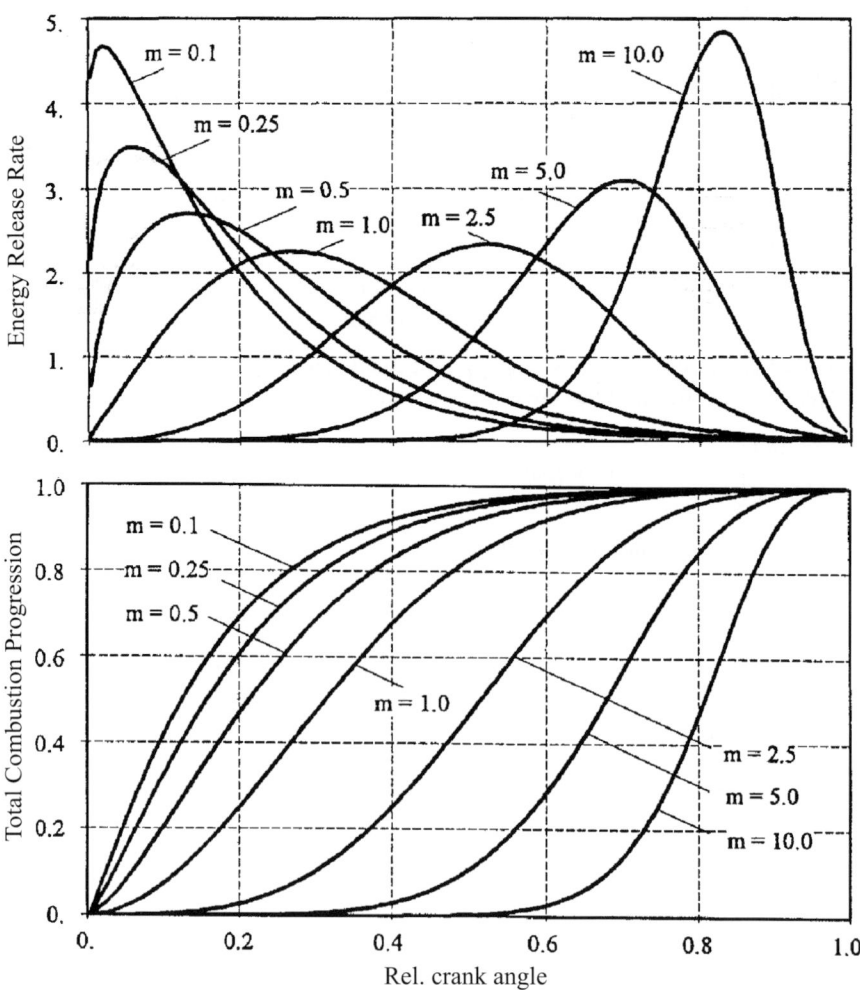

Fig. 7.12: The Vibe substitute heat release rate for different Vibe shape coefficients

- **The substitute heat release rate as a double Vibe function**

In engines or operating points with a clearly marked premixed combustion, reproduction with a simple Vibe substitute heat release rate is usually too imprecise. We thus often replace the simple Vibe substitute heat release rate with the superimposition of two Vibe functions, the so-called double Vibe function. If one describes an actual heat release rate with a double Vibe function, a subdivision of the energy shares of both functions is also necessary besides the two simple Vibe heat release rates. For a double Vibe heat release rate is valid

$$\frac{dQ_{f,1}}{d\varphi} = Q_{f,1} a(m_1 + 1)\left(\frac{\varphi - \varphi_{SOC,1}}{\Delta\varphi_{CD,1}}\right)^{m_1} e^{-a\left(\frac{\varphi - \varphi_{SOC,1}}{\Delta\varphi_{CD,1}}\right)^{m_1+1}} \quad (7.68)$$

with $\varphi_{SOC,1} \leq \varphi \leq \varphi_{SOC,1} + \Delta\varphi_{CD,1}$,

$$\frac{dQ_{f,2}}{d\varphi} = Q_{f,2} a(m_2 + 1)\left(\frac{\varphi - \varphi_{SOC,2}}{\Delta\varphi_{CD,2}}\right)^{m_2} e^{-a\left(\frac{\varphi - \varphi_{SOC,2}}{\Delta\varphi_{CD,2}}\right)^{m_2+1}} \quad (7.69)$$

with $\varphi_{SOC,2} \leq \varphi \leq \varphi_{SOC,2} + \Delta\varphi_{CD,2}$,

$$Q_{f,1} = x Q_{f,total} \text{ and } Q_{f,2} = (1-x) Q_{f,total}, \quad (7.70)$$

$$\frac{dQ_f}{d\varphi} = \frac{dQ_{f,1}}{d\varphi} + \frac{dQ_{f,2}}{d\varphi}. \quad (7.71)$$

The diagram on the left in Fig. 7.13 shows the reproduction of the heat release rate in the rated performance point of a high speed high performance diesel engine via a double Vibe substitute heat release rate. One can clearly see that the double Vibe function can not exactly reproduce combustion lasting until the exhaust phase. This has to do with the mathematical form of this substitute heat release rate. The diffusion phase is described by the central exponential term of the double Vibe function. This exponential term with its asymptotic course towards burning matter can under certain circumstances still not exactly describe large energy release rates at the burning location in the concrete case, which is why on occasion a threefold Vibe function is used instead of the double Vibe function.

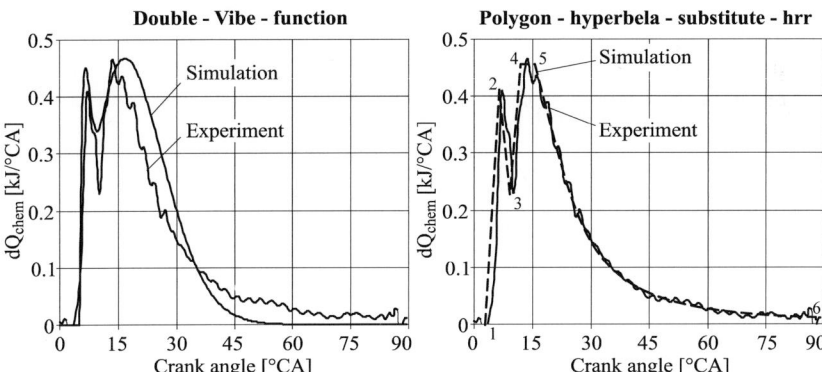

Fig. 7.13: The double-Vibe function and polygon-hyperbola heat release rate

The exact reproduction of premixed combustion is an important prerequisite for a good agreement of the working cycle calculation with measurements, which is why the adjustment and weighing of the first Vibe function should receive careful consideration. One problem in the double Vibe function is – as in the case of the simple Vibe function as well – the exact description of the burn-out phase, which is, especially in charged engines, responsible for the exhaust gas temperature level. Good indicator for this is the total heat release rate, in which deviations are most clearly visible. The method of the least square has proved to be the most suitable adjustment routine in the case of the double Vibe function as well.

- **The polygon-hyperbola substitute heat release rate**

In view of the calculation of NO_x formation by means of multi-zone models, a further substitute heat release rate, the polygon-hyperbola substitute heat release rate, was suggested by Schreiner (1993), which is represented in Fig. 7.13 on the right. It consists of a polygonal port (1-4-5) and a hyperbola connected to it (5-6). A triangle (1-2-3) is superimposed over the polygonal port, which serves in the description of premixed combustion. For the mathematical description of this polygon-hyperbola substitute heat release rate, we need nine parameters.

Premixed phase:

$$\frac{dQ_{f,pre}}{d\varphi} = y_2^* \frac{(\varphi - \varphi_1)}{(\varphi_2 - \varphi_1)} \text{ with } \varphi_1 \leq \varphi \leq \varphi_2 \tag{7.72}$$

$$\frac{dQ_{f,pre}}{d\varphi} = y_2^* \frac{(\varphi_3 - \varphi)}{(\varphi_3 - \varphi_2)} \text{ with } \varphi_2 \leq \varphi \leq \varphi_3 \tag{7.73}$$

Diffusion phase:

$$\frac{dQ_{f,diff}}{d\varphi} = y_4 \frac{(\varphi - \varphi_1)}{(\varphi_4 - \varphi_1)} \text{ with } \varphi_1 \leq \varphi \leq \varphi_4 \tag{7.74}$$

$$\frac{dQ_{f,diff}}{d\varphi} = y_4 \text{ with } \varphi_4 \leq \varphi \leq \varphi_5 \tag{7.75}$$

$$\frac{dQ_{f,diff}}{d\varphi} = h_3 + h_1 (\varphi - \varphi_1)^{h_2}$$

$$= y_4 - \frac{(y_4 - y_6)}{1 - \left[\frac{(\varphi_6 - \varphi_1)}{(\varphi_5 - \varphi_1)}\right]^{h_2}} +$$

$$+ \frac{(y_4 - y_6)}{(\varphi_5 - \varphi_1)^{h_2} - (\varphi_6 - \varphi_1)^{h_2}} (\varphi - \varphi_1)^{h_2} \tag{7.76}$$

mit $\varphi_5 \leq \varphi \leq \varphi_6$.

For the entire combustion

$$\frac{dQ_{f,total}}{d\varphi} = \frac{dQ_{f,pre}}{d\varphi} + \frac{dQ_{f,diff}}{d\varphi} \tag{7.77}$$

7.1 Single-zone cylinder model

is valid.

The quantity y_2^* sets the height of the peak in premixed combustion. The three hyperbolic parameters h_1, h_2 and h_3 result from the conditions that the hyperbola go through points 5 and 6 and that the integral under the substitute heat release rate has the value of 1. h_2 thereby fixes the path of the hyperbola. The maximum height of the substitute heat release rate in the diffusion phase y_4 results from the condition that the center of the substitute heat release rate agree with the experimentally determined heat release rate. The quantity y_6 describes the conversion at the end of combustion.

The agreement in Fig. 7.13 (right) is very good, such that there practically no disparity arises in the results of the actual working cycle calculation between the point-for-point provision and the approximation of experimental data with this substitute heat release rate.

Further details and an extensive analysis of the polygon-hyperbola substitute heat release rate can be found in Schreiner (1993).

- **Combinations of substitute heat release rates**

In order to describe the long burn-out phase and the variable pre-injection possible with the common-rail injection system, a combination of the Vibe substitute heat release rate and the hyperbola substitute heat release rate was suggested by Barba et al. (1999). However, the equations described and their conversion regularities are so complex that a general formulation is impossible, which is why we simply refer here to a corresponding source in the literature.

- **Heat release rate precalculation for diesel engines**

In order to take into account the quantity of fuel energy introduced into the combustion chamber required for the respective operating point and correctly to reproduce the resulting changes in the thermal state quantities of the engine, like exhaust gas temperature, a possibility must be created to pre-calculate the heat release rate contingent upon the respective operating point data in the simulation. This is above all inevitable for non-stationary calculations. The Vibe substitute heat release rate is determined – as described above—by the three parameters start of combustion, combustion duration, and the shape parameter. Conversion regularities for these parameters can be found contingent on the respective operating point. The foundations for the description of these regularities were laid, amongst others, by Woschni and Anisits (1973) and will here be represented in a universally valid form. A known initial operating point of the engine (index A) is used as the basis for the conversion. Partially constant values can be found in the literature for the parameters introduced in the following equations. Here, we will however represent the possible quantities of influence on the particular heat release rate quantities as parameters. The following equations describe the conversion of Vibe parameters for the diesel engine.

For the determination of start of combustion, proceeding from the geometrically set delivery start of the injection pump, the following relation is applicable

$$\varphi_{SOC} = \varphi_{DS} + \Delta\varphi_{IND} + \Delta\varphi_{IGD} \,. \tag{7.78}$$

For an injection pump without regulation of the start of injection, the injection delay between the geometrically fixed delivery start of the injection pump and the start of injection – i.e. the opening of the injection nozzle needle – must be considered, see Merker and Schwarz (2001)

$$\Delta\varphi_{IND} = \Delta\varphi_{IND,A} \left(\frac{n}{n_A}\right)^{a_{IND}}. \tag{7.79}$$

In engines with regulation of the start of injection, the geometrical delivery start is delayed until the start of injection, which can be determined via the needle lift signal, agrees with the desired value. The determination of the injection delay time and the geometrical delivery start are thereby left out. The equation for start of combustion is then

$$\varphi_{SOC} = \varphi_{SOI} + \Delta\varphi_{IGD}. \tag{7.80}$$

The ignition delay can be described with different, formulaic connections. According to Woschni and Anisits (1973), for the pre-chamber engine is valid

$$\Delta\varphi_{IGD} = a_{IGD}\, 6n10^{-3}\, 1.3\, e^{\frac{990}{T(\varphi_{SOI})}}\, p(\varphi_{SOI})^{-0.35}. \tag{7.81}$$

For direct injection diesel engines, the ignition delay can be calculated according to Sitkei (1963)

$$\Delta\varphi_{IGD} = 6n10^{-3}\left[a_{IGD} + b_{IGD}\, e^{\frac{7{,}800}{6.9167\,RT}}\left(1.0197\, p^{-0.7}\right)+ \right.$$

$$\left. + c_{IGD}\, e^{\frac{7{,}800}{6.9167\,RT}}\left(1.0197\, p^{-1.8}\right)\right]. \tag{7.82}$$

For pressure and temperature, the pressure and temperature values averaged from start of injection until start of combustion must be employed. The combustion duration depends on the air-fuel ratio and, in the case of some engines, on the speed as well

$$\Delta\varphi_{CD} = \Delta\varphi_{CD,A} \left(\frac{\lambda_A}{\lambda}\right)^{a_{CD}} \left(\frac{n}{n_A}\right)^{b_{CD}}. \tag{7.83}$$

For engines with exhaust gas recirculation, the description of the combustion duration via the air-fuel ratio is insufficient, since the air-fuel ratio in exhaust gas recirculation is contingent to a large extent on the mass of the recirculated gas. It is thus not a clear indicator of engine load. For this reason, the combustion duration for engines with exhaust gas recirculation is described contingent on the injected fuel mass. The equation then reads

$$\Delta\varphi_{CD} = \Delta\varphi_{CD,A} \left(\frac{\lambda_A}{\lambda}\right)^{a_{CD}} \left(\frac{n}{n_A}\right)^{b_{CD}} \left(\frac{m_{fuel}}{m_{fuel,A}}\right)^{c_{CD}}. \tag{7.84}$$

The form parameter describes the course of energy conversion and is thus contingent on the ignition delay (mixture preparation time), on the speed, and on the gas conditions or on the

7.1 Single-zone cylinder model

gas mass in the cylinder at "intake closes". In order to calculate shape parameters smaller than 0 as well, the equation is supplemented with an additional part. Shape parameters smaller than 0 describe heat release rates with very high heat release rate increase speeds, which can appear in direct injection diesel engines under weak loads and low speeds, when a load pressure is not yet available.

$$m = (m_A + \Delta m) \left(\frac{\varphi_{IGD,A}}{\varphi_{IGD}} \right)^{a_{VM}} \left(\frac{n_A}{n} \right)^{b_{VM}} \left(\frac{p_{IVC} V_{IVC} T_{IVC,A}}{p_{IVC,A} V_{IVC,A} T_{IVC}} \right)^{c_{IVC}} - \Delta m \quad (7.85)$$

Tab. 7.1 provides, besides the "basic" parameters known from the literature, an overview of usual parameters as they have been used by the authors for extensive calculations with good agreement with reality.

Tab. 7.1: Parameters for the pre-calculation of heat release rates

	Original equation	Large diesel engine	Commercial vehicle DI	Passenger car DI
a_{IND}	1.0	1.0	-	-
a_{IGD}	1*; 0.5	0.39	0.625*	0.1
b_{IGD}	0.135	0.105	-	0.135
c_{IGD}	4.8	3.12	-	4.8
a_{CD}	0.6	0.6	-0.3	0.0
b_{CD}	0.5	0.5	-0.65	0.0
c_{CD}	0.0	0.0	0.0	-0.1
Δm	0.0	0.0	0.4	0.3
a_{VM}	0.5	0.5	0.5	0.2
b_{VM}	0.3	0.3	-0.8	-0.4
c_{VM}	1.0	1.0	1.0	1.0

*Anisits

Conversion regularities for double Vibe heat release rates are much more costly, since another conversion for energy distribution of both individual portions is necessary for the second set of Vibe parameters. Conversion regularities for a high speed diesel engine are described by Oberg (1976).

For the polygon-hyperbola substitute heat release rate, the following contingencies are necessary for the conversion of the parameters for high performance diesel engines according to Schreiner (1993)

$$\varphi_{50\,mfb} = \int_{\varphi_1}^{\varphi_6} \frac{dQ_B}{d\varphi} \varphi \, d\varphi \,, \tag{7.86}$$

$$IND = IND_A \left(\frac{n}{n_A}\right), \tag{7.87}$$

$$\frac{\frac{ID}{ID_A} - e_1}{1 - e_1} = e_2 \left(\frac{n}{n_A}\right) + e_3 \left(\frac{m_{fuel}}{m_{fuel,a}}\right) + (1 - e_2 - e_3)\left(\frac{n}{n_A}\right)\left(\frac{m_{fuel}}{m_{fuel,a}}\right), \tag{7.88}$$

$$\frac{\tau_{IGD}}{\tau_{IGD,A}} = \left(\frac{e^{\frac{a}{T_{IGD}}}}{e^{\frac{a}{T_{IGD,A}}}}\right) \left(\frac{P_{IGD}}{P_{IGD,A}}\right)^b, \tag{7.89}$$

$$\varphi_{SOC} = \varphi_{SOI} + IND + IGD\,, \tag{7.90}$$

$$x_{IGD} = \frac{m_{fuel,IGD}}{m_{fuel}}\,, \tag{7.91}$$

$$\frac{x_{pre} - k_1}{x_{pre,A} - k_1} = \frac{x_{IGD} - 1}{x_{IGD,A} - 1}\,, \tag{7.92}$$

$$\frac{\frac{CD_{50\,mfb}}{CD_{50\,mfb,A}} - k_3}{1 - k_3} = \left(\frac{ID}{ID_A}\right)^{k_4} \left(\frac{\lambda}{\lambda_A}\right) \left(\frac{n}{n_A}\right)^{k_5}, \tag{7.93}$$

$$\frac{(\varphi_4 - \varphi_1)}{(\varphi_4 - \varphi_1)_A} = \left(\frac{n}{n_A}\right)^{k_6} \left(\frac{m_{fuel}}{m_{fuel,A}}\right)^{k_7} \text{ and} \tag{7.94}$$

$$\frac{(\varphi_5 - \varphi_4)}{(\varphi_5 - \varphi_4)_A} = \left(\frac{n}{n_A}\right)^{k_8} \left(\frac{m_{fuel}}{m_{fuel,A}}\right)^{k_9}. \tag{7.95}$$

In Tab 7.2, standard values for the nine parameters are given.

7.1 Single-zone cylinder model

Tab. 7.2: Standard values

Injection duration:	$e_1 = 0$	Center of comb. duration:	$k_3 = 0.3$
	$e_2 = 1 - e_3$		$k_4 = -0.3$
	$e_3 = 0.8$		$k_5 = 0$
Ignition delay:	$a = 1{,}500$ K	Diffusion combustion:	$k_6 = 0$
	$b = -0.8$		$k_7 = 1$
			$k_8 = 1$
Premixed portion:	$k_1 = 0.4$		$k_9 = 0$
	$k_2 = 0.125$		

- **Heat release rate conversion for SI engines**

A conversion of the Vibe parameters for varying operating points has been introduced for the SI engine as well, e.g. in Csallner (1981). Since the cylinder charge is externally ignited with the spark plug, the derivation of the start of combustion via injection/ignition delay is dropped in the case of the SI engine. In thermodynamic evaluations however, a time shift between ignition (ignition time) and a noticeable release of energy (increase of the heat release rate/5% mass fraction burned) is demonstrable in the SI engine as well. This is explained in that at first some time passes because of the point light ignition until a larger volume of the flame front is included. The time period until the rise of the heat release rate is designated as apparent ignition delay. Csallner (1981) has described the contingency of the Vibe parameters on the process quantities. The investigations were executed such that the individual operating parameters were varied independently of one another. Csallner thus chose a description by means of a multiplicative method. Proceeding from an referenced point (index A), for the ignition delay thus results

$$IGD = IGD_A \, f_{IT} \, f_n \, f_p \, f_T \, f_{x_{rg}} \, f_\lambda \, . \tag{7.96}$$

For the combustion duration, it can be described in the same way

$$\Delta\varphi = \Delta\varphi_A \, g_{IT} \, g_n \, g_p \, g_T \, g_{x_{rg}} \, g_\lambda \, . \tag{7.97}$$

For the form factor we obtain

$$m = m_A \, h_{IT} \, h_n \, h_p \, h_T \, h_{x_{rg}} \, h_\lambda \, . \tag{7.98}$$

The particular functions f, g and h are to be found in the following table (Tab. 7.3).

Tab. 7.3: Functions according to Csallner (1981)

	Ignition delay	Combustion duration	Shape parameters
Ignition time 25-50 ° b. TDC	$f_{IT} = \dfrac{430 - \varphi_{IT}}{430 - \varphi_{IT,A}}$	$g_{IT} = 1$	$h_{IT} = 1$
Speed 1,000-4,500 rpm	$f_n = \dfrac{1 + \dfrac{400}{n} - \dfrac{8 \cdot 10^5}{n^2}}{1 + \dfrac{400}{n_A} - \dfrac{8 \cdot 10^5}{n_A^2}}$	$g_n = \dfrac{1.33 - \dfrac{660}{n}}{1.33 - \dfrac{660}{n_A}}$	$h_n = \dfrac{0.625 + \dfrac{750}{n}}{0.625 + \dfrac{750}{n_A}}$
Cylinder-press. at 300 °CA	$f_p = \left(\dfrac{p_{300}}{p_{A,300}}\right)^{-0.47}$	$g_p = \left(\dfrac{p_{300}}{p_{A,300}}\right)^{-0.28}$	$h_p = 1$
Cylinder-temp. at 300 °CA	$f_T = 2.16 \dfrac{T_{A,300}}{T_{300}} - 1.16$	$g_T = 1.33 \dfrac{T_{A,300}}{T_{300}} - 0.33$	$h_T = 1$
Residual gas port. 0-10 %	$f_{x_{rg}} = 0.088 \dfrac{x_{rg}}{x_{rg,A}} + 0.912$	$g_{x_{rg}} = 0.237 \dfrac{x_{rg}}{x_{rg,A}} + 0.763$	$h_{x_{rg}} = 1$
Combus. air ratio 0.7-1.2	$f_\lambda = \dfrac{2.2\lambda^2 - 3.74\lambda + 2.54}{2.2\lambda_A^2 - 3.74\lambda_A + 2.54}$	$g_\lambda = \dfrac{2.0\lambda^2 - 3.4\lambda + 2.4}{2.0\lambda_A^2 - 3.4\lambda_A + 2.4}$	$h_\lambda = 1$

The range of validity for these conversions is confined to very minimal residual gas amounts. Modern combustion processes show higher amounts of residual gas, which is why Csallner's influence equations were adjusted for a throttled and an unthrottled the SI engine (fully variable valve train) by Witt (1999).

In Witt (1999), the following is valid:

$$IGD = IGD_A \, f_{IT} \, f_{x_{rg}} \, f_n \, f_{wi} \,, \tag{7.99}$$

$$\Delta\varphi = \Delta\varphi_A \, g_{IT} \, g_{x_{rg}} \, g_n \, g_{wi} \quad \text{and} \tag{7.100}$$

$$m = m_A \, h_{IT} \, h_{x_{rg}} \, h_n \, h_{wi} \,. \tag{7.101}$$

7.1 Single-zone cylinder model

Tab. 7.4: Functions according to Witt (1999)

	Ignition delay	Combustion duration	Shape parameters
Ignition time 17-57 ° BTDC	$f_{IT} = \dfrac{a + b\varphi_{IT}^2}{a + b\varphi_{IT,A}^2}$	$g_{IT} = \dfrac{a + b\varphi_{IT}^{-0.5}}{a + b\varphi_{IT,A}^{-0.5}}$	$h_{IT} = \dfrac{a + b\varphi_{IT}^{-2}}{a + b\varphi_{IT,A}^{-2}}$
Residual gas portion 10-26 %	$f_{x_{rg}} = \dfrac{a + bx_{rg}^2}{a + bx_{rg,A}^2}$	$g_{x_{rg}} = \dfrac{a + bx_{rg}}{a + bx_{rg,A}}$	$h_{x_{rg}} = \dfrac{a + bx_{rg}^2}{a + bx_{rg,A}^2}$
Speed 1,000 - 4,000 rpm	$f_n = \dfrac{a + b\ln(n)}{a + b\ln(n_A)}$ $f_n = \dfrac{a + bn^{-2}}{a + bn_A^{-2}}$	$g_n = \dfrac{a + bn^{-0.5}}{a + bn_A^{-0.5}}$	$h_n = \dfrac{a + bn^{1.5}}{a + bn_A^{1.5}}$
Indicated work 0.2-0.8 kJ/l	$f_{wi} = \dfrac{a + bwi^{1.5}}{a + bwi_A^{1.5}}$	$g_{wi} = \dfrac{a + bwi}{a + bwi_A}$	$h_{wi} = \dfrac{a + b\ln(wi)}{a + b\ln(wi_A)}$

The parameters for this can be taken from the following table.

Tab. 7.5: Parameters for the conversion of the heat release rate parameters

				IT	x_{rg}	n	wi
Ignition delay		throttled	a	0.678	0.879	0.992	1.112
			b	$2.383 \cdot 10^{-4}$	$3.648 \cdot 10^{-4}$	$-1.246 \cdot 10^{-4}$	-0.545
		unthrottled	a	0.638	0.914	-1.284	1.162
			b	$2.614 \cdot 10^{-4}$	$2.795 \cdot 10^{-4}$	0.292	-0.589
Combustion duration		throttled	a	0.596	0.429	1.355	1.115
			b	2.480	0.031	-18.49	-0.346
		unthrottled	a	0.477	0.690	1.701	1.295
			b	3.200	0.017	-34.50	-0.699
Shape parameters		throttled	a	0.964	1.076	1.046	1.007
			b	75.56	$-2.534 \cdot 10^{-4}$	$-4.075 \cdot 10^{-7}$	0.004
		unthrottled	a	1.000	1.061	1.016	1.053
			b	19.36	$-1.656 \cdot 10^{-4}$	$-1.206 \cdot 10^{-7}$	0.065

- **Neural networks for Vibe heat release rates**

Besides the analytic methods of pre-calculation of heat release rates already presented, there is still another possibility in determining the heat release rate contingent upon its influence quantities. For this, the Vibe parameters are determined by means of a neural network, which has previously been trained by a number of measured operating points evaluated with reference to the Vibe parameters. In principle, all types of neural networks are suitable for this task. In comparison to the analytic relations presented in the previous section, clear improvements can be attained by determining the Vibe parameters by means of a neural network, which makes possible a halving of the averaged error and thus positively influences the quality of the calculation results. The disadvantage of neural networks in this case is the impossibility of extrapolation and the lacking transparency with reference to the contingencies of particular influence quantities in comparison to the analytic approach. An advantage is to be seen in the possibility of obtaining statements about the contingencies of the Vibe parameters from a number of unsystematic experimental results. A more exact description of neural networks can be found at the end of this section.

- **Neural networks for discrete heat release rates**

In the previous section, the description of heat release rate functions via Vibe parameters pre-calculated with neural networks was introduced. However, because of the increasing complexity of combustion processes both in diesel and in SI engines, it is necessary to describe the discrete heat release rates. In the diesel engine, because of common-rail technology, the injection and thus within certain boundaries the combustion as well become freely formable (pre-/post-injection). In the SI engine also, the combustion is influenced in a lasting way by load control by means of fully variable valve trains or via direct injection. In gasoline direct injection, a clearly lengthened burn-out phase appears for the stratified area. For both cases, an identification of the combustion via a simple substitute heat release rate function is therefore no longer sufficient in describing correctly the details of the combustion process. Moreover, the number of influence quantities is becoming larger and larger in such complex combustion processes, to the extent that an experimental scanning of these quantities is practically impossible technically.

Fig. 7.14 shows a heat release rate for a stratified operating point in gasoline direct injection, in which the delayed burn-out phase is clearly recognizable. In the literature, a large number of methods are described for calculating heat release rates discretely with a neural network, see Zellbeck (1997). The method of Reulein et al. (2000) introduced in the following distinguishes itself from known methods and attempts to avoid the disadvantages of these methods. The experimentally determined heat release rates are first filtered, standardized to 1, and centered to their 50 mfb point. Through this, we can exclude significant deviations in the training data and make the network input regular. This process is represented in Fig. 7.15. In this process, it is necessary to train two neural networks, of which the first reproduces the path contingent on crank angle and the other contains the position of the center point. However, precisely this method shows clear advantages, since, on the one hand, the number of input parameters corresponding to the influence quantities to be described can be clearly enlarged as opposed to a functional description, and, on the other hand, the influence quantities for the shape and the 50 mfb point position can be set separately. In this way, the quality of the results is clearly improved. In most professional simulation tools, training algorithms with corresponding network topologies are now offered.

7.1 Single-zone cylinder model

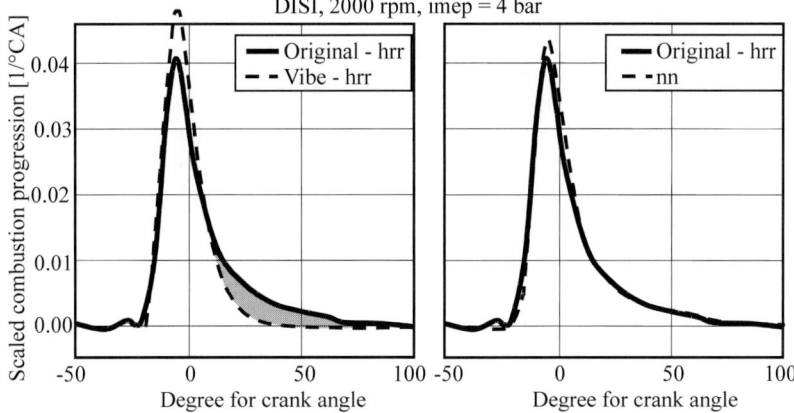

Fig. 7.14: Heat release rate for a DISI engine at 2,000 rpm and imep = 4 bar. Comparison of the original heat release rate with the Vibe heat release rate (left) and with the neural network heat release rate (right)

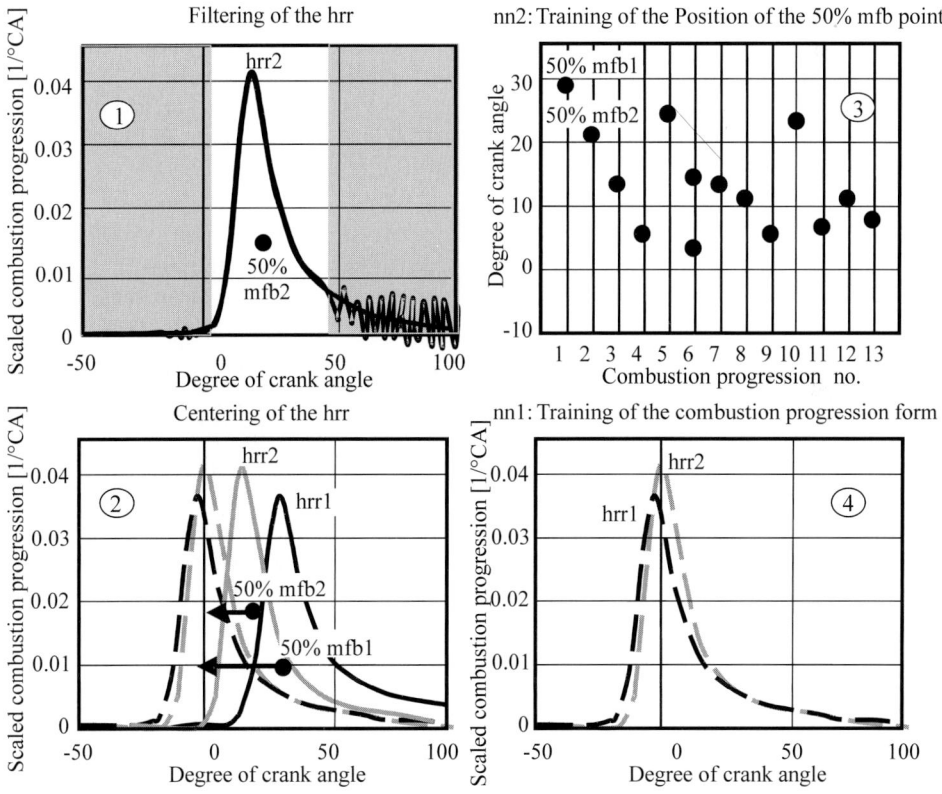

Fig. 7.15: Net training procedure

In Fig. 7.14 is presented a comparison between a measured heat release rate and one calculated with a neural network. As far as the utilized network is concerned, we are dealing with a dual-layered multi-layer perceptron network with 25 nodes/points of intersection per layer and sigmoid activation functions, which are trained with a back-propagation algorithm. Of course, the heat release rate to be compared was not used for the network training and determined within process calculation with the following input parameters:

- degree of crank angle
- speed
- air mass,
- fuel mass,
- ignition time, and
- residual gas amount.

On a critical note, we should note about neural networks that they are only valid within their trained range. An extrapolation is in principle impossible and must be made recognizable to the user, as otherwise the simulation results are considerably falsified.

- **Phenomenological models**

One fundamentally different method for modeling combustion is the use of phenomenological models. In the literature, several phenomenological models for the calculation of the heat release rate of a diesel engine from the injection rate have become known, see Stiesch (1999), Eilts (1993), or Chmela et al. (1998) and Constien (1991). For a description of these models, see chapter 5.

7.1.8 Knocking combustion

While in chap. 7.1.7 methods were described for calculating normal combustion, in this chapter we will go into the simulation of autoignition and into the criteria for the prediction of the appearance of knocking combustion.

Knocking combustion occurs in spark ignited engines (SI engines). Under knocking combustion is understood autoignition of parts of the final gas before it is covered by the flame front, which is initiated by the spark plug, see also chap. 4.1.4. A superimposition of several pressure waves in the combustion space then occurs, which leads to mechanical damage in the engine. The pressure curve of a knocking combustion is shown in Fig. 4.9.

The goal of the calculation of autoignition behavior is not the statement that a case of knocking combustion in the working cycle is concerned, but the predetermination of the degree of crank angle at which autoignition of the final gas appears.

In the literature, one distinguishes between methods with detailed or reduced reaction kinetics and empirical, phenomenological models. The first describe more or less extensively the reaction mechanisms of hydrocarbon oxidation in the low temperature range of the final gas, whereby so-called chain embranchments reinforce radical formation and chain breakage weakens it. The mechanisms are usually derived from shock wave pipes or rapid compression machines for a particular fuel type. An exponential increase in radical concentrations in the fresh air region over a certain limit value is assessed as an indicator for the commencement of knocking combustion. With this modeling approach there is thus the possibility of a prediction of the degree of crank angle, at which knocking begins. These modeling methods are, however, very complex, require a lot of calculation time, and would exceed the confines of this book. We therefore refer to the corresponding literature, e.g. Halstead et al. (1975, 1977), Li et al. (1992, 1994, 1996), Schreiber et al. (1994), and Kleinschmidt (2000).

The phenomenological methods view the processes in the final gas through an average reaction speed, which can be expressed by means of an Arrhenius relation. We will take a closer look at phenomenological approaches in the following.

- **Knocking criterion according to Franzke (1981)**

Since reaching a certain temperature level in the final gas area is not a sufficient criterion for the setting in of knocking combustion, the pressure-temperature history in the combustion chamber must be considered. For this, a so-called critical pre-reaction level is calculated as follows

$$I_k = \frac{1}{\omega} \int_{\varphi_{IVC}}^{\varphi_k} p^a \, e^{\left(\frac{b}{T_{ub}}\right)} d\varphi = \text{const.} \quad (7.102)$$

The parameter φ_k describes thus the degree of crank angle, at which the critical pre-reaction level is reached.

The temperature of the unburned T_{ub} is calculated via a polytrope state change from pressure and temperature at start of combustion and a polytrope exponent, which results from the conditions at compression start and start of combustion

$$T_{ub}(\varphi) = T_{SOC} \left(\frac{p(\varphi)}{p_{SOC}}\right)^{\frac{n_{SOC}-1}{n_{SOC}}} . \quad (7.103)$$

The parameters a and b are given in Tab. 7.6.

In addition, Franzke defines the appearance of knocking combustions by means of a constant parameter K specific for the combustion chamber space. This depends on different quantities (e.g. change motion level, combustion chamber space etc.) and represents a quotient of the combustion progress at the start of knocking and the entire combustion duration

$$K = \frac{\varphi_E - \varphi_{SOC}}{\Delta \varphi_{CD}} . \quad (7.104)$$

The difference of the 95 % mfb point and start of combustion (1 % mfb) is assumed as the combustion duration.

The parameter φ_E describes the degree of crank angle during the combustion progress, at which the critical pre-reaction level in the unburned gas must be reached, such that knocking combustion can appear.

Valid for the appearance of knocking combustion according to this is

$$\varphi_K < \varphi_E \ . \tag{7.105}$$

With the equation method of Franzke it is thus possible to determine the crank angle at which knocking combustion begins.

- **Modifications of the knocking criterion of Franzke**

Modifications of Franzke's approach were carried out by Spicher and Worret (2002). For the pre-reaction state I_K is valid

$$I_K = \frac{1}{6n} \frac{1}{c \, 10^{-3}} \int_{\varphi_{cs}}^{\varphi_K} p^a e^{\left(\frac{b}{T}\right)} d\varphi = 1 \ . \tag{7.106}$$

φ_{cs} thereby signifies the degree of crank angle at calculation start, i.e. 90 °CA before the TDC. The parameters a, b and c can be taken from tab. 7.6.

Tab. 7.6: Parameters for the knocking criterion

		Franzke	Spicher/Worret Simple 2-ZM, HTR Bargende	Spicher/Worret Actual 2-ZM, HTR Bargende	Spicher/Worret Simple 2-ZM, HTR Woschni
a	[-]	1.5	-1.299	-1.267	-1.262
b	[K]	-14,000	4,179	4,080	3,964
c	[-]	-	2.370	2.24	2.714
a_{IK}	[-]	-	-0.557	-0.449	-0.553
a_K	[-]	-	-0.236	-0.241	-0.231
b_K	[-]	-	1.292	1.395	1.275
c_K	[-]	-	0.251	0.313	0.244
a_{kp}	[-]	-	0.211	0.227	0.273
b_{kp}	[-]	-	0.288	0.277	0.233

Tab. 7.6 describes the parameters adjusted to the respective model. According to Spicher and Worret, a parameter sensitivity contingent on the chosen modeling is given. Methods for a simple and an actual two-zone model and heat transfer relations acc. to Bargende and Woschni were thereby considered. With the method introduced, it is possible to determine the start of knocking combustion in modern SI engines up to ±2 °CA.

In order to limit the range of deviations, the 75 % mfb point (φ_{75}) must be taken into consideration according to Spicher and Worret. We thereby utilize – as in the conversion of heat release rates – an additional reference point, for which all parameters must be known. The formal relationship reads

7.1 Single-zone cylinder model

$$I_K = I_{K,ref} \left(\frac{\varphi_{75} + 6}{\varphi_{75,ref} + 6} \right)^{a_{IK}}. \tag{7.107}$$

In the same way, for the k-value corrections, ($\varphi_{50\ mfb}$) results for the 50 mfb point position of combustion and λ for the air-fuel ratio. The following equation shows the correlation

$$K = K_{ref} \left(\frac{\varphi_{50\ mfb} + 8}{\varphi_{50\ mfb,ref} + 8} \right)^{a_K} \left(\frac{b_K - c_K \lambda}{b_K - c_K \lambda_{ref}} \right). \tag{7.108}$$

As opposed to Franzke, Spicher and Worret calculate the gas states in the fresh gas by means of a two-zone model (see chap. 7.2), which in the simplest case assumes an adiabatic calculation of temperature in the fresh gas zone.

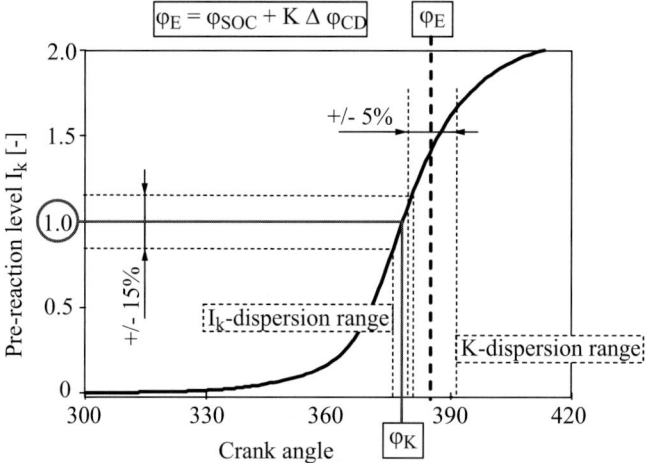

Fig. 7.16: Effect of the oscillation range of I_K and K

Spicher and Worret further provide an equation method for knocking probability

$$kp = 100 \left(a_{kp} + b_{kp} \ln \left(\frac{\varphi_{kbr,max}}{2} \right) \right)^{\varphi_{kbr}}. \tag{7.109}$$

In it, φ_{kbr} signifies the degree of crank angle, which results from a superimposition of the scatter bands of φ_E and φ_K (see Fig. 7.16).

7.1.9 Internal energy

The internal energy or enthalpy of the gas in the cylinder must be calculated as a further term of the 1st law of thermodynamics. The gas composition is usually assumed to be homogeneous.

- **Component model**

One possible approach in calculating the internal energy is the description via a mixture of single components of the gas, which are each considered as an ideal gas. The internal energy of each single component of the gas (initial products like O_2, N_2, etc. or reaction products CO_2, H_2O) can be calculated separately, since their standard formation enthalpies, the reaction enthalpies, as well as the molar heat are in tabular reference works, e.g. NIST JANAF (1993). Knowing the respective fraction of these individual components, one can then calculate the total internal energy of the gas. One thereby considers the cylinder charge as an ideal mixture of ideal gasses, for which the thermal state equation is valid

$$p_i V = n_i \widetilde{R} T \ . \tag{7.110}$$

Thereby is

$\left. \begin{array}{ll} p_i : & \text{partial pressure} \\ n_i : & \text{quanity of substance} \end{array} \right\}$ of component i

and the caloric state equation

$$\widetilde{u}_i = \widetilde{u}_i(\widetilde{v}_i, T) \ ,$$

$$\widetilde{h}_i = \widetilde{h}_i(p_i, T) \ .$$

For the total internal energy of the cylinder charge, we thus obtain the expression

$$U = \sum_{i=1}^{k} n_i \widetilde{u}_i(\widetilde{v}_i, T) \ . \tag{7.111}$$

The only difficulty is that the fractions of the single components must be determined via the chemical reactions. It is thereby at first irrelevant which fuel one considers and whether the reaction progresses with an excess of air (lean) or a shortage of air (enriched).

For the sake of simplicity, we will first limit ourselves to stoichiometric operation. For 1 kmol of any fuel $C_x H_y O_z$ and under the assumption that the combustion air consists of 21 % oxygen and 79 % nitrogen, the following mass amounts result

$$\begin{aligned} C_x H_y O_z + \left(x + \frac{y}{4} - \frac{z}{2} \right) \left(O_2 + \frac{79}{21} N_2 \right) \\ \rightarrow x CO_2 + \frac{y}{2} H_2 O + \frac{79}{21} \left(x + \frac{y}{4} - \frac{z}{2} \right) N_2 \end{aligned} \tag{7.112}$$

7.1 Single-zone cylinder model

$$n_{CO_2} = x \left[\frac{\text{kmol } CO_2}{\text{kmol fuel}} \right], \qquad (7.113)$$

$$n_{H_2O} = \frac{y}{2} \left[\frac{\text{kmol } H_2O}{\text{kmol fuel}} \right] \quad \text{and} \qquad (7.114)$$

$$n_{N_2} = \frac{79}{21} \left(x + \frac{y}{4} - \frac{z}{2} \right) \left[\frac{\text{kmol } N_2}{\text{kmol fuel}} \right]. \qquad (7.115)$$

The system can, on this basis, be constructed to an arbitrarily complex extent and can take into consideration the amount of water in the air in the same way as a changed composition of air with respect to existing inert gasses. In the case of the fuel, for example, the amount of sulfur can also enter into the chemical reaction. In super-stoichiometric operation, pure air must also be considered in addition to the stoichiometric combustion gasses. Contingent on the combustion process under investigation, the evaporated fuel must be calculated as a further component (e.g. during the compression phase or under sub-stoichiometric operation). This is especially the case in the SI engine, since according to the definition of diesel combustion the fuel is usually only added to the system just as it is burning.

Although, compared with polynomial methods, table values from data banks are no longer required considering the performance of contemporary computers, they are still in wide use because of their clarity. In Heywood (1988) we find polynomial methods for the most important species. These refer to the JANAF tables.

Valid for molar enthalpy is

$$\tilde{h}_i = \tilde{R}T \left(a_{i,1} + \frac{a_{i,2}}{2}T + \frac{a_{i,3}}{3}T^2 + \frac{a_{i,4}}{4}T^3 + \frac{a_{i,5}}{5}T^4 + \frac{a_{i,6}}{T} \right) \left[\frac{\text{kJ}}{\text{kmol}} \right]. \qquad (7.116)$$

For the molar internal energy results

$$\tilde{u}_i = \tilde{h}_i - \tilde{R}T . \qquad (7.117)$$

The specific internal energy is obtained by means of division of the molar mass of the respective component

$$u_i = \frac{\tilde{u}_i}{M_i} . \qquad (7.118)$$

The parameters for varying substances are given in Tab 7.7.

There is also a description by means of polynomial methods for fuel vapor according to Heywood (1988). A consideration of the components, as in the case of gas components, does not exist in the case of fuels. For this, the evaporation behavior of the particular components is too complex. Nevertheless, most of the usual fuels are included. Valid is

Tab. 7.7: Coefficients for the molar enthalpy of gases, acc. to Heywood (1988)

Component	Temp.-range [K]	$a_{i,1}$	$a_{i,2}$	$a_{i,3}$	$a_{i,4}$	$a_{i,5}$	$a_{i,6}$	Molar mass
CO_2	1.000-5.000	0.44608(+1)	0.30982(-2)	-0.12393(-5)	0.22741(-9)	-0.15526(-13)	-0.48961(+5)	-0.98636(0)
	300-1.000	0.24008(+1)	0.87351(-2)	-0.66071(-5)	0.20022(-8)	0.63274(-15)	-0.48373(+5)	0.96951(+1)
H_2O	1.000-5.000	0.27168(+1)	0.29451(-2)	-0.80224(-6)	0.10227(-9)	-0.48472(-14)	-0.29906(+5)	0.66306(+1)
	300-1.000	0.40701(+1)	-0.11084(-2)	0.41521(-5)	-0.29637(-8)	0.80702(-12)	-0.30280(+5)	-0.32270(0)
CO	1.000-5.000	0.29841(+1)	0.14891(-2)	-0.57900(-6)	0.10365(-9)	-0.69354(-14)	-0.14245(+5)	0.63479(+1)
	300-1.000	0.37101(+1)	-0.16191(-2)	0.36924(-5)	-0.20320(-8)	0.23953(-12)	-0.14356(+5)	0.29555(+1)
H_2	1.000-5.000	0.31002(+1)	0.51119(-3)	0.52644(-7)	-0.34910(-10)	0.36945(-14)	-0.87738(+3)	-0.19629(+1)
	300-1.000	0.30574(+1)	0.26765(-2)	-0.58099(-5)	0.55210(-8)	-0.18123(-11)	-0.98890(+3)	-0.22997(+1)
O_2	1.000-5.000	0.36220(+1)	0.73618(-3)	-0.19652(-6)	0.36202(-10)	-0.28946(-14)	-0.12020(+4)	0.36151(+1)
	300-1.000	0.36256(+1)	-0.18782(-2)	0.70555(-5)	-0.67635(-8)	0.21556(-11)	-0.10475(+4)	0.43053(+1)
N_2	1.000-5.000	0.28963(+1)	0.15155(-2)	-0.57235(-6)	0.99807(-10)	-0.65224(-14)	-0.90568(+3)	0.61615(+1)
	300-1.000	0.36748(+1)	-0.12082(-2)	0.23240(-5)	-0.63218(-9)	-0.22577(-12)	-0.10612(+14)	0.23580(+1)
OH	1.000-5.000	0.29106(+1)	0.95932(-3)	-0.19442(-6)	0.13757(-10)	0.14225(-15)	0.39354(+4)	0.54423(+1)
NO	1.000-5.000	0.31890(+1)	0.13382(-2)	-0.52899(-6)	0.95919(-10)	-0.64848(-14)	0.98283(+4)	0.67458(+1)
O	1.000-5.000	0.25421(+1)	-0.27551(-4)	-0.31028(-8)	0.45511(-11)	-0.43681(-15)	0.29231(+5)	0.49203(+1)
H	1.000-5.000	0.25(+1)	0.0	0.0	0.0	0.0	0.25472(+5)	-0.46012(0)

7.1 Single-zone cylinder model

$$\tilde{h}_f = \left(A_{f,1} \vartheta + \frac{A_{f,2}}{2} \vartheta^2 + \frac{A_{f,3}}{3} \vartheta^3 + \frac{A_{f,4}}{4} \vartheta^4 \right.$$

$$\left. - \frac{A_{f,5}}{\vartheta} + A_{f,6} + A_{f,8} \right) 4186.6 \left[\frac{kJ}{kmol} \right].$$
(7.119)

The reference temperature for this amounts to 273.15 K. The molar internal energy and the specific internal energy also amount to

$$\tilde{u}_f = \tilde{h}_f - \tilde{R} T ,$$
(7.120)

$$u_f = \frac{\tilde{u}_f}{M_f} .$$
(7.121)

The parameters for the fuels investigated can be found in Tab. 7.8.

Tab. 7.8: Coefficients for the molar enthalpy of fuels, acc. to Heywood (1988)

Fuel	$A_{f,1}$	$A_{f,2}$	$A_{f,3}$	$A_{f,4}$	$A_{f,5}$	$A_{f,6}$	$A_{f,8}$	Molar mass
Methane	-0.29149	26.327	-10.610	1.5656	0.16573	-18.331	4.3000	16.04
Propane	-1.4867	74.339	-39.065	8.0543	0.01219	-27.313	8.852	44.10
Hexane	-20.777	210.48	-164.125	52.832	0.56635	-39.836	15.611	86.18
Isooctane	-0.55313	181.62	-97.787	20.402	-0.03095	-60.751	20.232	114.2
Methanol	-2.7059	44.168	-27.501	7.2193	0.20299	-48.288	5.3375	32.04
Ethanol	6.990	39.741	-11.926	0	0	-60.214	7.6135	46.07
Regular	-24.078	256.63	-201.68	64.750	0.5808	-27.561	17.792	114.8
Premium	-22.501	227.99	-177.26	56.048	0.4845	-17.578	15.235	106.4
Diesel	-9.1063	246.97	-143.74	32.329	0.0518	-50.128	23.514	148.6

We should make brief mention at this point of the properties of diesel and gasoline fuels. The data for the molar fraction of carbon x and of hydrogen y can be taken from Tab. 7.9. From the molar mass of the fuel with the corresponding molar fractions of C and H results a mass fraction c for carbon and h for hydrogen for both fuels. As one can easily see, the mass fractions for both fuels are very close to each other.

Tab. 7.9: Properties of diesel and gasoline fuels

	Diesel		Otto (premium)	
	kmol / kmol fuel	kg / kg fuel	kmol / kmol fuel	kg / kg fuel
C	$x = 10.8$	$c = 0.874$	$x = 7.76$	$c = 0.877$
H	$y = 18.7$	$h = 0.126$	$y = 13.1$	$h = 0.123$
O	$z = 0$	$o = 0$	$z = 0$	$o = 0$
Molar mass	148.3 kg fuel / kmol fuel		106.2 kg fuel / kmol fuel	
Minimal air required	14.33		14.26	
Lower heating value	42,600 kJ / kg fuel		42,900 kJ / kg fuel	

The calculated molar mass of the fuel amounts to

$$M_{fuel} = x\,M_C + \frac{y}{2} M_{H_2} + \frac{z}{2} M_{O_2} \quad \left[\frac{\text{kg fuel}}{\text{kmol fuel}}\right]. \tag{7.122}$$

For the oxygen balance results

$$n_{O_2,min} = c \left[\frac{\text{kg C}}{\text{kg fuel}}\right] \frac{1}{M_C} \left[\frac{\text{kmol C}}{\text{kg C}}\right] 1 \left[\frac{\text{kmol } O_2}{\text{kmol C}}\right]$$

$$+ h \left[\frac{\text{kg } H_2}{\text{kg fuel}}\right] \frac{1}{M_{H_2}} \left[\frac{\text{kmol } H_2}{\text{kg } H_2}\right] \frac{1}{2} \left[\frac{\text{kmol } O_2}{\text{kmol } H_2}\right] . \tag{7.123}$$

$$- o \left[\frac{\text{kg } O_2}{\text{kg fuel}}\right] \frac{1}{M_{O_2}} \left[\frac{\text{kmol } O_2}{\text{kg } O_2}\right] 1 \left[\frac{\text{kmol } O_2}{\text{kmol } O_2}\right].$$

With the fraction of 21 % oxygen in the combustion air and the molar mass for air of 28.85 kg/kmol, the minimal air requirement amounts to

$$L_{min} = \frac{n_{O_2,min}}{0.21} M_{air} . \tag{7.124}$$

Since in sub-stoichiometric operation incomplete reactions occur, for the determination of these components the so-called water-gas equilibrium is utilized. It describes, contingent on the temperature, the equilibrium constant of the most important reaction (CO_2, CO) under oxygen deficiency and dissociation

$$CO_2 + H_2 \xleftrightarrow{K_p} CO + H_2O \tag{7.125}$$

$$K_p = \frac{p_{CO}\, p_{H_2O}}{p_{CO_2}\, p_{H_2}} . \tag{7.126}$$

Values between 3.5 and 3.7 can be set for the equilibrium constant, which approximately corresponds to a usual combustion temperature of 1,800 K.

7.1 Single-zone cylinder model

- **The Justi method**

In the 1930's, investigations were carried out in order to describe the internal energy and enthalpy of combustion gasses in terms of the quantities of temperature, pressure, and gas composition. The gas composition is thereby expressed as the so-called air-fuel ratio. In describing diesel engine combustion gasses, this procedure is correct as long as one stays within the super-stoichiometric range ($\lambda > 1$). The composition of the combustion air is assumed to be constant; a varying water content cannot be taken into consideration. Fuel mixtures and alternative fuels cannot be represented with these methods. The relation is valid, strictly speaking, only for a fixed C-H-ratio. This is however very similar to diesel and gasoline fuel. The dissociation of the gas can also not be considered. Under dissociation we understand the change in otherwise constant equilibrium constants of chemical reactions at very high temperatures, e.g. over 2,000 K. Since a single-zone combustion chamber model is used as a rule only for calculating the cylinder pressure and the caloric mean temperature and with that for further thermodynamic quantities like performance, efficiency, and heat flows, it is usually totally sufficient to represent the internal energy as a function of temperature and the air-fuel ratio, for which Justi (1938) gives the following empirical function

$$u(T,\lambda) = 0.1445 \left[1{,}356.8 + \left(489.6 + \frac{46.4}{\lambda^{0.93}} \right)(T - T_{ref})10^{-2} + \right.$$
$$+ \left(7.768 + \frac{3.36}{\lambda^{0.8}} \right)(T - T_{ref})^2 10^{-4} - \tag{7.127}$$
$$\left. - \left(0.0975 + \frac{0.0485}{\lambda^{0.75}} \right)(T - T_{ref})^3 10^{-6} \right] \text{ in } \left[\frac{kJ}{kg} \right].$$

We are dealing with a polynomial method, whereby for the reference temperature $T_{ref} = 273.15$ K is valid.

- **The Zacharias method**

Zacharias (1966) also suggests a polynomial method, but also takes into consideration the pressure of the combustion gas. Otherwise the above restrictions of Justi's relation are valid for this as well

$$u(T,p,\lambda) = \left[-A \frac{\pi}{\vartheta^2} e^{\frac{D}{\vartheta^2}} \left(1 + 2\frac{D}{\vartheta} \right) + \sum_{i=0}^{6} \left[FA(i) \vartheta^i \right] - 1 \right] R_0 T \text{ in } \left[\frac{kJ}{kmol} \right] \tag{7.128}$$

$$r = \frac{\lambda - 1}{\lambda + \frac{1}{L_{min}}} \tag{7.129}$$

$$R_0 = \frac{\tilde{R}}{28.89758 + 0.06021 r} \tag{7.130}$$

$$\pi = \frac{p}{0.980665} \tag{7.131}$$

$$\vartheta = \frac{T}{1{,}000\ \mathrm{K}} \tag{7.132}$$

$$A = 0.000277105 - 0.0000900711\, r \tag{7.133}$$

$$D = 0.008868 - 0.006131\, r \tag{7.134}$$

$$\begin{aligned}
FA(0) &= 3.514956 & &- 0.005026\, r \\
FA(1) &= 0.131438 & &- 0.383504\, r \\
FA(2) &= 0.477182 & &- 0.185214\, r \\
FA(3) &= -0.287367 & &- 0.0694862\, r \\
FA(4) &= 0.0742561 & &+ 0.016404110\, r \\
FA(5) &= -0.00916344 & &- 0.00204537\, r \\
FA(6) &= 0.000439896 & &- 0.000101610\, r
\end{aligned} \tag{7.135}$$

Despite the supposed increase in precision via the consideration of the pressure in determining the internal energy, in using the equations of Zacharias, exactly because of this pressure contingency, an iterative calculation of the internal energy becomes necessary, which costs additional time. The contingency of internal energy on the temperature and gas composition for the Justi method is shown in Fig. 7.17. If we elect a very high air-fuel ratio, we obtain the graphs for pure air.

We should briefly mention here the various ways of representing the composition of the combustion gas. In this case, clear differences result between diesel engine and SI engine model representations. While in the case of diesel engine models, the injected fuel normally plays no role, as it is added proportionally to the heat release rate, in the case of SI model representations the fuel must be taken into consideration because of the evaporation heat. Only the states "fuel burned" as well as "air burned" and "air unburned" exist, since in the diesel engine we always assume lean operation.

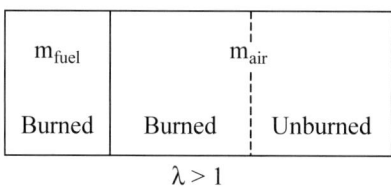

7.1 Single-zone cylinder model

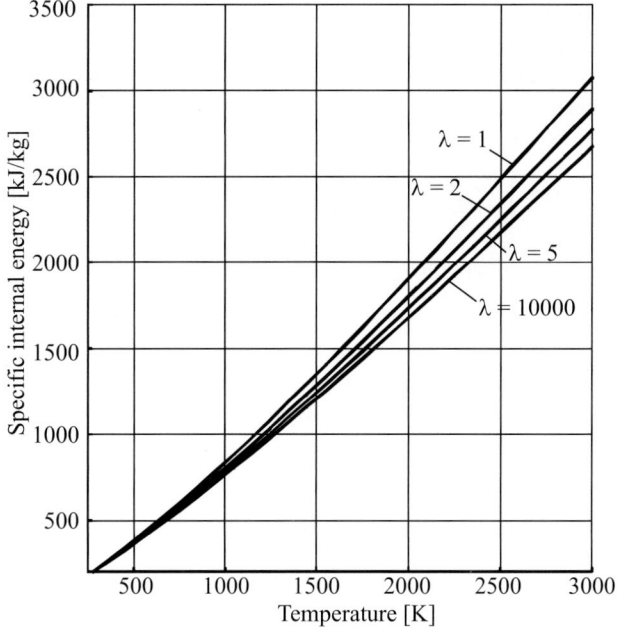

Fig. 7.17: Internal energy, acc. to Justi (1938)

If the internal energy is described according to the methods of Justi or Zacharias, the total differential for the internal energy and the partial differential for the specific internal energy must be formed. From this follows

$$\frac{dU}{dt} = \frac{d(mu)}{dt} = m\frac{du}{dt} + u\frac{dm}{dt} = m\left(\frac{\partial u}{\partial T}\frac{dT}{dt} + \frac{\partial u}{\partial p}\frac{dp}{dt} + \frac{\partial u}{\partial \lambda}\frac{d\lambda}{dt}\right) + u\frac{dm}{dt} \ . \quad (7.136)$$

The partial differentials can be calculated with the help of Justi's or Zacharias's relations. The air-fuel ratio is defined as

$$\lambda = \frac{m_{air}}{m_{fuel} L_{min}} \ . \quad (7.137)$$

The change in the air-fuel ratio amounts to

$$\frac{d\lambda}{dt} = \frac{1}{m_{fuel}^2 L_{min}}\left(m_{fuel}\frac{dm_{air}}{dt} - m_{air}\frac{dm_{fuel}}{dt}\right) \ . \quad (7.138)$$

The 1st law of thermodynamics is solved after the change in temperature and integrated. The term dp/dt can be calculated from the state equation for ideal gas in differential form

$$V\frac{dp}{dt} + p\frac{dV}{dt} = mR\frac{dT}{dt} + RT\frac{dm}{dt} + mT\frac{dR}{dt} \ . \quad (7.139)$$

Because of the fact that in the case of Zacharias the gas constant also depends on the quantities of temperature, pressure, and the air-fuel ratio, under complete differentiation of dR/dt, terms with dT/dt and dp/dt arise, which is why an iterative process or setting the differentials equal to zero becomes necessary.

ξ_{fuel}	ξ_{air}	$\xi_{exhaust\,gas}$
m_{fuel}	m_{air}	$m_{exhaust\,gas}$
Evaporated		Burned $\lambda = 1$

$$\xi_{air} + \xi_{exhaust\,gas} + \xi_{fuel} = 1$$

In the case of the SI engine, we subdivide in the same manner three ranges, as the sketch above shows. However, stoichiometrically burned exhaust gas, combustion air, and evaporated fuel exist here as components.

$$\xi_{exhaust\,gas} = \frac{m_{exhaust\,gas}}{m_{ent.}}; \quad \xi_{air} = \frac{m_{air}}{m_{total}}; \quad \xi_{fuel,evap.} = \frac{m_{fuel,evap.}}{m_{total}}, \qquad (7.140)$$

$$\xi_{exhaust\,gas} + \xi_{air} + \xi_{fuel,evap.} = 1. \qquad (7.141)$$

The total internal energy is to be calculated from the single portions of internal energy of the three areas

$$U = u_{exhaust\,gas}\,m_{exhaust\,gas} + u_{air}\,m_{air} + u_{fuel,evap.}\,m_{fuel,evap.} \cdot \qquad (7.142)$$

The differential of the internal energy becomes

$$\frac{dU}{dt} = \frac{d(u_{exhaust\,gas}\,m_{exhaust\,gas})}{dt} + \frac{d(u_{air}\,m_{air})}{dt} + \frac{d(u_{fuel,evap.}\,m_{fuel,evap.})}{dt}. \qquad (7.143)$$

Since we are concerned in the case of the single shares with "pure" components, the specific internal energies can be calculated either via classical polynomial methods or component for component. For the change in the mass of the single components results

$$\begin{aligned}
\frac{dm_{fuel,evap.}}{dt} &= \frac{dQ_{fuel}}{dt}\frac{1}{lhv} \\
\frac{dm_{fuel,evap.}}{dt} &= -\frac{dm_{fuel,evap.}}{dt} + \frac{dm_{fuel,evap.,new}}{dt} \\
\frac{dm_{air}}{dt} &= \frac{dm_{fuel,evap.}}{dt}L_{min} \\
\frac{dm_{exhaust\,gas}}{dt} &= \frac{dm_{fuel,evap.}}{dt}(L_{min}+1) \cdot
\end{aligned} \qquad (7.144)$$

For the port fuel injection SI engine, the evaporated fuel is already found in the cylinder. In the case of the direct injection SI engine, the fuel is injected during either charge changing or compression and has to evaporate.

7.2 The two-zone cylinder model

7.2.1 Modeling the high pressure range according to Hohlbaum

In the following, a two-zone model from the zero-dimensional model class will be considered in more detail. This model has gained a certain importance for the calculation of NO_x formation, whereby the heat release rate is given beforehand.

In this model, the combustion chamber is divided into two zones, which one should imagine as being divided by the flame front. Strictly speaking, the flame front itself represents a zone onto itself, i.e. the third zone. However, because simple assumptions are made about the reaction kinetics of the flame front and no balance equations are solved, this is usually not considered to be an independent zone, and the designation "two-zone model" has become customary.

An extensive description of this model can be found in Hohlbaum (1992) and in Merker et al. (1993). In the following, only the essential traits of the model will be explained. The basic idea of the model is schematically shown in Fig. 7.18.

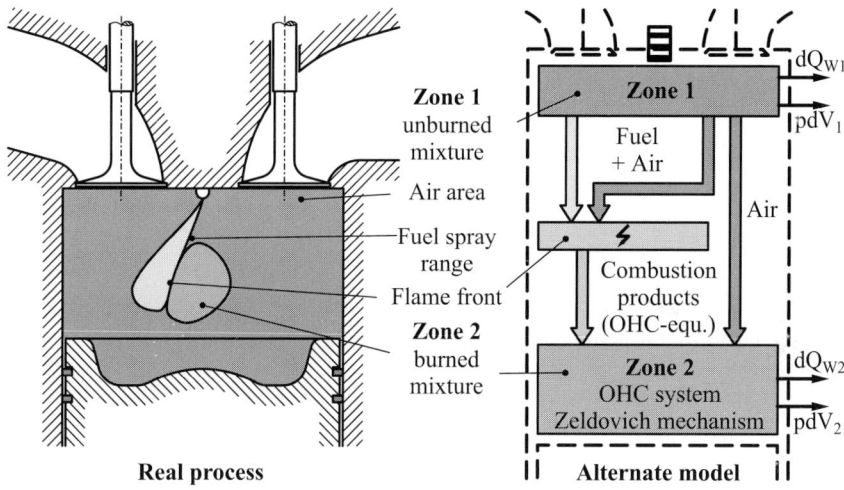

Fig. 7.18: Two-zone model for diesel combustion

Zone 1 should contain unburned mixture, i.e. air and fuel, which will in the following be designated as unburned zone and described with the state quantities $p, V_1, T_1, n_{1,i}$ and λ_1

Zone 2 should contain burned mixture, or more precisely speaking incompletely oxidized fuel, and will in the following be designated as burned zone and described with the state quantities $p, V_2, T_2, n_{2,i}$ and λ_2. In zone 2, "secondary oxidation" occurs. For this, reaction-kinetic models are necessary. Thermal NO_x formation also takes place in zone 2, which is described by the Zeldovich mechanism, see chap. 6.5.

The flame front separates both of these zones. It is assumed to be infinitely thin and without mass. "Primary oxidation" takes place in the flame front until OHC equilibrium, i.e. the OHC components $O^\bullet$, $H^\bullet$, O_2, H_2, H_2O, CO, CO_2 and $O^\bullet H$ are in chemical equilibrium in the flame front.

By means of a balancing of the relevant mass and energy flows, we obtain equations for the mass and energy in both zones. Valid thereby for the mass balance in zone 1 is

$$dm_1 = dm_{fuel} - dm_{1f} - dm_{12} \tag{7.145}$$

with

dm_{fuel} : injected fuel mass ,

dm_{1f} : mass added to the flame front (fuel + air) and

dm_{12} : air mass passing by the flame front and added "directly" to zone 2 (must appear because of $\lambda_1 > 1$); it can also go through the flame front, however without participating in the reactions occurring there and without heat absorption.

The energy balance can be written thus

$$dU_1 = dm_{fuel}\, h_{fuel} - dm_{1f}\, h_{1f} - dm_{12}\, h_{12} + dQ_1 - p\, dV_1 \;. \tag{7.146}$$

Analogously, for the mass balance of zone 2 is valid

$$dm_2 = dm_{f2} + dm_{12} \tag{7.147}$$

with dm_{F2} : components of the OHC equilibrium.

The energy law reads

$$dU_2 = dm_{f2}\, h_{f2} + dm_{12}\, h_{12} + dQ - p\, dV_2 \;. \tag{7.148}$$

Because the flame front is assumed to be without mass, it is furthermore valid

$$dm_{1f} = dm_{f2} = dm_f \;. \tag{7.149}$$

The specific enthalpy of the mass transported from the flame front to zone 2 dm_{F2} is reaction-enthalpy $\Delta_R h$ larger than that which is transported from zone 1 into the flame front, i.e.

$$h_{f2} = h_{1f} + \Delta_R h \;. \tag{7.150}$$

The terms dQ_1 and dQ_2 describe the energy losses of both zones via heat transfer as a result of radiation and convection to the wall limiting the combustion chamber. The total transferred heat

$$dQ = dQ_1 + dQ_2 = \alpha\, A\, (T_w - T)\, dt \tag{7.151}$$

can, for example, be calculated again according to Woschni's method, whereby T is the energetic mean temperature, which can be determined for the caloric mixture from the relation

$$(m_1 + m_2)\, u(T) = m_1 u_1(T_1) + m_2 u_2(T_2) \;. \tag{7.152}$$

7.2 The two-zone cylinder model

However, for the subdivision of the total transferred heat dQ into dQ_1 and dQ_2, we require a model, because the surface of the flame front and thus the size of the surface of both zones is not defined in the two-zone model. Hohlbaum (1992) proposes for this distribution the following relation

$$\frac{dQ_1}{dQ_2} = \left(\frac{m_1}{m_2}\right)^2 \frac{T_1}{T_2} \quad . \tag{7.153}$$

On the one hand, this approach takes into consideration that zone 2 of the burned zone contributes more to the total heat loss because of the higher temperature T_2 than unburned zone 1. On the other hand, the method considers the fact that at the beginning of combustion, the mass of zone 2 and thus its contribution to heat transfer is minimal. Finally, the temporal progression of the bypass air mass flow $\dot{m}_{12}$ must still be determined. The quantity designated as mixture stoichiometry λ^* is defined as

$$\lambda^* = \frac{dm_{1f,L} + dm_{12}}{L_{min} \, dm_{fuel,1f}} \quad . \tag{7.154}$$

In Fig. 7.19, the paths of the air-fuel ratios in the flame front and in zone 2, λ_f and λ_2, as well as the mixture stoichiometry λ^* are sketched over crank angle.

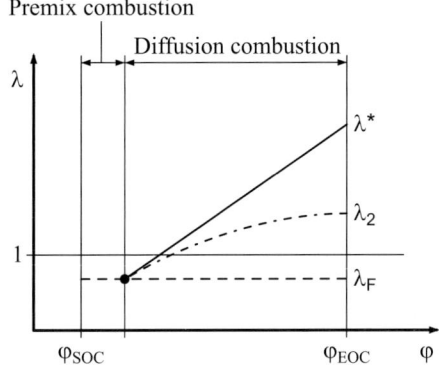

Fig. 7.19: Air ratio for premixed and diffusion combustion

Air ratio λ_F in the flame front is assumed to be <1 and temporally constant, thus $\lambda_F \neq f(\varphi)$, the mixture stoichiometry on the other hand is increasing in linear proportion to the crank angle, such that at the start of diffusion combustion $\lambda^* = \lambda_F$ and the excess air has at the end of diffusion combustion just completely intermixed with the unburned in zone 2.

Especially this assumption for the mixture stoichiometry makes clear the problematic nature of this simple zero-dimensional model; the lack of model depth (lacking physics) has to be substituted with more or less arbitrary assumptions.

7.2.2 Modeling the high pressure phase according to Heider

Heider (1996) has introduced another way to calculate the conditions in both zones. We distinguish thereby between two zones. In zone 1, which is designated as the reaction zone, energy conversion takes place. In the reaction zone, the air-fuel ratio λ_0 is assumed to be constant throughout the working cycle. The mass in the reaction zone is thus clearly fixed over the heat release rate. Zone 2 describes the unburned and thus sets the remaining volume of the combustion chamber. No combustion occurs.

As opposed to Hohlbaum's model, in the case of Heider, the results are based on the process calculation of the zero-dimensional single-zone model (see chap. 7.1). The paths of the swept volume, of pressure, and of mass mean temperature as well as the cylinder mass, compression air-fuel ratio, heat release rate, and wall heat losses are assumed to be known. The following assumptions are valid as conditions of compatibility

$$V_1 + V_2 = V(\varphi), \tag{7.155}$$

$$m_1 + m_2 = m(\varphi), \tag{7.156}$$

$$p_1 = p_2 = p(\varphi). \tag{7.157}$$

Provided we know the air-fuel ratio λ_0, the mass in the reaction zone can be calculated

$$\lambda_0 = \frac{m_{air1}(\varphi)}{L_{min}\, m_{fuel}(\varphi)} = \text{const.} \,. \tag{7.158}$$

For the fuel mass is valid in the case of a known heat release rate and a known residual gas mass, from which the burned fuel $m_{fuel,0}$ can also be calculated

$$m_{fuel}(\varphi) = \frac{1}{lhv} \int \frac{dQ_{fuel}}{d\varphi} d\varphi + m_{fuel,0}\,. \tag{7.159}$$

With this, for the mass of zone 1 we have

$$m_1(\varphi) = m_{air1}(\varphi) + m_{fuel}(\varphi) = (\lambda_0 L_{min} + 1)\, m_{fuel}(\varphi)\,. \tag{7.160}$$

The ideal state equation is valid for both zones

$$\begin{aligned} p_1 V_1 &= m_1 R_1 T_1 \\ p_2 V_2 &= m_2 R_2 T_2 \end{aligned}. \tag{7.161}$$

In the final analysis it must be determined which part of the energy released in the reaction zone is transferred to zone 2. This happens in the model representation essentially via turbulent mixing and less so via radiation and convection. For this, the following boundary conditions must be kept.

At the beginning of combustion, the temperature difference between both zones as a result of the high temperature difference between the flame and the unburned mass is maximal. Furthermore, this temperature difference is contingent on heat release via combustion. The turbulent mixing of both zones leads, with progressing combustion, to a lowering of temperature in the reaction zone and an increase of temperature in the zone with the unburned substance. At

7.2 The two-zone cylinder model

the end of the combustion, the temperature difference is around zero, since both zones are then completely intermixed. These considerations lead to the following empirical method for the temperature difference between both of these zones

$$T_1(\varphi) - T_2(\varphi) = B(\varphi) A^* . \tag{7.162}$$

For the function $B(\varphi)$ is valid

$$B(\varphi) = 1 - \frac{\int\limits_{\varphi_{SOC}}^{\varphi}[p(\varphi) - p_0(\varphi)]m_1\, d\varphi}{\int\limits_{\varphi_{SOC}}^{\varphi_{EVO}}[p(\varphi) - p_0(\varphi)]m_1\, d\varphi} . \tag{7.163}$$

As in the determination of the heat transfer coefficient according to Woschni (1970), here too, the difference between the cylinder pressure $p(\varphi)$ and the theoretical pressure of the motored engine $p_0(\varphi)$ is utilized for the consideration of the influence of combustion. A^* describes the temperature level in the reaction zone as the start of combustion. Detailed investigations have shown that minimal adjustments of the A^* value and the air-fuel ratio λ_0 are necessary for varying engines and combustion processes. For small to medium-sized diesel engines possessing an intake swirl

$$\lambda_0 = 1.0 \text{ and}$$

$$A^* = A \frac{1.2 + (\lambda_{gl} - 1.2)^{C_{gl}}}{2.2\,\lambda_0} \tag{7.164}$$

is applicable. A is an engine-specific factor, which has to be determined once for the respective engine. For C_{gl} is valid

$C_{gl} = 0.15$ for engines with 4-valve technology and central injection nozzle,

$C_{gl} = 0.07$ for engines with 2-valve technology and a side injection nozzle,

λ_{gl} describes the global air-fuel ratio.

Valid for large diesel engines without intake swirl is

$$\lambda_0 = 1.03$$

and

$$\lambda_0 = 1{,}03 - 0{,}24\frac{EGR}{100}$$

in the case of external exhaust gas recycling. In large diesel engines, the A^* value can be assumed to be constant

$$A^* = A = \text{const.}$$

Although this model was first developed only for the diesel engine, it can also be applied to SI engines with favorable results. Then

$$\lambda_0 = \lambda_{gl} \text{ and } A^* = \text{const.}$$

is valid.

Fig. 7.20 shows a characteristic temperature curve for a high speed diesel engine with approx. 4 l cylinder volume at a speed of 1,400 rpm and an actual load of 8 bar, as can be calculated with this model. In Tab. 7.10 below, typical A values for various engines are summarized.

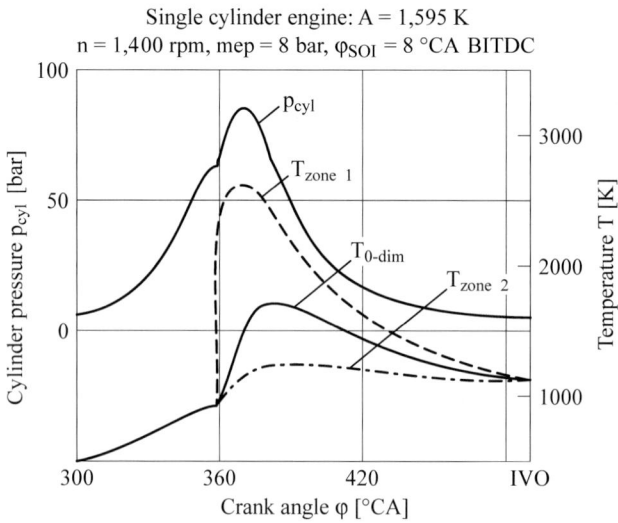

Fig. 7.20: Temperature curve acc. to Heider (1996)

Tab. 7.10: Typical A values

engine bore [mm]	stroke [mm]	cycle	compression ratio	rated speed [rpm]	A-value [K]
79.5	95.5	4	19,5	4,000	1,650
128	142	4	16	2,100	1,740
160	180	4	14	1,500	1,580
480	600	4	14	450	1,650
580	1,700	2	17	127	1,655

Despite the obvious empirical nature of this model, it offers a very good basis for nitrogen oxide calculation described in the following. Moreover, it is convincing in its simplicity. With

7.2 The two-zone cylinder model

this model, no assumptions must be made regarding the distribution of the wall heat losses to the two zones, which can only be determined as a whole. The calculation model is also appealing because of its short calculation times.

7.2.3 Results of NOx calculation with two-zone models

With the two-zone models of Hohlbaum and Heider described in the preceding sections, nitrogen oxide emissions in the zone of the hot combustion products can be calculated. For this, we utilize the description via the so-called Zeldovich mechanism, which is thoroughly described in chapter 6.5.1.

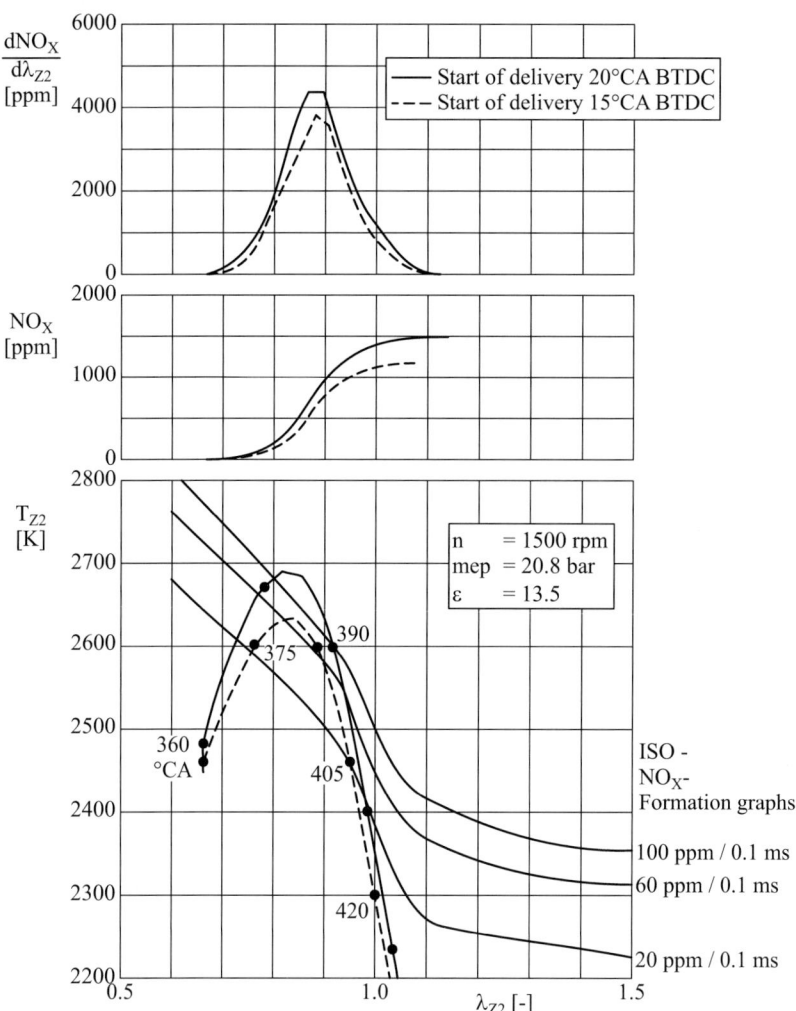

Fig. 7.21: Start of delivery variation, acc. to Hohlbaum (1992)

As an example, Fig. 7.21 shows the effect of the start of delivery timing from "early" to "late" on the rate of formation of NO calculated with the Hohlbaum model. At a start of delivery delay of 5 °CA from 20 °CA BTDC to 15 °CA BTDC the peak temperature sinks from about 2,680 K to 2,630 K, and the temperature as a whole reaches clearly lower values. This finally leads to a decrease in the amount of NO formed, which lowers from about 1,500 ppm to 1,200 ppm.

For the operating point described in Fig. 7.20, the temperature curve in the hot zone according to Heider (1996) is represented in Fig. 7.22 on the left. Via this temperature curve, a NO formation rate and the NO concentration in the combustion chamber is adjusted. We recognize that NO formation is over very quickly and that only a minimal reverse reaction occurs. Heider utilizes reaction constants of Pattas, organized in Tab. 6.1 (chap. 6.5.1), for the Zeldovich mechanism. For this engine, at a speed of 1,500 rpm, a comparison between measurement and calculation is given for injection time variation (Fig. 7.22, right).

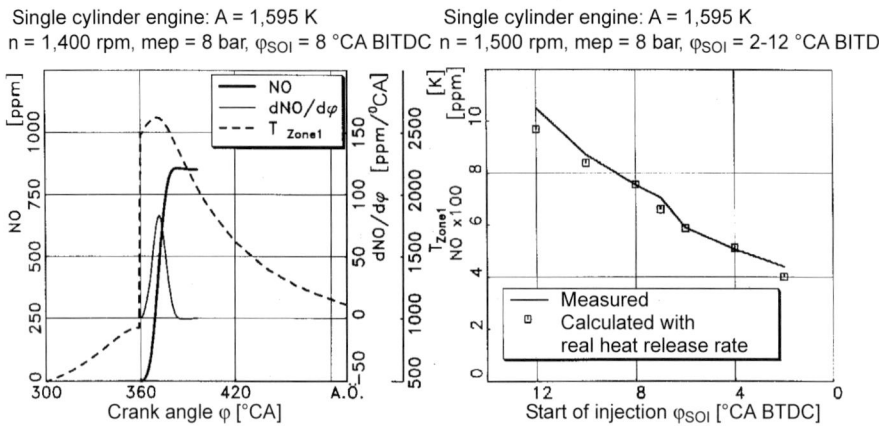

Fig. 7.22: Influence of start of injection on the NO formation rate

The coincidence is very good in this case, as it is in the case of variation of the charge air temperature (Fig. 7.23, left) and the exhaust gas recirculation rate (Fig. 7.23, right), which has a massive influence on NO formation.

We recognize that these very simple models are very much capable of describing reality not only according to tendency, but also quantitatively correctly. It is however of decisive importance that the heat release rate in the cylinder is exactly described. Results of transient calculations of nitrogen oxide emissions in a high speed diesel engine of a passenger car with pre-calculated heat release rates can be found in chap. 8.7.5.

7.2 The two-zone cylinder model

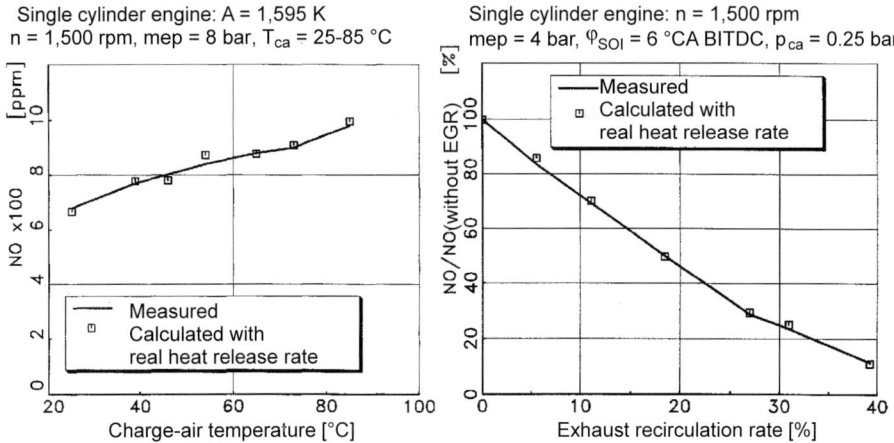

Fig. 7.23: Influence of the charge air temperature (left) and the exhaust gas recirculation rate (right) on the NO formation rate

7.2.4 Modeling the charge changing for a 2-stroke engine

Charge changing calculation for the 2-stroke engine is much more difficult than for the 4-stroke engine, since, on the one hand, only a small amount of time is available for charge changing and, on the other hand, the fresh gas flowing into the cylinder has to suppress the exhaust gas found in the cylinder without mixing with it. In literature a large number of models for loop scavenged as well as for longitudinally scavenged 2-stroke engines is described. We are hereby dealing with two or three-zone models. Differences arise thereby between zones, in which fresh gas, a mixture between exhaust gas and fresh gas, or pure exhaust gas is found. A more exact list and description of these models can be found in Merker and Gerstle (1997).

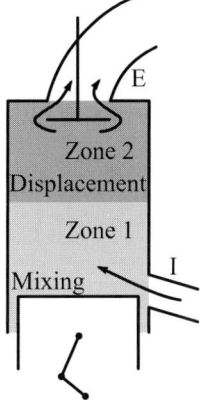

Fig. 7.24: Two-zone model for calculation of charge changing in a 2-stroke engine

The model described here proceeds from a two-zone approach. Since, practically speaking, two components – fresh gas and exhaust gas – are found in the cylinder during charge changing, this fact must be accounted for with two zones. Here, we will only go into the essential properties and descriptive equations of both zones and their interaction. In modeling, displacement scavenging with underlying intake ports and an upper exhaust valve usual today is used as a basis. Fig. 7.24 shows a diagram for the two-zone model for the charge changing calculation for a 2-stroke engine, in which the gas mass in the cylinder is subdivided by an infinitely thin, impermeable yet shiftable horizontal membrane. An exchange of gas between the zones is thus impossible, as opposed to the model of Streit and Bormann (1971).

At the beginning of charge changing at exhaust opens – i.e. after combustion and expansion – a homogeneous mixture is in the cylinder. The exhaust gas mass found in the cylinder is subdivided at exhaust opens into a displacement and a mixing zone via a so-called scavenging factor, which is characteristic for the charge changing properties of the engine. This scavenging factor can have values of 0 to 1. A scavenging factor of 0 sets a pure mixture scavenging, the success of which is, however, difficult to grade as a result of the constant ideal mixing of fresh gas and exhaust gas. In this case, the mixing zone has no mass at the start of the calculation, which is why, after the beginning of the intake process, only pure fresh gas is mixed. An ideal mixture is always assumed for this zone, as in the case of the entire charge changing of the 4-stroke engine. The displacement zone in a charge changing without mixing with the fresh gas flowing in through the intake is the first zone to be pushed out through the exhaust valve. The gas composition thus corresponds at every moment to the composition of the exhaust gas. According to the adjacent scavenging pressure ratio, the displacement zone can be only partially or completely expelled. In addition, part of the mixing zone can also be ejected. Towards the end of charge changing, both zones, insofar as they still exist and are not completely expelled, must be returned to a homogeneously mixed state. The descriptive equations of the two-zone model for the charge changing of a 2-stroke engine is represented in the following, whereby the index 1 is used for the mixing zone and 2 for the displacement zone. If the mass of the displacement zone is completely expelled and thus only the mixing zone still exists, this is treated like a 4-stroke engine. A subdivision of mass or heat flows to the single zones is then no longer necessary. The subdivision of cylinder mass at "exhaust opens" describes the following equation

$$m_1 = m(1 - SF) \quad m_2 = m\,SF \ . \tag{7.165}$$

For both zones, several basic couple conditions are applicable, which must be kept at every moment

$$m_1 + m_2 = m \ , \tag{7.166}$$

$$V_1 + V_2 = V \quad \text{and} \tag{7.167}$$

$$p_1 = p_2 = p = \frac{m_1 R_1 T_1 + m_2 R_2 T_2}{V} \ . \tag{7.168}$$

The mass balance for both zones reads

$$\frac{dm_1}{d\varphi} = \frac{dm_I}{d\varphi} \ , \tag{7.169}$$

$$\frac{dm_2}{d\varphi} = \frac{dm_O}{d\varphi} \; . \tag{7.170}$$

According to the 1st law of thermodynamics, for both zones results

$$\frac{dU_1}{d\varphi} = \frac{dW_1}{d\varphi} + \frac{dQ_{W,1}}{d\varphi} + \frac{dH_I}{d\varphi} \quad \text{and} \tag{7.171}$$

$$\frac{dU_2}{d\varphi} = \frac{dW_2}{d\varphi} + \frac{dQ_{W,2}}{d\varphi} + \frac{dH_O}{d\varphi} \; . \tag{7.172}$$

The enthalpy flow through that valves is formed – as in the case of the 4-stroke engine – according to the respective flow direction from the product of the mass flow through the valve and the specific enthalpy of the gas found in the flow direction before the valve.

Via a horizontal splitting of the system, the wall heat flow to the piston can be ascribed to the mixing zone and the wall heat flow to the cylinder lid to the displacement zone. Wall heat flow is distributed to both zones in accordance with the position of the imaginary membrane – i.e. proportional to volume. With that, all equations describing the two-zone model are at our disposal. At the closing of the last control unit, both zones can be ideally mixed again for the following high pressure part, so long as one has at an earlier degree of crank angle not already switched to a single-zone inspection after the complete expelling of a zone. For the mixing temperature in the cylinder is then valid

$$T = \frac{c_{v,1}\, m_1\, T_1 + c_{v,2}\, m_2\, T_2}{c_{v,1}\, m_1 + c_{v,2}\, m_2} \; . \tag{7.173}$$

7.3 Modeling the gas path

In order to build a complete engine model, we need besides the cylinder still other components for a description with the filling and emptying method, like the volume, orifice plates, or throttles as well as flow machines for charged engines. An exact description of these components is to be found in the following sections.

7.3.1 Modeling peripheral components

- **Volume**

A volume (intake manifold, etc.) is usually modeled as a cylindrical solid. Such a model is illustrated in the following sketch.

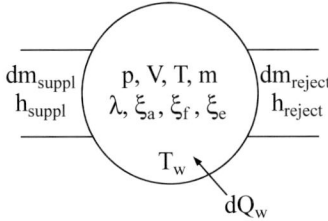

The cross-sectional surface and the volume and the volume itself thus amount to

$$A_{vol} = d_{vol}^2 \frac{\pi}{4} \quad \text{and} \quad V_{vol} = l_{vol}\, d_{vol}^2 \frac{\pi}{4}. \tag{7.174}$$

From a thermodynamic perspective, we are thereby dealing with an open, steadily flowed-through system. The resulting mass balance is

$$\frac{dm_{vol}}{dt} = \dot{m}_{suppl.} + \dot{m}_{removed}. \tag{7.175}$$

Furthermore, valid for the energy balance of the volume is

$$\frac{dU}{dt} = \frac{dQ_W}{dt} + h_{suppl.} \frac{dm_{suppl.}}{dt} + h_{remuved} \frac{dm_{removed}}{dt}. \tag{7.176}$$

The internal energy can be calculated by means for the relations introduced already in describing the cylinder (see chap. 7.1.9). Heat transfer is calculated with the Newtonian equation (see (7.10)), whereby we can utilize as heat transfer coefficient the Hausen relation (1976) assuming a turbulent pipe flow

$$\alpha = 0.024 \frac{\lambda_{vol}}{d_{vol}} \left[1 + \left(\frac{d_{vol}}{l_{vol}} \right)^{\frac{2}{3}} \right] Re^{0.786} Pr^{0.45}. \tag{7.177}$$

The Prandtl number is set at 0.731. For the Reynolds number and the viscosity we have

$$Re = \frac{\dot{m}_{mean}\, d_{vol}}{A_{vol}\, \eta}; \quad \dot{m}_{mean} = \frac{|\dot{m}_{suppl.}| + |\dot{m}_{removed}|}{2}; \quad \eta = 5.17791 \cdot 10^{-7}\, T^{0.62}. \tag{7.178}$$

The heat conductivity according to Woschni amounts to

$$\lambda_{vol} = 3.65182 \cdot 10^{-4}\, T^{0.748}. \tag{7.179}$$

It is thus possible to calculate a pipe volume which is switched between two succeeding throttle locations.

- **Orifice plate (throttle valves)**

For the simulation of switching flaps, throttle valves, engine bypass flaps, EGR valves, or wastegates in charges engines, we need trigger wheel vanes with constant or variable cross sectional surfaces. The modeling of these components is identical with that of a throttle location at the valves of the cylinder head. For this as well, the so-called flow equation is used.

7.3 Modeling the gas path

For the mass flow through a trigger wheel vane is valid (see (2.37))

$$\dot{m} = \alpha\, A_1 \sqrt{p_0\, \rho_0} \sqrt{\frac{2\kappa}{\kappa-1}\left(\pi^{\frac{2}{\kappa}} - \pi^{\frac{\kappa+1}{\kappa}}\right)}. \tag{7.180}$$

It must also be considered here that at the realization of the critical pressure ratio, the mass flow, as represented in (2.40), is limited. The flow coefficients are shown, contingent on the degree of the opening of the trigger wheel vane, in the form of lines on a characteristic map.

- **Flow machines**

In representing flow machines (compressor, turbine) in charged engines, a consideration with the help of characteristic maps is also possible and sufficient for unsteady processes. Because of the complexity of the presentation of these aggregates, we will take a special look at this in chap. 7.5.

7.3.2 Model building

Fig. 7.25 shows a simple example of modeling the gas path of a combustion engine. The model must in principle be built such that a "throttle building block" follows a "storage building block" and then a storage building block again etc. In the storage building block (e.g. the volume), the differential equations for the mass and energy balance are solved. From this results the mass determined for the actual integration step as well as the temperature and, via the general gas equation, the pressure in the storage building block. The mass and enthalpy flows coming in and out of the system are required as initial quantities for calculating the mass and energy balance. These can be calculated in the throttle building blocks when given the temperature and pressures determined in the previous integration step in the storage building blocks lying before and after the throttle location. In the case of a trigger wheel vane, the calculation of the mass flow takes place, for example, by means of the flow equation. The enthalpy flow is determined upon acquaintance with the gas states (esp. enthalpy) of the storage building block lying in the current flow direction before the throttle location. Fig. 7.26 shows this constantly repeating process.

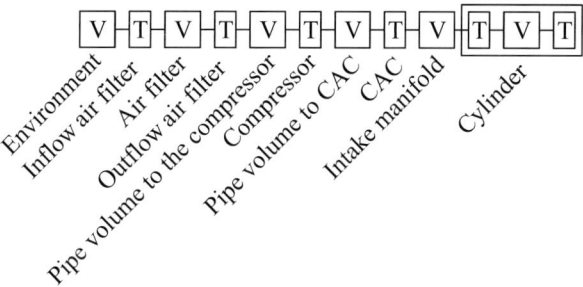

Fig. 7.25: Simple model of the gas path acc. to the filling and emptying method

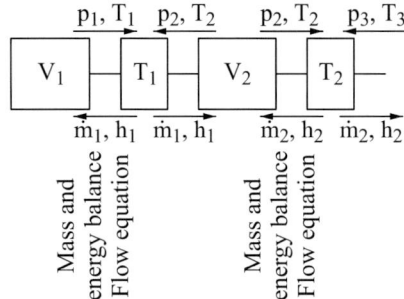

Fig. 7.26: Calculation scheme for the filling and emptying method

The cylinder has a special position, in which the calculation of the mass and enthalpy flows is usually integrated through the valves. Insofar as this is concerned, the cylinder represents a throttle location for the interconnection in the model, although – as shown in chap. 7.1 – the mass and energy balance is solved and the combustion chamber volume is of course a mass and energy storage space itself.

One additional special position in modeling is occupied by flow machines. While we can designate the flow turbine as a throttle location with energy emission, the flow compressor usually causes an increase in pressure. By means of the use of characteristic maps for describing operating behavior, which is itself dependent on the pressures and temperatures before and after the compressor (see chap. 7.5), this does not however play a role for the model described above.

7.3.3 Integration methods

The so-called Runge-Kutta method of the 4th order has proved itself to be a completely adequate method for solving the differential equations of mass and energy in the modules described in this section.

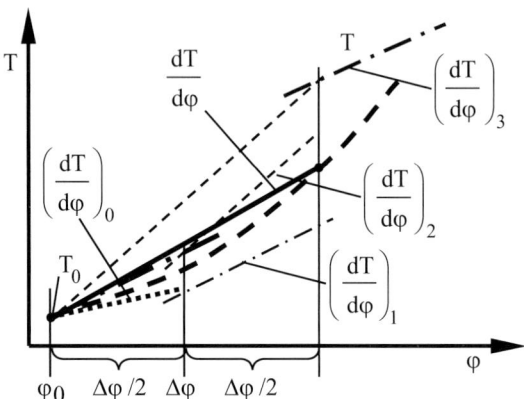

Fig. 7.27: Integration according to the 4th order Runge-Kutta process

The actual integration time is halved and four gradients are determined – as shown in Fig. 7.27 – which are finally differently weighed. These relations are shown in (7.181).

$$\left(\frac{dT}{d\varphi}\right)_0 = f(T_0, \varphi_0)$$

$$T_1 = T_0 + \left(\frac{dT}{d\varphi}\right)_0 \frac{\Delta\varphi}{2}$$

$$\left(\frac{dT}{d\varphi}\right)_1 = f\left(T_1, \varphi_0 + \frac{\Delta\varphi}{2}\right)$$

$$T_2 = T_0 + \left(\frac{dT}{d\varphi}\right)_1 \frac{\Delta\varphi}{2}$$

$$\left(\frac{dT}{d\varphi}\right)_2 = f\left(T_2, \varphi_0 + \frac{\Delta\varphi}{2}\right)$$

$$T_2 = T_0 + \left(\frac{dT}{d\varphi}\right)_2 \Delta\varphi$$

$$\left(\frac{dT}{d\varphi}\right)_3 = f(T_3, \varphi_0 + \Delta\varphi)$$

$$\frac{dT}{d\varphi} = \frac{1}{6}\left(\left(\frac{dT}{d\varphi}\right)_0 + 2\left(\frac{dT}{d\varphi}\right)_1 + 2\left(\frac{dT}{d\varphi}\right)_2 + \left(\frac{dT}{d\varphi}\right)_3\right)$$

$$T = T_0 + \frac{dT}{d\varphi}\Delta\varphi\;.$$

(7.181)

The Runge-Kutta method thus distinguishes itself from simple methods like the Euler-Cauchy method, in which only one gradient is formed and the integral thus falls short of reality. We find in the literature a number of other integration methods, but these are usually more complex and not as easy to manage as the Runge-Kutta method.

7.4 Gas dynamics

7.4.1 Basic equations of one-dimensional gas dynamics

In principle, the basic equations that describe a one-dimensional flow can be derived from the Navier-Stokes equations introduced in chap. 9.1 restricting to one dimension and neglecting gravity. Here we will however provide a simple and illustrative "derivative". For this, we will take into consideration the port section sketched in Fig. 7.28 with variable cross section along the x-coordinate.

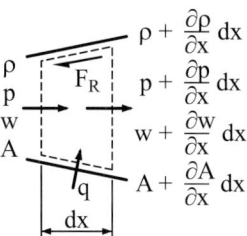

Fig. 7.28: Port length with variable cross section

We assume that the cross sectional alteration of the port along length dx is small, such that only contingencies on the x-coordinate and on time must be considered.

- **Mass balance**

The port section with volume dV and a medium with density $\rho = \rho_x$ contains the mass

$$dm = \rho_x \, dV = \rho_x \, A_x \, dx \; . \tag{7.182}$$

Through the cross sections A_x and A_{x+dx} the medium flows out with a speed of w_x and a speed of w_{x+dx}. Thus is valid for the mass flows

$$\dot{m}_x = w_x \, \rho_x \, A_x \; . \tag{7.183}$$

$$\dot{m}_{x+dx} = w_{x+dx} \, \rho_{x+dx} \, A_{x+dx} \; . \tag{7.184}$$

The mass balance for the port section of length dx then reads

$$\frac{\partial m}{\partial t} = \dot{m}_x - \dot{m}_{x+dx} \; , \tag{7.185}$$

$$\frac{\partial}{\partial t}(\rho_x \, A_x \, dx) = w_x \, \rho_x \, A_x - w_{x+dx} \, \rho_{x+dx} \, A_{x+dx} \; . \tag{7.186}$$

With the Taylor row development of the state quantities at location $x + dx$, we obtain

$$\frac{\partial \rho_x}{\partial t} dx = w_x \, \rho_x - \left(w_x + \frac{\partial w_x}{\partial x} dx \right)\left(\rho_x + \frac{\partial \rho_x}{\partial x} dx \right)\left(A_x + \frac{\partial A_x}{\partial x} dx \right)\frac{1}{A_x} \; . \tag{7.187}$$

Via multiplication and neglecting the terms of higher order, from this follows the continuity equation

$$\frac{\partial \rho}{\partial t} = -\frac{\partial (\rho w)}{\partial x} - \rho w \frac{d \ln (A)}{dx}$$

or

$$\frac{\partial \rho}{\partial t} + w \frac{\partial \rho}{\partial x} + \rho \frac{\partial w}{\partial x} + \rho w \frac{d \ln (A)}{dx} = 0 \; . \tag{7.188}$$

The index x can be left out thereby to keep things clear.

7.4 Gas dynamics

- **The law of conservation of impulse**

Change of the impulse I within the port section under consideration over time is equal to the sum of the impulse flows caused by the mass flow at its cross sections and the external forces that work upon the mass. The impulse is defined as

$$I = m w_x = \rho_x A_x \, dx \, w_x \, . \tag{7.189}$$

For the impulse flows is valid

$$\dot{I}_x = w_x \, w_x \, \rho_x \, A_x \quad \text{or} \tag{7.190}$$

$$\dot{I}_{x+dx} = w_{x+dx} \, w_{x+dx} \, \rho_{x+dx} \, A_{x+dx} \, . \tag{7.191}$$

The external forces are composed of the pressure forces F_x, which result from the various cross sectional surfaces, and the friction forces F_f of the fluid on the internal port wall,

$$F_x = p_x A_x \, , \tag{7.192}$$

$$F_{x+dx} = p_{x+dx} A_{x+dx} \, , \tag{7.193}$$

$$F_f = k_f \, \rho_x \, A_x \, dx \, . \tag{7.194}$$

With this, it follows for the impulse balance

$$\frac{\partial}{\partial t}(w_x \, \rho_x \, A_x \, dx) = w_x^2 \, \rho_x \, A_x - w_{x+dx}^2 \, \rho_{x+dx} \, A_{x+dx} + p_x \, A_x \\ - p_{x+dx} \, A_{x+dx} - k_f \, \rho_x \, A_x \, dx \tag{7.195}$$

With the Taylor row development analogous to the mass balance results

$$\frac{\partial(\rho w)}{\partial t} = -\frac{\partial(\rho w^2 + p)}{\partial x} - \rho w^2 \frac{\partial \ln(A)}{\partial x} - \rho k_f$$

or after transformation under consideration of the mass balance (7.185)

$$\frac{\partial w}{\partial t} + w \frac{\partial w}{\partial x} + \frac{1}{\rho} \frac{\partial p}{\partial x} + k_f = 0 \, . \tag{7.196}$$

The pipe friction coefficient k_f is determined by the pipe friction number λ_f and the internal pipe diameter d contingent on the density of the medium ρ and its speed w

$$k_f = \frac{\lambda_f}{d} \rho \frac{w^2}{2} \frac{w}{|w|} \, . \tag{7.197}$$

The pipe friction number is determined contingent upon the flow condition and wall roughness (here hydraulically smooth pipes) with the help of the equations of Blasius and Nikuadse or according to Prandtl (implicit equation), see Beitz and Grote (1997, "Dubbel").

$$\lambda_f = \frac{0.3164}{\sqrt[4]{Re}} \quad \text{for } 2{,}320 < Re < 10^5 \quad \text{(Blasius)} , \tag{7.198}$$

$$\lambda_f = 0.0032 + \frac{0.221}{\text{Re}^{0.237}} \quad \text{for } 10^5 < \text{Re} < 10^8 \quad \text{(Nikuradse)}, \tag{7.199}$$

$$\lambda_f = \frac{1}{\left[2\lg\left(\dfrac{\text{Re}\sqrt{\lambda_R}}{2.51}\right)\right]^2} \quad \text{for } 2{,}320 < \text{Re} \quad \text{(Prandtl)}. \tag{7.200}$$

- **The law of conservation of energy**

On the basis of the first law of thermodynamics, the change in energy in the port section under consideration over time in equal to the sum of the energy flows entering and leaving over the cross sectional surfaces and the heat flow supplied and removed from outside. For the energy and the energy flows is valid

$$E = m\left(u_x + \frac{w_x^2}{2}\right) = \rho_x A_x \, dx \left(u_x + \frac{w_x^2}{2}\right), \tag{7.201}$$

$$\dot{E}_x = w_x \rho_x A_x \left(h_x + \frac{w_x^2}{2}\right), \tag{7.202}$$

$$\dot{E}_{x+dx} = w_{x+dx}\, \rho_{x+dx}\, A_{x+dx} \left(h_{x+dx} + \frac{w_{x+dx}^2}{2}\right). \tag{7.203}$$

The heat flow amounts to

$$\dot{Q} = \dot{q}\, A_x \, dx \,. \tag{7.204}$$

With that follows for the energy balance

$$\frac{\partial}{\partial t}\left[\rho_x A_x \, dx \left(u_x + \frac{w_x^2}{2}\right)\right] = w_x \rho_x A_x \left(h_x + \frac{w_x^2}{2}\right) - w_{x+dx}\, \rho_{x+dx}\, A_{x+dx}\left(h_x + \frac{w_{x+dx}^2}{2}\right) + \dot{q}\, A_x\, dx \,. \tag{7.205}$$

Developed into a Taylor row, we obtain after a quick transformation

$$\frac{\partial}{\partial t}\left[\rho\left(u + \frac{w^2}{2}\right)\right] = -\frac{\partial\left[w\rho\left(h + \dfrac{w^2}{2}\right)\right]}{\partial x} - w\rho\left(h + \frac{w^2}{2}\right)\frac{\partial \ln(A)}{\partial x} + \dot{q} \,. \tag{7.206}$$

If we insert the mass and impulse balance ((7.187) and (7.195)) and transform further, the result is

7.4 Gas dynamics

$$\frac{\partial h}{\partial t} + w\frac{\partial h}{\partial x} - \frac{1}{\rho}\left(\frac{\partial p}{\partial t} + w\frac{\partial p}{\partial x}\right) - \frac{\dot{q}}{\rho} - wk_f = 0 \ . \tag{7.207}$$

The heat transfer in the pipe amounts with the help of the Newtonian method to

$$\dot{Q} = \alpha_w\, A(T_w - T_{gas}) \ . \tag{7.208}$$

In (7.209) is represented a semi-empirical approach of Gniellinski (see Stephan (1993)) based on the Prandt analogy for the mean Nußelt number contingent on the respective range of validity of the Prandtl and Reynolds numbers. The Prandtl analogy assumes a dual-layered model that subdivides the flow into a laminar boundary layer and a turbulent core flow, which is directly joined with the lower laminar stratum. It is assumed that in the fully turbulent flow, the speed, temperature, and concentration profiles depend only on the normal wall coordinate, while in the laminar boundary layer, the total value of thrust tension, heat, and diffusion flow density are independent of the wall normal. The approach is used for components like the charge air cooler or the exhaust manifold, in which heat transfer plays an essential role.

$$\mathrm{Nu}_{m,turb} = \frac{\lambda_f}{8} \frac{(\mathrm{Re} - 1{,}000)\mathrm{Pr}}{1 + (\mathrm{Pr}^{2/3} - 1)12.7\sqrt{\lambda_f/8}}\left[1 + \left(\frac{d}{l}\right)^{2/3}\right] . \tag{7.209}$$

Equation (7.209) is valid in the range $2{,}300 \leq \mathrm{Re} \leq 5\,10^5$, $0.5 \leq \mathrm{Pr} \leq 2{,}000$ and $l/d > 1$. From the definition of the Nußelt number, the heat transfer coefficient comes to

$$\alpha_w = \frac{\lambda}{d}\frac{\lambda_f}{8} \frac{(\mathrm{Re} - 1{,}000)\mathrm{Pr}}{1 + (\mathrm{Pr}^{2/3} - 1)12.7\sqrt{\lambda_f/8}}\left[1 + \left(\frac{d}{l}\right)^{2/3}\right] . \tag{7.210}$$

The determination of the heat transfer for $\mathrm{Re} < 2{,}300$ results via a quadratic averaging of these methods for the turbulent (7.209) and the laminar Nußelt number (7.211) according to (7.212)

$$\mathrm{Nu}_{lam} = 0.664\,\mathrm{Re}^{\frac{1}{2}}\,\mathrm{Pr}^{\frac{1}{3}} , \tag{7.211}$$

$$\mathrm{Nu}_m = \sqrt{\mathrm{Nu}_{m,turb}^2 + \mathrm{Nu}_{lam}^2} \ . \tag{7.212}$$

7.4.2 Numerical solution methods

In the previous section, equations were derived that describe one-dimensional gas dynamics. We are dealing here with a partial differential equation system, which is analytically not solvable. For this reason, a large multitude of graphic and numerical solution techniques have been developed, of which numerical solutions in the form of finite differences, supported by the steadily increasing calculation power of computers, has been quite successful, as it offers the necessary flexibility and precision. In this case, a discretization of the location with an approximation of the local gradients takes place. This renders possible the transformation of the partial differential equation system into a row of common differential equations. This process will be described in more detail in the following.

- **The single-step Lax-Wendroff method**

The Lax-Wendroff method offers the possibility of describing the gas dynamics in the pipes using finite differences. First the conservation equations are expressed in a form which corresponds to the vector form represented in (7.213)

$$\frac{\partial G(x,t)}{\partial t} + \frac{\partial F(x,t)}{\partial x} = -C(x,t) \ . \tag{7.213}$$

The vectors $G(x,t)$, $F(x,t)$ and $C(x,t)$ are easy to derive from the laws of conservation for mass, impulse, and energy,

$$G(x,t) = \begin{bmatrix} \rho \\ \rho w \\ \rho\left(u + \dfrac{w^2}{2}\right) \end{bmatrix} \ , \quad F(x,t) = \begin{bmatrix} \rho w \\ \rho w^2 + p \\ w\rho\left(h + \dfrac{w^2}{2}\right) \end{bmatrix} \quad \text{and}$$

$$C(x,t) = \begin{bmatrix} \rho w \\ \rho w^2 \\ w\rho\left(h + \dfrac{w^2}{2}\right) \end{bmatrix} \frac{d \ln(A)}{dx} + \begin{bmatrix} 0 \\ \rho k_f \\ \dot{q} \end{bmatrix} \ . \tag{7.214}$$

In the following, from reasons of improved clarity, the expression $G_t(x,t)$ will be used for $\partial G(x,t)/\partial t$ and the expression $F_x(x,t)$ for $\partial F(x,t)/\partial x$.

In the next step, the Taylor row of function $G(x,t)$ is developed around the point $(x, t + \Delta t)$ and a location discretization is carried out. From that results the following equation

$$G(x_i, t + \Delta t) = G(x_i,t) + \Delta t\, G_t(x_i,t) + O(\Delta t^2) \ . \tag{7.215}$$

Then $G_t(x,t)$ is substituted with (7.215) and the formation of the local gradient occurs with the central differences. With the approximation

$$G(x_i,t_j) \approx \frac{1}{2}\left[G(x_{i+1},t_j) + G(x_{i-1},t_j)\right] \tag{7.216}$$

we finally obtain the equation of the single-step Lax-Wendroff process

$$G(x_i,t_{j+1}) = \frac{1}{2}\left[G(x_{i+1},t_j) + G(x_{i-1},t_j)\right] - \frac{\Delta t}{2\Delta x}\left[F(x_{i+1},t_j) - F(x_{i-1},t_j)\right] - \Delta t\, C(x_i,t_j) \ . \tag{7.217}$$

7.4 Gas dynamics

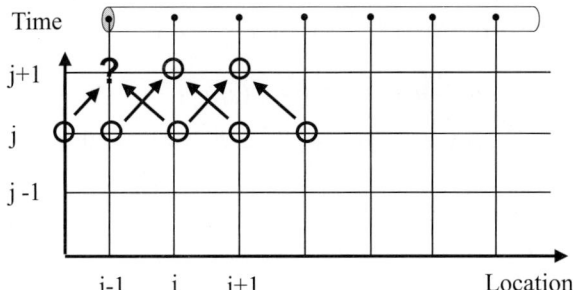

Fig. 7.29: Single-step Lax-Wendroff process with determination of the internal pipe nodes

In Fig. 7.29 we can recognize the method for the determination of the state of the particular internal pipe nodes. The condition of the node at location i at time j is known, as well as the conditions at location $i - 1$ and $i + 1$ at time j. From the states $i - 1$ and $i + 1$ at time j, we can determine with the help of the previously described equations the condition of the node at location i at time $j + 1$. However, the conditions at the external pipe nodes cannot be determined with this method, as even though the states at locations $i - 1$ and i at time j are known, the condition at location $i - 2$ is not. This must take place by means of the pipe-margin coupling, which we look at in more detail in chap. 7.2.3.

- **The two-step Lax-Wendroff method**

The two-step Lax-Wendroff method (Fig. 7.30), which follows from the Peyret-Lerat process, describes a method of finite differences, which consists of two steps and can heighten the stability though the use of further coefficients in comparison with the single-step process.

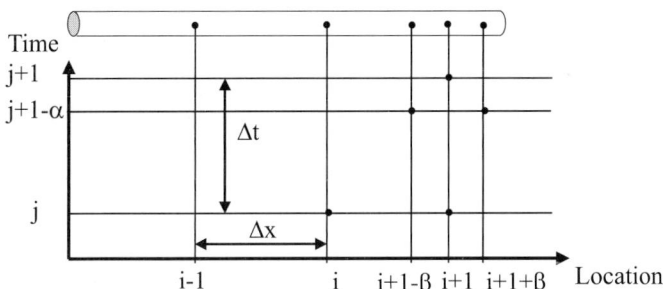

Fig. 7.30: Two-step Lax-Wendroff process

The Peyret-Lerat method can be represented in the following way.

1st step:

$$G(x_{i+\beta}, t_{j+1}) = (1-\beta)G(x_i, t_j) + \beta G(x_{i+1}, t_j)$$
$$- \alpha \frac{\Delta t}{\Delta x}\left[F(x_{i+1}, t_j) - F(x_i, t_j)\right] \qquad (7.218)$$
$$- \alpha \Delta t\left[(1-\beta)C(x_i, t_j) - \beta C(x_{i+1}, t_j)\right]$$

2nd step:

$$G(x_i, t_{j+1}) = G(x_i, t_j) - \frac{\Delta t}{2\alpha \Delta x}\left[(\alpha - \beta)F(x_{i+1}, t_j)\right.$$
$$+ (2\beta - 1)F(x_i, t_j)\right] - \frac{\Delta t}{2\alpha \Delta x}\left[(1-\alpha-\beta)F(x_{i-1}, t_j) \qquad (7.219)\right.$$
$$+ F(x_{i+\beta}, t_{j+\alpha}) - F(x_{i-1+\beta}, t_{j+\alpha})\right] - \Delta t\, C(x_i, t_j).$$

From the Peyret-Lerat method, we obtain via substitution of $\alpha = \beta = 1/2$ the two-step Lax-Wendroff method:

$$G\left(x_{i+1/2}, t_{j+1}\right) = \frac{1}{2}\left[G(x_i, t_j) + G(x_{i+1}, t_j)\right]$$
$$- \frac{\Delta t}{2\Delta x}\left[F(x_{i+1}, t_j) - F(x_i, t_j)\right] - \frac{\Delta t}{4}\left[C(x_{i+1}, t_j)\right], \qquad (7.220)$$

$$G(x_i, t_{j+1}) = G(x_i, t_j) - \frac{\Delta t}{\Delta x}\left[F(x_{i+1/2}, t_{j+1/2})\right.$$
$$- F(x_{i-1/2}, t_{j+1/2})\right] - \Delta t\, C(x_i, t_j). \qquad (7.221)$$

7.4.3 Boundary conditions

Previously, only one port/pipe section has been considered and the conservation equations and their solutions shown. A real system (intake system, exhaust gas system) consists however of a large number of single components (discontinuity points), which are connected with pipes. Only via methods for a coupling of various partial systems does such a complex system become calculable. For this, we must exchange between the partial systems, e.g. the energy and mass flows, thereby setting the bondary conditions of the pipe described above. The structure thereby is always the same and resembles the one described in chap. 7.3.2: – unsteady location – pipe – unsteady location – pipe – unsteady location. Consequently, the following components represent locations of unsteadiness from the standpoint of the pipe:

- pipe end
- pipe branching
- trigger wheel vane
- volume
- cylinder
- compressor
- turbine.

7.4 Gas dynamics

The coupling of a pipe with its neighboring component is described by:
- The method of characteristics, which calculates the conditions in the outlet cross section of the pipe (compatibility conditions).
- The general flow equation, which illustrates the influence of the throttle location. It describes the mass flow coming through the throttle location contingent on the conditions in the pipe border and the unsteady location.
- The conservation equations of the respective points of discontinuity, which describe changes in state via common differential equations contingent purely on time. These will be presented in chapters 7.1 to 7.3 and 7.5.

Since the equations are implicitly contingent upon each other, the solution must be determined iteratively. In the following, the iterative solution of the boundary conditions will be introduced, as it has been developed by Görg (1982) and Stromberg (1977). An exact derivation and arrangement can be found in Miersch (2003). Fundamentally, two cases of flow can be determined at the margin: in the case of elementary flow situation 1, the mass flow flows from the pipe into the margin, while in elementary flow situation 2 the mass flow runs from the border into the pipe.

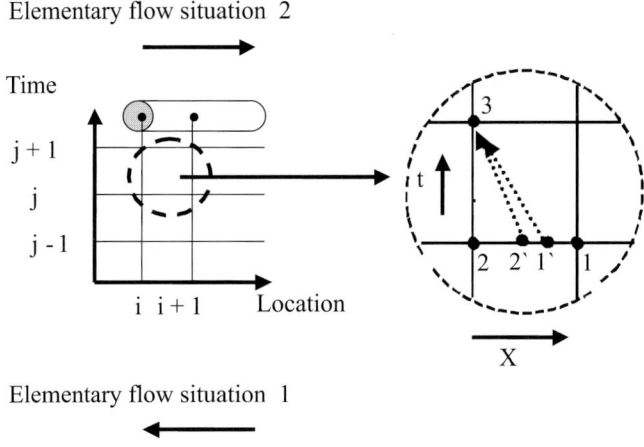

Fig. 7.31: Boundary coupling by of characteristic lines

Fig. 7.31 shows the time-location lattice at the pipe border. Point 3 represents the marginal state to be determined in the new point in time. It is determined from the known conditions of the states of rest (index 0) of the border node and from the conditions of the previous time in lattice points 1 and 2. The distance between points 1 and 2 corresponds to the local discretization, the distance between points 2 and 3 the temporal discretization. The location, from which in the former instant the new boundary conditions depart (point 1′ and point 2′), is called the foot of the Mach or particle path. These location coordinates are fixed such that at the new time the conditions starting from point 1′ reach the border (point 3) along the Mach path and the conditions starting from point 2′ the border (point 3) along the particle path. The foot points are set at the beginning of the iteration to the middle of the distance between points 1 and 2 and then again after every iteration loop as follows:

$$x_{f,\,new} = x_{f,\,old} - \Delta x_f$$

$$\Delta x_f = \frac{\Delta t}{2} \frac{v_f^* v_2^*}{v_2^* + v_2^*};$$

$$v_f = u_f - a_f ; \quad \text{for Mach line path (point 1'),}$$

$$v_f = u_f ; \quad \text{for particle path (point 2').}$$

The solution in elementary flow situation 1 is thus determined from the condition of compatibility along the particle path

$$\frac{2a}{\kappa-1}\left(\frac{\partial a}{\partial t} + w\frac{\partial a}{\partial x}\right) - \frac{1}{\rho}\left(\frac{\partial p}{\partial t} + w\frac{\partial p}{\partial x}\right) = 0 \tag{7.222}$$

and the condition of compatibility along the Mach line

$$\frac{\partial w}{\partial t} + (w \pm a)\frac{\partial w}{\partial x} \pm \frac{2}{\kappa-1}\left[\frac{\partial a}{\partial t} + (w \pm a)\frac{\partial a}{\partial x}\right] = \mp w(w \pm a)\frac{d}{dx}[\ln(A)] \tag{7.223}$$

the general flow equation

$$\dot{m} = \alpha\, A_2 \sqrt{2\, p_{01}\, \rho_{01}} \sqrt{\frac{\kappa}{\kappa-1}\left[\left(\frac{p_2}{p_{01}}\right)^{\frac{2}{\kappa}}\left[1 + \frac{(\kappa-1)}{\kappa R}\frac{\dot{W}}{\dot{m}_{is}\, T_{01}}\right] - \left(\frac{p_2}{p_{01}}\right)^{\frac{\kappa+1}{\kappa}}\right]} \tag{7.224}$$

and the conditions in the system connected to the pipe.

Fig. 7.21 shows the particle and Mach line paths on the flow level. These are described by

$$\frac{\partial x}{\partial t} = w \tag{7.225}$$

and

$$\frac{\partial x}{\partial t} = w \pm a . \tag{7.226}$$

The index α describes the propagation of a pressure wave along the Mach line in the flow direction, index β the propagation against the flow direction.

The formulation of the condition of compatibility occurs via transformation of the equation system (7.213) from the representation in the independent variables density ρ, pressure p, speed w to a representation in the independent variables speed of sound a, pressure p, and speed w.

The transformed system consists of coupled common differential equations, the solution of which is found through a purely temporal integration. The detailed derivation is described in Seifert (1962), Stromberg (1977), and Görg (1982) (see. Miersch (2003)).

7.4 Gas dynamics

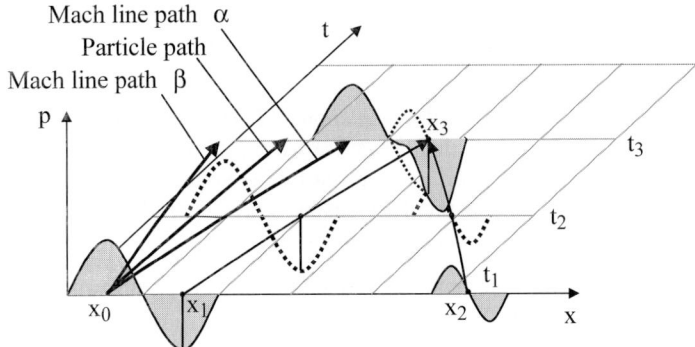

Fig. 7.32: Mach line trace and particle trace in the flow level (Miersch (2003))

The discretization of the conditions of compatibility tracing the Mach lines occurs via

$$dp = p_3 - p_{1'},$$
$$dw = w_3 - w_{1'}, \tag{7.227}$$
$$dt = \Delta t.$$

in the dimensions pressure p and velocity w. In this way, we obtain a linear relation between pressure and speed

$$p_3 = C_1 w_3 + C_2 . \tag{7.228}$$

In elementary flow situation 1, the condition of compatibility along the particle path is discretized according to

$$dp = p_3 - p_{2'},$$
$$da = a_3 - a_{2'}, \tag{7.229}$$
$$dt = \Delta t.$$

For the solution in elementary flow situation 2, instead of the condition of compatibility along the particle path, that of the quasi-stationary energy balance (7.231) is utilized. Furthermore, one obtains, via the quasi-stationary energy balance between state and state of rest at point 3, a linear relation between pressure and the speed of sound (7.232).

$$\left(\frac{a_2}{a_{01}}\right)^2 = \frac{\hat{A}\left(\dfrac{p_2}{p_{01}}\right)^{\frac{\kappa-1}{\kappa}}}{\alpha\sqrt{\dfrac{2}{\kappa-1}\left[1-\left(\dfrac{p_2}{p_{01}}\right)^{\frac{\kappa-1}{\kappa}}\right]}}\left(\frac{w_2}{a_{01}}\right), \tag{7.230}$$

$$a_0^2 = a_3^2 + \frac{1}{2}(\kappa-1)w_3^2 , \tag{7.231}$$

$$a_3 = \frac{p_3}{C_3} + C_4 \ . \tag{7.232}$$

In the context of iteration, the general flow equation (7.224) is solved in the implicit variables pressure or speed. This equation distinguishes itself from the known differential equation of Saint-Venant by the term for the work carried out by the system $\dot{W}$. In this way, there is a possibility of calculating mass flows that flow against a pressure ratio greater than one (e.g. in the case of a compressor).

It is thus possible to describe the boundary conditions of a trigger wheel vane and a flow machine in the same manner. In the case of the trigger wheel vane, the provision of the flow coefficient is necessary, while the flow machine can be seen as a trigger wheel vane for the sake of the solution of the boundary condition.

In the case of pipe branchings with three connections, a total of six flow situations appear (Fig. 7.33), for which the flow coefficients are given contingent on the branching angle and the particular cross sections, mostly in data banks.

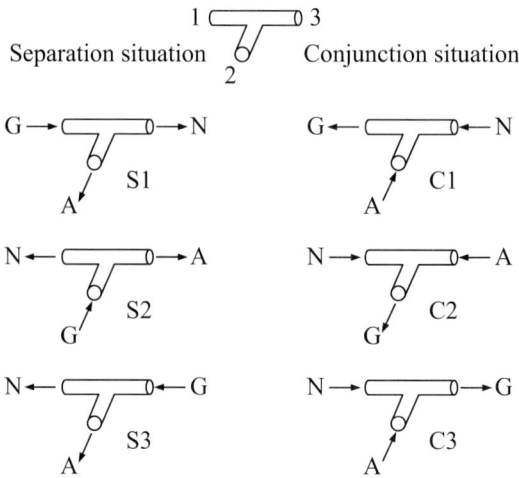

Fig. 7.33: Flow situations at a pipe branching

For one pipe branching, as a rule three separation situations and three conjunction situation must be taken into consideration, whereby in each case one pipe socket conducts the entire mass flow and the other two only a partial mass flow. In Fig. 7.34, the flow coefficients from the ratio from the branching mass flow and the total mass flow is applied to the main flow direction and the branching flow direction. The determination of the main flow directions serves only to improve the clarity of the system, having no effect on the calculation result.

The continuity equation must be fulfilled at the pipe embranchment. This is guaranteed by a corresponding iteration. Fig. 7.34 shows an example of the flow coefficient at a pipe em-

7.4 Gas dynamics

branchment with 120° between each pipe socket. The diameter of the branching pipe socket (2) is twice as large as the cross section of the other two pipe sockets.

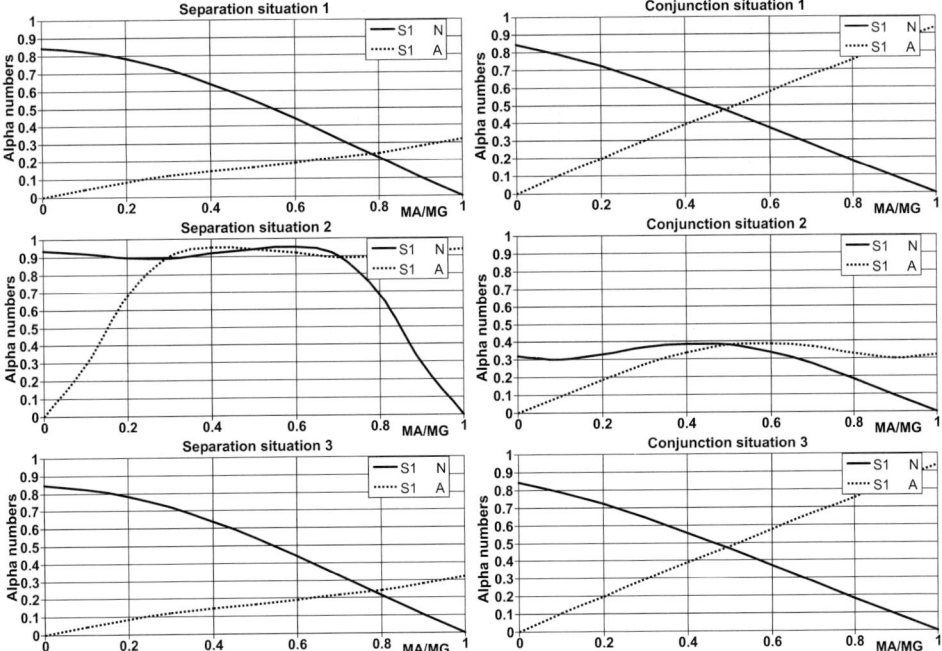

Fig. 7.34: Flow coefficients at a pipe branching

The entire iteration is successfully ended when the internal iteration loop converges over the flow equation and the external iteration loop according to the conditions of compatibility, the marginal equations, and the flow equation with a sufficiently small error.

Also of decisive importance for an exact calculation result is providing the calculation length, which, one the one hand, should of course not be too small, in order to maintain a stable solution and to save calculation time, and should, on the other hand, not be too large, since a secure convergence of the calculation time would thereby no longer be possible. The realization of the time allowance takes place according to the Courant-Friedrichs-Levy criterion of stability

$$\Delta t \leq \frac{\Delta x}{(|w| + a)} \ . \tag{7.233}$$

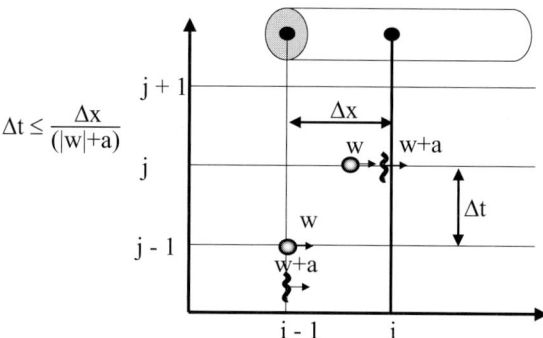

Fig. 7.35: Courant Friedrichs Levy stability criterium

These circumstances are shown with the help of Fig. 7.35. The conditions at locations $i-1$, i and $i+1$ at time j are known. In order now to draw a conclusion from the conditions $(i-1; j)$ and $(i+1; j)$ about the condition $(i; i+1)$, Δt must be chosen *at least as large*, such that time Δt passes, while the path Δx is traversed. From this results a need of minimal time length in the order of magnitude of 10^{-4} to 10^{-6} seconds.

7.5 Charging

Chap. 2.6 describes the foundations of charging in internal combustion engines. This chapter aims at the understanding of the simulation of charging in the context of zero- and one-dimensional process calculation.

7.5.1 Flow compressor

The simulation of charged engines requires, in addition to the process simulation in the cylinder described in chap. 7.1 and 7.2 as well as the simulation of pipe and secondary aggregate components, a detailed description of the charging components. The description of these components is based essentially on special, standardized characteristic maps, which, according to the maker, are usually of varying quality. A description of the operating behavior of flow machines by means of characteristic maps has proved to be sufficiently exact for unstable processes.

- **Reference quantities**

In order to utilize compressor maps measured under more or less arbitrary environmental conditions under altered environmental conditions as well, the quantities that are stored in a characteristic map have to be made independent of the actual environmental conditions. The map quantities are usually converted into standardized reference environmental conditions (e.g. ISA condition: 288 K, 1.013 bar; standard conditions: 293 K, 0.981 bar). This takes place via the introduction of so-called reference quantities with the help of fluid-mechanical laws of similarity. These will now be more thoroughly explicated.

7.5 Charging

In the case of the speeds for flow machines, we use the similarity of the dimensionless, so-called Mach number and relate the circumference speed of the rotor, which is directly proportional to the speed, to the sound velocity of the gas at the entry state. The gas constant is assumed to be constant in the following investigations. Furthermore, one must take into consideration in the calculations whether one is dealing with pure air in the compression, for which the isentropic exponent can be set to 1.4 and the gas constant to 287 J/kg K, or whether one has another medium with other physical properties, as for example a mixture or the like.

$$\mathrm{Ma} = \frac{u}{a} = \frac{\omega}{\sqrt{T}} \frac{r}{\sqrt{\kappa R}} = \frac{2n}{\sqrt{T}} \frac{\pi r}{\sqrt{\kappa R} \, 60[\mathrm{s \cdot min^{-1}}]} \, . \quad (7.234)$$

One recognizes that the Mach number is proportional to the speed and to the square root from the entry temperature of the gas. According to this is sufficient to define a reference speed or a reference angle speed for which the speed or angle speed is divided by the root of the entry temperature. Using the reference quantities, one can in the case of the flow compressor simultaneously convert to a certain standard entry condition, in order again to assign the original unit to the reference speed or angle speed.

$$n_{ref} = \frac{n\sqrt{T_{ref}}}{\sqrt{T}}, \quad \omega_{ref} = \frac{\omega\sqrt{T_{ref}}}{\sqrt{T}} \, . \quad (7.235)$$

For the mass flows as well, independent quantities can be defined from the respective environmental and entry conditions. If we consider at first the flow equation again, the result is

$$\dot{m}_{th} = A \frac{p}{\sqrt{RT}} \sqrt{\frac{2\kappa}{\kappa-1} \left(\pi^{\frac{2}{\kappa}} - \pi^{\frac{\kappa+1}{\kappa}} \right)} = A \frac{p}{\sqrt{RT}} \Psi \, . \quad (7.236)$$

With this equation, all components, which are in some way to be seen as throttle locations – i.e. flow turbines or charger aggregates – can be described for the filling and emptying method.

As we can recognize with the help of (7.236), the mass flow through a throttle location is contingent only on the pressure, the square root from the temperature before the throttle location, and the flow function, which for its part is determined practically only by the adjoining pressure ratio. If we rearrange the equation and move the pressure and temperature to the left side of the equal sign, the mass flow referring to these quantities is now practically contingent only on the adjoining pressure ratio. This mass flow is designated as the reference mass flow and can then be referred to for the sake of keeping the corresponding units for the mass flow in the flow compressor at standard conditions

$$\dot{m}_{ref} = \dot{m}_{th} \frac{\sqrt{T}}{p} \frac{p_{ref}}{\sqrt{T_{ref}}} = \frac{p_{ref}}{\sqrt{T_{ref}}} A \frac{1}{\sqrt{R}} \Psi \, . \quad (7.237)$$

Analogously to the mass flow, there is also a reference volume flow, which can be determined by division by means of the reference density of the gas and under consideration of the definition of the actual volume flow under environmental conditions over the related density at the entry into the components

$$\dot{V}_{ref} = \frac{\dot{m}_{ref}}{\rho_{ref}} = \frac{\rho \dot{V}_{th}}{\rho_{ref}} \frac{\sqrt{T}}{\sqrt{T_{ref}}} \frac{p_{ref}}{p} = \dot{V}_{th} \frac{\sqrt{T_{ref}}}{\sqrt{T}} \,. \qquad (7.238)$$

The performance admitted by the compressor has already been described in chap. 2.6 (2.69). The compressor torque results from a division with the angle speed. Since however quantities are stored in the compressor maps, which were converted according to the laws of similarity to the reference conditions, the compressor torque can also be written with the reference quantities

$$T_{c,ref} = \frac{1}{\omega_{c,ref}} \dot{m}_{c,ref} \frac{\kappa_{b.c.}}{\kappa_{b.c.} - 1} R T_{ref} \frac{1}{\eta_{ref,c}} \left(\pi_c^{\frac{\kappa_{b.c.} - 1}{\kappa_{b.c.}}} - 1 \right). \qquad (7.239)$$

The torque necessary for the propulsion of the compressor can thus at first be calculated for characteristic map considerations or for the extrapolation of characteristic maps as a reference torque. It can then be converted into the actual torque needed with the help of the definitions of the laws of similarity. This is given by

$$T_{c,ref} = \frac{1}{\omega_c} \frac{\sqrt{T_{b.c.}}}{\sqrt{T_{ref}}} \dot{m}_c \frac{p_{ref}}{p_{b.c.}} \frac{\sqrt{T_{b.c.}}}{\sqrt{T_{ref}}} \frac{\kappa_{b.c.}}{\kappa_{b.c.} - 1} R T_{ref} \frac{1}{\eta_{total,c}} \left(\pi_c^{\frac{\kappa_{b.c.} - 1}{\kappa_{b.c.}}} - 1 \right)$$

$$= T_c \frac{p_{ref}}{p_{b.c.}} \qquad (7.240)$$

and

$$T_c = T_{c,ref} \frac{p_{b.c.}}{p_{ref}} \,. \qquad (7.241)$$

We recognize that the reference compressor propulsion torque and the actual compressor propulsion torque differ only by the quotient from the actual pressure adjacent to the compressor entry and the reference pressure. The influence of an altered temperature has already been eliminated because of the theory of similarity via the reference quantities of angle speed and mass flow.

- **Characteristic map representation**

Fig. 7.36 shows the characteristic map of a flow compressor. The lines of constant reference compressor speeds or constant peripheral speeds and the isentropic efficiency of the compressor are entered in this characteristic map. The so-called surge line represents a fictitious boundary line and limits the range of validity of the characteristic map to the left. This boundary is however not only set by the compressor type, but results from the interaction of pipe volumes, pipes, and the compressor. Therefore, the characteristic map is at first considered without the surge line and if necessary extrapolated beyond it.

7.5 Charging

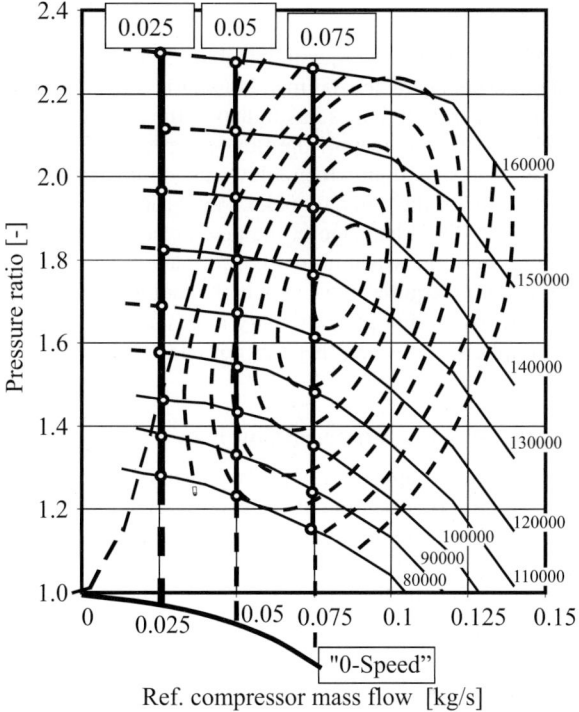

Fig. 7.36: Schematic map of a flow compressor

The determination of the isentropic efficiency, which is calculated because of its definition from the temperature difference between compressor intake and exhaust and the compressor pressure ratio, reaches its limits at low pressure ratios and small temperature differences, as is common at low speeds and mass flows. Therefore, a large area below a lowest reference speed is missing in most experimentally determined characteristic maps, see Fig. 7.36. This area is not of interest for an design of the engine under full load, since the interaction of engine and compressor under full load mostly takes place in the area of optimal efficiency at mid to high pressure ratios and speeds. For operating points at lower partial load or for the idle speed of the engine, as they typically appear in the cycles of motor vehicles or, at low speeds, at the propeller line in the diesel engines of large ships, the characteristic map does not provide reliable information. This leads to consequences which are not insignificant for the interaction of engine and the charger aggregate, as the calculations in chap. 8 show. While the real engine, even if in area with low efficiency, nonetheless continues to run, in a simulation calculation in these ranges on the other hand, it can come to discontinuations in the program, which makes a calculation of cycles with long partial load or idle phases practically impossible. For these reasons, one must attempt to extrapolate the characteristic map into areas up to a reference compressor speed of 0.

- **Extrapolation**

In this section we will introduce a method, with which the behavior of a flow compressor can be described beyond the usual characteristic map boundaries, in order to allow the clear representation of all possible operation ranges in simulating the interaction of the engine and the flow compressor. The characteristic map of the flow compressor must for these purposes be transported into another representation, which allows us, on the one hand, to extrapolate the map in a simple manner and, on the other, to make it accessible to calculations like the representation previously known. The advantage of the new presentation is a considerably broadened range of validity in areas, in which no more information was attainable in the previous representation. This procedure requires however a high amount of care and constant plausibility control. A reverse calculation into the customary representation after extrapolating is possible with certain assumptions, however, this can not take place in ranges of pressure ratio values smaller than 1.

- **Extrapolation in the pressure ratio-mass flow map**

The prerequisite for a secure extrapolation is the expansion of the characteristic map in the common and known way of representing it. The lines of constant reference compressor speeds can in most cases be easy expanded to the left and right. One must thereby consider that the lines lightly climb towards the surge line and beyond it, which is required in the case of interpolation in the characteristic map according to Münzberg and Kurzke (1997) as well. In some maps, the lines of a constant reference compressor speed first drop lightly at the surge line or to the left of it and then the rise again. This can be due to the specific events at the test bench during the measurement of the characteristic map (pipe lengths and volumes) and should not be considered in the extrapolation process. Operation left of the supposed surge line has to be indicated in the simulation calculation anyway, in order to leave the decision of whether or not the transgression should be tolerated to the user. For the extrapolation however, the surge line and all actual effects associated with it on the form of the characteristic map are at first ignored.

To the right as well, the lines of constant reference compressor speeds can be lead further to compressor pressure ratios of 1, since at a constant speed, the maximum mass flow is limited by the speed of sound in the compressor impeller and the speed lines slope down steeply (fill boundary). One can deal with partially existing fragments of lines of isentropic efficiency at the surge and fill line in the same way.

- **Determination of the zero-speed line as a characteristic map boundary**

A flow compressor with a speed of 0 rpm represents, from a fluid-mechanical perspective, a simple throttling. If a gas mass should flow through the flow compressor, a pressure ratio smaller than 1 must be adjacent to it according to the usual definition for the pressure ratio for flow compressors. In actual engine operation as well, e.g. turbocharged vehicle engines at idle operation, a pressure ratio smaller than 1 at the flow compressor can occur. At this operating point, the turbine performance available at the compressor is too low to deliver the fresh gas mass flow sucked by the engine at a corresponding speed and a resulting pressure ratio greater than 1. The speed of the flow compressor of passenger car diesel engines ranges – according to the respective condition of the engine and the bearing friction contingent on the oil viscosity – between 5,000 and 10,000 rpm. The so-called zero speed line of the compres-

7.5 Charging

sor is thus an important bottom limitation of the compressor map which one cannot fall beneath.

Fig. 7.37 shows the reference volume flow, determined contingently on the pressure ratio via the flow function with a constant actual throttle cross section, in comparison with the zero speed line on the compressor. The good agreement between both graphs permits us to conclude that the zero speed line of a flow compressor can be determined easily by means of the flow equation. In the case of the flow equations defined in (7.236), the effective cross-sectional surface must be used instead of the cross-sectional surface.

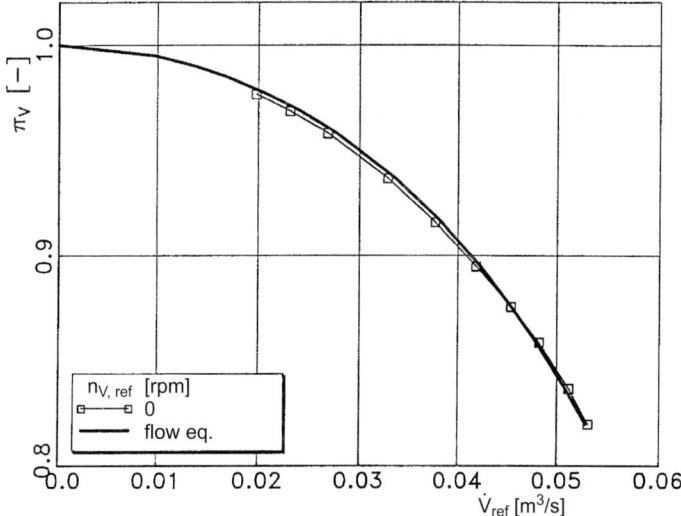

Fig. 7.37: Comparison of the zero speed line and the flow function

However, the isentropic efficiency at these speeds is not fixed by means of the interpolation, since in this area of the characteristic map an extrapolation contains large inaccuracies; the isentropic efficiency at a compressor pressure ratio smaller than 1 is not defined anyhow. In this context, it would be appropriate to cross over to a presentation of the reference compressor torque.

- **Extrapolation via coordinate transformation**

Proceeding from the known mode of representation of the characteristic map as a pressure ratio-mass flow map, the characteristic map is considered along constant reference mass flows and both the compressor pressure ratio and the reference compressor torque are plotted across the reference compressor speed with the parameter of reference mass flow. In the case of broad compressor maps as used in the context of passenger vehicles, this results in a huge amount of graphs, which are already defined across a wide speed range and are thus easy to extrapolate into low speed ranges. Fig. 7.38 shows characteristic maps determined with this method. The extrapolated range is dashed.

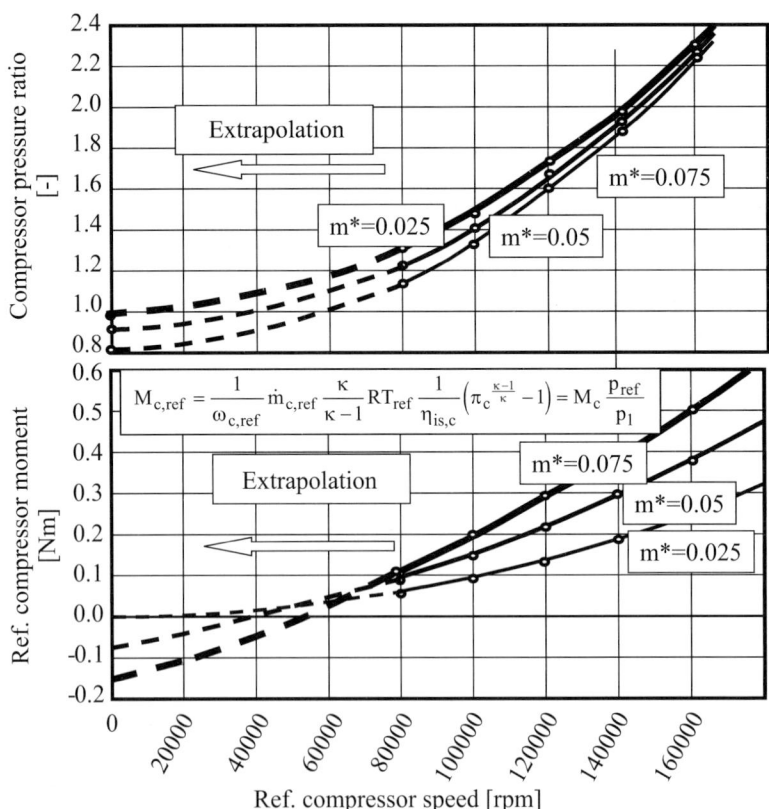

Fig. 7.38: Result of the extrapolation in the compressor map

In the case of narrow compressor maps for commercial vehicles or medium speed large diesel engines on the other hand, which are designed for optimum efficiency in a very specific operating point, problems arise in this method. From the surge line, which has a very small gradient and which therefore at high pressure ratios comes to a state of rest already at high reference volume flows, the reference speed lines runs nearly horizontally up to the fill boundary and then sink vertically. Such a characteristic map for a large diesel engine is presented in Fig. 7.39.

The extrapolation method described above along lines of constant reference volume flows is not suitable for this characteristic map type for describing behavior at low compressor speeds because of the steep decline of the speed lines. One remedial measure is the consideration of the characteristic map along lines that run approximately parallel to the surge line. The formal description for these "crooked" coordinates, seen in Fig. 7.39, reads

$$V_i^* = \dot{V}_{1,(\pi_c = 1)} + a(\pi_c - 1) . \tag{7.242}$$

7.5 Charging

As one can recognize, for positive values of a, it is a matter of parallel lines inclined to the right with a constant slope. These will be designated in the following as parameter lines and can be clearly identified by the respective i^{th} reference volume flow at a compressor ratio of 1 – as the start point of the line so to speak. If the slope a of these parameter lines takes on the value of 0, the consideration of the characteristic map on lines of constant reference volume flows results as a special case. The advantage of these parameter lines inclined dependently on slope a is clearly shown in Fig. 7.39. Almost all lines, at a suitable choice of a, intersect almost all the compressor speed lines and thus, at a later plotting of the compressor pressure ratio and the reference compressor torque over the speed, cover a large speed range beyond which can be easily extrapolated.

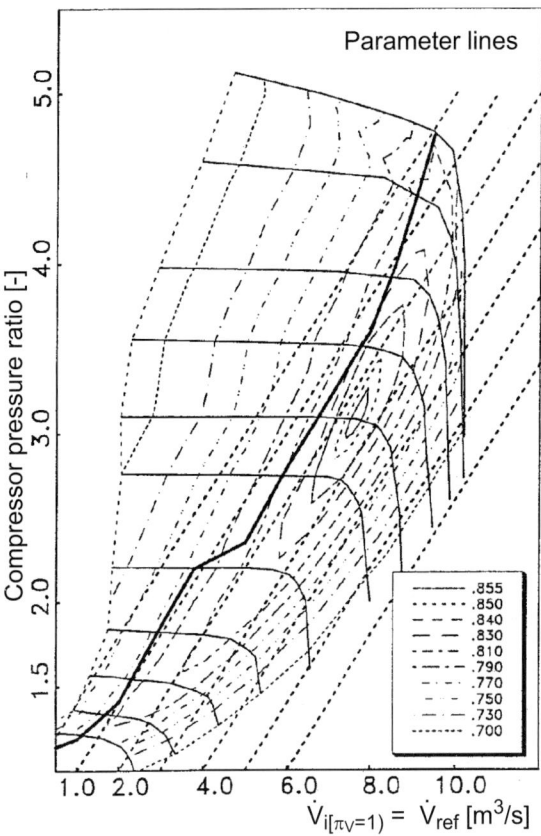

Fig. 7.39: Map of a flow compressor for a medium speed diesel engine

In the compressor map, the pressure ratios and the isentropic compressor efficiency are usually stored for numerical reasons along lines of constant reference compressor speeds. If the characteristic map is now considered along the parameter lines, the pressure ratios and the isentropic efficiency have to be determined at the points of intersection of the reference compressor speeds with the parameter lines. One can thereby proceed as represented in Fig. 7.40.

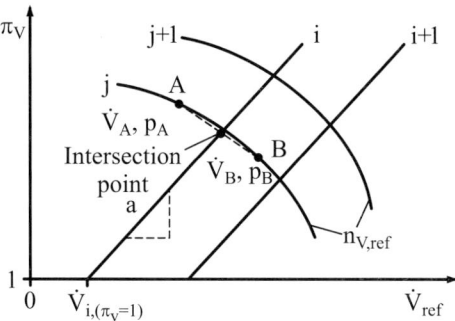

Fig. 7.40: Determination of the pressure ratio and the isentropic efficiency at the intersection points of sloping coordinates and lines of constant reference compressor speeds

For the intersection of the ith coordinate line with the jth compressor speed line is valid for a linear interpolation on the line of the constant compressor speed

$$\pi_{c,i,j} = \frac{\pi_{c,a,j} - \pi_{c,b,j}}{\dot{V}_{a,j} - \dot{V}_{b,j}} (\dot{V}_i^* - \dot{V}_{b,j}) + \pi_{c,b,j} \ . \tag{7.243}$$

Thereby is valid at the same time according to (7.242)

$$\dot{V}_i^* = \dot{V}_{i,(\pi_c = 1)} + a(\pi_{c,i,j} - 1) \ . \tag{7.244}$$

If one inserts both equations into each other and solves according to the compressor pressure ratio $\pi_{C,i,j}$, the result is

$$\pi_{c,i,j} = \frac{\dfrac{\pi_{c,a,j} - \pi_{c,b,j}}{\dot{V}_{a,j} - \dot{V}_{b,j}} \left(\dot{V}_{i,(\pi_c = 1)} - a - \dot{V}_{b,j} \right) + \pi_{c,b,j}}{1 - a \dfrac{\pi_{c,a,j} - \pi_{c,b,j}}{\dot{V}_{a,j} - \dot{V}_{b,j}}} \ . \tag{7.245}$$

With the known compressor pressure ratio at the intersection, the isentropic efficiency can be calculated analogously to (7.243)

$$\eta_{is,i,j} = \frac{\eta_{is,a,j} - \eta_{is,b,j}}{\dot{V}_{a,j} - \dot{V}_{b,j}} \left[\dot{V}_{i,(\pi_c = 1)} + a(\pi_{c,i,j} - 1) - \dot{V}_{b,j} \right] + \eta_{is,b,j} \ . \tag{7.246}$$

With the knowledge of the isentropic efficiency and the pressure ratio at the respective intersections with the compressor speed lines, the reference volume flow can be determined according to (7.244) and with that the reference compressor torque according to (7.240) along each parameter line. The result of this characteristic map transformation is shown in Fig. 7.41

7.5 Charging

for the compressor pressure ratio and the reference compressor torque. In place of the volume flow, the mass flow can of course be used in an analogous way for the parameter lines.

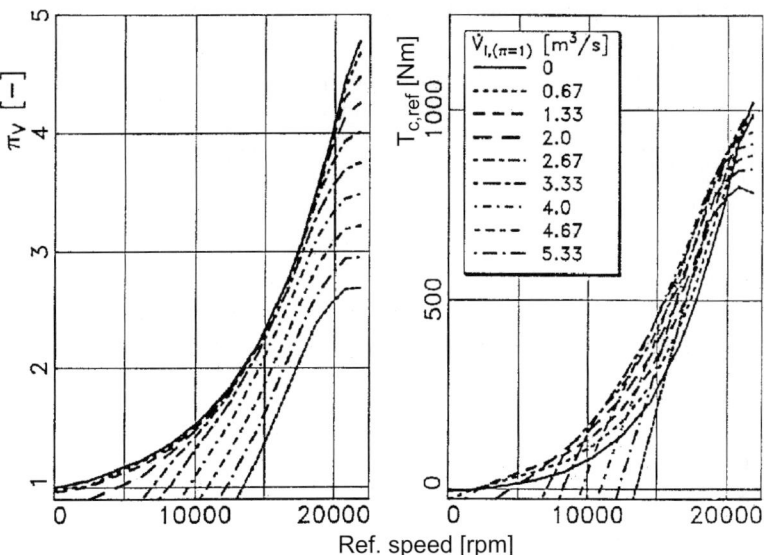

Fig. 7.41: Transformed compressor map

The interpolation in both of these characteristic maps for the calculation of the operating behavior of the compressor during process simulation can occur by means of a doubled linear interpolation in each characteristic map. This provides clear operating points. In this case, the volume flow parameter $\dot{V}_{(\pi_C = 1)}$ is first interpolated in the pressure ratio-speed map. This parameter can be transformed to the reference volume flow parameter with the help of the known slope a and the compressor pressure ratio according to (7.242). Simultaneously, the interpolated volume flow parameter serves, together with the reference speed in the compressor torque-speed map, as an input value for the interpolation of the reference compressor torque. By means of the actual pressure at the compressor entry, the actual compressor torque is calculated according to (7.241). For the parameter lines, the mass flow can of course also be used in place of the volume flow in an analogous manner.

- **Behavior at the surge line**

The surge line is displayed separately in the characteristic map as a graph of the reference mass flow contingent on the pressure ratio. At every computed operating point is examined whether the reference mass flow lies left of the surge line. If this is the case, instead of the mass flow, its negative value or 0 is used for further calculations. By means of this simple assumption, which takes into account the outline of the flow and even a reverse flow through the compressor, the pressure drops after the compressor via the reduction of delivered mass, which is why the pressure ratio sinks at the compressor, and the compressor can again leave the area of the pump. Usually, the graph given in the pressure ratio-volume flow map is used

7.5.2 The positive displacement charger

In the case of positive displacement chargers, the lines of constant charger speeds slope down considerably, similarly to the speed lines near the fill limit in flow compressors. This slope is grounded however in the working method of positive displacement chargers.

The characteristic map, contingent on the pressure ratio and the reference mass flow, the lines of constant speed and constant isentropic efficiency are represented. Now and then are found presentations of lines of constant temperature differences between the outlet and inlet temperature in the charger, which are however relatively easy to transform into isentropic efficiency or to utilize directly in the determination of the required compressor torque.

$$\eta_{is,c} = \frac{T_1 \left(\pi_c^{\frac{\kappa-1}{\kappa}} - 1 \right)}{\Delta T}, \tag{7.247}$$

$$T_{is,c} = \frac{1}{\omega_c} \dot{m} c_p \Delta T. \tag{7.248}$$

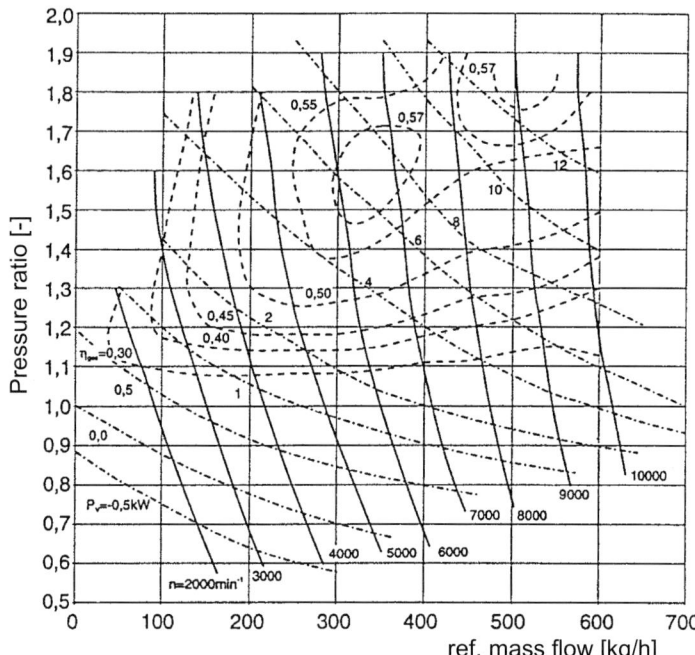

Fig. 7.42: Extrapolation at the positive displacement charger

In positive displacement chargers, the friction of the aggregate plays a significant role. While friction in the case of the turbocharger is usually contained in the total efficiency of the turbine, in the case of positive displacement chargers it has to be considered independently. This can take place either by means of the mechanical efficiency or directly with the aid of the lines of constant propulsion performance. These lines can also be extrapolated in areas of pressure ratios smaller than one with good approximation, in order, for example, to make possible a description of the interaction of the engine and the positive displacement charger at a load regulation by means of a change of the transmission ratio also in areas below the full load of the uncharged engine. With that, an extrapolation provides positive displacement charger performances as well, in which a performance is delivered due to the expansion of a gas mass to a pressure below the intake pressure in the charger. Thus the charger possesses thereby, as we would say in the case of the internal combustion engine, a positive charge changing loop. Fig. 7.42 shows such an extrapolation.

7.5.3 The flow turbine

The adjustment of flow turbines to the required performance range usually takes place experimentally and extends from an alteration of the spiral housing to an alteration of the nozzle ring or to an alteration of the trim. For these alterations, fixed characteristic maps result, with which the operating behavior of these turbines can be described clearly, also in the context of a simulation computation. However, the influence possibilities on the alteration of turbine performance in running operation (variable turbine geometry) partially cause more or less considerable changes in the form of the characteristic maps and in the efficiency behavior of the turbines. These changes can no longer be covered by the basic maps, which is why several maps – as long as they are available – must be superimposed. It is partially also possible with suitable theories of similarity to describe in sufficient exactness the operating behavior of the turbine being used proceeding from a basic characteristic map. For this and for the basic processing of characteristic maps for the simulation calculation from usually quite limited measurement data for a turbine, laws of similarity for a turbine must first be drawn up.

- **Reference quantities**

For the flow turbine, the laws of similarity described for the flow compressor are valid as well, in order to prepare the measurement results and with that finally the map characteristic for the entry conditions that deviate from the entry conditions from the determination of the characteristic map. In determining the maps, most turbocharger manufacturers use standardized entry quantities, which are often very helpful in checking the plausibility of extrapolation results. Thus, the reference entry temperature is set in the turbine, for example at 873 K (600 °C) by some manufacturers.

Similarity regularities can be derived from the similarity via the Mach number and the fluid-mechanical similarity. For the sake of clarity, we have to do without a reference to standard turbine conditions because of the large range of possible entry conditions in the turbine. Only an additional reference to the desired constant reference entry temperature is possible, which is used by some manufacturers. The similarity with respect to the Mach number will be represented here however only with the help of the reference to the actual entry temperature in (7.249)

$$n_{t,ref} = \frac{n}{\sqrt{T_{b.t.}}}; \quad \omega_{t,ref} = \frac{\omega}{\sqrt{T_{b.t.}}} . \tag{7.249}$$

The mass flow through the turbine can be determined by means of the flow equation from the adjoining pressure ratio and a specific flow coefficient, which considers the dependence on a fabricated specific cross sectional surface, which is of course contingent on the turbine speed as opposed to a normal throttle. For this reason, the opportunity presents itself for the turbine to relate the actual mass flow to the entry quantities of pressure and temperature in order to be independent of these quantities at least in the characteristic map representation. Corresponding to (7.236), the following relation is valid for the reference mass flow at the turbine

$$\dot{m}_{ref} = \dot{m} \frac{\sqrt{T_{b.t.}}}{p_{b.t.}} . \tag{7.250}$$

A transformation to standard conditions for the reference mass flow of the turbine is not sensible for the above mentioned reasons.

The reference mass flow through the turbine thus amounts to

$$\dot{m}_{t,ref} = \mu_t A_t \frac{1}{\sqrt{R}} \sqrt{\frac{2\kappa}{\kappa - 1} \left(\pi_t^{*\frac{2}{\kappa}} - \pi_t^{*\frac{\kappa+1}{\kappa}} \right)} . \tag{7.251}$$

The flow coefficient can be interpolated from the characteristic map described in Fig. 7.44.

In chap 2.6, the basic equations of the turbine performance is presented (2.72). For the isentropic torque of the turbine results from this

$$T_{t,is} = \frac{1}{\omega_t} \dot{m}_t \frac{\kappa_{v.t.}}{\kappa_{v.t.} - 1} R_{v.t.} T_{v.t.} \eta_{is.t} \left(1 - \pi_t^{*\frac{\kappa_{v.t.} - 1}{\kappa_{v.t.}}} \right) . \tag{7.252}$$

The isentropic turbine efficiency can thereby be determined from a characteristic map also described in Fig. 7.44. If we insert (7.251) into (7.252) and convert, we obtain the isentropic turbine torque related to the pressure after the turbine contingent on the reference turbine speed and the reference turbine pressure ratio. We should thereby consider that the reciprocal value of the actual turbine pressure ratio is utilized

$$\frac{T_{t,is}}{p_{a.t.}} = \mu_t A_t \eta_{is,t} \frac{60 \left[\frac{s}{\min} \right] \sqrt{2R}}{2\pi} \left(\frac{\kappa}{\kappa - 1} \right)^{1.5} \frac{1}{n_{t,ref}} \pi_t^{*\frac{\kappa+1}{\kappa}} \left(1 - \pi_t^{*\frac{\kappa-1}{\kappa}} \right)^{1.5} \tag{7.253}$$

with

$$\pi_t^* = \frac{p_{a.t.}}{p_{b.t.}} .$$

Precisely in the context of simulation computation, the turbine cross sectional surface must often be adjusted for the adjustment of the turbine and the fine tuning of possibly existing

7.5 Charging

measurement results. In the case of these alterations, not the size of the turbocharger but only its cross sectional surface is changed, which is why also the reference turbocharger speeds need not hereby be transformed. In an alteration of the cross sectional surface through a change of the nozzle ring, the in-flow of the impeller, but not its size is changed. For this reason, in this case as well, a conversion of the reference turbine speeds can be left out. The changes in efficiency and flow coefficients caused by the varying in-flow of the impeller have to be estimated on a individual case basis. In comparison to the errors made in the preparation of the characteristic maps in the context of extrapolation, they can, however, usually be ignored at minimal changes in quantity.

A further characteristic quantity for a flow turbine is the so-called type number, which is expressed in terms of the quotients from the peripheral speed of the impeller and the theoretical flow speed, which can be calculated from the adjoining static enthalpy slope at the turbine. The formal relation for the type number reads according to this

$$\frac{u}{c_0} = \frac{u}{\sqrt{\frac{2\kappa R}{\kappa - 1} T_{b.t.} \left(1 - \pi_t^{*\frac{\kappa-1}{\kappa}}\right)}} \quad . \tag{7.254}$$

For the peripheral speed at the mean diameter of the turbine d_m is valid

$$u = \frac{\omega_t \, d_m}{2} = \frac{\pi n d_m}{60 [\text{s} \cdot \text{min}^{-1}]} \quad . \tag{7.255}$$

With that, the result for the type number of the turbine is the following relationship with the reference speed. Here we already see the advantage of the use of laws of similarity

$$\frac{u}{c_0} = \frac{\frac{\pi d_m}{60[\text{s} \cdot \text{min}^{-1}]} \frac{n}{\sqrt{T_{b.t.}}}}{\sqrt{\frac{2\kappa R}{\kappa - 1}} \sqrt{1 - \pi_t^{*\frac{\kappa-1}{\kappa}}}} \quad . \tag{7.256}$$

Fig. 7.43 shows the relationships between the turbine pressure ratio, the type number, and the reference turbine speed according to (7.256) in the form of a diagram. The reference speeds increase towards the top right. For the creation of a diagram, a few simplifying assumptions must be made. The isentropic exponent κ is set at 1.34 and the gas constant R to 289 J/kg K. Furthermore, a constant mean turbine diameter is taken as a basis.

We recognize in Fig. 7.43 that the type numbers shift to higher values with increasing reference turbine speeds at a realistic turbine pressure ratio (< 5). This determination has effects on the extrapolation of turbine maps. The larger the reference speeds, the less important the course of the characteristic map quantities at smaller type numbers and the more important the course at larger type numbers. One further aspect of the diagram analysis is the fact that the type numbers can take on values larger than 1 at larger type numbers at corresponding combinations of turbine pressure ratios and reference speeds, which is also of importance for an extrapolation.

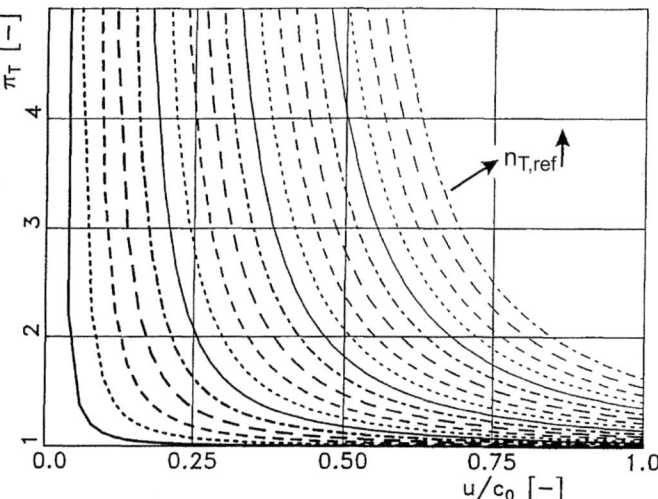

Fig. 7.43: Relation of the referenced turbine speed, type number and turbine pressure ratio

- **Extrapolation**

Most turbine maps are measured in the so-called dry cycling of the turbocharger at the combustion chamber test bench. The turbine is thereby admitted with hot gas, which – as mentioned above – usually possesses a constant temperature. The load on the turbine is adjusted via a throttling of the compressor and returned to equilibrium through a heightened mass flow at the turbine. Due to these facts, the working range of the turbine is very limited and the measured characteristic map range of the turbine is very small. For stationary approaches, this range is usually totally sufficient, while in the unsteady consideration of the interaction of engine and turbocharger or in partial or light load ranges, considerable problems result. The same effect makes itself known in this case as in the compressor maps: that precisely the characteristic map range of low reference speeds is lacking in the case of the turbine as well. This range is however of special interest for an exact simulation of the turbocharger.

The characteristic maps measured at the test bench are usually represented with the help of two diagrams. The first diagram contains the reference mass flow as a function of the pressure ratio with the reference turbine speed parameter, and the second diagram the total turbine efficiency again as a function of the pressure ratio with the reference turbine speed parameter. Fig. 7.44 (left) shows a characteristic map for a small radial turbine.

In this presentation, the characteristic maps can only be applied in a limited fashion for a sure interpolation of the values necessary for a simulation computation. The characteristic maps must therefore be transferred to a mode of representation, which, on the one hand, allows an extrapolation into the margin areas of the characteristic map, and on the other hand, guarantees a sure interpolation. With the aid of (7.256), the type number of the turbine can be calculated from the reference turbine speed and the turbine pressure ratio. The mean diameter of the turbine must be estimated in a suitable way for this, if it is not known. An inaccurate estimation of the mean diameter does not necessarily lead to an error in turbine calculations in

7.5 Charging

the context of simulation calculation, as long as this value remains tightly fixed to the turbine and the calculation of the type number always takes place with this value. The isentropic exponent κ and the gas constant R have to be determined with the help of the knowledge of the turbine entry temperature (873 K) and of the air-fuel ratio in the combustion chamber of the turbocharger test bench by means of a corresponding relation between temperature, the air-fuel ratio, and the condition quantities. For a diesel-run combustion chamber, an air-fuel ratio of approx. 6 can be assumed. With it, the isentropic efficiency can be represented as a function of type number with the reference turbine speed parameter. With (7.251), it is possible to determine the flow coefficient of the turbine with the help of the turbine cross sectional surface. Alternatively, the mass flow can be directly plotted here as well. The success of this transformation, as Fig. 7.44 (right) shows by means of the thick solid lines, is unfortunately usually rather disillusioning, since the area of such a transformed characteristic map originates in a narrow range of type number.

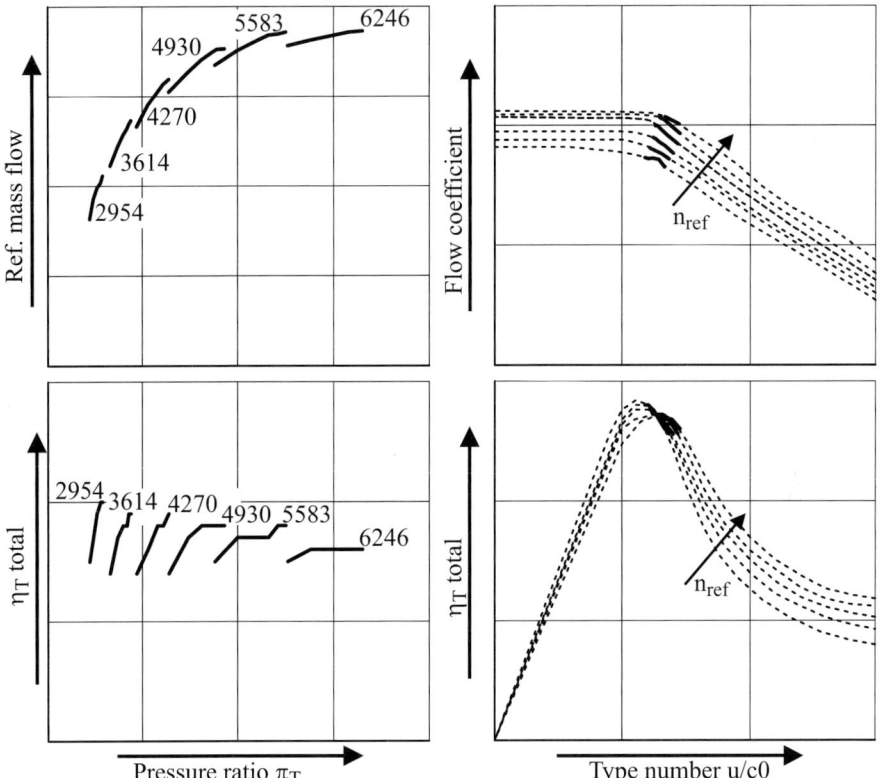

Fig. 7.44: Map for a small radial turbine (left), map using the type number (right)

The characteristic map has to be extrapolated appropriately for unsteady processes, for which the range of the type number extends from 0 at a sudden switching-on of the turbine in the context of register charging from a state of rest ($n_{t,ref} = 0$) to very high values for the type number at a sudden switching-off of the turbine ($\pi_t = 1$) at high speeds. The type number range of the turbine, according to the momentarily adjoining turbine pressure ratio, can deviate strongly also in an impact admission of the turbine at a practically constant turbine speed. The turbine pressure ratio thereby can also take on values of approx. 1, which at least theoretically would cause an infinitely large type number. As has been mentioned, an extrapolation into ranges of small type numbers is not necessary, since this would make necessary more or less infinitely high turbine pressure ratios. Thus at these high speeds, the path of efficiency can be extrapolated from the optimum efficiency towards the left to the coordinate source. At low reference turbine speeds, the type numbers also lie in lower areas. Usually, the optimum efficiency also shifts at these speeds to low type numbers and lower absolute values. This path as well can be extrapolated without much error to the coordinate source. Naturally, the boundary value problem, as the type number approaches the value of 0 at a reference turbine speed other than 0, must be considered here as well, which also allows for an extrapolation to isentropic efficiencies that are minimally larger than 0.

An extrapolation to type numbers right of the optimum efficiency is more difficult. As a basis, one can as a first approximation describe the course of efficiency as a parabola, which has its vertex in the optimum position of efficiency. This extrapolation method provides for most cases satisfactory results, since the range for the type number takes on values that are far right of the optimum efficiency – i.e. at higher type numbers – only in a few extremely unsteady cases.

It should be noted in conclusion that there is no patent formula for the extrapolation of turbine maps. Usually, the only thing that helps is a large number of the most varying plotting types for the most varying turbine parameters, the paths of which then, amongst other things, allow for a further extrapolation. These extrapolations must however then be set in relation to each other again and again by means of the equations described above in order to test their plausibility. A greater progress in the describing and extrapolation of turbine maps can be seen, as in the case of the compressor maps, in the use of the zero-speed line, which clearly limits the range of the characteristic map and thus contains valuable information regarding the usually missing intermediate area of the characteristic map.

- **Characteristic map representation**

Because of the relations between the type number, the reference turbine speed, and the turbine pressure ratio, there are differing ways of representing the parameters of the turbine map. The representational type of contingent quantities of the turbine, like efficiency or the flow coefficient, across the type number with the parameter of the reference speed allows for a good interpolation of these quantities, but it is not very effective in describing the interaction of the engine and the turbocharger or exhaust gas turbine. The plotting type shown in Fig. 7.45 has proven itself to be the most effective, whereby instead of the torque of the turbine the turbine efficiency is represented. This is much more common in association with the representation of the flow compressor in the pressure ratio-mass flow map, in which in the same way the isentropic compressor efficiency is drawn in as a parameter. The pressure ratio-mass flow field is too narrow in the case of turbines to be capable of representing significant alterations. In the case of the chosen mode of representation, both main influence quantities of the turbine, i.e. the pressure ratio and the speed, appear directly and thus do not require a transformation.

7.5 Charging

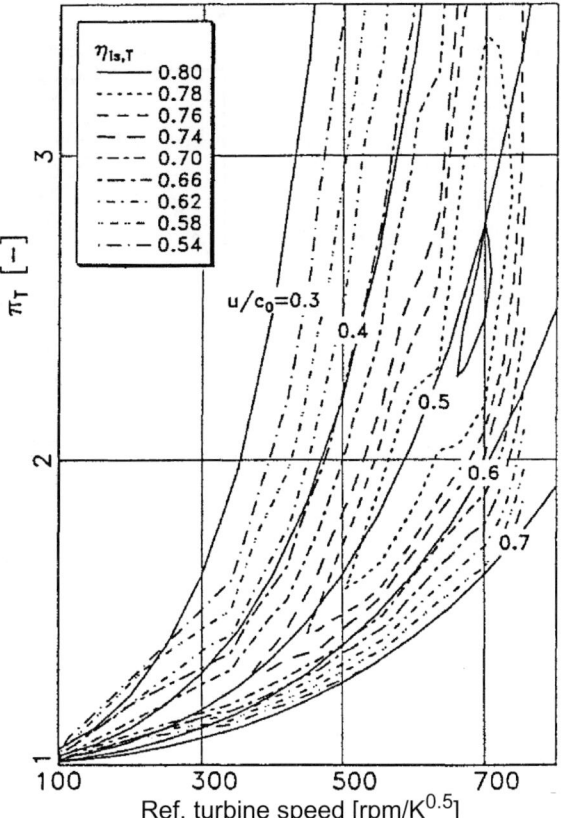

Fig. 7.45: Turbine map for description of operation condition

- **Variable turbine geometry**

In the case of a turbine with an adjustable diffuser, the guide vane position appears as a further parameter. For discrete guide vane positions between both end positions, a complete separate characteristic map is usually measured, which has to be prepared and extrapolated according to the above regularities. Fig. 7.46 shows the characteristic map of such a variable turbine. At a known shifting position, the map quantities for both guide vane positions, between which the actual position lies, must first be interpolated. Finally, the characteristic map values for this shifting position can be determined, for example, by means of a linear interpolation.

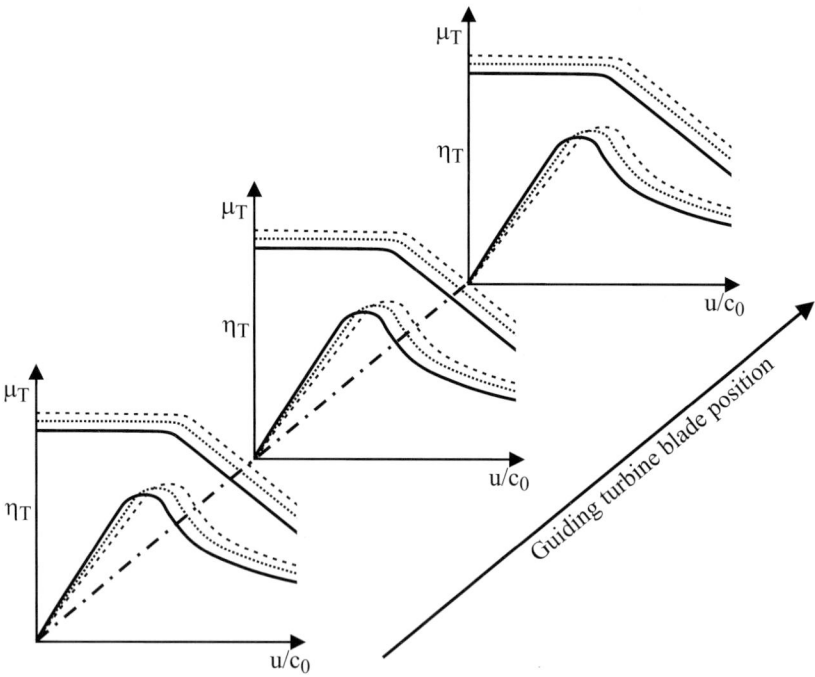

Fig. 7.46: Map of a turbine with an adjustable diffuser

- **Modeling a twin-scroll turbine**

Dual-flow turbines – also called twin-scroll turbines – are employed in engine concepts, in which one is trying to achieve a charge changing which is as frictionless as possible for the exploitation of the maximal potential of charge changing in the internal combustion engine and for the utilization of the exhaust gas pulses at the turbine. Possible areas of use are in-line four or six cylinder engines equipped with a turbine or V8 and V12 engines with a turbocharger for each side. Since in the 4-stroke four cylinder engine, for example, the exhaust impulses arrives shifted by 180 °CA in the exhaust gas system, the exhaust gas impulse of the following igniting cylinder influences the valve intersection phase of the cylinder proceeding it in the order of ignition considerably at conventional valve timing of approx. 240 °CA (see chap. 8.7.6). Such a charge changing leads to increased amounts of residual gas, which cause in the diesel engine a clearly reduced filling and thus losses in performance. In the case of the SI engine, on the other hand, the increased residual gas amounts lead to both losses in filling as well as a considerable increase in knocking potential, since a higher final compression temperature results from higher amounts of residual gas. These effects are shown and briefly described in chap. 7.5.5 and Fig. 7.53.

For this reason, in the case of the four cylinder engine, we summarize by means of so-called ignition sequence manifold two cylinders each igniting shifted by 360 degrees. With this, one creates two separate flows, the further summary of which can only occur further downstream in order to avoid crosstalk. The circumstances for the six cylinder engine are similar consider-

7.5 Charging

ing the other ignition intervals. For emission, response, and package reasons, one must arrange the turbocharger however usually very close to the engine so that one can lead the separate flows without a cross flow location through the turbocharger, so to speak.

Represented in Fig. 7.47 are a sectional view through the turbine scroll (left) and a layout of the turbine scroll (right) for a twin-scroll turbine.

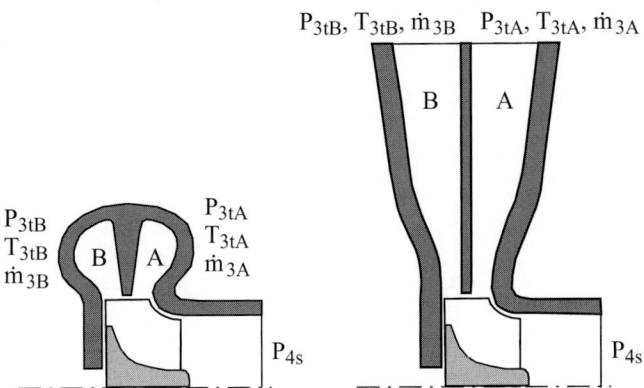

Fig. 7.47: Section through a twin-scroll turbine (Abdel Hamid et al. (2002))

We recognize that two inflow conveyors are arranged beside on another, which admit the turbine wheel directly. The separating wall in the turbine housing reaches almost as far as the turbine wheel, so that because of the high gas speeds there, a cross-flow is avoided similarly to a pulse converter. The flow off after the impeller takes place jointly. An influence upstream the turbine is not to be expected.

For simulation, several demands result with reference to the modeling of the gas path and the turbine, which we will go into in more detail in the following. Through the twin-scroll control of the exhaust gas pipes, the kinetic energy of the exhaust impulse is used directly at the turbine. A complete conversion into pressure and then into kinetic energy again, complemented with efficiency losses, is not necessary. In order to describe reality exactly, it is essential to model the gas path with the one-dimensional gas dynamics presented in chap. 7.4. Since there are varying pressures and mass flows in both flows of the turbine, one must already take this into consideration in the experimental determination of the characteristic map. Contingent upon the so-called branch pressure ratio

$$\Pi_{branch} = \frac{p_{3t,A}}{p_{3t,B}}, \qquad (7.257)$$

which describes the ratio between the pressure in branch A to that in branch B, several characteristic maps are recorded.

In this matter, a mass flow emerges in correspondence with the level of pressure in the branch at a given reference turbine speed. Fig. 7.48 describes these relations for branch pressure ratios of 0.9 (left) and 1.1 (right). Plotted is the reference mass flow across the pressure ratio for the corresponding branch.

$$\pi_{t,A} = \frac{p_{3t,A}}{p_{4s}}$$
$$\pi_{t,B} = \frac{p_{3t,B}}{p_{4s}}$$
(7.258)

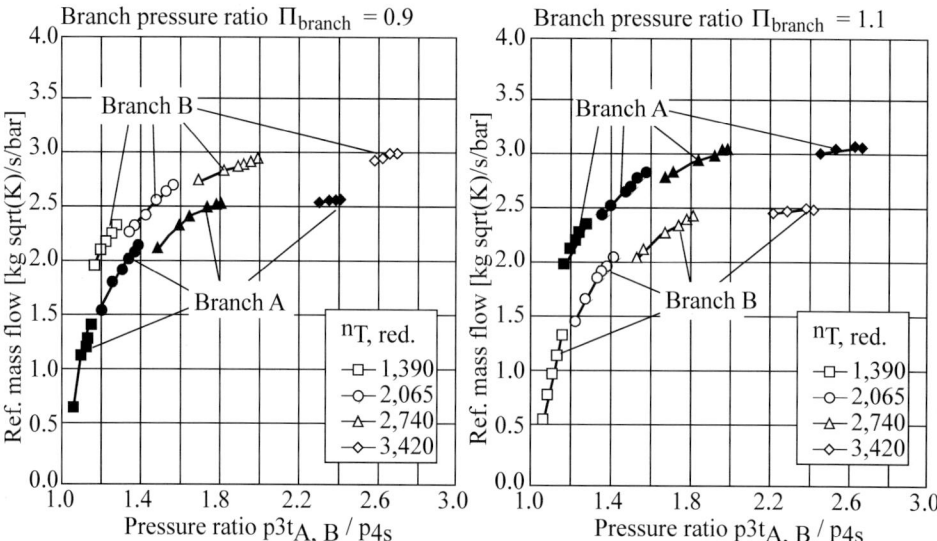

Fig. 7.48: Mass flow map of a twin-scroll turbine for branch pressure ratios 0.9 and 1.1 (Adbel Hamid et al. (2002))

For a given branch pressure ratio, the contingency known from the single-flow turbine shows itself, so that in these diagrams, two turbine maps are recognizable. In this example, the branch pressure ratios of 0.9 and 1.1 were chosen such that the mass flow characteristics of branch A in the right diagram corresponded to those in branch B in the left diagram and vice versa. In actuality, this is not the case: in the chosen example, the reference mass flow in branch A is higher than that in branch B. This becomes clear for a branch pressure ratio of 1 also in Fig. 7.49, in which the graphs for both branches should completely match up. This behavior does not have to be characteristic of twin-scroll turbines, but it does represent the difficulty of laying out twin-scroll turbines symmetrically. As one can see in Fig. 7.47, both flows admit different areas of the impeller, from which in principle a different behavior results. Via an adjustment of the cross sectional surface at the separation point, one can achieve a largely symmetrical behavior.

In the simulation of twin-scroll turbines, we can in principle proceed similarly as in the calculation of turbines with variable turbine geometry, for which we interpolate between several characteristic maps. In the case of twin-scroll turbines, separate characteristic maps are created for the particular branches contingently upon the varying branch pressure ratios. According to the engine's cylinder number and the expected height of pressure pulses, a further range for the branch pressure ratio must be covered that can take on orders of magnitude of 0.5 to 2. Resulting for the efficiency of the turbine is one characteristic map per branch pressure ratio. We do not distinguish hereby between the particular branches, since this is not possible in the experimental determination of the characteristic maps.

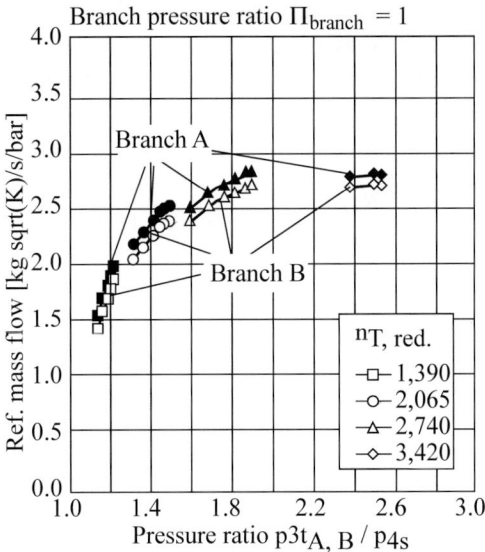

Fig. 7.49: Mass flow map of a twin-scroll turbine for a branch pressure ratio of 1.0 (Adbel Hamid et al. (2002))

We hereby make use of a mean turbine pressure ratio according to

$$\pi_{T,m} = \frac{p_{3t,A} + p_{3t,B}}{2\, p_{4s}} \,. \tag{7.259}$$

The maps for efficiency can also be prepared according to the simulation regularities described in this section.

The calculation of a twin-scroll turbine, roughly speaking, follows this pattern: first, the branch pressure ratio Π_{branch} is determined. Then, one interpolates linearly with this branch pressure ratio in the various branch pressure maps. With the reference turbine speeds for both branches, which can differ because of varying temperature levels, the reference mass flow and the efficiency are calculated, from which the turbine performance for the determination of the turbocharger speed can be calculated with the mean turbine pressure ratio. Because of the calculation in the framework of one-dimensional gas dynamics, the process takes place in a relatively iterative fashion and resembles that of a pipe embranchment for a conjunction situation without a crossing of the flows.

As one can see, the calculation of twin-scroll turbines demands the greatest complexity in turbine modeling. Moreover, a plethora of characteristic maps must be prepared for calculation, which implies a cost which is hardly insignificant.

7.5.4 Turbochargers

- **The core equation for turbochargers**

For the steady turbocharger operation, an equilibrium results between the torque given by the turbine and the torque received by the compressor. The frictional torque should be taken into consideration in turbine efficiency. Solved according to the compressor/turbine pressure ratio, the following relation result

$$\pi_c = \left[\frac{\dot{m}_t}{\dot{m}_c} \frac{\kappa_{b.c.}-1}{\kappa_{b.c.}} \frac{\kappa_{b.t.}}{\kappa_{b.t.}-1} \frac{R_{b.t.} T_{b.t.}}{R_{b.c.} T_{b.c.}} \eta_{total,c} \eta_{total,t} \left(1 - \pi_t^{*\frac{\kappa_{b.t.}-1}{\kappa_{b.t.}}} \right) + 1 \right]^{\frac{\kappa_{b.c.}}{\kappa_{b.c.}-1}} \tag{7.260}$$

$$\pi_t^* = \left[\frac{\dot{m}_c}{\dot{m}_t} \frac{\kappa_{b.c.}}{\kappa_{b.c.}-1} \frac{\kappa_{b.t.}-1}{\kappa_{b.t.}} \frac{R_{b.c.} T_{b.c.}}{R_{b.t.} T_{b.t.}} \frac{1}{\eta_{total,c} \eta_{total,t}} \left(\pi_c^{\frac{\kappa_{b.c.}-1}{\kappa_{b.c.}}} - 1 \right) + 1 \right]^{\frac{\kappa_{b.t.}}{\kappa_{b.t.}-1}} . \tag{7.261}$$

Employed in dynamic propulsion systems, a high measure of unsteady behavior is demanded of engines and thus of the charger aggregates as well. This effect must also be reproducible in the framework of simulation calculations.

- **Angular-momentum conservation law**

Whilst in the case of mechanically propelled supercharger aggregates, the propulsion performance comes directly from the engine and the momentum of inertia of the supercharger aggregate is added to the momentum of inertia of the engine corresponding to the square of the transmission ratio, the unsteady simulation of the turbocharger depends primarily on the difference torque of the turbine and the compressor and on the polar momentum of inertia of the mechanism of the turbocharger.

The formal relationship necessary for the simulation of the run-up of the turbocharger for the change in the angle speed of the turbocharger mechanism is

$$\frac{d\omega}{dt} = \frac{(T_t - T_c - T_f + T_{add.})}{\Theta_{TC}} . \tag{7.262}$$

In the case of electrically supported turbochargers, an additional torque, e.g. via an electric motor, is introduced to the turbocharger shaft in order to speed up the run-up. For the warm engine, the frictional torque can be considered in the form of a total mechanical efficiency in the case of the turbine torque, since the frictional torque occurs reproducibly in contingence on the operating point or the turbocharger speed. For the separate determination of the friction torque in the turbocharger, costly measurements of the entire turbocharger are required. Furthermore, an exact attribution and division to both flow machines is neither possible nor use-

7.5 Charging

ful, since we are essentially concerned with the frictional torque of the mutual connection shaft of the turbine and the compressor. At high speeds, the frictional torque of the mechanism take on an order of magnitude of 5 to 10 percent of the turbine torque.

The goal of an optimization of the mechanism for unsteady operation must therefore be, on the one hand, to minimize the frictional torque in order to increase the surplus torque between the turbine and the compressor, and, on the other hand, to minimize the polar torque of inertia in order to obtain a larger gradient of angle speed.

The reduction in size of one component of the turbocharger causes a clear change in the polar torque of inertia. This is clear from the formal relationship in (7.263). The polar torque of inertia is determined in geometric similarity, on the one hand, by means of the mass of the mechanism, which is proportional to the third power of a geometrical length (e.g. of the diameter). On the other hand, the polar momentum of inertia is determined in addition by square of the diameter

$$\Theta \sim d^3 \, d^2 = d^5 \ . \tag{7.263}$$

The same relation can be utilized for simulation purposes in the necessary size adjustment of the turbocharger, when the turbocharger as a whole or components thereof are geometrically enlarged or reduced.

- **Dependence of the frictional torque on the oil temperature**

In simulating transient warm-up behavior of engines and turbochargers, knowledge of the dependence of frictional losses on the working medium temperatures is of decisive importance.

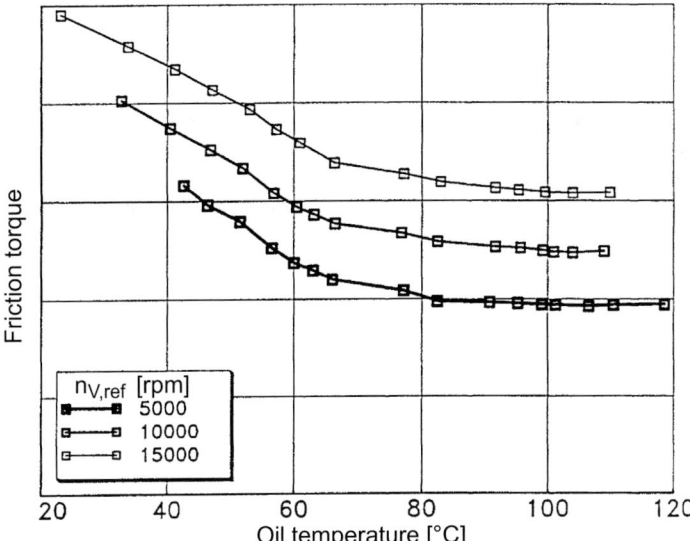

Fig. 7.50: Dependence of the friction torque in a passenger car turbocharger on the oil temperature

Fig. 7.50 shows the experimentally determined dependence of the frictional torque on the oil temperature for a turbocharger supported in babbit bearings for the passenger car range at low speeds, as they might appear in idle speed operation of the engine.

The graphs show an almost linear fall in the frictional torque up to an oil temperature of approx. 80 °C. After this temperature, the frictional torque remains practically constant. This effect remains at higher speeds as well and can be taken into consideration by means of a suitable superimposition of an additional frictional torque contingent on the oil temperature.

- **Total thermal behavior at high turbine temperatures**

In the case of the turbocharger, the heat release of the turbine influences the efficiency behavior of the compressor enduringly, since between the particular subsystems of turbine, bearing housing, and compressor heat flows are exchanged either via heat conduction or radiation/convection. Since the turbine is usually admitted with a constant gas temperature for the characteristic map measurement and is even occasionally isolated, stationary heat flows between turbine and compressor arise that do not occur in reality. With this, only one aspect of the real behavior of a turbocharger is covered in the adiabatic approach with corresponding isentropic efficiencies of compressor and turbine.

If in real operation a higher gas temperature as in the characteristic map measurement now arises, a greater heat flow runs from the turbine to the compressor/into the bearing housing or into the environment. This influences in a sustained way the energy balance of the subsystems and thus also the efficiency behavior. Fig. 7.51 shows schematically the heat flows for a water-cooled turbocharger, in which the compressor, bearing housing, and turbine are seen as their own subsystems.

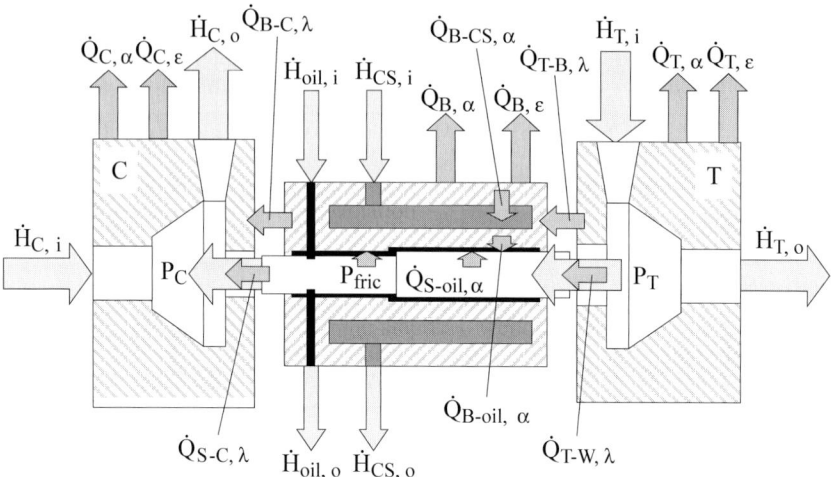

Fig. 7.51: Heat flow model for a water-cooled turbocharger

At high loads, the turbine releases heat to the environment primarily through radiation, which is partially taken in again by the compressor and bearing housing. According to the current

7.5 Charging

flow around of the turbocharger in the engine package, the three subsystems release heat to the surrounding air via free or forced convection. A large part of the heat is transported via heat conduction from the usually very hot turbine into the bearing housing despite baffle plates and decoupling measures. In the same way, a heat conduction takes place in the turbocharger shaft from the turbine into the compressor. A portion of these heat flows is lead away by the lubricant from the bearing housing or the turbocharger shaft. In the case of a water-cooled housing, as is used for thermally heavily burdened turbochargers in SI engines, heat is lead away from the bearing housing by coolant as well.

In order to allow for the description of the individual heat flow terms, we need extensive measurements of the surface temperatures and costly CFD calculations, which leads to unacceptable calculation times in the context of a transient simulation as introduced in chap. 8.7.

Above all at very high temperatures of up to 1,050 °C and the resultantly high wall temperatures, the share of radiation in heat removal becomes more significant. It should be remarked that a description of the efficiency behavior of the compressor and above all of the turbine by means of the characteristic maps measured at the stationary turbocharger test bench is flawed in the simulation with the more errors, the greater the difference in temperature between the real turbine entry temperature and the gas entry temperature at the turbocharger test bench in the determination of the characteristic map. As a simple measure for the simulation, a measurement of the maps of the compressor and the turbine with varying turbine entry temperatures is a possibility. This requires however a very high experimental cost and raises the cost drastically precisely in the case of these turbine models. In the simulation, we must then interpolate between these ancillary maps with the real exhaust gas temperature. The thermally inert behavior can however not yet be considered in such an investigation. This requires in addition heat conduction, convection and radiation models for the description of thermal behavior. A detailed formal modeling of these relations would however go beyond the scope of this book.

7.5.5 Charge air cooling

- **Foundations**

In charged internal combustion engines, the charger aggregate causes an increase in the temperature of the fresh gas as well, due to the change in state occurring in an increase of the pressure of the fresh gas. Via the isentropic state equation and the isentropic efficiency of the charger aggregate, both states are in close relation

$$T_2 = T_1 \left[\frac{1}{\eta_{is,c}} (\pi_c^{\frac{\kappa_1 - 1}{\kappa_1}} - 1) + 1 \right] . \qquad (7.264)$$

The purpose of charging, i.e. a density increase of the charge mass taken in by the engine via pressure increase, is thereby partially thwarted by the increase in temperature of the charge mass. The increase in density is set in the case of a engine that is not charge air cooled by the isentropic exponent, the compressor pressure ratio, and the isentropic efficiency

$$\rho_2 = \frac{p_2}{R_2 T_2} = \frac{p_1}{R_2 T_1} \frac{\pi_c}{\frac{1}{\eta_{is,c}}(\pi_c^{\frac{\kappa_1-1}{\kappa_1}} - 1) + 1} = \rho_1 \frac{\pi_c \, \eta_{is,c}}{\pi_c^{\frac{\kappa_1-1}{\kappa_1}} - 1 + \eta_{is,c}} . \quad (7.265)$$

Fig. 7.52 shows a characteristic map, in which the quotients from the density after the charger aggregate and the density before the charger aggregate are represented according to (7.265) contingently upon the isentropic efficiency and the pressure ratio at an isentropic compression with an isentropic exponent of 1.4. Even in the case of a possible increase of the isentropic efficiency to the value of 1, which is in praxis unachievable, the increase in density is not directly proportional to that of the charge pressure. This effect is all the more pronounced that higher the pressure ratio is, and is grounded in the isentropic change of state. The influence of the isentropic efficiency over the charge air temperature represents an additionally important potential in the increase of density.

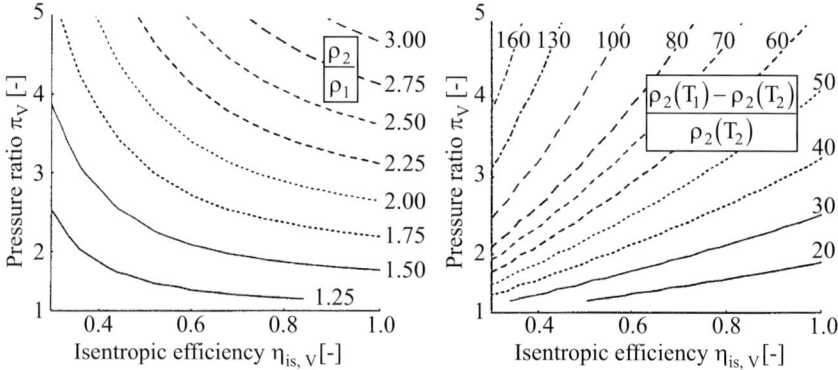

Fig. 7.52: Density ratio due to isentropic compression (left) and percental density factor due to recooling (right)

In Fig. 7.52 (right), the effect is represented of a possible recooling of the charge mass on the temperature before the compressor, again contingent on the isentropic efficiency and the pressure ratio. The theoretically maximum attainable density is thereby directly proportional to the compressor pressure ratio and is thus independent of the isentropic efficiency. The state changes in recooling differ at varying isentropic efficiencies only in the varyingly high amounts of heat that must be removed for the recooling of the charge mass to the state before the charger aggregate. In Fig. 7.52 (right) is represented the percentual increase in density after the charger aggregate at complete recooling as opposed to the density increase shown in Fig. 7.52 (left).

The charge air cooling thus represents a comparatively simple possibility to increase the charge mass and with that the performance of the engine. However, a lowering in temperature has a favorable effect not only with respect to the possibility of performance increase, but also – above all by means of a global lowering of the process temperature level, as well as by means of a lower thermal load on the components – in terms of pollutant emissions (nitrogen oxide) and possible knocking danger in the SI engine, see Dorsch (1982).

Thus, the necessity of charge air cooling also depends on the type of engine. Whilst in diesel engines charge air cooling is only needed for the increase of the charge mass and thus for performance increase or the reduction of emissions, in charged SI- or gas engines, it is absolutely necessary for the avoidance of knocking and for the achievement of a good level of efficiency (optimal position of the 50 mfb). Via the reduction of the temperature of the charge mass, we shift the combustion process away from the knocking boundary. The free space thus created can now be made use of either via a further raising of the pressure ratio, of the compression ratio of the engine, or though an efficiency-improving of ignition timing. One can see from this how much charge air cooling, especially in the case of SI- or gas engines, has an effect on the attainable values of performance and efficiency and thus also on fuel consumption, such that the advantages of an increase of the charge mass are not spoiled by a retarding ignition delay or a lowering of the compressor ratio.

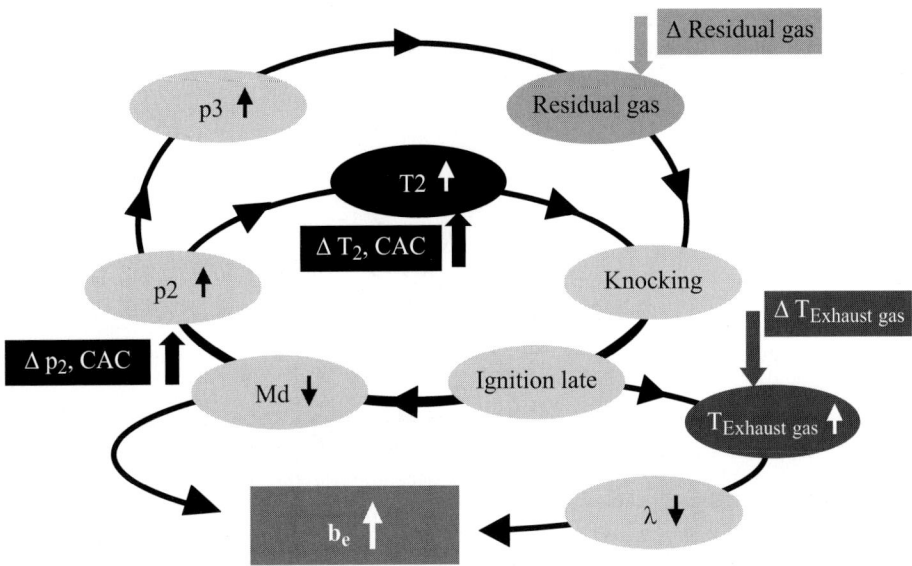

Fig. 7.53: Influence on consumption of a turbocharged SI engine

The relations for a turbocharged SI engine are summarized in Fig. 7.53. A higher loss in pressure in the charge air cooler causes under identical engine performance both a higher exhaust gas counter-pressure as well as a higher inlet manifold temperature. Both cause an increased knocking tendency, which has a ignition delay as a consequence. Through this, the exhaust gas temperature rises and, at the attainment of a value critical for the turbine or the catalyst, creates an enrichment demand, i.e. a higher fuel consumption. At the same time, the torque usually decreases because of the ignition delay, for which can be compensated by a higher charge pressure with the known consequences for exhaust gas pressure and the recooling rate. The spiral is run through once more. We recognize that already a small disturbance of the system can have disastrous consequences. Charge air cooling is thus one of the most decisive keys to efficient turbo engines.

The use and efficiency of charge air cooling depend however primarily on the availability of a corresponding coolant, in order to make possible the reduction of a sufficiently large quantity of heat. The temperature of the charge mass can however be lowered at most to the temperature of the coolant and is moreover still contingent on the heat transfer surface and the mass flow of the coolant. Therefore, various constructive solutions for charge air cooling result for different areas of use in charged engines. For application in ships or in stationary plants, relatively cold water in partly unlimited amounts is available for the cooling of the charge air, while for application in motor vehicles one can only fall back upon the surrounding air or the coolant of the vehicle. The design of the charge air cooler must in addition be taken into consideration the available mass flow of the coolant and its heat capacity.

- **Numerical treatment**

The most common representation of coolers is shown in Fig. 7.54. In this case, the specific cooling output Φ is plotted in contingence on the mass flow of the cooling medium and the medium to be cooled. The heat flow $\dot{Q}_{CAC}$ removed in the cooler can be calculated with

$$\dot{Q}_{CAC} = \Phi(T_{chrg,i} - T_{cool,i}) \ . \tag{7.266}$$

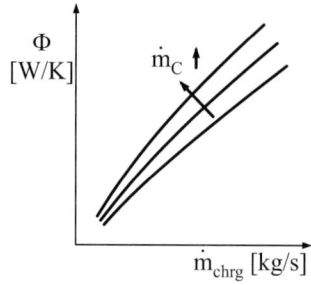

Fig. 7.54: Map of a charge air cooler

The influence of the construction form (cross-flow, counter-flow, or equal flow coolers), the heat transfer conditions, and the heat transfer surface are taken into consideration by the varying paths of the specific cooling performance. The effect of cooling is that much greater, the greater the adjoining difference in temperature between the entry temperature of the charge mass $T_{chrg,i}$ and the entry temperature of the coolant $T_{cool,i}$.

Another possibility for describing the charge air cooler is the cooling digit ε, which gives the quotient from the actual temperature lowering and the maximum possible temperature lowering

$$\varepsilon = \frac{T_{chrg,i} - T_{chrg,o}}{T_{chrg,i} - T_{cool,i}} \ . \tag{7.267}$$

7.5 Charging

For this as well, a representation in a characteristic map can be selected. With the cooling digit ε results for a rough estimate of the density $\rho_{chrg,o}$ of the charge mass after the charge air cooler under consideration of (7.265)

$$\rho_{chrg,o} = \rho_1 \frac{\pi_c \, \eta_{is,c}}{\left(\pi_c^{\frac{\kappa_1-1}{\kappa_1}} - 1 + \eta_{is,c} \right)(1-\varepsilon) + \varepsilon \, \eta_{is,c} \frac{T_{cool,i}}{T_1}} \quad (7.268)$$

The heat flow removed in the charge air cooler can thus be calculated contingently on the temperature difference of the charge air at the inlet and outlet of the charge air cooler

$$\dot{Q}_{CAC} = \dot{m}_{chrg} \, c_p \, \Delta T_{CAC} = \dot{m}_{chrg} \, c_p \left(T_{chrg,i} - T_{chrg,o} \right) . \quad (7.269)$$

With this, we obtain the following relation between the cooling digit and the specific cooling performance:

$$\varepsilon = \frac{\Phi}{\dot{m}_{chrg} \, c_p} . \quad (7.270)$$

In the case of the filling and emptying method, the heat flow taken from the fresh gas in the charge air cooler is calculated according to (7.266) and removed as wall heat flow in a specific volume module.

In reality, a pressure loss appears in the charge air cooler as in every pipe, which partially thwarts the effect of the increase of density of the charge mass via cooling. Therefore, in the constructive design of the charge air cooler, one must strive for a pressure decrease that is as small as possible. In existing engine vehicle construction types pressure loss values of up to 150 mbar appear under rated power conditions. Such pressure losses can no longer be neglected in the simulation.

The consideration of the pressure losses of the charge air cooler in the simulation is however not possible in a volume module, since the pressure in the volume is calculated from the thermal state equation and is thus set for both volumes exits to an equal amount.

The pressure loss in a charge air cooler can thus only be simulated by means of a specific trigger wheel vane, which is arranged before or after the charge air cooler volume. Since the description by means of a simple trigger wheel vane with a constant flow coefficient is insufficient because of the complexity of the charge air cooler, here too we must fall back upon a characteristic map for the description of pressure losses. The losses in the charge air cooler are composed of the losses in the inflow and outflow cases and of the frictional losses in the actual cooling grid. In the characteristic map, these total pressure losses are represented over mass flow. If we know the pressures before and after the charge air cooler trigger wheel vane, we can very easily determine the mass flow through the trigger wheel vane from the characteristic map. The charge air cooler trigger wheel vane thus provides a mass and enthalpy flow, with which the condition quantities in the volumes before and after the trigger wheel vane can be calculated according to the procedure described in chap. 7.3.2. Consequently, a new pressure difference arises, which under stationary conditions corresponds to the pressure loss of the charge air cooler.

The calculation of pressure losses in the charge air cooler using one-dimensional gas dynamics is much more complex. In this case, the charge air cooler is modeled as a pipe-bundle of a number of small tubes corresponding to the cooling shafts, into which the air to be cooled flows. The inflow and outflow cases are represented in this modeling as volumes with the corresponding volumes of the cases. The tubes serve as connections between the volumes. With this method, which is derived from the simulation of catalyst flow, the pressure loss in the charge air cooler can be approximated very well.

However, the heat removal of the charge air cooler is problematic. This has to occur by means of the walls of the pipes.

For this, the heat flow from the charge air to the wall of the tube is determined with the help of methods for the wall heat transfer at the internal pipe wall. The heat exchange of the external pipe wall with the environment is described by the charge air map shown in Fig. 7.54. In order to limit the cost of describing the charge air cooler, we assume that an equal heat flow is removed from each tube. That means that an equal in-flow of the tubes and an equal distribution of the air to be cooler is assumed for all tubes. With the number i of tubes results

$$\dot{Q}_{tubes} = \frac{\dot{Q}_{CAC}}{i} . \tag{7.271}$$

In the case of a cross flow heat exchanger, roughly equal temperatures of the cooling medium exist across the entire pipe course. In the direction of flow however, the temperature of the cooling medium decreases. From this results increasingly small temperature differences in the direction of flow of the cooling medium, which can be considered by means of the length discretization of the tube, necessary anyway for the calculation, with l segment. These affect the heat flow to be removed per segment as follows

$$\dot{Q}_{tubes,j} = \frac{\Delta T_j}{\sum_{k=1,...,l} \Delta T_k} \dot{Q}_{tubes} . \tag{7.272}$$

In this case, ΔT_k represents the temperature difference between the internal pipe wall and the charge air temperature for one segment. The charge air cooler is viewed as being without mass. A storage effect in the component cannot be considered in this modeling.

8 Total process analysis

8.1 General introduction

We understand under the term total process analysis the simulation of the total engine configurations under stationary and transient operating conditions. As a basis, the modeling builds upon the basic building blocks of the engine, like the cylinder, container, pipes, trigger wheel vanes or throttle valves, and charger aggregates, which are described in chap. 7. Since the calculation of the working process in the cylinder only provides indicated quantities, suitable methods for engine friction must be integrated into the total simulation model. In addition, the behavior of the oil and cooling cycle must be calculated in order to permit a calculation of the frictional torque in thermally unsteady processes, since the influence of oil and coolant temperatures on the frictional torque is considerable. In addition, one also needs models, with which a simulation of the loads connected to the internal combustion engine is possible, in order to provide an as exact as possible description of the boundary conditions for the combustion machine.

The behavior of the propulsion system and, for example, the adjustment of the charging unit in reproducible driving cycles can thus be replicated in a fashion which is true to reality. Moreover, the influence of single components within the total system can be worked out by means of an alteration in the model or a change of the parameters or of the characteristic maps for every individual component. This is applicable especially for the selection of charger components and for a possible comparison of varying charging concepts in unsteady processes, for which a description of the individual components as true to reality as possible with a correspondingly large model depth is a requirement. Such a model depth, without which particular influences can no longer be exactly worked out, itself presupposes considerable calculation times. The description of these engine peripherals as well as chosen examples for stationary and transient calculation results for the internal combustion engine are the object of this chapter.

8.2 Thermal engine behavior

8.2.1 Basics

In order to describe the thermal behavior of an engine, e.g. the warm-up, both the oil cycle and the cooling cycle must be modeled. Moreover, both cycles are connected to each other by means of an exchange of heat in the cylinder head, in the engine block, and in (existing if necessary) oil-water heat exchangers. The principal heat supply occurs by means of the wall heat flow originating in combustion. This heat flow is distributed in the engine block, whereby the largest amount is lead away by the wall of the cylinder liner and the cylinder head to the coolant. A small amount is lead away by the combustion chamber walls and via the cooling of the piston to the engine oil. In a turbocharged engine, an additional heat flow is transferred from the turbocharger to the oil. Since the frictional output in the bearings of the engine are also lead away to a large extent by the oil in the form of heat, yet another heat flow joins the oil.

The heat is primarily transferred from the oil cycle to the environment by means of the oil pan and, if present, by the oil radiator and the coolant. Besides that, the heat is emitted by means of the vehicle radiator from the coolant to the surroundings. Considering an unsteady situation, the water and oil masses contained in the particular cycles and the heat storage behavior of the walls must be taken into consideration. Thus, in a cold engine, only a little heat flows from the combustion chamber wall to the coolant cycle, as the engine block must first warm up. The heat removal from the coolant to the environment only begins when the engine itself has reached its normal operation temperature.

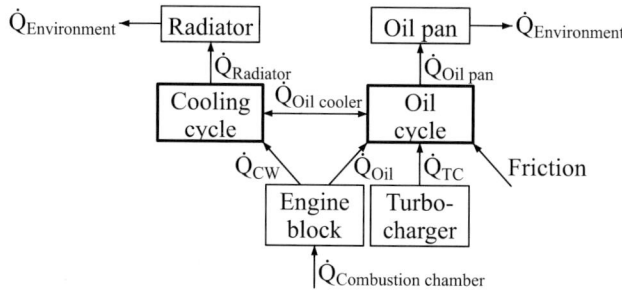

Fig. 8.1: Heat flows in an internal combustion engine

8.2.2 Modeling the pipeline system

Fig. 8.2 shows an analogy consideration (Reulein, 1998) for the calculation model of pipelines. This model has the advantage that no iterations are necessary for the calculation of the mass flows, since the pipe wall is viewed as elastic. Due to the elasticity of the pipe walls, the pipe expands under pressure. One can also analogously consider a container with inflexible walls, which is weighed down by a spring force (spring stiffness k_c). In this way, a certain pressure is initiated in the container. Every change in mass in the container thus alters the pressure. A mass change in the container amounts to

$$\frac{dm}{dt} = \sum \dot{m}_i + \sum \dot{m}_o . \tag{8.1}$$

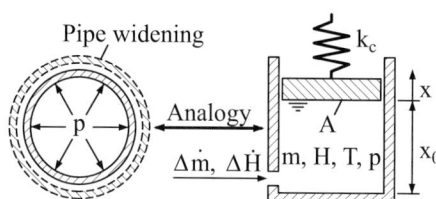

Fig. 8.2: Computational model of a pressure pipe

Under these assumptions, the pressure change can be determined to be:

8.2 Thermal engine behavior

$$\frac{dp}{dt} = \frac{dm}{dt} \frac{k_c}{\rho} . \tag{8.2}$$

The container stiffness k_c thereby corresponds physically to the elasticity module of the pipeline materials. The change of enthalpy in the pressure container can be written as

$$\frac{dH}{dt} = \dot{H}_i + \dot{H}_o + \dot{Q}_w . \tag{8.3}$$

In addition, another wall heat flow $\dot{Q}_w$ is considered in this case, which makes possible a heat removal to the environment. The temperature of the container can be calculated under consideration of the total container mass m and the specific heat capacity c

$$\Delta T_{cont.} = \frac{\Delta H_{cont.}}{m\,c} . \tag{8.4}$$

The mass flow of a fully developed turbulent flow through a pipe with constant cross section A can be calculated as follows

$$\dot{m} = k_{corr}\, A \frac{1}{v} \frac{1}{\sqrt{\zeta}} \sqrt{2\rho(p_i - p_o)} . \tag{8.5}$$

In this case p_i is the pressure on the inlet side and p_o the pressure on the outlet side of the pipe and ρ the pressure of the medium. Friction losses in the pipe are given by the factor ζ, which can be determined in dependence on the geometry and length of the pipe as well as the Reynolds number Re. Additionally, with the correction factor k_{corr}, the behavior of a valve can be simulated. For representing a closed valve, the value $k_{corr} = 0$ must be used, while a completely open valve has the value $k_{corr} = 1$ (modeling of the pressure control valve and of the bypass in the oil cycle as well as the simulation of the thermostat in the cooling cycle).

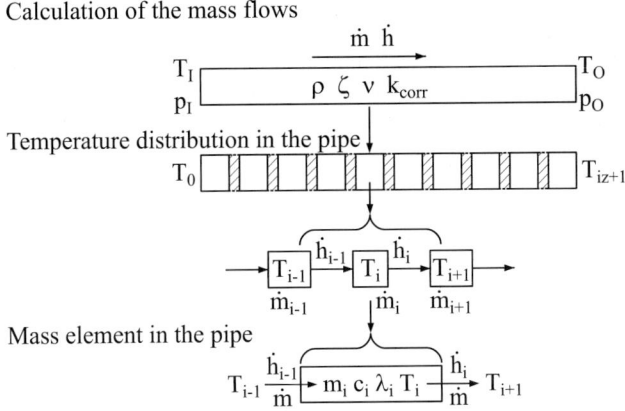

Fig. 8.3: Computational model of a pipeline

In order to render possible a temperature equalization within the pipe section, if the flow comes to a halt, heat conduction must also be considered in addition to heat transport (enthalpy transport). In accordance with finite element models, the conduction is subdivided according to Fig. 8.3 into a number of equally large conduction segments each with homogenous temperature. Then both the heat transport and the heat conduction within the pipe can be calculated.

With iz conduction segments, the temperatures T_0 and T_{iz+1} of the adjoining pressure container, the mass flow $\dot{m}$ through the pipe, and the fluid mass of an element m_i are known as boundary conditions for the entire pipe. From these assumptions, the following equation can be set up for the temperature change of the mass element i:

$$m_i\, c_i(T)\frac{dT_i}{dt} = \left(\dot{H}_i + \dot{H}_o + \dot{Q}_{i,w} + \dot{Q}_L\right) . \tag{8.6}$$

Besides the ingoing and outgoing enthalpy flows, another amount of heat supply $\dot{Q}_{i,w}$ from outside and heat conduction $\dot{Q}_L$ along the pipe are also considered. Valid for heat conduction $\dot{Q}_L$ is

$$\dot{Q}_L = \frac{\lambda A}{l_i}\left(T_{i-1} - T_i\right) + \frac{\lambda A}{l_i}\left(T_{i+1} - T_i\right) . \tag{8.7}$$

For the temperature change of the conduction segment i, the differential equation then results

$$\frac{dT_i}{dt} = \frac{iz}{m\, c_i(T)}\left[\dot{m}\left(c_{i-1} T_{i-1} - c_i T_i\right) + \frac{\dot{Q}_w}{iz} + \frac{2m\,\lambda\, iz}{\rho l^2}\left(T_{i-1} + T_{i+1} - 2T_i\right)\right] . \tag{8.8}$$

In this case, m is the mass contained in the pipe, l the total length of the pipe, and $\dot{Q}_w$ the heat flow added to the pipe.

8.2.3 The cooling cycle

Fig. 8.4 shows, as a modular mimic display, the coolant cycle of a turbocharged diesel engine. The coolant is delivered from the water pump (1) through the engine block (2). Finally, the mass flow divides itself up, contingently on the position of the thermostat, (8) into the cycle segments of heating (7), radiator (6), and short circuit (3).

8.2 Thermal engine behavior

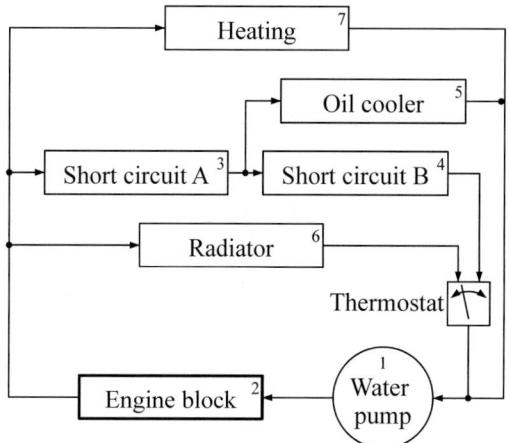

Fig. 8.4: Scheme of the coolant cycle

- **The water pump**

As a rule, single-stage centrifugal pumps are utilized for the delivery of the coolant.

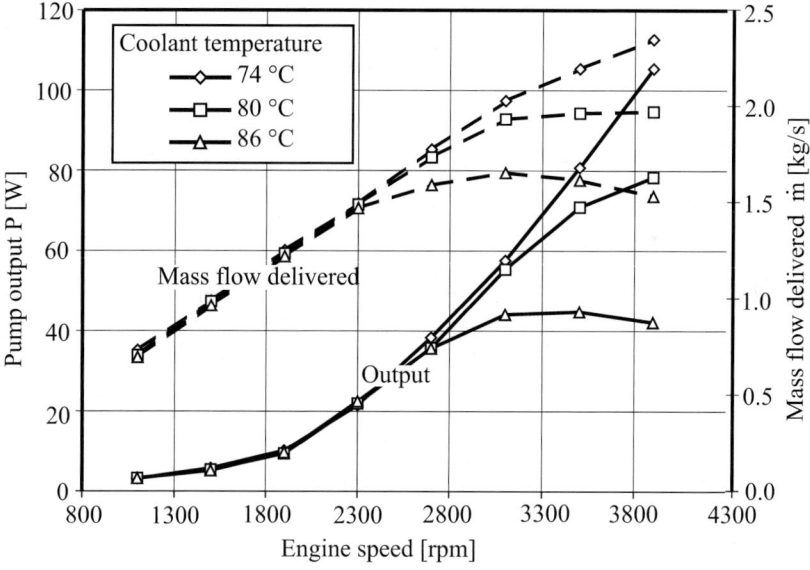

Fig. 8.5: Water pump map

The delivery flow of the pump is contingent on its speed and the flow circumstances at the entry and exit of the rotor. The delivery output of the pump can be determined with the following equation under consideration of the density of water ρ

$$P = \frac{\dot{m}_{cm}}{\rho}(p_o - p_i) \ . \tag{8.9}$$

The paths of the delivered coolant mass flow and the pump output are shown in Fig. 8.5.

- **The thermostat**

The control of the coolant temperature takes place by means of a thermostat. Besides two-way thermostats, which only influence the mass flow through the radiator, three-way thermostats are also employed. In the case of elastic material thermostats, the elastic material is surrounded by the flow of the coolant. In contingence on its thermal expansion, the valve openings are either opened or closed (see Fig. 8.6).

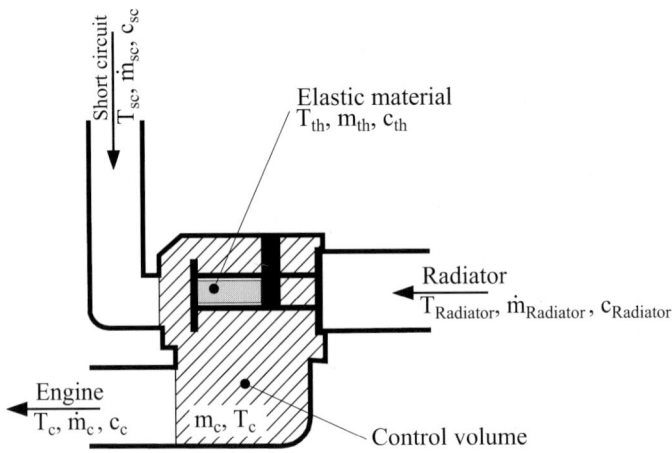

Fig. 8.6: Computational model of a two-way thermostat

For the reference opening path a of the thermostat, the following equation is valid

$$a = \begin{cases} 0 & \text{for} \quad T_{cm} < T_l \\ \dfrac{T_{cm} - T_l}{T_u - T_l} & \text{for} \quad T_l < T_{cm} < T_u \\ 1 & \text{for} \quad T_{cm} > T_u \end{cases} \ . \tag{8.10}$$

A linear approximation is sufficient for most cases.

The valve of three-way thermostats is simulated by the pipe calculation model, such that the reference opening path a is used for the conduction of the radiator and the complimentary value $(1 - a)$ is used as a correction factor k_{corr} for the conduction of the short circuit in (8.5).

This is adequate for the pre-calculation of mass flows. However, problems arise in unsteady considerations, since in this case, besides the opening characteristics, the temporal reaction behavior of the thermostat must be considered. This plays a particularly large role in the first

8.2 Thermal engine behavior

opening of the thermostat. In this case, large temperature difference arise between the mass flows of the radiator and short circuit cycles, while on the other hand the elastic material requires a certain amount of time to adjust to the adjoining temperatures. Thus, in order better to reproduce the behavior of the thermostat, its heat storage capacity is also considered.

For calculating thermostat reaction behavior, a simple model can be used, which is portrayed in Fig. 8.6. In this case, we assume that the total surface of the thermostat is surrounded by the flow of the control volume of coolant. Within the control volume, the coolant has a homogeneous temperature, which can be determined according to the energy balance

$$m_c c_c \frac{dT_c}{dt} = \dot{m}_{sc} c_{sc} T_{sc} + \dot{m}_{cooler} c_{cooler} T_{cooler} - \dot{m}_c c_c T_c . \tag{8.11}$$

If we assume a homogeneous temperature T_{th} of the thermostat, it can be determined – in correlation with its surface A, mass m, heat capacity c, and the heat transfer coefficient between the coolant and the thermostat – with the differential equation

$$m c \frac{dT_{th}}{dt} = \alpha A (T_c - T_{th}) . \tag{8.12}$$

- **Heat exchangers**

For the heat removal to the environment and for vehicle heating, water-air heat exchangers are employed, which are either finned tube radiators or gilled radiators. Engines of higher output and turbocharged engines additionally need an oil radiator. These are usually built in a round pipe or flat pipe fashion. Heat removal takes place thereby as a rule to the coolant. The procedure for the simulation of charge air radiators has already been described in chap. 7.5.5, for which reason we will not go into any more detail here.

8.2.4 The oil cycle

The lubrication of vehicle engines customarily takes place in the form of a pressure circulation lubrication. This is represented schematically in Fig. 8.7 for a turbocharger passenger car. In this case, the engine oil is delivered from the oil pump from an oil pan. It then flows through the oil filter and the oil radiator. The cooled engine oil is lead from the main oil bore ending in the engine block across branch canals to the single bearings. The lubrication of the turbocharger takes place by means of an additional oil lead.

To protect against excessive oil pressure in a cold engine, an excess pressure valve is built into the oil pump. Besides the sections engine block and turbocharger, the lubricant cycle has yet another bypass line. The control of the oil pressure in a operationally warm engine takes place via this pipe. Heat flows get into the oil cycle from both the engine block and the turbocharger. The direction of the heat flow transferred in the oil radiator is contingent on the present operating state of the engine. In a warm engine, heat is delivered to the coolant. During the warm-up phase, the engine oil takes pick up heat from the coolant. An additional cooling of the oil takes place by means of the heat removal from the oil pan to the surrounding air.

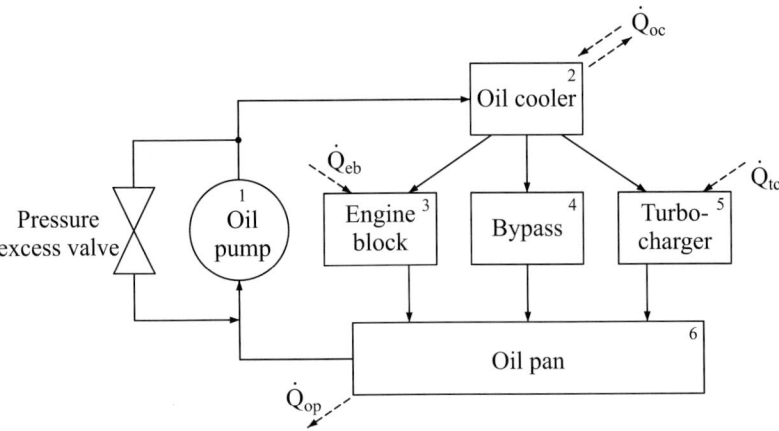

Fig. 8.7: Scheme of the oil cycle

- **The oil pump**

For the delivery of engine oil, gear pumps are used as a rule. According to Schlösser (1961) and Findeisen and Findeisen (1994), the delivered volume flow q of the pump consists of the geometrical delivery flow q_{theo} minus loss oil flows, which result due to the influence of viscosity $q_{V\eta}$ and the influence of density $q_{V\rho}$

$$q = q_{theo} - q_{V\eta} - q_{V\rho} \ . \tag{8.13}$$

The theoretical delivery flow results from the theoretical delivery volume V_i and the propulsion speed n of the pump

$$q_{theo} = V_i \, n \ . \tag{8.14}$$

The loss oil flow $q_{V\rho}$ can be derived from the law of Hagen-Poisseuille. According to it, the leakage flow through a crack of height h, breadth b, and length l is calculated in contingence on the pressure difference adjacent to the pump and the viscosity of the oil

$$q_{V\eta} = \frac{b\,h^3\,\Delta p}{12\,\eta\,l} = k_\eta \frac{\Delta p}{12\,\eta} \ . \tag{8.15}$$

The leakage flow $q_{V\rho}$ caused by the influence of density can be calculated with the following equation

$$q_{V\rho} = k_\rho \sqrt{2\frac{\Delta p}{\rho}} \ . \tag{8.16}$$

The factor k_ρ is dependent on the form of the crack. With it, the mass flow through the oil pump can be determined

$$\dot{m} = \rho \left(V_i\,n - 3k_\eta \frac{\Delta p}{12\,\eta} - k_\rho \sqrt{2\frac{\Delta p}{\rho}} \right) \ . \tag{8.17}$$

8.2 Thermal engine behavior

As an example, values for a turbocharged diesel engine with an output of approx. 66 kW is given in the table below (Reulein, 1998).

Tab. 8.1: Values for the diesel engine of a passenger car

V_i	k_p	k_η
$8 \cdot 10^{-6} \; \dfrac{m^3 \; min}{s \; rot.}$	$1 \cdot 10^{-7} \; m^2$	$1.7 \cdot 10^{-11} \; m^3$

The enthalpy flow delivered by the oil pump amounts to

$$\dot{H}_{oil} = c_{oil}(T) \, \dot{m}_{oil} \, T_{oil} \, . \tag{8.18}$$

Fig. 8.8 shows a delivery amount map for a spur gear pump.

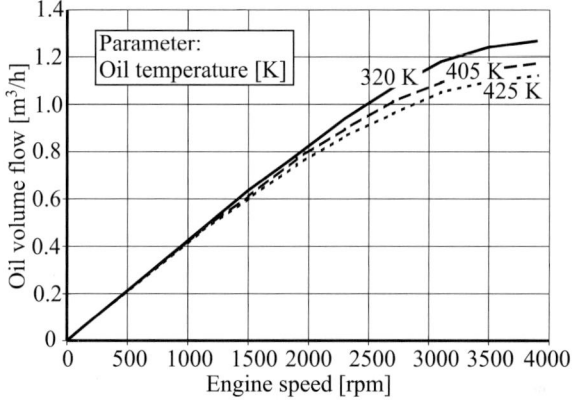

Fig. 8.8: Map for oil pump delivery (spur gear)

- **The oil pressure controller**

The delivery output of the oil pump must be designed for the lubricant requirements of the warm engine. Since the delivery output of the oil pump is nearly independent of the viscosity of the lubricant, a very high oil pressure occurs at engine start because of the high flow resistance of the bearings. This oil pressure is reduced by the excess pressure valve, in order to protect the oil pump and the oil filter. The excess pressure valve is built into the engine block very close to the oil pump as a rule, or it is directly integrated into the oil pump. usually, a valve ball is used for this, which is pressed against its valve seat with the help of a coil spring.

The opening characteristics of this valve can be given by the factor k_{corr} in (8.5). The oil mass flow is lead back into the oil pan after the oil pump and thus does not influence the warm-up behavior of the engine block. This valve is designed such that a pressure sufficient for the lubrication of the engine exists even at high oil temperatures and low engine speed. The bypass can be modeled as a pipe, similarly to the pressure control valve.

- **Oil cooling**

The heat received by the engine oil is removed by the oil pan in engines of lower performance. Engines of higher performance are considerably more thermally loaded and need additional oil radiators. Especially charged engines can reach critical oil temperatures, since a relatively large additional amount of heat is imparted to the engine oil by means of the lubrication of the turbocharger shaft besides the higher thermal load. In vehicle engines, the heat removal takes place as a rule from the oil radiator to the coolant.

The model for the oil pan can be considered in analogy to the wall model introduced in chap. 7.1.6, which described the thermally unsteady behavior of a wall. For the heat flow entering and leaving the wall is valid

$$\dot{Q}_i = \alpha_{oil}\ A\ (T_{oil} - T_w)$$
$$\dot{Q}_o = \alpha_{env}\ A\ (T_{env} - T_w)$$
(8.19)

The heat flow lead away from a heat exchanger can be determined fundamentally according to Küntscher

$$\dot{Q} = \Phi (T_c - T_{tc})\ .$$
(8.20)

In this case, T_c is the temperature of the coolant, T_{tc} the temperature of the medium to be cooled and Φ the so-called cooling digit. This procedure was already introduced in the modeling of charge air radiators and extensively described (see chap. 7.5.5). Fig. 8.9 shows a map for an oil-water heat exchanger.

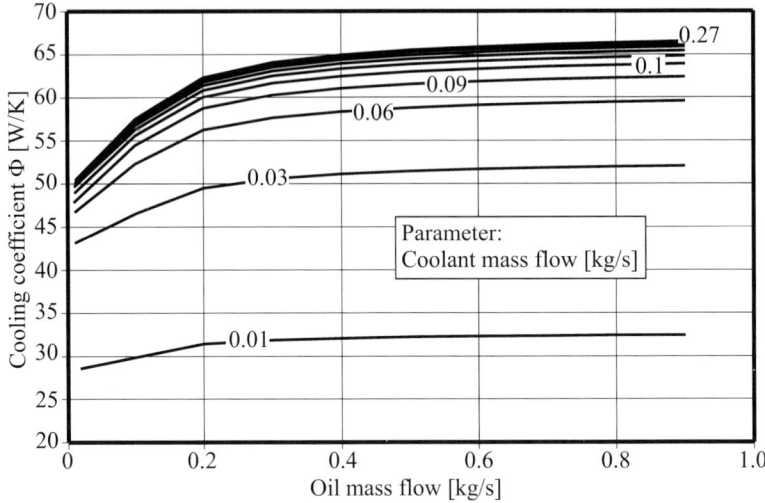

Fig. 8.9: Oil radiator map, measured

The heat flow transferred from the turbocharger to the oil is composed of the friction output at the turbocharger shaft and the heat flow given off of the hot turbocharger housing to the oil.

8.2 Thermal engine behavior

The modeling of the turbocharger with a simple model is not trivial, since hot exhaust gas, cool fresh air, as well as hot oil and, with a water-cooled housing, coolant come into close quarters with each other. For the bearing of the turbocharger, we can assume according to Niemann (1981) that the frictional performance P_f is independent of the load on the bearing. It can be determined by the speed n and the dynamic viscosity η of the lubricant

$$P_f = k_b \, n^2 \, \eta \frac{1}{3{,}600} \qquad (n \text{ in [rpm]}) \, . \tag{8.21}$$

The constant k_b is defined dependently on the diameter of the bearing pin d, the width of the bearing b, and the relative clearance ψ as

$$k_b = 3 d^2 \, b \, 2\pi^2 \, \frac{1}{\psi} \, . \tag{8.22}$$

For the calculation of the heat flow from the turbine housing to the oil, several simplifying assumptions must be made. A water-cooled housing will not be considered here. At first, the compressor side of the turbocharger is neglected in calculating the heat transfer, since the temperature of the compressor housing lies in the same order of magnitude as the temperature of the engine oil. For the temperature of the internal turbine wall, the temperature of the exhaust gas manifold can be employed with sufficient accuracy. Thermally unsteady behavior can again be described with the wall temperature model, whereby the relation of Hausen (1976) for a pipe can be used for the gas-side heat transfer coefficient, see (7.177).

If we combine the amounts of friction and heat transfer, we obtain for the heat flow transferred from the turbocharger to the engine oil

$$\dot{Q}_{oil} = \alpha_{oil} \, A \left(T_{w,tc} - T_{oil} \right) + k_b \, n_{Tl}^2 \, \eta \, . \tag{8.23}$$

The heat transfer coefficient α_{oil} for the ring gap of the bearing can be determined according to the VDI heat atlas (Verein Deutscher Ingenieure, 1991) to be

$$\alpha_{oil} = Nu \frac{\lambda}{d_o - d_i} \frac{0.86 \left(\dfrac{d_i}{d_o} \right)^{0.84} + \left[1 - 0.14 \left(\dfrac{d_i}{d_o} \right)^{0.6} \right]}{1 + \dfrac{d_i}{d_o}} \, , \tag{8.24}$$

with

$$Nu = \frac{\left(\dfrac{\xi}{8} \right)(Re - 1{,}000)Pr + \left[1 - 0.14 \left(\dfrac{d_i}{d_o} \right)^{0.6} \right]}{1 + 12.7 \sqrt{\dfrac{\xi}{8}} \left(Pr^{0.67} - 1 \right)} \left[1 + \left(\dfrac{d_o - d_i}{b} \right)^{0.67} \right] \tag{8.25}$$

and

$$\xi = \left(1.82 \, log_{10} \, Re - 1.64 \right)^{-2} \, . \tag{8.26}$$

In this case, d_o is the diameter of the bearing bore, d_i the diameter of the shaft and b the width of the bearing.

8.2.5 Physical properties of oil and coolant

In modeling the oil and coolant cycles, knowledge of the physical properties of the media utilized is necessary. Of particular importance hereby are heat capacity, density, and viscosity. These physical characteristics are usually contingent on temperature. In the following, we will briefly provide the conversion equations of the particular material quantities.

- **The cooling medium**

The course of the heat capacity of water is described according to Stephan and Mayinger (1992) by the following equation

$$c_{cm} = 3.175991 \cdot 10^{-6} T^4 - 0.004251556 T^3 + 2.1382 T^2 - 478.36 T + 44{,}310. \quad (8.27)$$

The density of water is also contingent on temperature according to Kuchling (1988). Its path is calculated as

$$\rho_{cm} = -0.0036 T^2 - 1.8843 T + 752.61. \quad (8.28)$$

The influence of temperature on the kinetic viscosity of water can be ignored. According to Truckenbrodt (1980), a constant value is utilized

$$\nu = 1.791 \cdot 10^{-6} \left[\frac{m^2}{s}\right]. \quad (8.29)$$

By means of the addition of glysantin to the coolant, the physical properties change slightly. The following material values refer to the indications of Küntscher (1987) for a mixture ratio of 50 %

$$c_{gl} = 0.0452 T^2 - 24.045 T + 6{,}361.1. \quad (8.30)$$

Valid for density is

$$\rho_{cm} = -0.0036 T^2 + 1.6665 T + 893.46. \quad (8.31)$$

- **Engine oil**

The physical quantities for engine oil differ quite considerably according to the type. The physical characteristics given in the following and the conversion equations refer to a multi-grade oil of type HDC 15W40.

The heat capacity of the engine oil is contingent on temperature. The conversion takes place according to Große (1962) with the equation

$$c_{oil}(T) = 4{,}459.05 \left[0.414824 + \frac{(T - 273.15)}{1{,}037.34}\right]. \quad (8.32)$$

The density of the engine oil can again according to Küntscher (1987) be determined contingently on temperature with

8.3 Engine friction

$$\rho_{oi}(T) = [\rho_{15} - (T - 288.15)0.64]\left[\frac{kg}{m^3}\right] \quad (8.33)$$

In this case, ρ_{15} is the density of the oil used for the investigations at a temperature of 15 °C with

$$\rho_{15} = 885 \left[\frac{kg}{m^3}\right].$$

The kinetic viscosity of the oil is also contingent on temperature. According to Chaimowitsch (1965), in the temperature range between 250 K and 420 K, the relation

$$v(T) = v_{20°C}\left(\frac{20°C}{T[°C]}\right)^K, \quad (8.34)$$

with

$$v_{20°C} = 3.6 \cdot 10^{-4} \left[\frac{m^2}{s}\right]$$

and $K = 2.0257$, can be used with sufficient precision. The kinetic viscosity $v_{20°C}$ is hereby valid only for the engine oil mentioned above. A pressure dependence of the dynamic viscosity η can be neglected.

8.3 Engine friction

8.3.1 Friction method for the warm engine

In Schwarzmeier (1992), a method is developed for the friction of an internal combustion engine with the help of experimental investigations of particular friction systems. The calculation of the friction mean effective pressure of the engine takes place according to the following equation

$$\begin{aligned}
fmep = fmep_x &+ C_1\left(\frac{c_m}{T_{cyl.w}^{1.66}} - \frac{c_{mx}}{T_{cyl.wx}^{1.66}}\right) \\
&+ C_2\left(\frac{mep}{T_{cyl.w}^{1.66}} - \frac{mep_x}{T_{cyl.wx}^{1.66}}\right) + C_3\left[\frac{(d \cdot n)^2}{T_{oil}^{1.66}} - \frac{(d \cdot n_x)^2}{T_{oilx}^{1.66}}\right], \\
&+ C_4\left[(1 + 0.012c_m)mep^{1.35} - (1 + 0.012c_{mx})mep_x^{1.35}\right] \\
&+ C_5\left(n^2 - n_x^2\right)
\end{aligned} \quad (8.35)$$

with the constants

$$C_1 = \frac{64.0}{z}$$

$$C_2 = 12$$

$$C_3 = \frac{30}{z} \cdot 10^{-3}$$

$$C_4 = 15 \cdot 10^{-3}$$

$$C_5 = \frac{P_{rated} \, xk \, z \, 0.6}{V_d \, n_{rated}^3} + cf \, ifa^2 \, z \, df^2$$

$$xk = 0.1 - 0.07 \, P_e^{0.04}$$

$$cf = 0.14 \cdot 10^{-6}.$$

In more detail, the first part of the method (the term with the constant C_1) considers the friction of the piston group in contingence on the glide speed and temperature of the oil film between the piston and the cylinder wall. The second part (C_2) takes the friction amount of the piston group into consideration contingent on engine load and oil film temperature between the piston and the cylinder wall. The third term of the equation (C_3) considers the friction behavior of the main and connecting rod bearings contingently on the oil temperature, engine speed and bearing geometry, the oil temperature-dependent oil pump work, and the ventilation losses of the crankshaft drive. The fourth part of the method reproduces the load- and speed-dependent influence on the injection pump and the last part the output need of the ancillary components, the coolant pump, and the cooling-air fan, which is determined above all be the speed and the geometry of the fan.

For the pre-calculation of the friction mean effective pressure at an arbitrary operating point, knowledge of the friction mean effective pressure $fmep_x$, the mean effective pressure mep_x, the mean piston speed c_{mx}, the friction-relevant cylinder wall temperature $T_{cyl.\,wx}$, the engine speed n_x, the lubricant temperature T_{oilx}, and the coolant temperature T_{cmx} at a reference point is necessary (index x).

The friction-relevant cylinder liner temperature is calculated according to

$$T_{cyl.\,w} = T_{cyl.\,wx} + f_1 (c_m - c_{mx}) + f_2 (mep - mep_x) + f_3 (T_{cm} - T_{cmx}), \qquad (8.36)$$

with $f_1 = 1.6$; $f_2 = 1.5$ and $f_3 = 0.8$. The range of validity of the method is indicated for oil temperatures over 40 °C.

8.3.2 Friction method for the warm-up

For the calculation of the friction of an engine beneath 40 °C, as is necessary for statement about warm-up behavior, the method of Reulein (1998) is been extended.

- **Friction method for low temperatures**

Already Schwarzmeier (1992) has suggested to include the influence of engine pressure on the bearing friction at low oil temperatures by means of

8.3 Engine friction

$$fmep_b = \frac{C}{T_{oil}^a} mep.$$ (8.37)

With this, we obtain for the friction method under consideration of thermal behavior

$$fmep = fmep_x + C_1\left(\frac{c_m}{T_{cyl.w}^{1.68}} - \frac{c_{mx}}{T_{cyl.wx}^{1.68}}\right) + C_2\left(\frac{mep}{T_{cyl.w}^{1.68}} - \frac{mep_x}{T_{cyl.wx}^{1.68}}\right)$$

$$+ C_3\left[\frac{(dn)^2}{T_{oil}^{1.49}} - \frac{(dn_x)^2}{T_{oilx}^{1.49}}\right]$$

$$+ C_4\left[(1 + 0.012c_m)mep^{1.35} - (1 + 0.012c_{mx})mep_x^{1.35}\right]$$

$$+ C_5\left(n^2 - n_x^2\right) + C_6\left(\frac{mep}{T_{oil}^{1.49}} - \frac{mep_x}{T_{oil}^{1.49}}\right).$$

(8.38)

In accordance with these changes, the coefficients of the remaining terms must be adjusted as well

$$C_1 = \frac{44}{z}$$

$$C_2 = 31$$

$$C_3 = \frac{22}{z}\cdot 10^{-3}$$

$$C_4 = 6\cdot 10^{-3}$$

$$C_5 = \frac{P_{rated}\, xk\, z\, 0.96}{V_d\, n_{rated}^3}$$

$$C_6 = 1.9$$

$$xk = 0.13 - 0.07\, P_e^{0.03}$$

The friction-relevant cylinder surface temperature $T_{cyl.w}$ is calculated according to equation (8.36). For a speed of 2,300 rpm and a torque of 40 Nm is represented in Fig. 8.10 the influence of coolant and oil temperature on the friction mean effective pressure for a 1.9 l turbocharged, direct injection diesel engine. One can see clearly that the influence of the oil temperature is much greater than the coolant temperature.

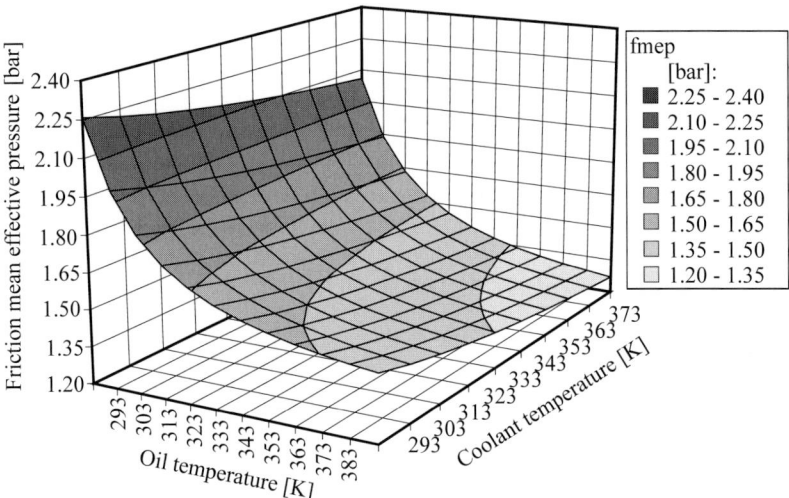

Fig. 8.10: Friction map calculated for a 1.9 l diesel engine at 2,300 rpm, 40 Nm

With the calculation model introduced here it is possible, contingent on the operating point-specific and thermal states of the engine, to pre-calculate the friction mean effective pressure in transient operation. However, the thermal states and the friction are in close connection, since a significant amount of friction is led away to the engine oil as heat. This amount can be determined from the Schwarzmeier (1992) method, if one investigates the single components of the method for their contribution to the warming up of the engine oil. The friction-relevant oil heat flow originates for the most part through bearing friction at the camshaft and crankshaft, the friction of the piston group and the warming up of the oil due to the loss-ridden delivery through the oil pump. Thus

$$\dot{Q}_{fric} = \frac{V_H \, n 10^5}{120} \left\{ k_{oil} \, fmep_x + C_1 \left(\frac{c_m}{T_{cyl.\,w}^{1.68}} - \frac{c_{mx}}{T_{cyl.\,wx}^{1.68}} \right) \right.$$

$$+ C_2 \left(\frac{mep}{T_{cyl.\,w}^{1.68}} - \frac{mep_x}{T_{cyl.\,wx}^{1.68}} \right) + C_3 \left[\frac{(d\,n)^2}{T_{oil}^{1.49}} - \frac{(d\,n_x)^2}{T_{oilx}^{1.49}} \right] \qquad (8.39)$$

$$\left. + C_6 \left(\frac{mep}{T_{oil}^{1.49}} - \frac{mep_x}{T_{oil}^{1.49}} \right) \right\}$$

can be utilized for the calculation of the friction-relevant oil heat flow. In this case, the fraction k_{oil} of the heat flow transferred to the engine oil at the reference point of the total friction output must still be known. A value for k_{oil} of 0.33 can, according to Reulein (1998) be chosen as a first approximation for a high speed diesel engine for a passenger vehicle. With that, it is possible to determine the heat entry resulting from the friction of the engine for a warm-up simulation.

8.4 Engine control

In modern engines, the influence of electronics is becoming more and more important. Closely associated with this is the attempt to influence certain engine or aggregate quantities via electronic control. For pragmatic reasons, PID controller adjusted to the respective employment case are used primarily for the description of control behavior. Especially in the interaction of the engine and the charger aggregate, the controller help to keep to a given rated pressure by means of valves functioning as control elements or throttle valves, or, as in the case of exhaust gas recirculation, to keep to a given rated fresh gas mass. Controller can also be employed in the context of active pump protection control in turbocharged SI engines etc.

A further important function for the simulation of transient behavior of complex vehicle systems is taken over by a so-called driver controller, which realistically imitates the behavior of a driver and also takes over control tasks regarding speed or the sequence of gears.

8.4.1 PID controller

The formal relationship of a PID controller reads

$$Y = K_p X_d + K_i \int X_d \mathrm{d}t + K_d \frac{\mathrm{d}X_d}{\mathrm{d}t} . \qquad (8.40)$$

The so-called control deviation X_d is formed from the difference between the command signal W (rated value) and the control variable X (actual value) and added to the controller. This calculates from that the new correcting variable Y and adds it to the control route, whereupon a new control variable is initiated under consideration of the interference quantity Z. In the case of the correcting variable, we can have to do with a valve adjustment as already mentioned, with which a desired pressure path can be adjusted.

8.4.2 Load control

Both varying load control concepts for engines, quality and quantity control, must also be considered in modeling and simulation of the entire system. While in the calculation of boundary conditions, in the intake pipe for example, a provision of the air-fuel ratio can already take place, in reality a corresponding control of the injected fuel mass is needed for the preservation of the pre-set air-fuel ratio by means of a λ-probe in the exhaust tract. This process is overlaid with the fuel mass evaporating from the intake pipe wall and the control of the throttle valve in order to maintain the pre-set engine load. The gas dynamics mistuned by the opening of the throttle valve creates further problems, since the fresh air and the residual gas amount can change because of this. From these complex relationships, we already recognize that the provision described above of the air-fuel ratio as a marginal value can clearly simplify several control problems for simulation and, for the stationary case example, makes an evaporation model for the intake pipe superfluous in the modeling. Finally, we also save calculation time to a considerable extent, since a build up of the controller is not necessary. The effects of non-homogeneity of the air-fuel mixture (droplet transport etc.) can only be grasped in an extremely rudimentary manner in zero- to one-dimensional calculations anyway.

The control of the load for direct injection engines is much simpler, since their load is primarily proportional to the injected fuel mass. With the exception of $\lambda = 1$ concepts in the case of the SI engine, for which again a dependence on the in-taken air mass is given, the demand on the exact determination of fresh gas or air mass is clearly lower than in the SI engine with quantity control. In quantity control, an error of 5 % in air mass almost directly causes a 5 % change in the load, while in quality control only the air-fuel ratio is altered by 5 % and the last remains for the most part uninfluenced. For this reason, for many years the process calculation of diesel engines was clearly more developed than those of SI engines. Only a relatively exact calculation possibility of the fresh gas mass via one-dimensional gas dynamics made possible a simulation of the SI engine cycle. In engines with electromechanical valve operation, the load is controlled as a rule by the helical groove of the intake valve. In this case, a control time deviation of only a few tenths of a degree of crank angle is permissible for precise load adjustment.

Under transient operation however, a modeling of evaporation in the intake pipe must be carried out, since a clear influence on calculation results from transient leaned mixture and/or enrichment is to be expected. These effects are also less strongly marked in direct injection engines, since in this case a cycle by cycle control of the engine last by means of the injected fuel mass is possible.

In most diesel engines, the load at full load acceleration is controlled in dependence on the build-up of the fresh air mass flow, which is usually delayed by the turbocharger. For mechanical injection pumps, this function is taken over by a boost pressure-dependent fuel limitation, which mechanically limits the injection amount by means of the helical groove of the injection pump until the corresponding boost pressure has been built up. In the case of electronically controlled injection pumps, this function is taken over by a smoke limitation map in the engine control, which electronically limits the injection amount contingent on the measured air mass.

8.4.3 Combustion control

Another great advantage of simulation is the possibility of controlling energy conversion in the cylinder such that, for example, the 50 mfb point is located in a consumption-optimal position (approx. 8 °CA after the ITDC). This can take place by means of either a provision of the position of corresponding substitute combustion functions, if these are already known before the calculation of the operating point, or a simple correction of the ignition time or the start of injection. In the cycle simulation, as opposed to reality, the temporal conversion of fuel energy and thus also the position of the 50 mfb point is known at the end of combustion at the latest, so that it can be reacted upon already at the beginning of the next cycle. This is important for an effective, calculation time-efficient use of simulation.

8.4.4 Control of exhaust gas recirculation

A recirculation of already burned exhaust gas into the combustion chamber of the internal combustion engine is possible in two ways: via internal and external recirculation, or by means of a combination of both. The amount of residual gas can be calculated in the simulation calculation at intake closes by means of the amounts of fresh gas mass taken in and exhaust gas found in the cylinder. These amounts are defined in chap. 7.1.9. In port fuel injection engines, this is the component ξ_{rg} directly, while in direct injection engines the residual

8.4 Engine control

gas amount can be calculated from the air-fuel ratios at "intake closes" and "exhaust opens". The following relationship is applicable

$$x_{rg} = \frac{\lambda_{EVO} L_{min} + 1}{\lambda_{IVO} L_{min} + 1}. \tag{8.41}$$

Usually, nothing can be said about the criteria of ignition conditions and residual gas compatibility of the combustion method in the simulation calculation, so that the results must always be checked for plausibility from these points of view. The situation is the same in the case of possible valve opening clearance in internal exhaust gas recirculation.

- **Internal exhaust gas recirculation**

If in the context of a simulation calculation the residual gas amount should be adjusted to a certain value by means of an internal exhaust gas recirculation, the spreads of intake and exhaust valves as well as the amount of intake pipe and exhaust gas counter-pressure are available as parameters for this. In principle, different types of exhaust gas recirculation can be adjusted across the valve control times. A lowering of the exhaust spread at constant load point causes, as a rule, an increase in residual gas, since with increasing reduction of the exhaust gas spread, the exhaust valve is open longer, by means of which the piston can again take in exhaust gas into the combustion chamber. In the case of a reduction of the intake spread, the intake valve already opens before the exhaust gas has been fully pushed out to the exhaust valve. Thus the exhaust gas is pushed back into the intake tract, in order then to be taken in again. The amount of intake pipe/exhaust gas counter-pressure against the cylinder pressure increases the described effects in the respective direction. A further possibility of raising the amount of residual gas is a choice of control times which impede a complete expulsion of exhaust gas (combustion chamber recirculation).

- **External exhaust gas recirculation**

In the case of external exhaust gas recirculation, a certain exhaust gas mass flow is lead through a separate port in front of the intake tract again. Fig. 8.11 shows this schematically for a direct injection engine

A mass flow can only flow, if a corresponding pressure slope exists between the exhaust and intake sides. The most important parameters for the control of the exhaust gas recirculation rate in external exhaust gas recirculation are thus the intake pipe pressure, the exhaust gas counter-pressure at the respective entry or exhaust location, and the position of the control valve in this pipe. The exhaust gas recirculation rate is determined by the quotient from the recirculated exhaust gas mass flow and the total mass flow taken in through the intake valve into the cylinder. If we consider in Fig. 8.11 the paths for the air and fuel mass flows though the engine separately of one another, we will recognize that the exhaust gas flowing out of the cylinder, the exhaust gas which has flowed back, as well as the exhaust gas after the exhaust gas recirculation location each possess the same air-fuel ratio λ_A for a stationary operating point.

For step-wise cycle calculation, e.g. with the filling and emptying method, the influence of exhaust gas recirculation described above is reproduced automatically, since a mixing of the exhaust gas and the fresh gas occurs in the manifold before the intake valve and the mixture ratio is known at every time.

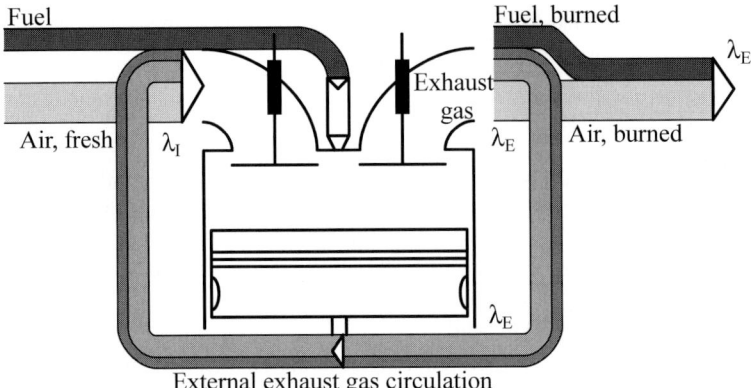

Fig. 8.11: Diagram of external exhaust gas recirculation

While it is thus possible for process calculation to determine the residual gas amount or the exhaust gas recirculation rate at any time, in the real engine, these quantities cannot be determined. In the application therefore, the following procedure is utilized to regulate the exhaust gas recirculation rate: the exhaust gas recirculation valve is adjusted to a rated value with the deviation of the fresh gas mass determined by the air mass meter. If the measured fresh air mass is too high, the valve is opened. With this, under identical engine intake behavior, more exhaust gas mass would be taken in by the engine, which again leads to a reduction of the fresh air mass flow.

Since we are dealing in the case of external exhaust gas recirculation over a throttle location with a continuous process, while the engine process proceeds discontinuously, it can last some working cycles until a stable equilibrium sets in. This should also be considered in the simulation.

8.4.5 Charger aggregate control

One of the most crucial tasks of charger aggregates for internal combustion engines is a provision of boost pressure in as efficient a manner as possible as well as the maintenance of operating boundaries for the charger and the engine. Since the characteristics of the engine and the charger aggregate do not usually match in an optimal fashion, their interaction must be tuned by means of corresponding control strategies.

- **Boost pressure control**

Boost pressure control plays the main role in this case. Other demands are made upon a vehicle engine as, for example, on a large diesel engine with a generator. The latter is designed for the highest and most effective rated performance, the first for the most optimal full load line, which provides a high engine torque already at very low speeds. For this reason, the turbine of a turbocharger for a vehicle engine must be designed smaller than would be necessary for those made for operation at the rated output point. With this, already at very low engine speeds a high boost pressure can be created, which would however climb quite powerfully with increasing engine speed and exceed both the valid range of operation of the compressor as well as the permissible mechanical strength of the engine. For this reason, the exhaust gas

8.4 Engine control

before the turbine of the turbocharger must be diverted and bypassed it, so that the boost pressure can be adjusted in a roughly constant way. In addition, a bypass valve (wastegate) is switched parallel to the turbine, which takes over this task. By means of a control (PID controller), this valve is adjusted such that a predetermined rated boost pressure can be set. The positioning of the bypass valve takes place these days, as a first approximation, without delay and free of overshooting, since the valve is operated either electronically or by means of a modulated boost or vacuum pressure. This is sufficiently fast, so that the behavior of the position controls need not be reproduced in the simulation. In the case of older "controls", in which the adjustment of the bypass took place by means of a pressure-sensitive capsule, the boost pressure was imprinted, should the wastegate be modeled for transient inspections as mechanical single mass oscillator in consideration of all influential pressure, spring, and mass forces.

In the control of the boost pressure by means of the adjustment of a turbine with variable geometry, the adjustment speed is large enough in comparison with the change of thermodynamic quantities to render a detailed modeling of the position controller superfluous. The maximum adjustment speed can also be limited by corresponding parametering in the PID controller.

- **Dump control (surge protection)**

In the case of a sudden closing of the throttle valve of a throttled SI engine, the compressor delivers, because of the inertia of the mechanism of the turbocharger at a high turbocharger speed, against a nearly closed volume, which is why the pressure after the compressor does not drop quickly enough at a, in the first instant, constant compressor speed. The operation line of the compressor moves parallel to the lines of constant compressor speeds in the direction of the surge line. This can thereby be briefly exceeded until the mass flow and the speed have assumed lower values again. Most flow compressors used in SI engines therefore have a dump valve, which is either integrated in the compressor housing or separately arranged. By means of the opening of this valve arranged parallel to the compressor, a mass flow is diverted after the compressor and blown back in front of the compressor. In this way, the pressure can be quickly reduced so that a surpassing of the surge line is avoided.

A further possibility is an active control of the dump valve in connection with the throttle valve. The path of the line for the rated boost pressure can be taken from Fig. 8.12. The rated boost pressure is in this case parallel to the compressor intake line shifted to the left and has to preserve a sufficient distance from the surge line. As long as the rated boost pressure is larger than the actual boost pressure, the dump valve stays closed, which means that the dump valve can always remain completely closed in stationary operation. Control with the dump valve is therefore only activated in a sudden lowering of the rated boost pressure, for example, in the case of a dropping of load.

The compressor operation line wanders, due to the opening of the damp valve in the compressor map, not towards the left to the surge line, but rather down towards the fill limit. The boost pressure may however drop at most to the value which corresponds to the boost pressure in steady operation after the load dropping. For this purpose, the damp valve must be punctually closed. Both the opening speed and the maximum opening path of the damp valve must thereby be adjusted to the closing of the throttle valve and the compressor map, so that one does not fall below the fill limit of the compressor.

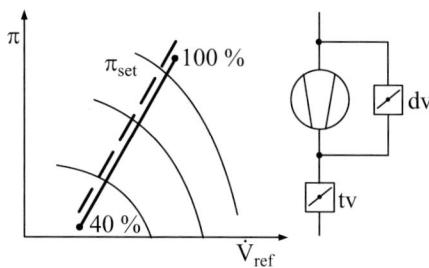

Fig. 8.12: Active control for surge protection

- **Exhaust gas temperature control**

For the reduction of the enrichment need for short-term performance peaks in turbocharged SI engines, thermal inertias of the exhaust gas manifold and the turbine are utilized and for unsteady operation exhaust temperatures allowed, which are not possible in the stationary continuation state. In order to allow for the description of this effect in the simulation as well, the components (exhaust gas manifold and turbine) have to be illustrated as thermally inert masses via corresponding wall temperature models. These component temperatures provide the input for the control of the enrichment need in combination with the exhaust gas temperature.

8.4.6 The driver controller

The calculation of the operation behavior of engines in the context of a simulation of a complete vehicle requires a realistic description of the driver. Its task is not only the reproducible preservation of pre-set speed profiles, but also the independent execution of starting or shifting processes, as described in chap. 8.7 in the investigation of unsteady processes. Precisely these marginal areas can substantially influence the decision in driving cycles for or against a corresponding engine concept, which is why they must be made accessible to simulation. For a vehicle with a gear box (commercial vehicles or passenger cars), the driver must be capable of taking control of the clutch and choice of gear in addition to that of the gas pedal and the brake. All possible operation states of the vehicle must be able to be portrayed for this, which range from a standstill with a floored brake and opened clutch to clutching with a gliding clutch to completely clutched operation. At the same time, the driver must be able to distinguish between a full load acceleration and a part load acceleration and implement the clutching process correspondingly faster or slower.

The starting and shifting processes have a small place in the simulation of driving cycles, but can influence considerably the interaction of the engine and the charger aggregate. The calculation of reproducible driving cycles is important, such that a previous choice of concepts can be made with the help of the simulation.

In addition, the driver must be provided with a special predicting speed controller that approximates the driving speed in the rated speed profile. For speed control for a PID controller, the current rule deviation between rated and actual speed is not used, but rather a rule deviation which run ahead of the current time by a so-called prediction time period. For the determination of this rule deviation, a fictitious actual speed is calculated from the current accel-

8.5 Representing the engine as a characteristic map

eration and speed of the vehicle with the prediction time. From the rated speed at the prediction time and this fictitious actual speed, the rule deviation is determined and added to the PID controller. Fig. 8.13 clarifies this procedure.

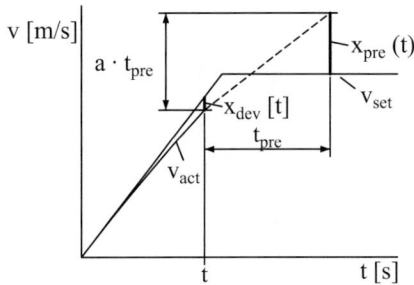

Fig. 8.13: Predicting vehicle controller

8.5 Representing the engine as a characteristic map

In the calculation of total systems in transient operation, an extremely high modeling cost is required. The model depth has to be selected very high at many places, in order to make possible the differentiation of various concepts. Because of this high number of single module and the necessarily great model depth, the computation time cost increases considerably. Even in the representation of the engine cycle according to the filling and emptying method, a very small calculation step length is necessary, which must then conceivably be impressed upon the entire model.

8.5.1 Procedure and boundary conditions

For this reason, in the calculation of the internal processes in the cylinder, we fall back not on the calculation-intensive step-wise working process calculation, but on a multi-dimensional engine map, by means of which the calculation times can be reduced by factors of 10 to 100, see Schwarz et al. (1992). Although the construction and use of a so-called engine map is extensively described there, we will nonetheless briefly describe the basics here.

The engine map is pre-calculated with the help of the working process calculation described in chap. 7 contingently on several entry parameters (Fig. 8.14). The entry parameters are quantities, on which the results of the working process calculation, so-called output quantities, depend. These are set via the pressure and temperature before the intake valve and the pressure after the exhaust valve. The pressure after the exhaust valve is, in the case of the turbocharger engine, in the order of magnitude of the charge pressure and differs according to the respective engine type from it only by the varying scavenging pressure ratio level. In the systematic variation of parameters in the pre-calculation of the engine map, in order to exclude unfavorable parameter constellations with a high pressure before the intake valve and a very low pressure after the exhaust valve or vice versa, one usually varies the scavenging pressure ratio instead of the pressure after the exhaust valve. In that case, we can go into the engine type and the expected scavenging pressure ratio in relation to a turbocharger, by means of which the engine map range can also be restricted to the range which we would expect in reality.

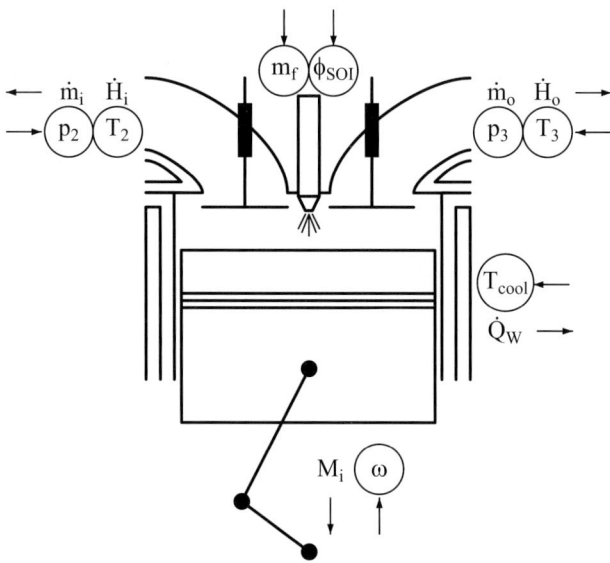

Fig. 8.14: Representation of the engine as an n-dimensional map

From this results an improvement of the interpolation result in the engine map at an identical support point number or a lowering of the support points to be calculated at an equal interpolation quality. Further entry parameters are the engine speed, the fuel mass injected per working cycle, and the start of injection/delivery as well as the coolant temperature as a descriptive quantity for the thermal state of the engine. In an engine with exhaust gas recirculation, the external exhaust gas recirculation rate is taken into consideration as the eight quantity. The entry parameters are systematically varied with certain support points within their range of validity, and every operating point is calculated until a stationary condition is initiated. The thus determined working process quantities for this operating point, which are needed for the further computation of the connected components, are stored in the engine map. For the quantities stored in the characteristic map (here $\dot{H}_A$ as an example) is thus valid

$$\dot{H}_A = f(p_2, T_2, \frac{p_3}{p_2}, n_e, m_{fuel}, \varphi_{SOI}, T_{cm}, EGR) \; . \tag{8.42}$$

In the context of a calculation of the total system, the previously stored output quantities are then interpolated in correspondence with the entry parameter bordering the respective time and the connected components made available. We are dealing in this case with quantities like mass and enthalpy flows through the intake and exhaust valves for the volume attached to the engine; the indicated engine torque, which is available for the propulsion branch after the withdrawal of the friction torque at the clutch entrance; the wall heat flow for possible warm-up processes; and the air-fuel ratio. With these quantities, the linked mechanical or thermody-

8.5 Representing the engine as a characteristic map

namic components or the control unit of the engine are provided for possible control interferences. The exact procedure in setting up a characteristic map and interpolating in the map is described in Schwarz et al. (1992). Östreicher (1995) also uses a similar method for describing the internal processes in the cylinder.

In modeling the internal combustion engine via the description of gas-dynamic effects (see chap. 7.4), this procedure is eliminated. The boundary conditions for charge changing cannot in this case be assumed to be constant, which is way the step-wise working process calculation cannot be replaced.

8.5.2 Reconstruction of the torque band

In the simulation of stationary or transient operating behavior by means of a n-dimensional characteristic map, it is still sometimes necessary to be able to consider the torque path within a working cycle. If the cylinder process is calculated with the help of the step-wise thermodynamic cycle calculation, the path of the indicated cylinder torque can be determined by means of the pressure in the cylinder and the kinetics of the engine. Knowledge of the crank angle-solved torque band is the prerequisite for, e.g., torsional vibration analyses, simulations of clutch processes, or for the development of rule algorithms for the avoidance of jolting in the vehicle diesel engine. Using a pre-calculated engine map, the torque is averaged over one working cycle and thus only represents a constant value. In order to be able nevertheless to represent the dynamics of the cylinder torque, we can use the method of Gerstle (1999).

The torque path within the working cycle is composed of the two components of effective torque and drag torque.

$$T_{cyl,ind.}(\varphi) = T_{cyl,eff}(\varphi) + T_{cyl,drag}(\varphi) \ . \tag{8.43}$$

The effective torque originates essentially from the combustion of the introduced fuel and is independent of the drag torque, which results from gas and mass forces, which originate by the compression and expansion of the cylinder charge as well as piston and connecting rod movement. If the blow-by losses and the heat transfer between the gas in the cylinder and the combustion chamber walls are ignored, the integration of the drag torque amounts to zero for one working cycle – the drag torque thus does not contribute to the propulsion of the engine. For the charge changing period, we simplify and assume that the pressure in the cylinder is constant and corresponds to the pressure at the start of the compression phase.

$$p_{cyl,drag}(\varphi) = \begin{cases} p_{cyl,0} & \text{charge changing} \\ p_{cyl,0}\left(\dfrac{V_{cyl,0}}{V_{cyl}(\varphi)}\right)^n & \text{high pressure .} \end{cases} \tag{8.44}$$

In the high pressure component, the drag pressure path can with good approximation be modeled by a polytropic state change. In order to determine the drag momentum from the drag pressure path, the geometry and kinetics of the crankshaft drive must be considered. According to Maas and Klier (1981), the torque adjoining the crankshaft can be derived from the tangential pressure path as follows

$$T_{cyl,drag}(\varphi) = \frac{1}{2} V_d \ p_T(\varphi) \ . \tag{8.45}$$

The tangential pressure is thereby composed of two components. On the one hand, the gas forces cause a so-called gas tangential pressure amount, which results from the drag pressure path. On the other hand, the mass forces cause a so-called mass tangential pressure component. For a conventional crankshaft drive, we obtain in consideration of (2.7)

$$p_T(\varphi) = p_{Tg}(\varphi) + p_{Tm}(\varphi) = \left[\sin(\varphi) + \frac{\lambda_s \sin(\varphi)\cos(\varphi)}{\sqrt{1-\lambda_s^2 \sin^2(\varphi)}} \right] \cdot$$

$$\left\{ p_g(\varphi) - \frac{m_{osc.}}{V_d} \frac{\pi^2}{2} c_m^2 \left[\cos(\varphi) + \frac{\lambda_s \cos^2(\varphi) - \lambda_s \sin^2(\varphi) + \lambda_s^3 \sin^4(\varphi)}{\left(\sqrt{1-\lambda_s^2 \sin^2(\varphi)}\right)^3} \right] \right\}. \quad (8.46)$$

The gas and mass tangential pressures are depicted in Fig. 8.15 for a medium speed diesel engine.

The effective torque is approximated with a Vibe function in accordance with the Vibe hrr.

$$T_{cyl,drag}(\varphi) = \begin{cases} \dfrac{f\,\overline{T}_{cyl,ind}\, a(m+1)\left(\dfrac{\varphi-\varphi_{TDC}}{\varphi_{BDC}-\varphi_{TDC}}\right)^m e^{-a\left(\dfrac{\varphi-\varphi_{TDC}}{\varphi_{BDC}-\varphi_{TDC}}\right)^{m+1}}}{\text{expansion stroke}} \\ \underset{\text{residual process}}{0} \end{cases}. \quad (8.47)$$

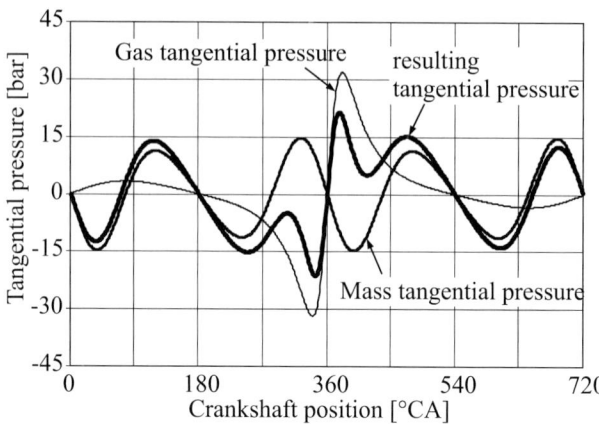

Fig. 8.15: Tangential pressure in a medium speed diesel engine

Since we can assume for the effective torque with good approximation that it only appears in the expansion stroke, the mean cylinder torque has to by referred to the duration of the expansion stroke. For the parameter a, $a = 6.905$. The parameter f helps distinguish between two

8.5 Representing the engine as a characteristic map

and four-stroke, assuming values of 2 or 4. To set the Vibe function, the parameter m is still missing. If one assumes that the position of the maximum of the effective torque is known, valid for the maximum of the function for the effective torque is

$$\frac{T_{cyl,eff}(\varphi)}{d\varphi} = f\,\overline{T}_{cyl,ind.}\,a(m+1)e^{-a\left(\frac{\varphi-\varphi_{TDC}}{\varphi_{BDC}-\varphi_{TDC}}\right)^{m+1}}$$

$$\left[m\left(\frac{\varphi-\varphi_{TDC}}{\varphi_{BDC}-\varphi_{TDC}}\right)^{m-1} - a(m+1)\left(\frac{\varphi-\varphi_{TDC}}{\varphi_{BDC}-\varphi_{TDC}}\right)^{2m}\right] = 0. \qquad (8.48)$$

Excluding the trivial solution $m = -1$, for the shape parameter m we obtain:

$$\frac{m+1}{m}\left(\frac{\varphi_{max}-\varphi_{TDC}}{\varphi_{BDC}-\varphi_{TDC}}\right)^{m+1} = \frac{1}{a}. \qquad (8.49)$$

This expression can be solved iteratively according to m. The function for the effective torque is thus set. The effective torque maximum can be determined in a step-wise construction of the engine map and stored in the map. In transient calculations, it can then be interpolated again from this characteristic map.

By means of the superimposing of the drag and effective torques, the path of the indicated torque can be reconstructed according to (8.43). The result of the reconstruction is compared in Fig. 8.16 with the indicated cylinder momentum, as determined by the cycle calculation for the full load point of a moderately fast-running large diesel engine. In this case, the effective torque maximum, which has been determined for the characteristic map representation in the cycle calculation, is utilized for the reconstruction. In Fig. 8.17, the entire procedure is summarized once again.

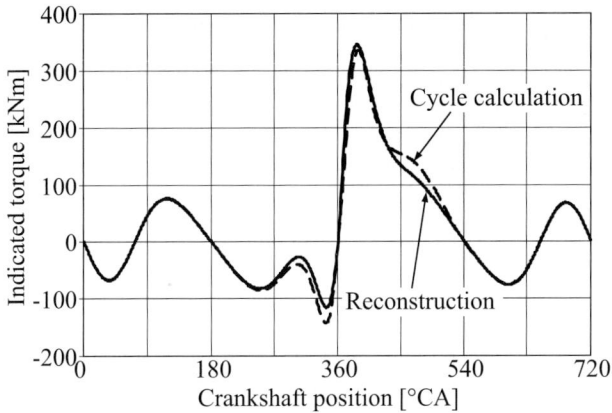

Fig. 8.16: Indicated torque of a medium speed 6-cylinder diesel engine

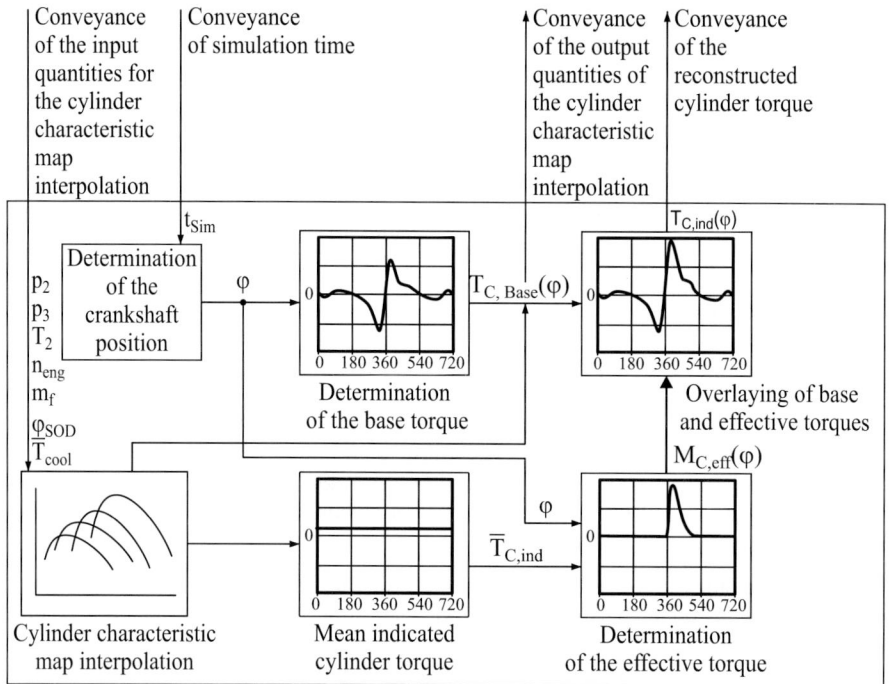

Fig. 8.17: Procedure for determination of the indicated torque

8.6 Stationary simulation results (parameter variations)

Stationary process calculation is used primarily for parameter variation. With it, explications for phenomena can be found or purposeful statements about the influence of particular parameters made, which can be varied only with difficulty in isolation from other parameters. This is one of the most important advantages of the real working process calculation. One should be warned however against all too uncritical parameter studies, since not all entry quantities are physically completely independent of each other. For example, a change in the compression ratio usually results in a change in the heat release rate as well.

In terms of result quantities, we distinguish between those averaged over a working cycle – like the indicated mean effective pressure or the indicated specific fuel consumption – and those, the path of which is represented in degrees of crank angle. In addition to these are quantities that indicate a maximum value – like peak pressure or temperature.

8.6 Stationary simulation results (parameter variations)

8.6.1 Load variation in the throttled SI engine

As a first example, a variation of load in the SI engine will be investigated. The engine under scrutiny is a 6-cylinder engine with 3 l cylinder capacity, which was calculated with the methods described in chap. 7.4 for the description of gas dynamics. 2,000 rpm was chosen as a speed for this operating point. The valve control times for this are conventional.

Fig. 8.18 shows the charge changing loops in the p, V diagram for the first cylinder proceeding from an indicated mean effective pressure of 2 bar, of 5 bar, and at full load, which corresponds to an indicated mean effective pressure of approx. 11 bar. We see that with decreasing load, the charge changing losses clearly increase, since a negative pressure must be constructed for the load control of the intake system of the engine via the throttle valve. This pressure impacts the cylinder shortly after the opening of the intake valve as well.

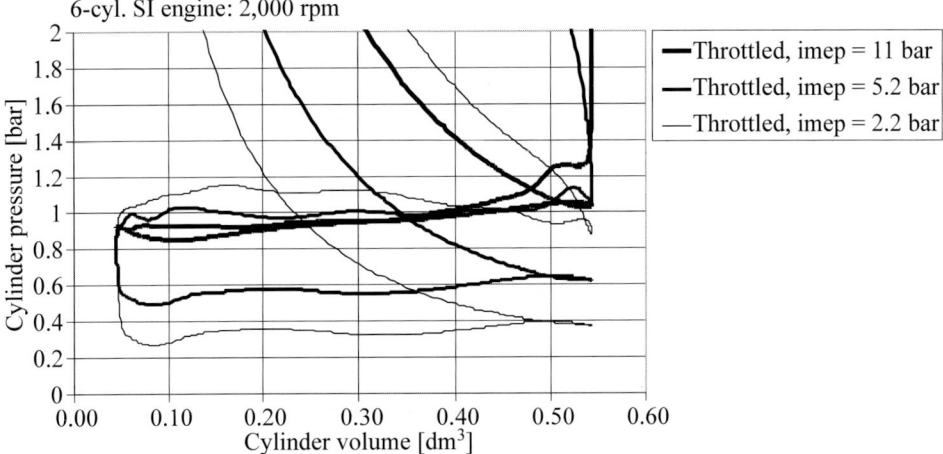

Fig. 8.18: Charge changing loops under different operating conditions

Depicted in Fig. 8.19 above are, for the indicated mean effective pressure of 2 bar, the calculated cylinder pressure, as well as the calculated pressures in the intake and exhaust ports of the cylinder under investigation. We can recognize clearly the pressure vibrations in the intake and exhaust ports of the engine. In Fig. 8.19 below, the mass flows through the valves are shown. As we have already seen in the pressure paths, a reverse flow of exhaust gas into the cylinder occurs shortly before the closing of the exhaust valve.

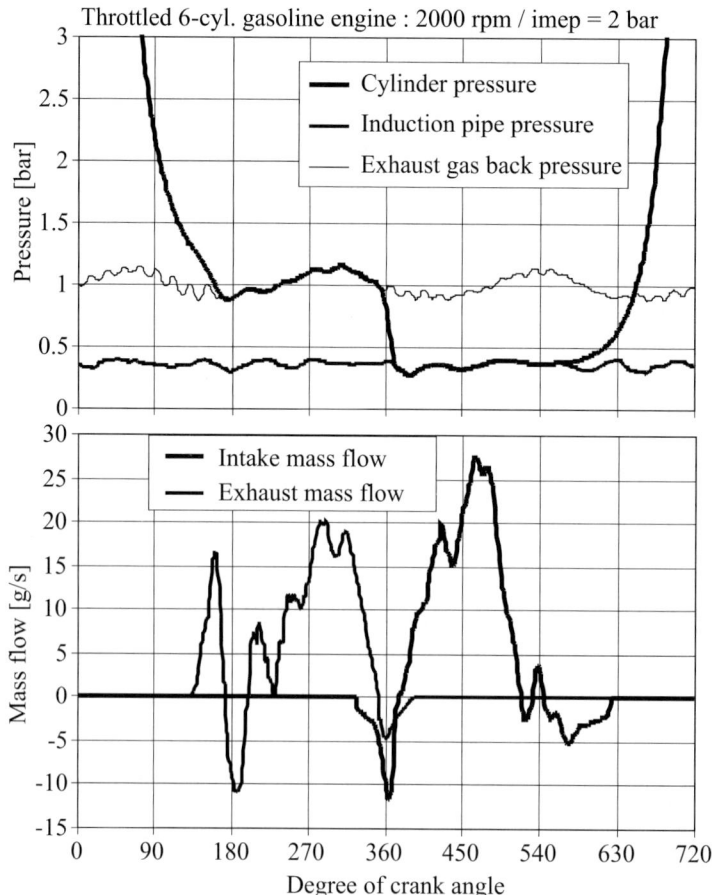

Fig. 8.19: Pressures and mass flows for the operating point *imep* = 5 bar

8.6.2 Influence of ignition and combustion duration

For the SI engine presented in chap. 8.6.1, variations of the start of combustion (ignition time) and the combustion duration were carried out at an indicated mean effective pressure of about *imep* = 5 bar. The combustion duration cannot usually be altered in an experiment, but is determined rather by the combustion process. Nevertheless, a quantitative investigation can supply important information regarding how a possible shortening of the combustion duration affects the process quantities for the combustion process concerned. Fig. 8.20 shows the influence of a start of combustion shifted earlier by 9 °CA and one shifted later by 9 °CA.

The form of the heat release rate and the combustion duration were modeled in this case as Vibe substitute heat release rates and held constant (combustion duration: 58°CA, shape factor: 2.3). The indicated mean effective pressure clearly drops at an earlier start of combustion, while at a combustion set later than the optimal initial position, it only drops slightly. Because of the fact that the fuel mass for this investigation was left constant, the exact reciprocal be-

8.6 Stationary simulation results (parameter variations)

havior is shown under the indicated specific fuel consumption. At an earlier start of combustion, the fuel energy is released so early before the TDC. As Fig. 8.20 further shows, at an earlier start of combustion the peak pressure, and with it the entire pressure level in the cylinder ascends. The result of this is that the heat transfer in the cylinder also increases, leading to higher wall heat losses, as one can also recognize in Fig. 8.20. Accordingly, because of the higher process losses, the exhaust gas temperature drops much more at an earlier start of combustion and goes up at a later start of combustion over-proportionally, since the fuel energy released very late can only make a small contribution to the work of the piston.

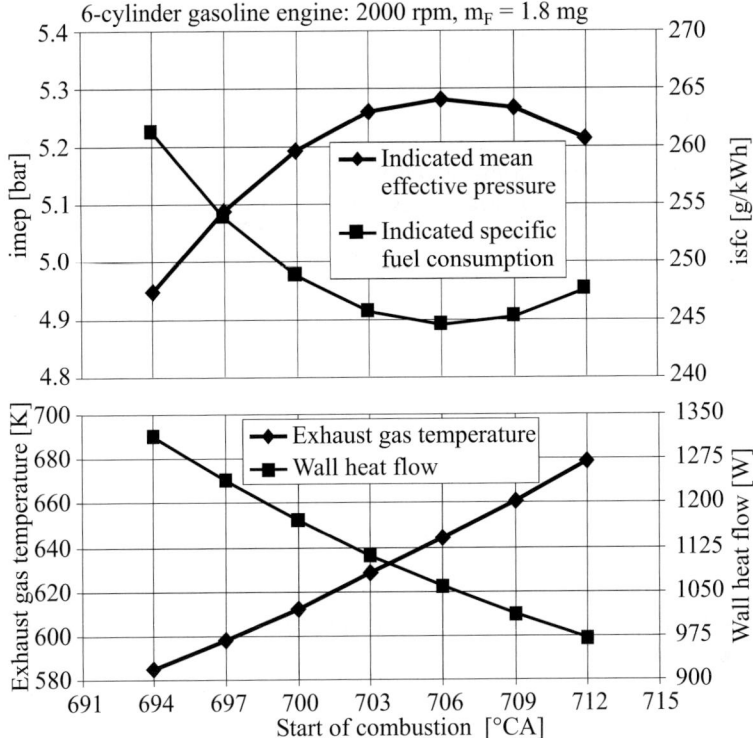

Fig. 8.20: Variation of the start of combustion (SI engine)

On the other hand, the influence of a variation of the combustion duration is much less. Proceeding from the operating point at the consumption-optimal start of combustion, a variation of a combustion duration altered by -10 to +30 degrees was carried out. The 50 mfb point was retained and the shape factor adopted by the start of combustion variation. In Fig. 8.21, one can clearly recognize that the indicated mean effective pressure drops with a longer combustion duration and the indicated consumption increases. Because of the retarded combustion, the exhaust gas temperature goes up with increasing combustion duration.

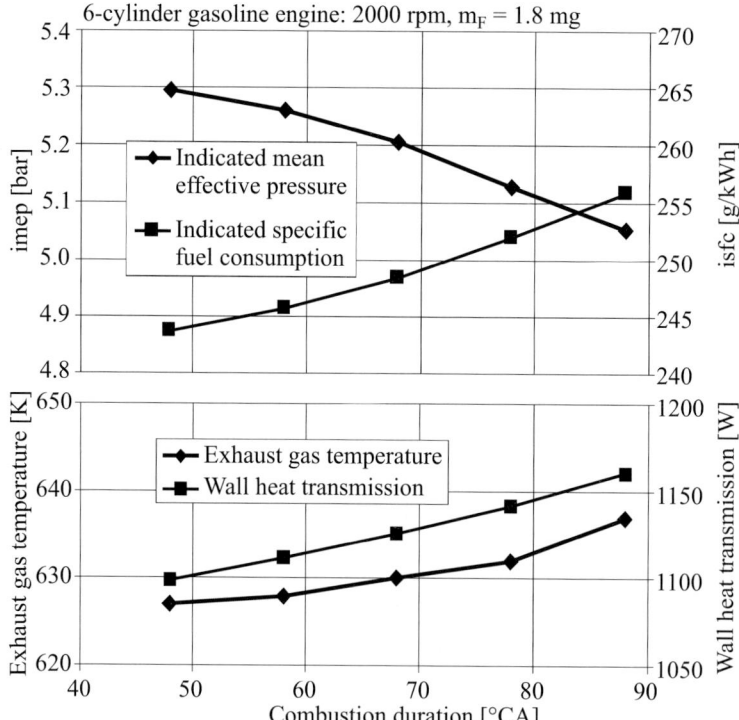

Fig. 8.21: Variation of the combustion duration (SI engine)

With these examples, it becomes clear that, as opposed to engine experiments, the influence of particular parameters can be separately investigated in simulation calculations.

8.6.3 Variation of the compression ratio, load, and peak pressure in the large diesel engine

Fig. 8.22 shows a study of a variation of the compression ratio for varying load-peak pressure-ratios for a large diesel engine with a capacity of 113 l per cylinder.

The variation is carried out at the rated speed of the engine of 450 rpm and for a combustion duration of 72 °CA at a air-fuel ratio of 2.2. The mechanical efficiency is set to a value of 0.92, and the total turbocharger efficiency constantly amounts to 0.65. The description of the turbocharger takes place by means of the so-called 1st fundamental equation of turbocharger equation, see (7.261). The influence of the compression ratio on the specific fuel consumption is clearly recognizable. For every graph for a constant compression ratio, there is a pronounced minimum for the fuel consumption. This minimum is valid respectively for a quotient from the mean effective pressure and for the peak pressure, which represents a limiting factor with respect to the mechanics. With this, at a pre-set peak pressure, the mean effective pressure for the consumption-optimal operating point can be easily determined. For example, an engine with a peak pressure of 180 bar and a compression of 12 is designed for optimal consumption at a mean effective pressure of approx. 20 bar. Interestingly, most engines are

8.6 Stationary simulation results (parameter variations)

designed more in the direction of higher mean effective pressure and less in the area of optimal consumption. That which envelops the individual curves can be seen as the design graph of these engines, which represents the respective compromise between consumption and output. Especially in the case of diesel engines, simulation is an important aid in designing engines at an early phase.

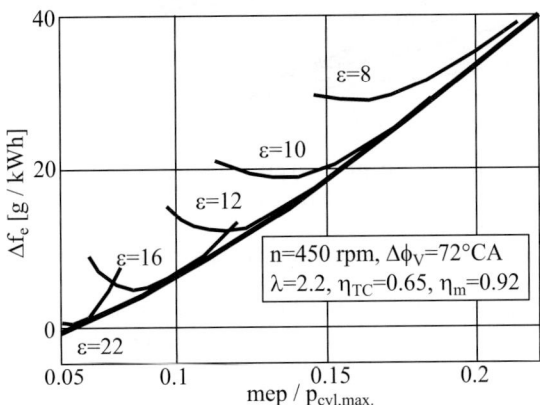

Fig. 8.22: Influence of compression ratio, load, and peak pressure on the specific fuel consumption (medium speed diesel engine)

8.6.4 Investigations of fully variable valve trains

Fig. 8.23 shows two charge changing loops for a 6-cylinder SI engine with 3 l capacity, which is equipped with a fully variable mechanical valve train. At this operating point, the engine has a speed of 2,000 rpm and an indicated mean effective pressure of 2 bar. Juxtaposed, both of these graphs of a charge changing loop for a conventional throttled engine are at the same operating point. Since the maximum valve lift for this operating point lies below one millimeter, practically no residual gas can be expelled from the intake side and taken in again later. An intake of residual gas is only possible by means of the exhaust valve. Both of the operating points represented differ by a varying amount of residual gas of approx. 17 % and 35 %. This is possible due to an exhaust spread of 90 °CA for the 17 % residual gas point and 50 °CA for the 35 % point. The throttled operating point has a residual gas amount of 17 % as well. We clearly see the decrease in charge changing work at fully variable operation in both of the operating points of equal residual gas. This has the effect of a 6-7 % improvement of the indicated consumption in this operating point. With further unthrottling with residual gas, a larger consumption potential results. We should in this case notice however that a portion of the charge changing potential is compensated again by a very much retarded combustion.

A completely different kind of load control results in electromechanical valve trains. In this case, at full valve lift, the timing of "intake valve closes" is set such that only the desired filling remains in the cylinder. For this, 2 kinds of load control are available – the "early intake valve closes" (EIVC) and "late intake valve closes" (LIVC). While an EIVC prevent more charge from arriving in the cylinder, in the case of LIVC, a portion of the cylinder charge taken in is expelled again.

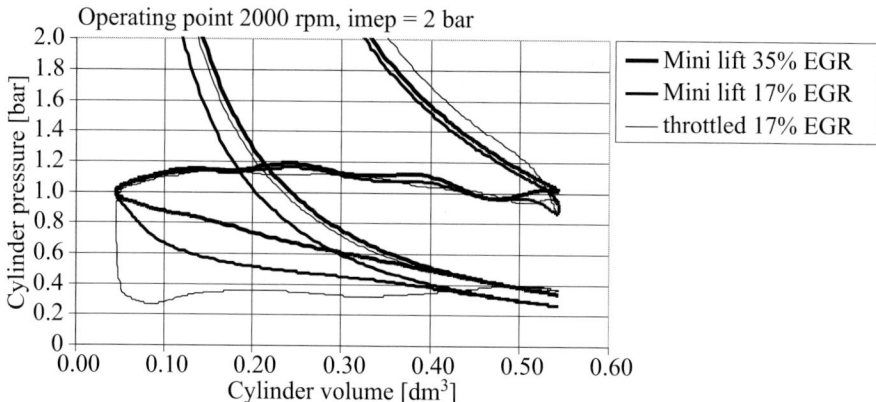

Fig. 8.23: Charge changing loops of a fully variable mechanical valve train

Fig. 8.24 shows a comparison of these two load control methods with each other and with a throttled engine with a conventional valve train for an operating point of 2,000 rpm and a load of 2 bar. Both methods indicate a residual gas amount of approx. 20 %, so that they can be compared with the throttled engine. The improvement of the total indicated specific fuel consumption in comparison with the throttled engine ranges from 7 % in LIVC to 9 % in EIVC. Part of the existing charge changing potential is compensated by the system-contingent retarded consumption in the high pressure range. We recognize from this again the necessity of an exact modeling of all components of the working process calculation. Since electromechanical valve trains provide the highest flexibility with reference to the control times and an unimagined number of combination possibilities, process simulation represents a valuable aid in systematic variation and thus in the evaluation of this valve train concept.

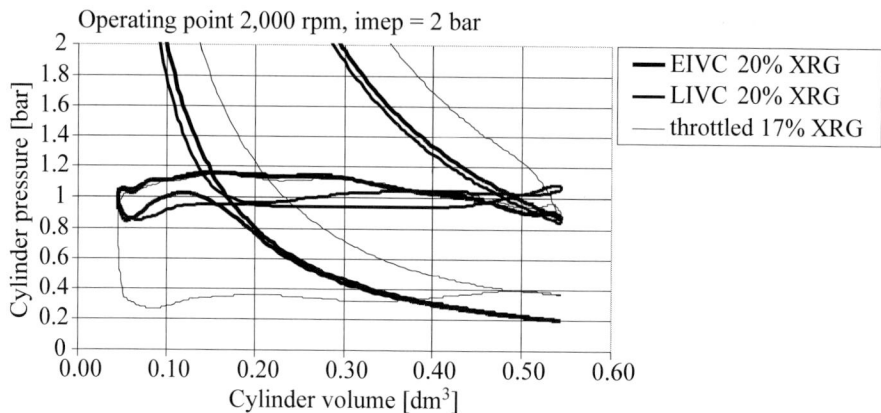

Fig. 8.24: Charge changing loops of an electromechanical valve train

8.6 Stationary simulation results (parameter variations)

8.6.5 Variation of the intake pipe length and the valve control times (SI engine, full load)

For a 4-cylinder SI engine, in Fig. 8.25 a variation of the intake pipe length is carried out in two steps of 350 mm and 600 mm for the optimization of the full load torque and the output.

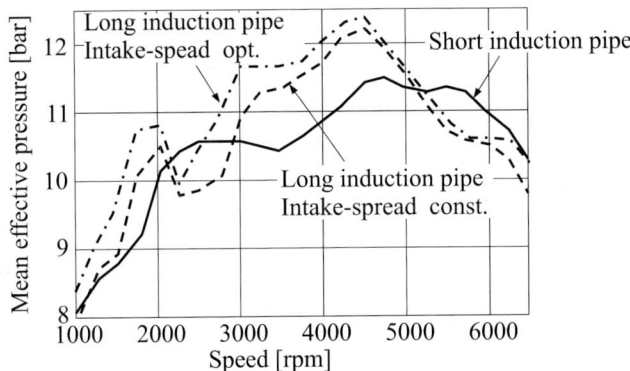

Fig. 8.25: Intake pipe variation in a SI engine

Both intake pipes procure from the intake manifold. In the case of the long intake pipe, an additional optimization for the lower speed range of the control times for the achievement of lower amounts of residual gas and for the adjustment of the pressure wave timing for the highest possible volumetric efficiency is carried out. The control times are thereby varied by the intake spread. Modeling takes place by means of the regularities for one-dimensional gas dynamics here as well.

We clearly recognize the marked course of the effective specific work for the short intake pipe at high speeds (rated performance point) and, for the long intake pipe, at low speeds as well as the decline of each curve in the other respective speed ranges. The use of variable valve control times once again heightens the medium pressure for long intake pipe lengths considerably. We see that the quantitative design of an intake system should always be parallel to the calculation process. We will not go further in this context into the exact modeling and consideration of diameters, inlet funnels, valves, etc.

8.6.6 Exhaust gas recirculation in the turbocharged diesel engine of a passenger car

Fig. 8.26 shows the connection diagram of an engine with charge air cooling and exhaust gas recirculation. In the case of the exhaust gas recirculation – as long as the scavenging pressure is larger than 1 – exhaust gas is led back in front of the intake valve by means of a specially controlled valve between the engine exhaust and intake. The maximum amount of recirculated exhaust gas thereby depends on the scavenging pressure ratio of the engine. A recirculation of the exhaust gas to the front of the compressor or the charge air radiator is not possible, since, on the one hand, the entry temperatures in the compressor become very high and, on the other, the oil components carried along in the exhaust gas and the like would stick to the compressor or the charge air radiator.

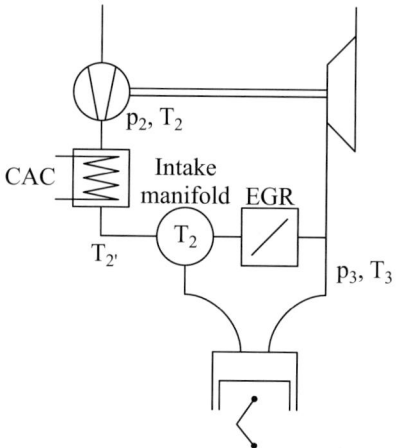

Fig. 8.26: Scheme of an engine with exhaust gas recirculation

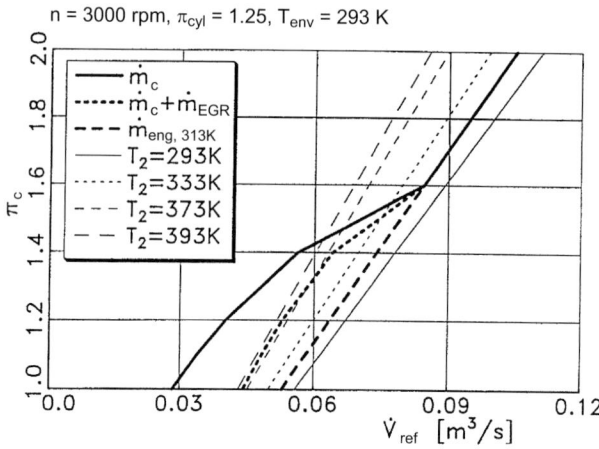

Fig. 8.27: Differences between engine and compressor suck lines under exhaust gas recirculation conditions

Fig. 8.27 shows the engine suck line for an average speed of 3,000 rpm and a temperature of 313 K, which should correspond to the temperature after the charge air cooler and is held constant. This line (roughly dashed) runs between the lines drawn in additionally for constant temperature before the intake valve of 293 to 393 K. The scavenging pressure ratio is assumed to be a constant 1.25. Below a compressor pressure ratio of 1.6, the exhaust gas recirculation rate increases linearly from 0 % to 36 % at a compressor pressure ratio of 1. The

8.6 Stationary simulation results (parameter variations)

mass flow taken into the engine decreases however at first due to the increase in mixture temperature with exhaust gas recirculation. The effect of pure temperature increase is depicted by the finely dashed line. In addition however, the mass flow delivered from the compressor is lessened by the recirculated exhaust gas mass flow. The solid line describes the mass flow and the reference volume flow through the compressor. The difference between the engine suck line at the respective mixture temperature (fine dashes) and the compressor absorption line corresponds to the recirculated exhaust gas mass flow. We recognize a clear shift of the compressor suck line to the left towards the surge line, which is caused on the one hand by the increase in temperature and on the other by the additional recirculated exhaust gas mass.

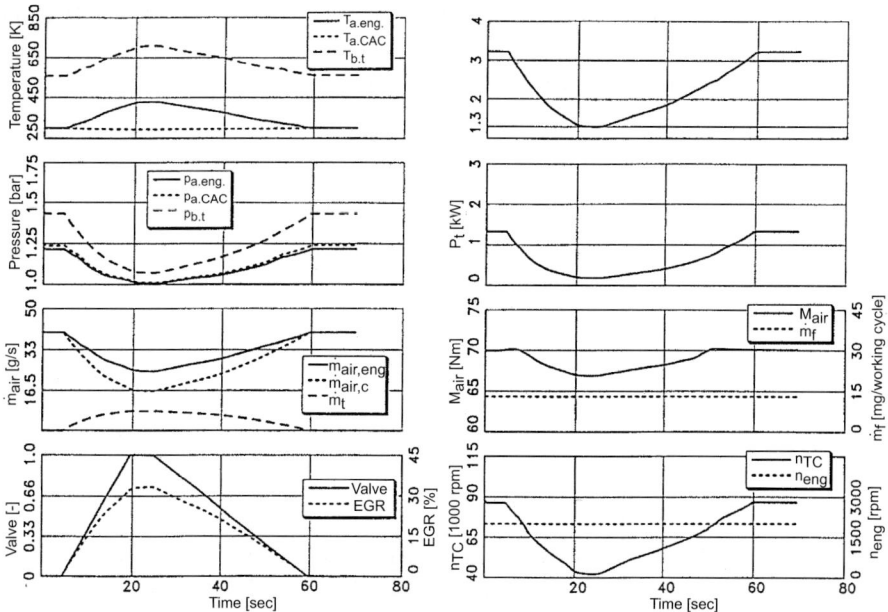

Fig. 8.28: Effect of the lift of the exhaust gas recirculation valve to engine operating conditions

For the clarification of the statements made above, the effect of exhaust gas recirculation is shown in Fig. 8.28 for a 4-cylinder diesel engine with 1.9 l capacity and turbocharging. The results are calculated by a simulation of the entire engine. The fuel mass injected per working cycle and the engine speed were held constant by a specially adjusted speed controller. The diagrams show the effects of an adjustment of the exhaust gas valve for the temporal course of engine-specific quantities, proceeding from a stationary engine operation state at a speed of 2,000 rpm and an exhaust gas recirculation rate of 0 %. After 5 s, the exhaust gas recirculation valve, as Fig. 8.28 below shows, is linearly opened, until at approx. 20 s it is fully open. After 25 s, the valve is linearly closed until approx. 60 s. This process goes very slowly, which is why the conditions can be seen as quasi-stationary. With the increase of the valve opening, the pressures before the turbine and after the compressor conform to each other, but drop as a whole, since the turbine output lowers because of the pressure decrease over-proportionally to the increase in exhaust gas temperature. The intake temperature before the engine rises with increasing exhaust gas recirculation, which is why the mass flow into the engine with the known effects on exhaust gas counter-pressure and the turbocharger speed goes back additionally. The air-fuel ratio was limited to 1.3 minimally by the maximum throttle cross-sectional surface of the exhaust gas recirculation valve. Fig. 8.28 (left) shows the path of the mass flows through the compressor and into the engine as well as the recirculated exhaust gas mass. The indicated mean effective engine torque T_i (Fig. 8.28, right) also collapses in the expected way because of the reduction of the air-fuel ratio caused by exhaust gas recirculation with the known effects on the compression path and the heat release rate as well as on the condition quantities in the cylinder at a constant speed and injected fuel mass. Clearly recognizable as well is the collapse of the compressor speed n_{TC} associated with the collapse of the compressor pressure ratio. Very important in this context is the exact knowledge of the compressor map in the lower speed regions.

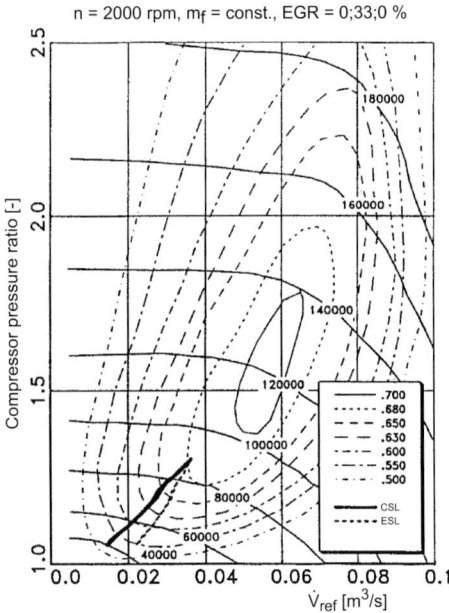

Fig. 8.29: The exhaust gas recirculation process in the compressor map

Fig. 8.29 shows the extrapolated map of the compressor used for the engine under investigation. Entered into this characteristic map is the path of the compressor and engine suck lines for the exhaust gas recirculation path shown in Fig. 8.28. We recognize that the compressor absorption line is shifted to the left by the amount of recirculated exhaust gas flow and clearly drops with an increasing recirculation rate to lower pressure ratios and compressor speeds. An unsteady acceleration process must therefore begin at a much lower initial boost pressure level, which leads partially to a considerable delay of the reaction behavior.

8.6 Stationary simulation results (parameter variations)

8.6.7 Engine bypass in the large diesel engine

Similarly to exhaust gas recirculation in a small, high speed diesel engine, which only proceeds independently when the scavenging pressure slope is greater than 1, the process of engine bypass in engines can only be exploited at a scavenging pressure ratio smaller than 1. For this reason, engine bypass is primarily employed in large, medium speed engines, in which the scavenging pressure ratio is automatically smaller than 1 because of the favorable turbocharger-total efficiency.

Fig. 8.30 shows the connection diagram of an engine with engine bypass. In a turbocharged engine, the compressed fresh air is lead past the engine to the turbine and mixed with the exhaust gas. One distinguishes thereby between so-called "warm" and "cold" engine bypass processes. In cold engine bypass, the air is diverted after the charge air radiator and lead to the turbine, while in warm engine bypass the air is directly diverted after the compressor. By means of the mixing of fresh air with the exhaust gas, the temperature in part clearly drops in front of the turbine, which is why, from a thermodynamic perspective, warm engine bypass is preferable to cold. From a technical perspective on the other hand, cold engine bypass often involves a shorter and simpler pipe construction. The effect of engine bypass, in which the temperature is lowered in front of the turbine, is at first difficult to understand.

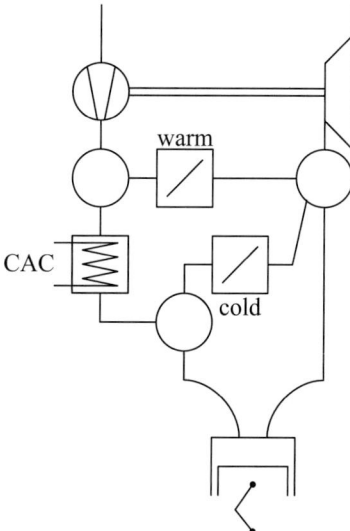

Fig. 8.30: Scheme of a turbocharged engine with engine bypass

Fig. 8.31 shows the compressor operation lines for various load points at a speed of 310 rpm and warm engine bypass.

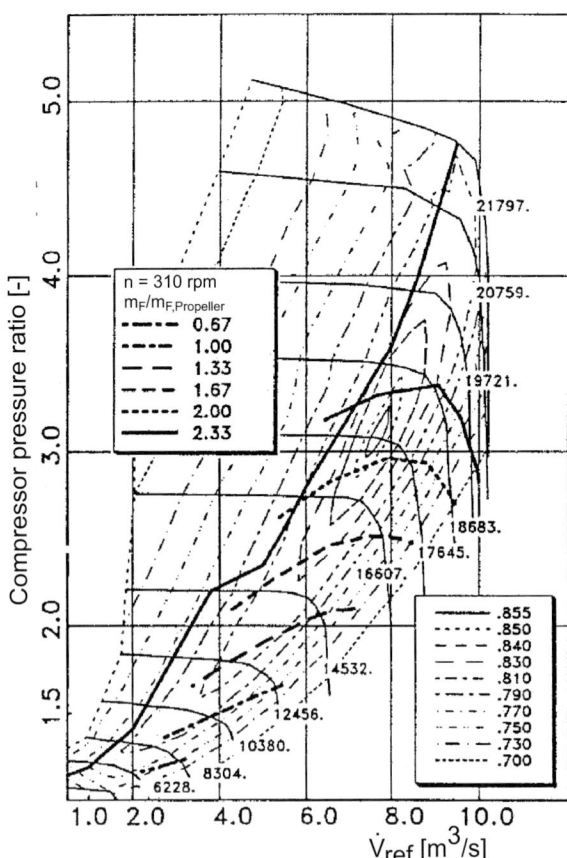

Fig. 8.31: Compressor operating lines for different loads at 310 rpm

Represented here are the load points of 66 to 233 % load, proceeding from the load at the propeller line in respective steps of 33 %. The left starting position of the respective compressor operation line thereby corresponds to a engine bypass rate of 0, and the right final position of the operation line the maximum engine bypass rate. The starting positions of the solid compressor operation lines lie on the engine or compressor suck lines, as they would result without engine bypass for the standard turbine cross-sectional surface. With an increasing engine bypass rate, the operating points for the compressor shift toward the right, away from the surge line to higher reference mass flows and to partially higher compressor pressure ratios, while the reference volume flow taken in by the engine remains roughly on the engine suck line. The effect of engine bypass can be explained as follows. Because of engine bypass, the mass flow through the turbine at first increases. In this way, a larger output can be created in the turbine and placed at the disposal of the compressor. The latter reacts to this for its part with a higher pressure ratio and with a higher delivered mass flow. In this way, the enthalpy at the turbine is further increased. However, this increase is restricted for several reasons. On the one hand, an increase of the engine bypass rate causes an ever-increasing drop in temperature in front of the turbine; on the other hand, the engine bypass rate, and with it the engine bypass mass through the opening of the engine bypass valve, is limited because of the increase in the scavenging pressure ratio.

A further reason is the possible shifting of the operating point of the compressor to regions with low compressor efficiency, which also has a negative effect on the potential scavenging pressure ratio. For this reason, the operation lines decline to the right again from a maximum at high engine bypass rates. We recognize that an engine operation at a doubled load would not be possible without engine bypass, since this operating point would already lie to the left of the surge line of the compressor. By means of engine bypass however, the operating point can be shifted towards the right away from the surge line.

8.7 Transient simulation results

In the following chapters, several results of the simulation of transient processes in the internal combustion engine will be shown. We will consider thereby processes of power switching for stationary engines, the simulation of unsteady processes in vehicle engines, and the calculation of entire driving cycles.

8.7.1 Power switching in the generator engine

A large, medium speed diesel engine is operated not only on the propeller line, but serves as an energy generator in power stations. In this case, the engine is operated on the generator line – i.e. at a constant engine speed. In this chapter, we will investigate the effects of power switching on engine operation and particularly on the transient behavior of the compressor and turbine of the turbocharger.

Fig. 8.32 shows the effects of power switching of approx. 20 %, beginning with a basic load of approx. 30 % at a rated speed of 450 rpm. Each consecutive power switching is carried out after 20 s, so that the previous power switching has reached a stationary position. As one can see in Fig. 8.32 on the left. the actual speed only deviates noticeably from the rated speed in the first power switching at 5 s, which is to be attributed to the restriction of the injection amount for the avoidance of a lowering of the air-fuel ratio below a value of approx. 1.3. Clearly recognizable is the increase of the scavenging pressure ratio and the exhaust gas temperature shortly after the power switching, since the release of a larger fuel mass at an at first equal fresh gas mass in the engine causes a higher exhaust gas temperature. Because of the inertia of the mechanism, the turbocharger cannot immediately react upon the higher enthalpy supply, which is why there is at first a damming up in front of the turbine. This process again worsens for a short term the charge changing in the engine and, in addition, reinforces the dropping of the air-fuel ratio in the engine, as one can see in Fig. 8.32 on the right. A torque difference of up to 50 Nm is sufficient to accelerate the turbocharger shaft after approx. 7 s to the desired speed.

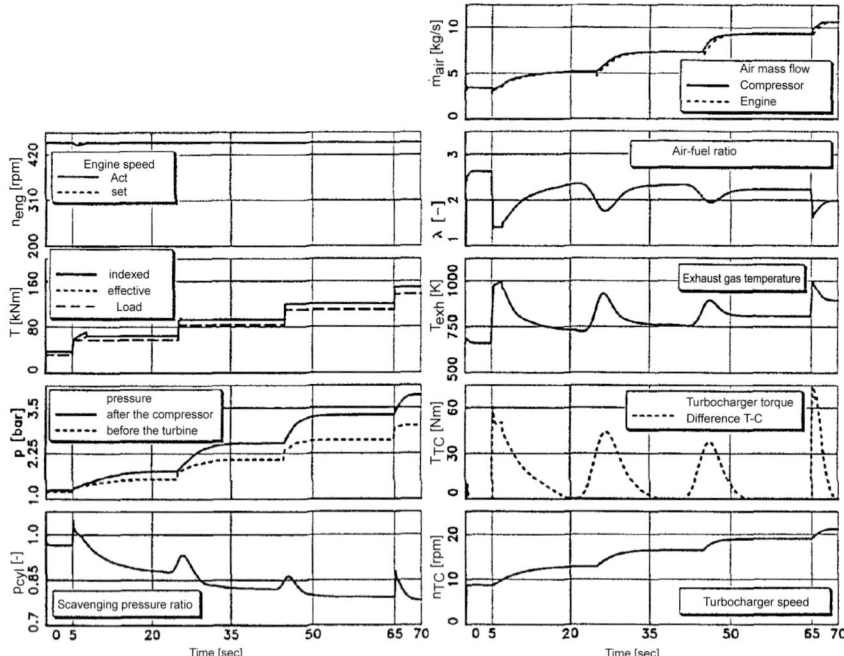

Fig. 8.32: Power switching

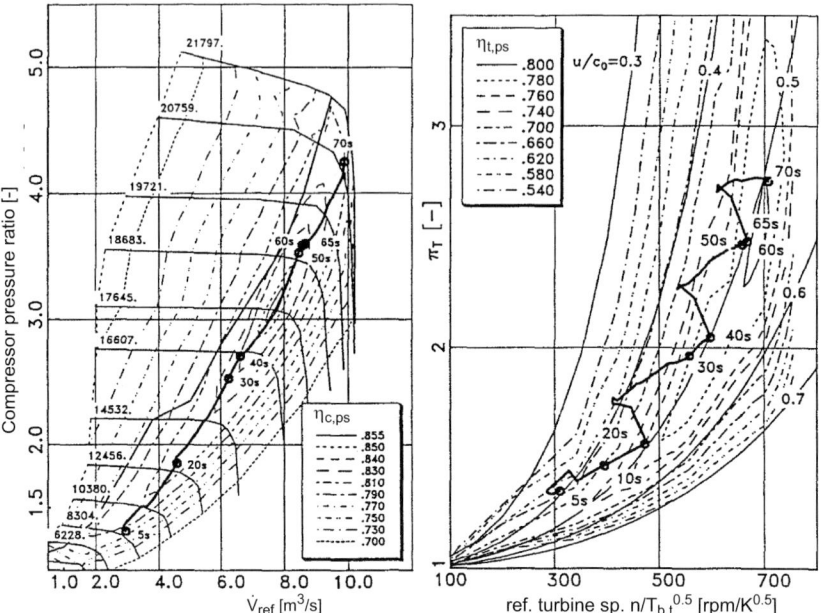

Fig. 8.33: Power switching, compressor and turbine operating lines

8.7 Transient simulation results

One should pay special attention however to the operational behavior of the turbocharger in the respective maps for the compressor and the turbine, which are shown in Fig. 8.33. While the power switching in the case of the compressor only expresses itself in lightly arched curves, the operation line of the turbine is quite jagged. The cause of this is the plotting over the reference turbine speed. Therefore, in the case of a sudden increase in the exhaust gas temperature, the reference turbine speed drops. The turbine is operated briefly in regions of very poor efficiency at lower speeds. This process can only be made clear by means of characteristic map representations as introduced in chap. 7.5, and again corroborates the necessity of extrapolation into map regions as the basis for an exact simulation of transient behavior.

8.7.2 Acceleration of a commercial vehicle from 0 to 80 km/h

As a basis for investigating the diesel engine of commercial vehicles, we are using a 6-cylinder in-line engine with approx. 12 l capacity. This engine propels a commercial vehicle with a total mass of 40 tons. The measured values forming the basis for the comparison between measurement and calculation were determined on a dynamic engine test bench. Comparisons with the respective measurement (solid) and calculation (dashed) have been illustrated. The propulsion torque of the vehicle and the gear shifting are thereby put upon the engine by a controlled brake. For simulation calculation, the complete power train was illustrated such that the propulsion torque in the engine corresponds to the brake torque for the simulation calculation of the measurement.

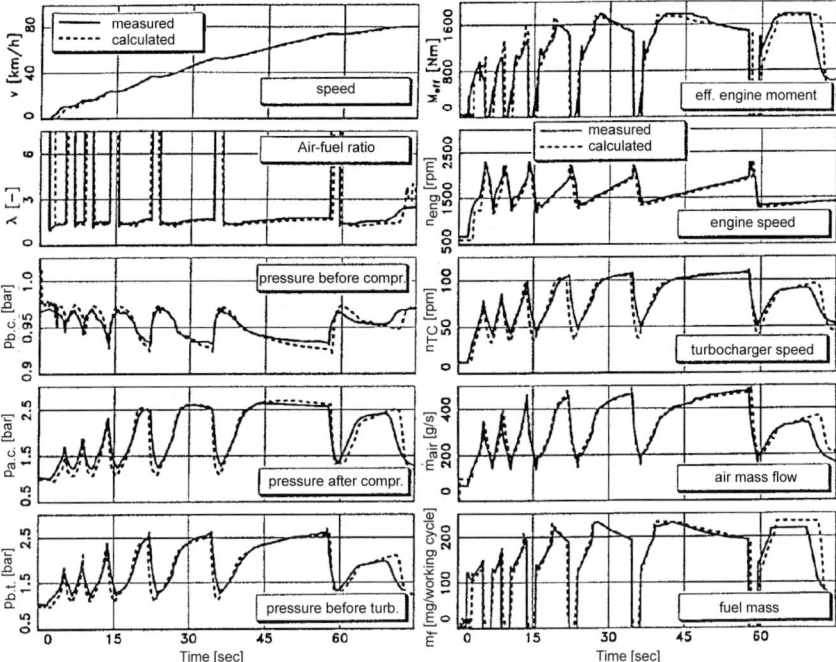

Fig. 8.34: Free acceleration, 40 ton truck

Fig. 8.34 shows a free acceleration of the truck from standstill with start-up and gear-shifting until a speed of 80 km/h is reached. The speed for the gear-shift is set at exactly 1,900 rpm. For the simulation, at the beginning of the calculation, a target speed of 80 km/h is given. The start and gear shifting at 1,900 rpm are in the further course of the calculation carried out independently by the driver controller. For this reason, the agreement of the measurement and the calculation should receive special attention, since the gear shifting times in the measurement and in the simulation calculation are reached at practically the same time, which testifies to a high simulation quality in terms of a firmly pre-set shifting speed and the asymptotic approximation of these speeds at higher gears. The agreement of all quantities of the measurement-calculation comparison in Fig. 8.34, of the fuel mass injected per working cycle, of the air mass flow through the compressor, of the air-fuel ratio, of the effective torque, and of the engine speed already dealt with, confirm the informational value of the simulation. Quantities like exhaust gas temperature, pressure before and after the compressor and before the turbine, turbocharger speed, and the achieved driving speed also show few discrepancies between measurement and calculation, which verifies a correct and realistic procedure in the simulation of the charger aggregate and of the entire engine.

In Fig. 8.35, this acceleration process is depicted in the compressor map. For the sake of clarity, only the compressor operation line is drawn from the start-up to the shift into fourth gear. The circulation direction for the operation lines is clockwise. The low boost pressure build up in the 1st and 2nd gear and the considerable distance of the operation line from the surge line are clearly recognizable in the characteristic map.

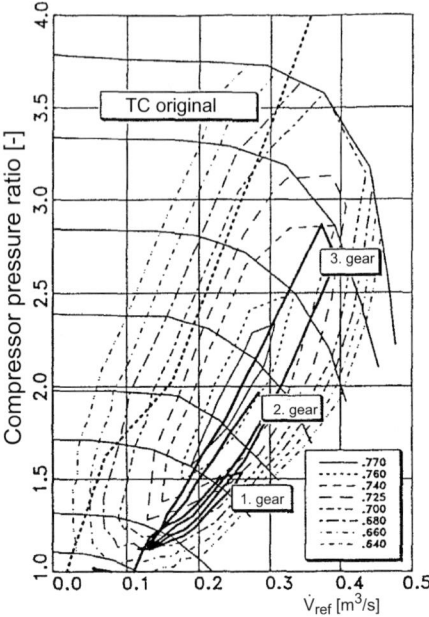

Fig. 8.35: Acceleration process in the compressor map

8.7.3 Turbocharger intervention possibilities

For a turbocharger diesel engine, different possibilities of improving unsteady behavior at low speeds are being investigated.

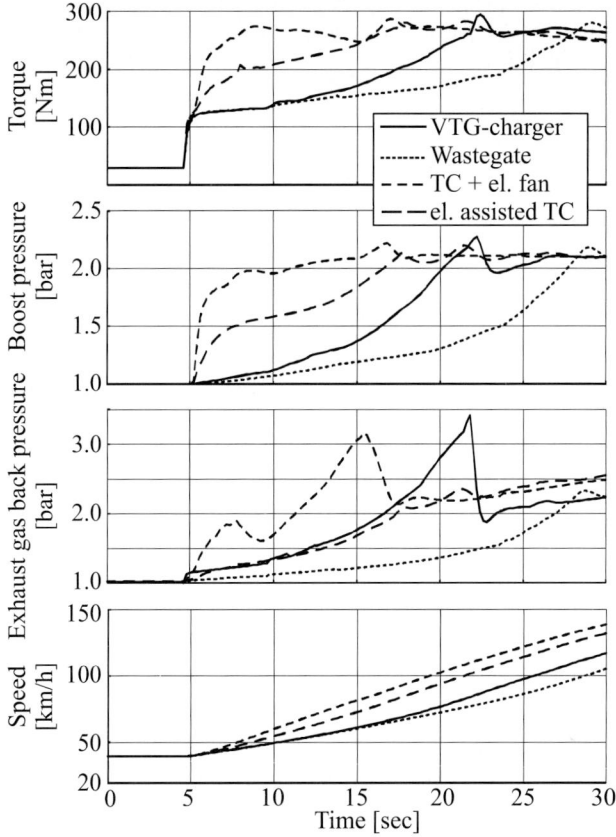

Fig. 8.36: Comparison of different operating strategies of the turbocharger

As an initial basis, a turbocharger configuration with control of the boost pressure by means of a wastegate has been selected. Under investigation are a configuration with a variable turbine geometry (VTG), an electrically assisted turbocharger, and an additional compressor switched before the turbocharger, which is electrically driven. Both electrical propulsions have a peak output of approx. 2 kW. The base engine is a 2-liter, 4-cylinder engine, which is utilized in a 1,350 kg vehicle. In order to workout the effect of the particular measures effectively, a full load acceleration process of approx. 40 km/h in a direct course was selected, in which an engine speed of approx. 1,000 rpm results. This is an extreme acceleration process, since the engine and turbocharger speeds are very low. As is shown in Fig. 8.36, in the case of the conventional engine with a wastegate, it takes approx. 25 s to build up a corresponding boost pressure. With the utilization of a turbine with a variable geometry, the time until complete boost pressure build-up is shorted by approx. 7 s. The electrically supported turbocharger reaches the target pressure approx. 11 s before the engine with the wastegate. The fastest boost pressure build-up is made possible by the additional, electrically propelled compressor. In this case, the boost pressure is reached in a few seconds.

In Fig. 8.36, the engine torque and the speed of the vehicle are additionally shown. The path of the operation lines for the particular concepts in the compressor map is also interesting. This is portrayed in Fig. 8.37.

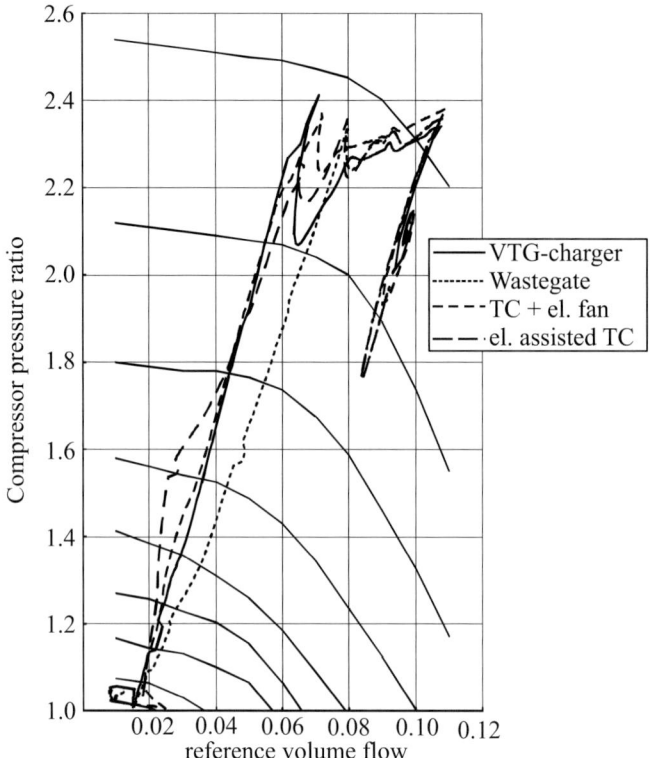

Fig. 8.37: Operating lines in the compressor map

8.7.4 Part load in the ECE test cycle

Part of the ECE test cycle for a turbocharger 4-cylinder diesel engine in a medium-class vehicle is represented in Fig. 8.38. In this case, we are dealing with operation in the lower part load of the engine, in which exhaust gas recirculation is activated with the corresponding effects on the unsteady behavior of the turbocharger.

The measured values plotted in the images were taken from an electronically controlled engine test bench, in which both the vehicle with all driving resistance as well as the driver (gas pedal, gears) can be simulated. Since the speed of the ECE cycle is adjusted by the driver

8.7 Transient simulation results

controller, the paths for the effective torque and the speed of the engine must correspond exactly to the measured values (Fig. 8.38, left, 2nd and 3rd diagrams). The fuel volume flow and the air mass flow taken in by the engine can adjust themselves freely. Both of these quantities are represented in the 4th and 5th diagrams and display a fine agreement between measurement (solid) and calculation (dashed). The situation is similar in the case of the quantities of boost pressure and exhaust gas counter-pressure as well as the turbocharger speed (Fig. 8.38, right). Thus all quantities characteristic for charging proceed under part load operation congruently to the measurement results. In the case of exhaust gas temperature, represented in the 4th diagram of Fig. 8.38 right, the thermal inertia of the thermocouple that was used for the data acquisition is simulated as well. These paths are also nearly congruent. A decent agreement in the case of the paths for carbon dioxide values results as a necessity from the good agreement of the air and fuel masses, since the values for carbon dioxide emissions can be calculated from the air-fuel ratio.

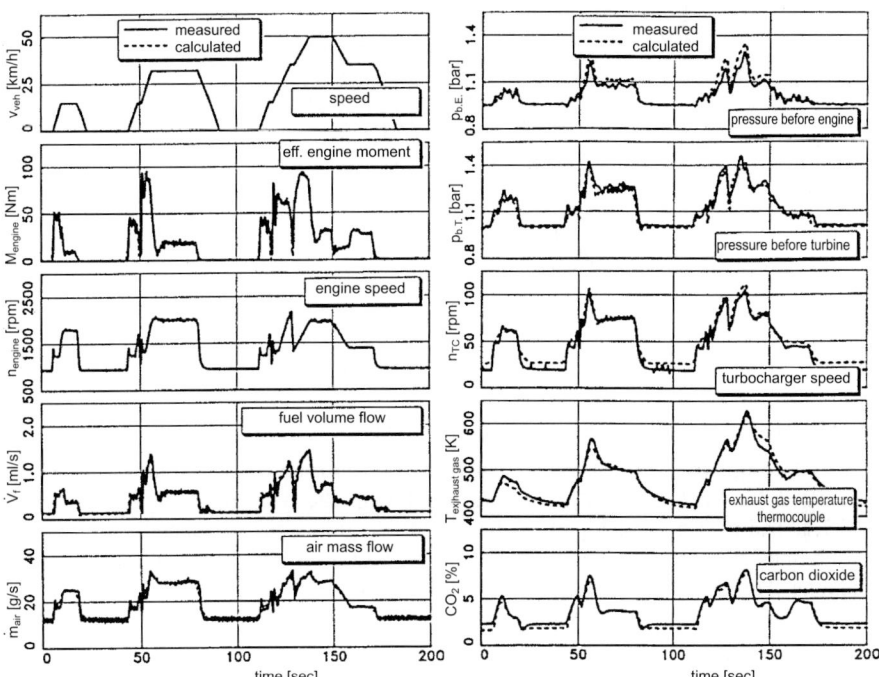

Fig. 8.38: Part of the ECE test cycle

In Fig. 8.39, the operation line of the compressor for the part of the ECE cycle under consideration is plotted into the compressor map. We recognize that precisely for the quantitative simulation of the part load region, an exact extrapolation into area of lower pressure ratios and speeds in the compressor and turbine maps is essential. Aggravatingly, in the case of this configuration, because of exhaust gas recirculation, the operation range is additionally shifted into these ranges.

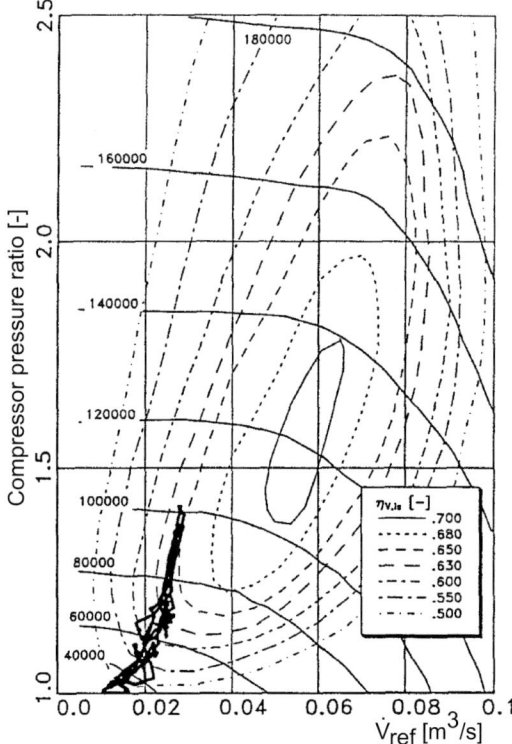

Fig. 8.39: Operating line in the compressor map, part of the ECE test cycle

8.7.5 The warm-up phase in the ECE test cycle

In Fig. 8.40, results of a simulation calculation of the warm-up process in the ECE test cycle, as calculated by Reulein (1998), is represented with the 4-cylinder engine described in chap. 8.7.4. The measured paths are dashed in the diagrams, the calculated paths illustrated with solid lines. All represented curves reveal favorable agreement between measurement and calculations. This makes it clear that the warm-up phase of the internal combustion engine can be realistically simulated with the assumptions made in chap. 8.2 and 8.3.

A high temperature-dependence is shown by the path of the friction torque, shown in Fig. 8.40 in the top right. Within the first 300 seconds, a clear increase in engine friction occurs, especially in the higher engine speed range. The paths of the temperatures of the coolant and the oil are shown in the two following diagrams. In this case as well, a very good agreement results between measured and calculated temperature paths. The next diagram shows the paths of the measured and calculated cylinder head temperature. Again, a favorable agreement between calculation and measurement can be seen.

8.7 Transient simulation results

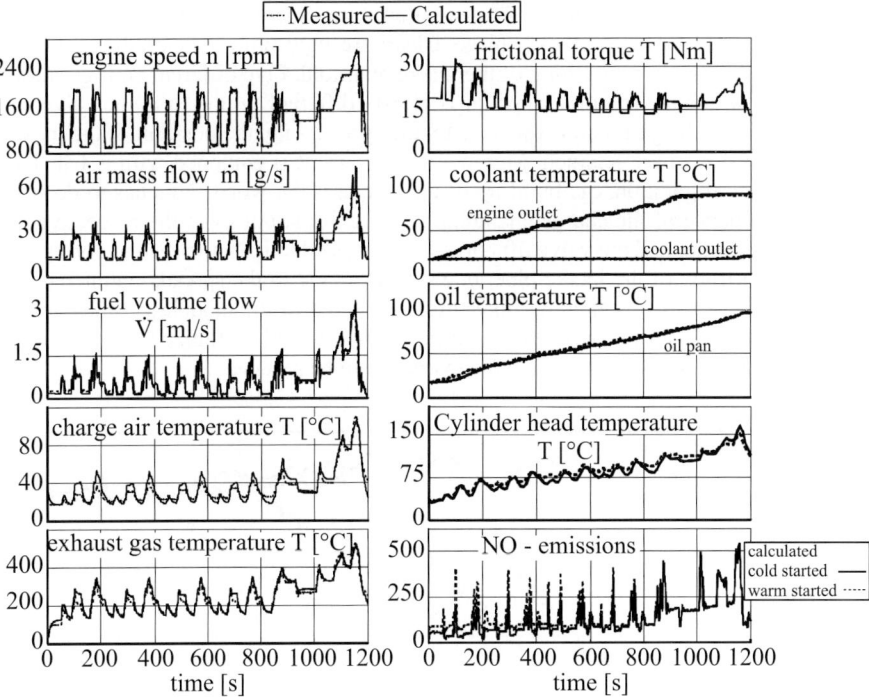

Fig. 8.40: Warming-up in the ECE test cycle

The diagram on the bottom right shows the effect of the engine temperature on the nitrogen oxide concentration in the exhaust gas, which were calculated according to the method of Heider (1996) introduced in chapter 7.2.2. The path for the engine started at operationally warm temperatures is shown as a dashed line, while the solid line shows in comparison to this the calculated path during the warm-up phase described above. The path of the operationally warm-started engine corresponds very well with measurements, which are however not shown here. Very easy to recognize in the first half of the driving cycle are the initially lower and then gradually increasing nitrogen oxide emissions of the engine in warm-up operation. The cause of these effects are the initially very low combustion chamber wall temperatures, which lower the process temperature and thus work against the formation of nitrogen oxide.

8.7.6 Full load acceleration in the turbocharged SI engine

In the following section, a calculation of the acceleration behavior of a 4-cylinder engine with a capacity of 2 l, direct injection, and turbocharging in a medium-class vehicle is represented. The engine has a cam phase with a large shifting range for both camshafts and can be equipped for modeling both with a "4-in-1 exhaust gas manifold" with a mono-scroll turbine as well as with a "4-in-2 manifold" with a twin-scroll turbine. Since we are concerned with a concept with stoichiometric combustion, the engine has a throttle valve arranged after the charge air cooler.

In Fig. 8.41, acceleration paths in the 4th gear at an initial speed of 1,500 rpm are shown. At approx. 1 s, the full load demand takes place. Clearly recognizable is the relatively fast build-up of the basic torque after the filling of the intake manifold. Five different configurations are represented, in which the reaction behavior is clearly different. In the first, the valve control times optimized for the part load (expressed by the spread – i.e. the position of the maximum of the valve lift curve) are retained. Moreover, the concept is equipped with a 4-in-1 manifold. With this configuration, the build-up of the torque is the worst. This has to do with the unfavorable influence of the charge changing by the exhaust pulses of the respective, previously igniting cylinder and the valve lift curves not adjusted to it. In the second configuration, shortly after the filing of the intake pipe, one switches over to the spreads optimized for full load. In this case, realistic switching times are considered in the modeling. One can clearly recognize a fast build-up of torque, since filling improves with optimized control times and the amount of residual gas remaining in the cylinder can be reduced.

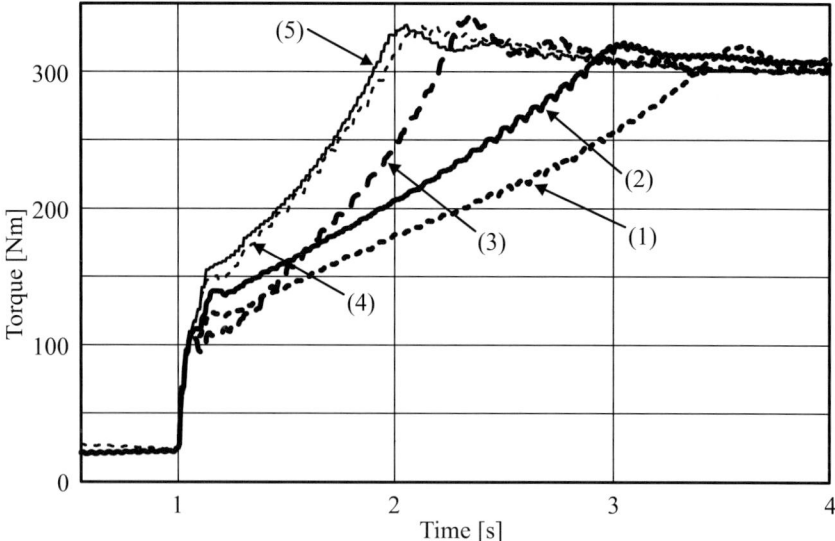

Fig. 8.41: Concept-comparison for a charged 4-cylinder engine

In chap. 8.6.7, the principle of engine bypass for the large diesel engine was described. This procedure entails a shifting of the operating point in the compressor map towards the right away from the surge line, so that at an equal engine speed (absorption line), a larger pressure ratio in the compressor and thus a better filling in the cylinder is made possible. At the same time, the compressor efficiency improves significantly. In the SI engine, we can also use this method, since, according to the design, it possesses a positive scavenging slope in the range up to 3,500 rpm. However, the possibility presents itself of passing the mass flow, but rather loading through the combustion chamber. As a result, both the residual gas is minimized and the cylinder is additionally cooled, which in both cases clearly increases the filling. This is made possible in the SI engine by a cam phaser. With it, the engine absorption line in the compressor map tilts to the right to the higher pressure ratios and efficiencies already described. In the case of the direct injection SI engine, pure air is scavenged thoroughly during the valve overlap phase, which is very favorable with respect to HC emissions. With this, the possibility arises of forming a sub-stoichiometric air-fuel ratio in the combustion chamber, which with high burning velocities makes possible the conversion of the higher filling without knocking problems. The unburned fuel reacts in the best case before the turbine with the air mass previously scavenged, thus raising the enthalpy of the exhaust gas before the turbine. In the unfavorable case, the fuel first reacts with the air in the catalyst converter.

In the third configuration in Fig. 8.41, we switched to a large valve overlap after the filling of the intake pipe. After a short stagnation of the torque build-up, one can indeed see a clearly larger gradient in the torque build-up, but this behavior is difficult for a driver to administer. At the beginning of the load change process, at first the system "swallows" regularly. The positive scavenging slop breaks down briefly because of the control times, and the amount of residual gas climbs massively, until the system begins again after approx. 1 s to show the expected behavior.

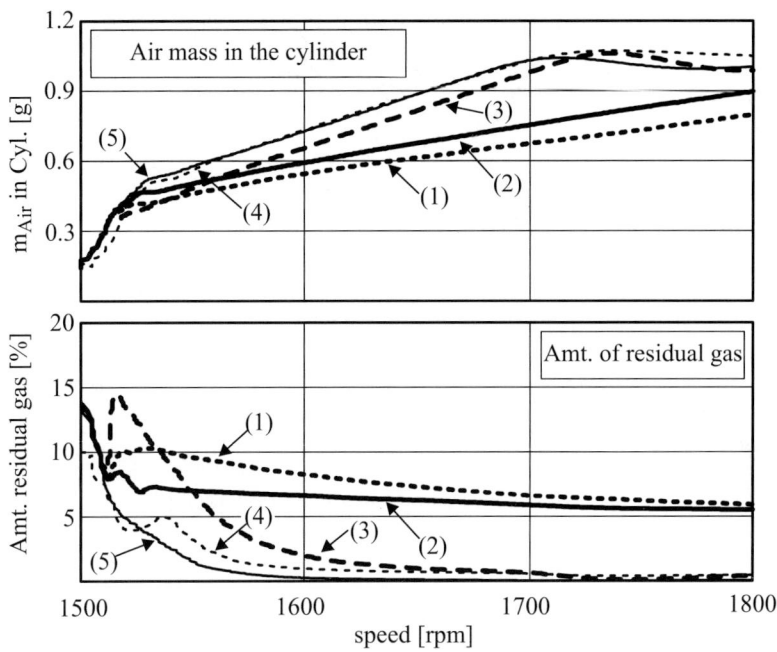

Fig. 8.42: Amounts of fresh and residual gas for different charging concepts

Fig. 8.42 shows the fresh gas mass remaining in the cylinder after the closing of the intake valve and the fraction of residual gas in the total cylinder mass for the five configurations. These are for reasons of comparison plotted over the engine speed. An exact analysis of the processes during charge changing for the third concept results in a large amount of residual gas of up to 20 %, which is caused by a superimposing of the exhaust pulses of successively igniting cylinders and the influence of charge changing behavior resulting from this. By means of a shortening of the exhaust control times to approx. 200 degrees of crank angle, this behavior can be considerably avoided, as curve four in Fig. 8.41 shows. In this case, the interfering pressure wave of the previously igniting cylinder is "hidden" by the shortening of the exhaust control time. The residual gas is thus minimized, by means of which the necessary positive scavenging slope then quickly constructs itself.

The cleanest technical solution is represented by the fifth configuration, in which a reciprocal influence of the charge changing of the cylinder can be fundamentally avoided through a 4-in-2 combination in connection with a twin-scroll turbine with a preservation of the exhaust control time. In this configuration, the best unsteady outcome results with a very pleasant behavior.

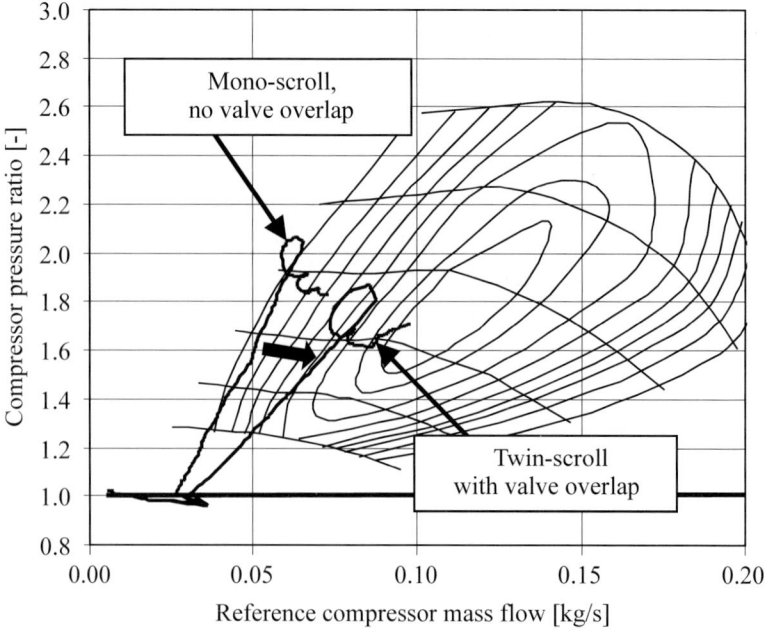

Fig. 8.43: Operating lines for concept 2 and 5 in the compressor map

Fig. 8.43 shows the operation lines in the compressor map for the 2nd and 5th concepts – thus an operation optimized with respect to valve control times with and without scavenging of the fresh air. We recognize in the scavenging arrangement the compressor operation line lying further right, which has a clear distance from the surge line. The higher output in the compression of the larger fresh air mass is partially overcompensated by the increased compressor efficiency. With this process, the low end torque speed can be lowered far below 1,500 rpm in order to achieve the maximum stationary torque.

9 Fluid mechanical simulation

Fluid mechanical or CFD simulation (CFD: computational fluid dynamics) is playing an increasingly important role in the simulation of engine processes, as it makes possible the most detailed physical description of the relevant processes. The most diverse processes in the engine field are considered, like charge changing, in-cylinder flow, exhaust gas recirculation, outflow processes, secondary reactions in the exhaust tract, catalyst converter flow, turbine and compressor flow in the turbocharger, internal nozzle flow, or coolant flow.

A relatively costly process is still necessary: firstly, computational meshes must be generated, and after a definition of the (often extensive) initial and boundary conditions, the actual calculation can finally be started. Evaluation is also typically costly due to the large amount of data. Finally, the theoretical requirements are also quite considerable. We are therefore devoting an entire chapter to the topic of "CFD".

In chapter 9.1, the foundations of numerical fluid mechanics are first explicated. Calculations of internal engine processes like injection and combustion demand a detailed treatment, which we will explore in chapters 9.2 and 9.3.

9.1 Three-dimensional flow fields

9.1.1 Basic fluid mechanical equations

- **Mass and momentum transport**

In the following, the basic equations of fluid mechanics will be briefly recapitulated, whereby the components will be expressed in sum convention, i.e. summation over twice-appearing indexes has to be performed. For an exhaustive derivation, see Merker and Baumgarten (2000), Cebeci (2002) and White (1991). The local equation of mass conservation, called the *continuity equation*, reads

$$\frac{\partial}{\partial t}\rho(x,t) + \frac{\partial}{\partial x_i}(\rho(x,t)v_i(x,t)) = 0 \ . \tag{9.1}$$

The dependency of the field quantities on x or t is usually simply left out. The momentum equation (*Navier-Stokes equation*) then reads

$$\rho\left(\frac{\partial}{\partial t} + v_j \frac{\partial}{\partial x_j}\right) v_i - \frac{\partial}{\partial x_j}\left(\tau_{ij}\left[\frac{\partial v_k}{\partial x_l}\right]\right) = -\frac{\partial p}{\partial x_i} + f_i \ , \tag{9.2}$$

whereby

$$\tau_{ij} = \mu\left(\frac{\partial v_i}{\partial x_j} + \frac{\partial v_j}{\partial x_i}\right) + \xi \frac{\partial v_k}{\partial x_k}\delta_{ij} \qquad \mu, \xi: \text{1. and 2. viscosity coefficient} \tag{9.3}$$

designates the *stress tensor* and f_i the external force density (e.g. gravity). Often,

$$\xi = -\frac{2}{3}\mu \qquad (9.4)$$

is assumed, i.e. τ_{ij} is traceless.

So far the general compressible case was considered, i.e. ρ is variable, a function of location and time. For the case of incompressible flow (typically for liquids), i.e. ρ is constant, the continuity and Navier-Stokes equations are simplified drastically, and we obtain for the incompressible fluid

$$\frac{\partial v_i}{\partial x_i} = 0 \qquad (9.5)$$

$$\rho\left(\frac{\partial}{\partial t} + v_j \frac{\partial}{\partial x_j}\right)v_i - \mu \Delta v_i = -\frac{\partial p}{\partial x_i} + f_i , \qquad (9.6)$$

whereby

$$\Delta = \frac{\partial^2}{\partial x^2} + \frac{\partial^2}{\partial y^2} + \frac{\partial^2}{\partial z^2}$$

designates the *Laplace Operator*. Differentiation of the second term in the continuity equation (9.1) according to the product rule provides

$$\left(\frac{\partial}{\partial t} + v_i \frac{\partial}{\partial x_i}\right)\rho + \rho \frac{\partial}{\partial x_i} v_i = 0 . \qquad (9.7)$$

The operator

$$\frac{\partial}{\partial t} + v_i \frac{\partial}{\partial x_i}$$

in equation (9.7) is called the *convective* or *substantive derivative*. It also appears in the Navier-Stokes equation (9.2) and expresses the temporal change of local fluid quantities in the external, spatially fixed laboratory coordinates. The transition from a local coordinate system carried along with the fluid (*Lagrangian coordinates*) to a global, spatially fixed coordinate system (*Euler coordinates*) thus corresponds to the substitution

$$\left.\frac{\partial}{\partial t}\right|_{Lagrange} \longrightarrow \left(\frac{\partial}{\partial t} + v_i \frac{\partial}{\partial x_i}\right)_{Euler} . \qquad (9.8)$$

If we now assume that the 2nd Newtonian Axiom

$$m\frac{dv_i}{dt} = F_i$$

is valid in the local coordinate system, carried along with the fluid, whereby force F is composed of an external component and the pressure gradient in the flow, from this follows then the *Euler equation*

9.1 Three-dimensional flow fields

$$\rho\left(\frac{\partial}{\partial t}+v_j\frac{\partial}{\partial x_j}\right)v_i = -\frac{\partial p}{\partial x_i}+f_i \ . \tag{9.9}$$

This is valid for an ideal, frictionless fluid and distinguishes itself from the Navier-Stokes equation (9.2) by the term of viscosity. That is, in a real flow, an additional force, the force of friction, is in effect, for which commonly the Newtonian approach

$$f_{i,friction} = \frac{\partial}{\partial x_j}\left(\tau_{ij}\left[\frac{\partial v_k}{\partial x_l}\right]\right) \tag{9.10}$$

is used. This additional term also means that the Navier-Stokes equation is now a 2nd order differential equation in the space coordinates and therefore requires additional boundary conditions. The physical equivalent is the phenomenon of wall friction, because of which a Navier-Stokes fluid rests directly on the wall relative to it and thus forms a *boundary layer*, while a Euler equation-based fluid flow along the wall frictionlessly, i.e. with finite speed.

Partial differential equations are classified according to their properties. We distinguish between *elliptic*, *parabolic*, and *hyperbolic* equations. The incompressible Euler equation (9.9) is of the hyperbolic type with reference to the variables v[1]. For such a differential equation, *characteristics* exist, i.e. a set of curves, along which temporal development is given by means of a system of ordinary differential equations. Thus, the Euler equation is simplified along a spatial curve $\chi(t)$ with

$$\frac{d\chi_i}{dt} = v_i \tag{9.11}$$

to

$$\rho\frac{dv_i}{dt} = -\frac{\partial p}{\partial x_i}+f_i \ ; \tag{9.12}$$

i.e. one can imagine, that the solution field "propagates"[2] along the set of curves determined by the characteristics (9.11) and (9.12). Thus, in order to define the boundary conditions, only the initial values $v(t=t_0, x_i=x_{0,i})$ and $v(t=t_0, x_i=x_{o,i})$ at starting time t_0 must be given for various starting positions $x_{0,i}$. One typical elliptic differential equation is the *Poisson equation*

$$\Delta\phi = \left(\frac{\partial^2}{\partial x^2}+\frac{\partial^2}{\partial y^2}+\frac{\partial^2}{\partial z^2}\right)\phi = 4\pi\gamma \ . \tag{9.13}$$

It can be shown, that the solution in a certain area $\phi(x,y,z)$ depends on all ϕ-values of the boundary of this area; there are no preferable propagation directions nor characteristics.

The *heat conduction equation* or *Helmholtz equation* (for the description of heat conduction in a solid)

---

[1] For that it is, together with the incompressible continuity equation, elliptic in pressure!
[2] This is valid under the specification of an external pressure and force field

$$\rho c_V \frac{\partial}{\partial t} T - \lambda \Delta T = 0 \tag{9.14}$$

as well as the Navier-Stokes equation are parabolic equations. They indeed possess a definite propagation direction in time, but not in space. One typical initial condition consists therefore in the specification of all values for the boundary of a certain area at starting time t_0.

The above suggests that these differential equation properties are also decisive for the numerical solution. With reference to the Navier-Stokes equation, we should especially keep in mind that, by adding the viscosity term, an alteration of the type of differential equation is caused; the incompressible Euler equation is hyperbolic, while the Navier-Stokes equation is parabolic with corresponding consequences for the solution method. The propagation behavior of the compressible equations is once again different, since now the phenomenon of sound appears as well. Typical engine 3D-flows like cylinder flows are necessarily compressible, but this compressibility is so weak (a measure for this is the Mach number $a = v/c$), that in fact the properties of the incompressible equations are valid.

- **Transport of internal energy and species**

The equation set must still be completed. In the incompressible case, the continuity equation (9.5) and the Navier-Stokes equation (9.6) (a vector equation, i.e. three component equations), four equations in total, are already complete to determine the four unknowns velocities (vector!) and pressure. In the compressible case however, the density must also be determined. For a single-component or a homogeneously mixed gas (which can again be treated in a single-component fashion) we obtain the density from the pressure by means of the thermal equation of state.

$$p = \frac{\rho \tilde{R} T}{M} . \tag{9.15}$$

This equation now contains the temperature in addition, which by means of the caloric equation of state

$$u = \int_{T_0}^{T} c_V(\vartheta) d\vartheta + u_0 \tag{9.16}$$

is linked with the (specific) internal energy. The internal energy is an extensive quantity, for which a transport equation can be formulated (similarly as for the momentum)

$$\rho \left(\frac{\partial}{\partial t} + v_j \frac{\partial}{\partial x_j} \right) u - \frac{\partial}{\partial x_i} \left(\lambda \frac{\partial T}{\partial x_i} \right) = -p \frac{\partial v_i}{\partial x_i} + \tau_{ij} \frac{\partial v_i}{\partial x_j} + q , \tag{9.17}$$

see Merker and Eiglmeier (1999). The second term on the left side is a diffusion term and corresponds to the viscosity term in the Navier-Stokes equation (λ designates the heat conductivity). The first two terms on the right side represent energy sources and sinks, the first term $-p \, \partial v_i / \partial x_i$ can assume both signs and corresponds to the reversible mechanical compression work, which is performed on the volume element. The second term $\tau_{ij} \, \partial v_i / \partial x_j$ describes the heat released by internal friction; this term is always positive (2nd law of ther-

9.1 Three-dimensional flow fields

modynamics). The third term q describes further heat sources, be it via evaporation or via combustion.

With this, our equation system is also complete for the incompressible case – for the seven unknowns, velocity (3), pressure, density, temperature, and internal energy, we have seven equations, (9.1), (9.2), (9.15), (9.16) and (9.17). For the case that the fluid is an inhomogeneous mixture of several components, further equations are required for the determination of the material composition, see below.

We should point out that the energy equation required here is the equation for internal energy and by no means the equation for kinetic energy. The latter results from the Navier-Stokes equation and is therefore not an independent quantity. Of course, instead of the internal energy, the total energy (internal energy + kinetic energy) can be transported, or the thermal enthalpy $(w = u + p/\rho)$. This is all equivalent, because when one knows the quantities v, p, ρ, all forms of energy can be converted into each other. It is very popular, especially amongst chemists, to transport the total enthalpy (thermal enthalpy + chemical energy). Assuming a known material composition, this is also convertible into the other energy quantities, making the process equivalent.

Finally, we should also consider the case that the fluid under investigation is given as an inhomogeneous mixture of several species. In this case, transport equations for the concentrations $c_{(k)}$ of the particular species must be formulated,

$$\rho\left(\frac{\partial}{\partial t} + v_i \frac{\partial}{\partial x_i}\right) c_{(k)} - \frac{\partial}{\partial x_i}\left(D_{(k)}\rho \frac{\partial}{\partial x_i} c_{(k)}\right) = Q_{(k)}(c_{(j)}, p, T) \qquad (9.18)$$

with

$$\sum c_{(k)} = 1 ,$$

which are constructed quite analogously to the Navier-Stokes and energy transport equation. On the left side are a convective derivative and the diffusion term and on the right a source term, which is only other than zero when chemical reactions are occurring, which is especially the case in combustion. Also, an additional diffusion term appears in the energy equation (9.17), so that it now reads

$$\rho\left(\frac{\partial}{\partial t} + u_j \frac{\partial}{\partial x_j}\right) u - \frac{\partial}{\partial x_i}\left(\lambda \frac{\partial T}{\partial x_i} + \rho D_{(k)} \sum_{(k)} h_{(k)} \frac{\partial c_{(k)}}{\partial x_i}\right) = -p\frac{\partial v_i}{\partial x_i} + \tau_{ij}\frac{\partial v_i}{\partial x_j} . \qquad (9.19)$$

All current CFD codes use the equations of state of ideal gases. Yet this assumption is not particularly suitable for diesel engine conditions (peak pressure of more than 200 bar); in this case, one should rather make use of the equations of state of real gases. An implementation is currently unavailable commercially.

- **Passive scalars and the mixture fraction**

Frequently, transport equations of other, formal scalars are defined, like progress variables or the flame surface density. These transport equations follow essentially the pattern of (9.18) (i.e. convection, diffusion, source term). Those quantities however make no contribution to the thermodynamic equations (9.15) and (9.16) and are therefore called *passive scalars* as opposed to the genuine chemical species, called *active scalars*.

One important scalar is the so-called *mixture fraction Z*. The mixture fraction field describes the local mixture states of two gases, whereby each of these gasses may be given as a homogeneous mixture of different species; it assumes values between 0 and 1. As long as there is no reaction, one can express it as

$$Z = \frac{\rho_{gas\,I}}{\rho_{gas\,I} + \rho_{gas\,II}} \ . \tag{9.20}$$

We now establish the fact that Z is a linear function in an arbitrary element mass fraction. The element mass fraction c_X of the element X (e.g. C, O, or H) is defined in this case as

$$c_X = \frac{\rho_X}{\rho_{total}} \ . \tag{9.21}$$

Be $c_{X,I}$, the X-mass fraction in gas I, correspondly $c_{X,II}$ the X-mass fraction in gas II and c_X the X-mass fraction in the local I-II mixture state. We then find the following dependence:

$$Z(c_X) = \frac{c_X - c_{X,II}}{c_{X,I} - c_{X,II}} \ . \tag{9.22}$$

The basic idea is to use this relation for the definition of the mixture fraction, because this definition is not influenced by chemical reactions, as it is defined on an elemental basis. In this way, we obtain a quantity suitable for describing mixtures independently of reactions (combustion!). The mixture fraction is thus an essential concept for the sake of describing diffusion flames. The transport equation of the mixture fraction corresponds to that of a species

$$\rho \left(\frac{\partial}{\partial t} + v_i \frac{\partial}{\partial x_i} \right) Z - \frac{\partial}{\partial x_i} \left(D \rho \frac{\partial}{\partial x_i} Z \right) = 0 \ , \tag{9.23}$$

except that no chemical source term appears[3]. For the diffusion constant D, a "mean" value of the diffusion constant of the species involved must be employed[4]. Since the mixture fraction can be used to calculate the mixture composition, it can function as an active scalar.

- **Conservative formulation of the transport equations**

Finally, it shall be mentioned that transport equations (for energy, momentum, and scalars) can also, with the help of the continuity equation, be represented in the so-called *conservative formulation*. For the scalar transport, this reads

$$\frac{\partial}{\partial t}\left(\rho c_{(k)}\right) + \frac{\partial}{\partial x_i}\left(\rho v_i c_{(k)}\right) - \frac{\partial}{\partial x_i}\left(D_{(k)} \rho \frac{\partial}{\partial x_i} c_{(k)}\right) = Q_{(k)}(c_{(j)}, p, T) \ , \tag{9.24}$$

---

[3] Evaporation source terms may however appear.
[4] In the case of turbulent flow, this problem of varying laminar diffusion constant is reduced.

9.1.2 Turbulence and turbulence models

- **The phenomenology of turbulence**

In the Navier-Stokes equation, the relative order of magnitude of the viscous term has a large influence on the character of the flow. In order to understand this, we will consider a typical flow problem like the flow over of a cylinder[5] (for reasons of simplification we will consider the incompressible case) and introduce characteristic scales for length L (e.g. the cylinder diameter) and velocity v (e.g. in-flow velocity). With this, we obtain in reference to these scales the standardized variables

$$x = x^* L, \qquad v = v^* V, \qquad t = t^* L/V, \qquad p = p^* \rho V^2 . \tag{9.25}$$

With the standardized variables x^*, v^*, t^* and p^*, the problem can be formulated in a scale-invariant manner. We finally obtain the equations

$$\frac{\partial v_i^*}{\partial x_i^*} = 0 \text{ and} \tag{9.26}$$

$$\left(\frac{\partial}{\partial t^*} + v_j^* \frac{\partial}{\partial x_j^*}\right) v_i^* - \frac{1}{\text{Re}} \frac{\partial^2 v_i^*}{\partial x_i^{*2}} = -\frac{\partial p^*}{\partial x_i^*} , \tag{9.27}$$

whereby

$$\text{Re} = \frac{\rho V L}{\mu} \tag{9.28}$$

designates the *Reynolds number*. In the Reynolds number, all scale influences are now included. Flows with the same Reynolds number can be formed into each other via rescaling of variables; we say that they are similar to each other. Thus, the Reynolds number classifies flows. In addition, it describes the relative size of the viscous term. If the Reynolds number is small and the viscous term large, then we have the case of a "honey-like", "viscous" flow. In the other limit case, i.e. in the boundary value of an infinitely large Reynolds number, one could at first suppose that the viscous term simply vanishes and that the Navier-Stokes equation would be reduced to the Euler equation. This is however not the case, for, as opposed to the Euler equation, a (indeed increasingly thin) viscous wall boundary layer (property of the Navier-Stokes equation, see above) exists as before, in which the velocity is reduced to zero (relative to the wall velocity). In this way, high velocity gradients exist close to the wall, which lead to vortex formation in the wake of the cylinder.

The flow over of our cylinder resembles for Re = 10^{-2} approximately as represented in Fig. 9.1 a – a laminar, viscous flow. With an increasing Reynolds number, more vortices forms we find behind the cylinder, which also detach, but first still show periodical structures

---

[5] the so-called Karman vortex street

(Fig. 9.1 b and c). With further increase of the Reynolds number, the flow finally fluctuates chaotically and three-dimensionally – it is now *turbulent* (Fig. 9.1 d and e). Large eddies disintegrate into smaller ones, these in turn into still smaller ones, a eddy spectrum forms down to a very small length scale, the Kolmogorov scale, on which the flow becomes viscous (laminar) again. Such a chaotic process can no longer be calculated deterministically, even with an arbitrarily large computer, because the minutest cause may have the largest effects.

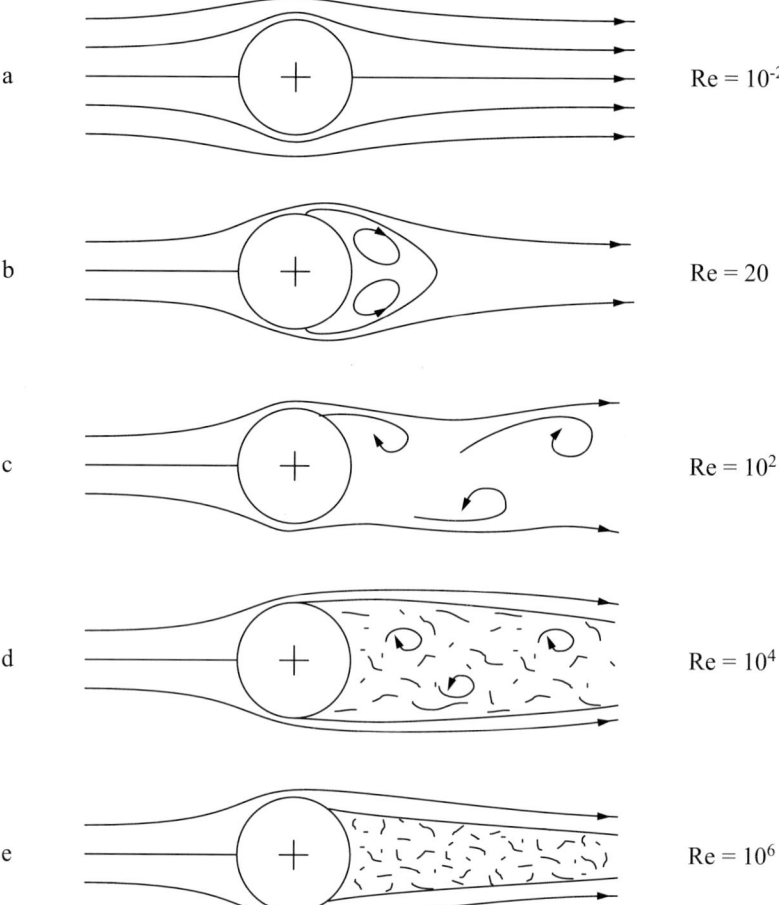

Fig. 9.1: Karman vortex street (flow over a cylinder) at varying Reynolds' numbers

- **Modeling turbulence**

However, statistical quantities can be provided for the description of turbulent phenomena; the spectrum of turbulent fluctuations is also accessible to theoretical considerations. Typical quantities are the *turbulent length scale* l_t, the *turbulent time scale* τ_t, the *turbulent velocity*

9.1 Three-dimensional flow fields

scale v_t, as well as the *turbulent viscosity* μ_t. Assuming spatially homogeneous turbulence, there are two independent quantities, most commonly the (specific) *turbulent kinetic energy* k and the *turbulent dissipation* ε are used. With these we obtain

$$v_t = \sqrt{k}, \quad \mu_t = c_\mu \rho \frac{k^2}{\varepsilon}, \quad l_t = c_l \frac{k^{\frac{3}{2}}}{\varepsilon}, \quad \tau_t = \frac{l_t}{v_t}, \tag{9.29}$$

whereby c_μ and c_l represent proportionality constants. The mathematical method in introducing these turbulence quantities consists in an ensemble averaging of fluid mechanical equations, i.e. not one single realization is calculated, but the mean value from all possible realizations under identical measurement-technical boundary conditions[6]. This Reynolds averaging will be carried out for the case of a source-free scalar transport equation (9.24).

If $\langle v_i \rangle$ and $\langle c \rangle$ are chosen as the ensemble-averaged quantities, we then obtain for the boundary instantaneous values v_i and c the following reductions

$$v_i = \langle v_i \rangle + v_i' \quad \text{with} \quad \langle v_i' \rangle = 0, \quad c = \langle c \rangle + c' \quad \text{with} \quad \langle c' \rangle = 0, \tag{9.30}$$

whereby v_i' and c' now describe the turbulent velocity and scalar fluctuations. The (specific) turbulent kinetic energy introduced above thus amounts to

$$k = \frac{1}{2} \sum_i (v_i')^2. \tag{9.31}$$

For all terms linear in v and c the averaging of (9.24) can be simply carried out, but the non-linear convection term leads to an additional term

$$\frac{\partial}{\partial t}(\rho \langle c \rangle) + \frac{\partial}{\partial x_j}(\rho \langle v_j \rangle \langle c \rangle) + \frac{\partial}{\partial x_j}(\rho \langle v_j' c' \rangle) - \frac{\partial}{\partial x_j}\left(D_c \rho \frac{\partial}{\partial x_j} \langle c \rangle\right) = 0. \tag{9.32}$$

If we now assume that the turbulence behaves like a spatially isotropic fluctuation process uncorrelated at different times ("white noise"), we can show that this additional term takes on the structure of an additional diffusion term with the diffusion constant

$$D_t = \text{const.} \cdot k^2/\varepsilon = \mu_t/(\rho\, Sc_t).$$

This implies the so-called *eddy diffusivity approach*

$$\langle v_i' c' \rangle = -\frac{\mu_t}{\rho\, Sc_t} \frac{\partial \langle c \rangle}{\partial x_i}. \tag{9.33}$$

Sc_t designates the turbulent Schmidt number, i.e. the ratio of turbulent viscosity and turbulent diffusion. A similar term, $\langle v_i' v_j' \rangle$, appears in the averaged Navier-Stokes equation. It takes over the role of an additional tension stress, the so-called *Reynolds stress* $\tau_{R,ij}$. In analogy to the eddy diffusivity approach, for this the *eddy viscosity approach* is used (we are working as before on an incompressible basis)

---

[6] For engines it is often said that this corresponds to an averaging over several cycles. One should however be wary of such a "descriptive" interpretation. Engine cycle fluctuations are to a large extent caused by for fluctuations of the boundary conditions, e.g. of the injection spray or the residual gas content of the previous cycle.

$$\langle v'_i v'_j \rangle = -\frac{\mu_t}{\rho} \cdot \left(\frac{\partial \langle v_j \rangle}{\partial x_i} + \frac{\partial \langle v_i \rangle}{\partial x_j} \right) + \frac{2}{3} \delta_{ij} k \ . \tag{9.34}$$

This means that in addition to the original, laminar diffusion/viscosity in the averaged equations, a usually much larger turbulent diffusion/viscosity appears ($D_t \gg D_c$). One can now easily understand the essential effect of turbulent averaging; the viscous, diffusive character of the differential equations is now more in the foreground and produces a *well posed* problem (alterations in the solution of a given order of magnitude demand a sufficiently large alteration of the initial conditions, there is no longer any chaos). Frequently, because of the dominant turbulent quantities, laminar viscosity and diffusion constants are simply ignored (the same is valid of course for all other transported quantities).

For the compressible case however, we only obtain formally a very similar result, if we carry out the averaging *density-weighed*, i.e. using the so-called *Favre averaging*

$$\langle \Phi \rangle_F = \langle \rho \Phi \rangle / \langle \rho \rangle \qquad \Phi = \langle \Phi \rangle_F + \Phi'' \ , \tag{9.35}$$

which will be utilized in the following.

For the calculation of k and of ε, we still need additional equations. From the Navier-Stokes equation, a transport equation for the turbulent kinetic energy, the k-equation, can be derived. For this, we only need to assume that pressure correlations are being neglected. The k-equation (in the Favre average) reads

$$\langle \rho \rangle \frac{\partial}{\partial t} k + \langle \rho \rangle \langle v_i \rangle_F \frac{\partial}{\partial x_j} k - \frac{\partial}{\partial x_i} \left(\frac{\mu_t}{\Pr_k} \frac{\partial k}{\partial x_i} \right) = \tau_{R,ij} \langle S_{ij} \rangle_F - \langle \rho \rangle \varepsilon - \frac{2}{3} \langle \rho \rangle k \nabla \cdot \langle \vec{v} \rangle_F \ , \tag{9.36}$$

with the compressible Reynolds tension stress $\tau_{R,ij}$

$$\tau_{R,ij} = \mu_t \left(\frac{\partial \langle v_j \rangle_F}{\partial x_i} + \frac{\partial \langle v_i \rangle_F}{\partial x_j} - \frac{2}{3} \delta_{ij} \left(\nabla \cdot \langle \vec{v} \rangle_F \right) \right) \tag{9.37}$$

and the shear tensor S_{ij}

$$S_{ij} = \frac{1}{2} \left(\frac{\partial v_j}{\partial x_i} + \frac{\partial v_i}{\partial x_j} \right) \ . \tag{9.38}$$

The fact that the turbulent diffusion constant of the k transport equation can be quite different from μ_t is taken into account by the introduction of a proportionality constant, the turbulent Prandtl number for k and $\Pr_k$. The turbulent dissipation ε amounts into

9.1 Three-dimensional flow fields

$$\varepsilon = v \left\langle \frac{\partial v_i''}{\partial x_j} \left(\frac{\partial v_i''}{\partial x_j} + \frac{\partial v_j''}{\partial x_i} \right) \right\rangle_F \quad . \tag{9.39}$$

A transport equation for ε can in principle also be derived from the Navier-Stokes equation, however it must be modeled in several terms in accordance with the k-equation. Its usual form reads

$$\langle \rho \rangle \frac{\partial}{\partial t} \varepsilon + \langle \rho \rangle \langle v_i \rangle_F \frac{\partial}{\partial x_j} \varepsilon - \frac{\partial}{\partial x_i} \left(\frac{\mu_t}{\Pr_\varepsilon} \frac{\partial \varepsilon}{\partial x_i} \right) =$$

$$c_{\varepsilon,1} \frac{\varepsilon}{k} \tau_{R,ij} \langle S_{ij} \rangle_F - c_{\varepsilon,2} \langle \rho \rangle \frac{\varepsilon^2}{k} - \left(\frac{2}{3} c_{\varepsilon,1} - c_{\varepsilon,3} \right) \langle \rho \rangle \varepsilon \nabla \cdot \langle \vec{v} \rangle_F \quad . \tag{9.40}$$

According to the standard, the following coefficient set is used in the k-ε-model.

c_μ	$c_{\varepsilon 1}$	$c_{\varepsilon,2}$	$c_{\varepsilon,3}$	$\Pr_k$	$\Pr_\varepsilon$
0.09	1.44	1.92	-0.33	1.0	1.3

Finally, we still require a turbulent transport equation for the internal energy. This differs from the laminar equation only by turbulent transport coefficients and ε as an additional source term.

With that, a closed equation system has been derived. With the help of k and ε, the diffusion constants for the additional turbulent diffusion terms can be calculated in the transport equations. Since we will actually always be looking at Favre averaged equations and quantities in the following, the symbol for the Favre averaging $\langle \ \rangle_F$ will usually be left out.

- **The turbulent law of the wall**

However, there is still a problem concerning the boundary conditions. On the walls, a boundary layer is formed, in which the velocity decreases because of friction to zero, i.e. the flow becomes laminar. Thus in a turbulent flow the boundary layer consists typically of a laminar sub-layer and a turbulent zone. The k-ε-model can not be applied across the entire boundary layer. Moreover, the boundary layers are often so thin (especially in engines) that they are hardly numerically solvable anyway.

The usual way to overcome this problem is by deriving the so-called law of the wall with the help of the boundary layer equations, because the shear stress is constant over the boundary layer, i.e. it describes a tangential momentum flow constant over the wall distance and flowing off onto the wall (this is to speak the definition of the boundary layer). We now need a turbulent law of the wall, i.e. an analytical boundary layer model, in order to calculate shear stress from the local velocities in the cells closest to the wall.

At a given wall shear stress τ_w, only one velocity scale is available in the turbulent boundary layer, the shear stress speed v_τ, with $\tau_w = \rho v_\tau^2$. It should be proportional to the turbulent velocity scale, while the turbulent length scale should be proportional to the wall distance y. Under these assumptions, we obtain for the turbulent viscosity

$$\mu_t = \kappa \rho y v_\tau ,$$

whereby κ designates a proportionality constant, the *von Karman constant*. For the shear stress τ_w we now obtain the following relation

$$\tau_w = \rho v_\tau^2 = \rho \kappa y v_\tau \frac{\partial v_w}{\partial y} , \qquad (9.41)$$

from which follows under the assumption of constant density the *logarithmic law of the wall*

$$v_w = \frac{v_\tau}{\kappa} \ln y + C \qquad (9.42)$$

with an integration constant C. In the standardized coordinates y^+ and v^+:

$$v^+ = \frac{v_w}{v_\tau} \text{ and } y^+ = \frac{\rho v_\tau y}{\mu} \qquad (9.43)$$

the logarithmic law of the wall takes on its universal form

$$v^+ = \frac{1}{\kappa} \ln y^+ + \tilde{C} \qquad (9.44)$$

with

$$\kappa = 0.4 \text{ and } \tilde{C} = 5.5 .$$

In this form, a universal range of validity of the law of the wall can now also be given, namely

$$20 < y^+ < 150 . \qquad (9.45)$$

For the numerical solution, the node nearest to the wall must remain within this boundary layer. The law of the wall is employed in CFD calculation insofar as v_τ and consequently τ_w are calculated from v^+ and y^+ in the cell closest to the wall. This τ_w then supplies a momentum source term (negative) as a boundary condition for the Navier-Stokes equation.

A law of the wall can be derived quite analogously for the temperature as well. From the heat flux

$$q_w = \rho \kappa y v_\tau \frac{c_p}{\Pr} \frac{\partial T}{\partial y} \qquad (9.46)$$

follows

$$T - T_w = \frac{q_w \Pr}{\kappa c_p \rho v_t} \left(\ln y^+ + \text{const.} \right) . \qquad (9.47)$$

Finally, one still needs boundary conditions for the turbulence quantities k and ε. Because of the constancy of the shear stress τ_w and the velocity scale v_τ in the boundary layer, a $k_w = \text{const.}$ approach is plausible. Such that this can be valid however, production and dissipation of k must be in equilibrium in the boundary layer

9.1 Three-dimensional flow fields

$$\mu_t \left(\frac{\partial v_w}{\partial y}\right)^2 = \kappa \rho y v_\tau \left(\frac{v_\tau}{\kappa y}\right)^2 = \rho \varepsilon_w \ . \tag{9.48}$$

Moreover, for the viscosity

$$\mu_t = \kappa \rho y v_\tau = c_\mu \rho \frac{k_w^2}{\varepsilon_w} \tag{9.49}$$

has to be valid. From both of these equations, it is not difficult to arrive at

$$k_w = \frac{v_\tau^2}{\sqrt{c_\mu}} \quad \text{and} \quad \varepsilon_w = \frac{v_\tau^3}{\kappa y} \ , \tag{9.50}$$

i.e. ε_W diverges towards the wall! For the validity of this relation however, diffusion and source terms of the ε-equation must also be in equilibrium

$$\frac{\partial}{\partial y}\left(\frac{\mu_t}{\Pr_\varepsilon}\frac{\partial \varepsilon_w}{\partial y}\right) + c_{\varepsilon,1}\frac{\varepsilon_w}{k_w}\mu_t\left(\frac{\partial v_w}{\partial y}\right)^2 - c_{\varepsilon,2}\rho\frac{\varepsilon_w^2}{k_w} = 0 \ . \tag{9.51}$$

Insertion of the already obtained results provides a relation between the model constants

$$\kappa = \sqrt{\sqrt{c_\mu}\ \Pr_\varepsilon (c_{\varepsilon,2} - c_{\varepsilon,2})} \ . \tag{9.52}$$

The derivation of these equations occurred under the assumption of stationary, flows parallel to the wall and constant density. From this it becomes clear that these laws of the wall can only have limited validity. In engines especially, the flows are transient, stagnation points appear (e.g. as the injection jets hit the wall), and the density of the boundary layer is not constant due to higher temperature gradients towards the wall. There are therefore methods for deriving law of the walls for variable density, e.g. the formulation of Han and Reitz (1995). In this case, in the equation for the heat flux (9.46), density ρ is considered as given by the ideal gas equation of state. The integration finally provides the relation

$$T \ln\left(\frac{T}{T_w}\right) = \frac{q_w \Pr}{\kappa c_p \rho v_t}\left(\ln y^+ + \text{const.}\right) \tag{9.53}$$

instead of (9.47). Of course, the influences of a variable density on k and ε distribution are not hereby considered. Nevertheless, the use of such a formulation deserves strong recommendation especially for combustion calculations.

- **Modeling the turbulent mixture state**

The mixture fraction was introduced in chap. 9.1.1 as an important reference quantity in describing mixing processes. Its transport equation should also undergo a turbulent averaging process; the result corresponds to the scalar transport equation (9.32) incl. (9.33). However, it must be kept clear that a mean value of Z no longer contains the total information about the local mixture states. Thus, a mean value of $Z = 0.5$ can indicate a fluctuation-free, perfect mixture from equal parts of "flow 1" and "flow 2", but it can also indicate a static superimposing of two totally unmixed states, "only flow 1" and "only flow 2", whereby both states appear with equal probability. Another helpful quantity in describing the local mixture states is the (Favre) variance of the mixture fraction

$$\langle Z''^2 \rangle_F = \langle [Z - \langle Z \rangle_F]^2 \rangle_F = \langle Z^2 \rangle_F - \langle Z \rangle_F^2 . \tag{9.54}$$

From the mixture fraction transport equation a transport equation for the variance can be derived

$$\langle \rho \rangle \frac{\partial}{\partial t} \langle Z''^2 \rangle_F + \langle \rho \rangle \langle v_j \rangle_F \frac{\partial}{\partial x_j} \langle Z''^2 \rangle_F - \frac{\partial}{\partial x_j} \left(\frac{\mu_t}{Sc_t} \frac{\partial}{\partial x_j} \langle Z''^2 \rangle_F \right) = $$

$$2 \frac{\mu_t}{Sc_t} (\nabla \langle Z \rangle_F)^2 - \underbrace{2D \langle \rho \rangle \langle (\nabla Z'')^2 \rangle_F}_{\chi} . \tag{9.55}$$

To model the last term χ of this equation, the so-called scalar dissipation rate, the commonly used approach is

$$\chi = c_\chi \frac{\varepsilon}{k} \langle Z''^2 \rangle_F . \tag{9.56}$$

Based on its character, the term χ corresponds to the turbulent dissipation ε in the k-equation (k is also a variance, that one of velocity). We thus see that gradients of the Z mean value (inhomogeneities) incite Z variance formation, while the latter decays with the turbulent time scale if further production is suppressed. The constant c_χ is usually set to 2 (see Peters, (2000)).

One can now attempt to reconstruct, by means of structural knowledge of the distribution function, the information about the local mixture states from Z and Z''^2. Such a distribution function f is often named a *pdf* (probability density function). It assigns a rate of occurrence to each mixture state (Z values between 0 and 1). For this reason, it must be standardized to 1

$$\int_0^1 f(Z) dZ = 1 . \tag{9.57}$$

Because of the definition of the mean value and the variance

$$\langle Z \rangle = \int_0^1 Z f(Z) dZ \quad \text{and}$$

$$\langle Z''^2 \rangle = \int_0^1 (Z - \langle Z \rangle)^2 f(Z) dZ = $$

$$\int_0^1 \left(Z^2 - \langle Z \rangle^2 \right) f(Z) dZ \le \int_0^1 \left(Z - \langle Z \rangle^2 \right) f(Z) dZ \tag{9.58}$$

we obtain the relation

9.1 Three-dimensional flow fields

$$\langle Z''^2 \rangle \le \langle Z \rangle (1 - \langle Z \rangle) . \tag{9.59}$$

As an example, the statistical superposition of completely unmixed states ($Z = 0$ and $Z = 1$) is described by the linear combination of two Dirac distributions

$$f(Z) = a\,\delta(Z) + b\,\delta(Z-1) \text{ with } a + b = 1 . \tag{9.60}$$

In this cases, the sign of equality is valid in (9.59). In Fig. 9.2, two distribution functions are shown, on the one hand, for the case of minimal mixing, and on the other for a case of considerable mixing. All Z distribution functions should be placed more or less between these limiting cases. In the literature, it has become common to use the so-called beta function[7] for this. It is a simple function with two free parameters, which can be related without difficulty to the mean value and the variance of the distribution.

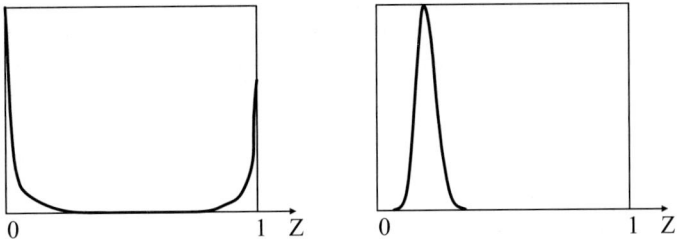

Fig. 9.2: Distribution functions for a slightly mixed state (left) and a highly mixed state (right)

It holds[8]:

$$f(Z) = \frac{\Gamma(a+b)}{\Gamma(a)\Gamma(b)} Z^{a-1}(1-Z)^{b-1}, \quad \int_0^1 f(Z)\,dZ = 1, \quad a,b > 0 , \tag{9.61}$$

$$\langle Z \rangle = \frac{a}{a+b} \text{ and } \langle Z'^2 \rangle = \frac{\langle Z \rangle (1-\langle Z \rangle)}{1+a+b} . \tag{9.62}$$

For $a,b < 1$, we obtain the "bath tub form" of Fig. 9.2 (left), for $a,b > 1$ a form like that in Fig. 9.2 (right). With the parameters $\langle Z \rangle$ and $\langle Z'^2 \rangle$, which are obtained from the transport equations, the equation system is now closed, and a and b can be determined from it. Although the total integral over the beta function is normalized to 1, the individual factors of (9.61) are very large for large or very small values of a and b. In the calculation of f, it is therefore recommendable to logarithmize e.g. (9.61) at first:

---

[7] In mathematical literature, the integral of this function as a function of the parameter a and b is called beta function:
$$B(a,b) = \int_0^1 (1-Z)^{a-1} Z^{b-1}\,dZ = \frac{\Gamma(a+b)}{\Gamma(a)\Gamma(b)} .$$

[8] The Gamma function is given as $\Gamma(x) = \int_0^\infty e^{-t} t^{x-1}\,dt$

$$\ln f = \ln \Gamma(a+b) - \ln \Gamma(a) - \ln \Gamma(b) + (a-1)\ln Z + (b-1)\ln(1-Z) \,, \tag{9.63}$$

in order to obtain *f* by means of exponentiation of the logarithm.

- **The validity of turbulence models – alternative approaches**

The k,ε-turbulence model and turbulent law of the wall thus far represent the current standard in the calculation of turbulent flow processes, and they have proven to be quite successful in most applications. This is of course also true for engine purposes. Nevertheless, one should always be aware of the fact that this is, in the final analysis, a rather simple model, which can never exactly reproduce the complete Navier-Stokes equations because of its simplified closure approaches, not even in the statistic average. This problem expresses itself in the non-universality of the model constants and the introduction of additional terms for the treatment of particular flow situations.

In certain cases, such modifications can also be of great relevance for engine calculations too, for example the Pope correction (Pope, 1978) for the calculation of free jets, a type of flow that occurs in direct injection. In many practical cases however, the influence of turbulence model modifications is rather minimal in comparison to the influence of the mesh resolution.

Reynolds stress models conceptually go beyond the k-ε-model. In this case, the Reynolds stress $\langle v_i' v_i' \rangle$ is not attributed to the turbulent kinetic energy

$$k = \frac{1}{2} \sum_{i=1,2,3} \langle v_i' v_i' \rangle$$

by means of the approach (9.37), but all six independent components of the Reynolds stress are described with their own transport equation. In this way, anisotropies in the turbulence structure, among other things, can be described (unfortunately, one usually works with only one single ε). However, such models demand intricate boundary conditions and a high calculation effort. For this reason, Reynolds stress models just start to be introduced to engine related problems.

One completely different method is the *large-eddy simulation* (LES). In this case, turbulent flow structures larger than a pre-given scale are directly modeled, while smaller ones are modeled. Such an approach should have a more general validity. However, here as well additional problems with the boundary values appear. Moreover, individual LES calculations no longer contain the total statistical information, bearing elements of randomness. In order to compensate for this, several realizations must be calculated. Finally, very high demands are made on numerical scheme and mesh quality. For this reason, it will still take some time before practical and workable LES applications will be available, even if they would also be very desirable in the sense of universal turbulence modeling.

Finally, we would however still like to point out that LES *cannot provide* the resolution of engine cycle-to-cycle fluctuations, even if that is typically given as the main argument for LES in the engine context. Turbulence models can only describe fluctuations caused by the turbulent flow itself, which reflect the chaotic, non-deterministic behavior of the Navier-Stokes equations. Variations or fluctuations that are created by variations of the boundary conditions cannot be described. In an engine, however, these are the dominant effects by far (injector, ignition, last cycle, throttle, ...). Thus, the stochastic case-to-case fluctuations of a

9.1 Three-dimensional flow fields

LES simulation (under unchanged, perfect boundary conditions) should be much smaller than the cycle-to-cycle fluctuations in the real engine.

Direct numerical simulation (DNS) will not be an option for the simulation of engine flows in the foreseeable future because of the small length and time scales that must be resolved directly. Moreover, individual calculations would only have a stochastic character, thus making several realizations again necessary. Still, DNS can also be used today profitably in the sense of a "numerical experiment", for example for the verification or calibration of models for turbulent flow processes (besides the actual turbulence models already discussed, this is valid also for spray and combustion models). For this purpose however, one must work in especially simple geometries at low Reynolds numbers. In principle, small, exemplary pieces of larger problems are investigated.

9.1.3 Numerics

In the following, some basic concepts of numerical fluid mechanics will be explained, in order to become acquainted with its most essential concepts, which are also of importance for practical work. For a detailed presentation, see Ferziger and Perić (1996) or Patankar (1980).

- **The finite volume method**

Customarily, CFD codes work in the finite volume method. This approach guarantees the numerical preservation of conservative quantities for incompressible flows (unfortunately, this is not obvious). However, the computational mesh should ideally be built hexahedrally. We proceed from a transport equation in the general form

$$\frac{\partial}{\partial t}\Phi + \frac{\partial}{\partial x_i}\Psi_i = \Xi , \qquad (9.64)$$

i.e., Ψ_i designates convection and diffusion flows of size Φ, Ξ the corresponding local source. With the help of the Gaussian law follows:

$$\frac{\partial}{\partial t}\int_V \Phi \, dV + \oint_{\partial V} \vec{\Psi} \, d\vec{S} = \int_V \Xi \, dV \qquad (9.65)$$

or for a hexahedron of the computational mesh

$$[\Phi(t+\Delta t) - \Phi(t)]\Delta V = -\sum_{l=1}^{6} \vec{\Psi}_{(l)} \Delta \vec{S}_{(l)} \Delta t + \Xi \Delta V \Delta t , \qquad (9.66)$$

whereby the sum runs over the six sides of the hexahedron.

$\vec{\Psi}_{(l)}$ designates the flux vector on the hexahedral side (l), $\vec{\Psi}_{(l)} \Delta \vec{S}_{(l)} \Delta t$ is the (oriented) flux, which in the time Δt leaves the hexahedral volume across side (l) with surface $\Delta \vec{S}_{(l)}$. The same term, but with the opposite sign, appears in the neighboring cell bordering the side (l). Thus insofar as the source Ξ is equal to zero, only fluxes between individual cells of the computational mesh are exchanged. The total amount

$$\sum_{z:\text{sum total}} \Phi_z \Delta V_z$$

remains constant over time, assuming that the flow over the border of the calculation range is zero.

- **Discretization of the diffusion term – central differences**

For the discretization of the diffusion term, we will at first look at a (purely elliptic) stationary diffusion equation in one dimension of space

$$\frac{\partial}{\partial x}\left[D\frac{\partial}{\partial x}\Phi\right] = 0 \ . \tag{9.67}$$

The Gaussian law states for a "one-dimensional cell" i

$$D_{i,+}\frac{\partial}{\partial x}\Phi(x_{i,+}) - D_{i,-}\frac{\partial}{\partial x}\Phi(x_{i,-}) = 0 \ , \tag{9.68}$$

whereby $x_{i,+}$ describes the location of the right cell border, $x_{i,-}$ the location of the left cell border (see Fig. 9.3); for equidistant cells of length Δx is valid

$$x_{i,+} = x_i + \frac{\Delta x}{2} \text{ and } x_{i,-} = x_i - \frac{\Delta x}{2} \ . \tag{9.69}$$

The quantities Φ are only known at the cell centers, i.e. at the locations x_{i-1}, x_i, x_{i+1},... and we have to attribute the gradients to these. The most obvious solution is the so-called *central difference scheme*, which provides for equidistant cells

$$\frac{\partial}{\partial x}\Phi(x_{i,+}) \cong \frac{\Phi(x_{i+1}) - \Phi(x_i)}{\Delta x} \ , \tag{9.70}$$

which finally (for constant D) leads to the following relation

$$\Phi(x_i) = \frac{1}{2}\Phi(x_{i+1}) + \frac{1}{2}\Phi(x_{i-1}) \ . \tag{9.71}$$

Since in this equation every Φ value is given as the mean value of its neighboring value, this equation system is not directly solvable. Rather, the equations provide, formulated for all cells $i = 1,..., N$, a linear equation system, which should be handled with an appropriate solution technique for linear equations. This equation system corresponds exactly to the character of an elliptic equation, in which all boundary values have to be prescribed for the problem definition.

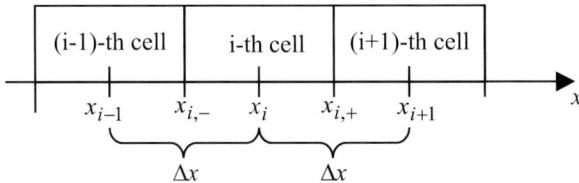

Fig. 9.3: Graphic representation of cells and cell boundaries

9.1 Three-dimensional flow fields

In general, every source-free transport equation can be attributed to such a pattern, so that the value of a field Φ results from the values of Φ in the temporally and spatially neighboring cells

$$\Phi(x_i, y_j, z_k; t_l) = \sum_{|\tilde{i}-i|+|\tilde{j}-j|+|\tilde{k}-k|+|\tilde{l}-l|=1} \alpha_{ijkl;\tilde{i}\tilde{j}\tilde{k}\tilde{l}} \Phi(x_{\tilde{i}}, y_{\tilde{j}}, z_{\tilde{k}}; t_{\tilde{l}}) . \tag{9.72}$$

Obviously (to guarantee that $\Phi = \text{const.}$ is also a solution of the discretized transport equation) the condition

$$\sum_{|\tilde{i}-i|+|\tilde{j}-j|+|\tilde{k}-k|+|\tilde{l}-l|=1} \alpha_{ijkl;\tilde{i}\tilde{j}\tilde{k}\tilde{l}} = 1 \tag{9.73}$$

has to be fulfilled. An essential criterion in this formulation is now the positivity of the coefficients

$$\alpha_{ijkl;\tilde{i}\tilde{j}\tilde{k}\tilde{l}} \geq 0 . \tag{9.74}$$

Only when this has been achieved does a stable, monotone numerical scheme exist. Physically, this corresponds to the fact that only via processes of diffusion and convection no local maximum in the value of a field can be generated (which had not already existed).

- **Discretization of the convection term – the upwind scheme**

If we extend the diffusion equation (9.67) by the convection terms, one then obtains

$$\rho \frac{\partial}{\partial x}[v\Phi] - \frac{\partial}{\partial x}\left[\rho D \frac{\partial}{\partial x} \Phi\right] = 0 . \tag{9.75}$$

For constant ρ and D, we obtain after integration

$$v_{i,+} \Phi(x_{i,+}) - v_{i,-}\Phi(x_{i,-}) - D\frac{\partial}{\partial x}\Phi(x_{i,+}) + D\frac{\partial}{\partial x}\Phi(x_{i,-}) = 0 . \tag{9.76}$$

The gradients

$$\frac{\partial}{\partial x}\Phi(x_{i,\pm})$$

can again be described by means of central differences. For $\Phi(x_{i,\pm})$ one could evidently use the difference system

$$\Phi(x_{i,\pm}) = \frac{1}{2}(\Phi(x_i) + \Phi(x_{i\pm 1})) . \tag{9.77}$$

This states in all

$$\left(\frac{2D}{\Delta x} + \frac{v_{i,+}}{2} - \frac{v_{i,-}}{2}\right)\Phi(x_i) = \left(\frac{D}{\Delta x} - \frac{v_{i,+}}{2}\right)\Phi(x_{i+1}) + \left(\frac{D}{\Delta x} + \frac{v_{i,-}}{2}\right)\Phi(x_{i-1}) . \tag{9.78}$$

For reasons of simplicity, assume from $v_{i,+} = v_{i,-} = v_i$ (this is exactly true in one dimension by the validity of the continuity equation, in several dimensions still roughly). With this, the positivity condition (9.74) leads to the following inequality[9]:

$$\frac{|v|\Delta x}{D} \leq 2 \ . \tag{9.79}$$

If this inequality is violated, which can easily be the case with less fine meshes or high convection velocities, then the difference system (9.77) is no longer stable. Thus, another discretization $\Phi(x_{i,\pm})$ must be introduced. The so-called upwind scheme[10] provides a stable difference scheme

$$\Phi(x_{i,+}) = \theta(v)\Phi(x_i) + \theta(-v)\Phi(x_{i+1}) \ , \quad \Phi(x_{i,-}) = \theta(v)\Phi(x_{i-1}) + \theta(-v)\Phi(x_i) \ , \tag{9.80}$$

i.e. the grid point lying upstream (where the flow is coming from) determines the local Φ value. The difference scheme now leads to

$$\left(\frac{2D}{\Delta x} + |v|\right)\Phi(x_i) = \left(\frac{D}{\Delta x} + \theta(-v)|v|\right)\Phi(x_{i+1}) + \left(\frac{D}{\Delta x} + \theta(v)|v|\right)\Phi(x_{i-1}) \ , \tag{9.81}$$

i.e. every coefficient is positive. In principle, this unsymmetrical pattern reflects the fact that a pure convection equation (without diffusion) is of the hyperbolic type and this does not "get along" with elliptic boundary conditions (the value in a cell results from its bordering e.g. all its neighboring cell values). It would rather have only one initial value (at the "inflow position"), from which it propagates with the flow. This property is passed down to convection-dominated flows.

If one develops $\Phi(x_{i,\pm})$ into a Taylor series, one then recognizes that the central difference scheme (9.77) is more precise than the upwind scheme (namely, of the second order in the discretization parameter Δx, while the upwind scheme is only of the first order). Still, numerical stability is a property one cannot do without for tasks with complex geometries and deformed grids. Therefore, in numerical fluid mechanics, we cannot avoid an upwind-type scheme (in the literature, there is a large number of modifications, including for higher orders of accuracy), or a so-called hybrid method, which still contains a substantial amount of "upwind".

- **Discretization of the time derivation – implicit scheme**

If we extend the equation (9.75) additionally by the time derivative, we obtain

$$\frac{\partial}{\partial t}\Phi + \frac{\partial}{\partial x}[v\Phi] - \frac{\partial}{\partial x}\left[D\frac{\partial}{\partial x}\Phi\right] = 0 \ . \tag{9.82}$$

After integration and insertion of the previously discussed discretization systems, we obtain

---

[9] The quantity $Pe = v\Delta x / D$ is also called the Peclet number.

[10] $\theta(x)$ designates the Heaviside function $\theta(x) = \begin{cases} +1 & x \geq 0 \\ -1 & x < 0 \end{cases}$

9.1 Three-dimensional flow fields

$$\frac{\partial \Phi(x_i)}{\partial t} \Delta x + \left(\frac{2D}{\Delta x} + |v|\right) \Phi(x_i) - \left(\frac{D}{\Delta x} + \theta(-v)|v|\right) \Phi(x_{i+1})$$
$$- \left(\frac{D}{\Delta x} + \theta(v)|v|\right) \Phi(x_{i-1}) = 0. \tag{9.83}$$

In the time variables, the differential equation is clearly hyperbolic, i.e. the state of time $t + \Delta t$ is calculated from the state of time t; one only has to provide a one-sided boundary condition (initial condition). Thus, the following time discretization (the so-called *explicit scheme*) is at first apparent

$$\frac{\Phi(x_i, t + \Delta t) - \Phi(x_i, t)}{\Delta t} \Delta x + \left(\frac{2D}{\Delta x} + |v|\right) \Phi(x_i, t) -$$
$$\left(\frac{D}{\Delta x} + \theta(-v)|v|\right) \Phi(x_{i+1}, t) - \left(\frac{D}{\Delta x} + \theta(v)|v|\right) \Phi(x_{i-1}, t) = 0. \tag{9.84}$$

This leads however to the following equation

$$\Phi(x_i, t + \Delta t) = \left[1 - \left(\frac{2D}{\Delta x} + |v|\right)\frac{\Delta t}{\Delta x}\right] \Phi(x_i, t) + \frac{\Delta t}{\Delta x}\left(\frac{D}{\Delta x} + \theta(-v)|v|\right) \Phi(x_{i+1}, t)$$
$$+ \frac{\Delta t}{\Delta x}\left(\frac{D}{\Delta x} + \theta(v)|v|\right) \Phi(x_{i-1}, t). \tag{9.85}$$

With the Courant number

$$Cou = \frac{|v|\Delta t}{\Delta x}$$

follows the stability condition in accordance with (9.74)

$$\left(\frac{2}{Pe} + 1\right) Cou \leq 1. \tag{9.86}$$

This condition states that the time advance Δt should not be too large, and must be the smaller, the finer the mesh is. In praxis, this condition is hardly maintainable for all mesh cells. Alternatively therefore, an *implicit scheme* is formulated, in which the amounts of "spatial" differential operators are set for the "new" time $t + \Delta t$

$$\left[1 + \left(\frac{2D}{\Delta x} + |v|\right)\frac{\Delta t}{\Delta x}\right] \Phi(x_i, t + \Delta t) =$$
$$\Phi(x_i, t) + \frac{\Delta t}{\Delta x}\left(\frac{D}{\Delta x} + \theta(-v)|v|\right) \Phi(x_{i+1}, t + \Delta t) + \frac{\Delta t}{\Delta x}\left(\frac{D}{\Delta x} + \theta(v)|v|\right) \Phi(x_{i-1}, t + \Delta t). \tag{9.87}$$

This system now has purely positive coefficients again, and is thus monotone and stable. However, it has the disadvantage that it is implicitly formulated, i.e. it represents a coupled system of linear equations in the $\Phi(x_i, t + \Delta t)$ with $i = 1, ..., N$ (from (9.85) on the other hand, the quantities $\Phi(x_i, t+\Delta t)$ can be directly calculated). Nevertheless, quite all CFD codes use implicit (or semi-implicit, hybridized) systems for reasons of stability.

- **Discretization of the source term**

For the complete discretization of the transport equation, we now still need the treatment of the source term. The previously discussed discretization systems lead for the equation

$$\frac{\partial}{\partial t}\Phi + \frac{\partial}{\partial x}[v\Phi] - \frac{\partial}{\partial x}\left[D\frac{\partial}{\partial x}\Phi\right] = Q \tag{9.88}$$

to

$$\left[1 + \left(\frac{2D}{\Delta x} + |v|\right)\frac{\Delta t}{\Delta x}\right]\Phi(x_i, t+\Delta t) = \Phi(x_i, t) + \ldots + Q(x_i, t+\Delta t)\Delta t \;. \tag{9.89}$$

Insofar as the term $Q(x_i, t+\Delta t)$ is unavailable, it has to be attributed to the term $Q(x_i, t)$. A special case exists if Q is directly depending on Φ. In this case, Q can be developed into a Taylor series up to the linear term in Φ

$$Q(\Phi(x_i, t+\Delta t)) \cong Q(\Phi(x_i, t)) + \underbrace{\frac{\partial Q}{\partial \Phi}(\Phi(x_i, t))}_{\alpha}[\Phi(x_i, t+\Delta t) - \Phi(x_i, t)] \;. \tag{9.90}$$

Inserted in (9.89), this leads to

$$\left[1 + \left(\frac{2D}{\Delta x} + |v|\right)\frac{\Delta t}{\Delta x} - \alpha\Delta t\right]\Phi(x_i, t+\Delta t) = (1 - \alpha\Delta t)\Phi(x_i, t) + \ldots + Q(\Phi(x_i, t))\Delta t \;. \tag{9.91}$$

The stability criterion is surely fulfilled for

$$\alpha = \frac{\partial Q}{\partial \Phi}(\Phi(x_i, t)) \leq 0 \;. \tag{9.92}$$

In this case, the employment of (9.91) is indeed highly recommendable. If the criterion is offended however, one should rather resort to (9.89) plus substitution $Q(x_i, t+\Delta t) \Rightarrow Q(x_i, t)$, even if this redness precision (e.g. convergence speed). The precision can namely be raised again, if one iterates (9.89) (or (9.91)). Proceeding from (9.89), one calculates the approximate values $Q_{(k+1)}(x_i, t+\Delta t)$ from $Q_{(k)}(x_i, t+\Delta t)$ via insertion into the source term, until the value changes only to an inessential extent. One obtains

$$\left[1 + \left(\frac{2D}{\Delta x} + |v|\right)\frac{\Delta t}{\Delta x}\right]\Phi_{(1)}(x_i, t+\Delta t) = \Phi(x_i, t) + \ldots + Q(\Phi(x_i, t))\Delta t \quad \text{i.e.} \tag{9.93}$$

$$\left[1 + \left(\frac{2D}{\Delta x} + |v|\right)\frac{\Delta t}{\Delta x}\right]\Phi_{(k+1)}(x_i, t+\Delta t) = \Phi(x_i, t) + \ldots + Q(\Phi_{(k)}(x_i, t+\Delta t))\Delta t \;.$$

9.1 Three-dimensional flow fields

- **The operator split method**

Especially in internal engine processes, injection and combustion processes appear in addition to pure gas flow, which demand the solving of new transport equations, but also create new source terms in the equations existing already. In this case, a typical transport equation contains terms for convection, diffusion, spray (e.g. evaporation), and combustion. It is recommendable to treat the various effects separately. This takes place by use of the operator split method. M and N designate two operators in this case, e.g. convection/diffusion and a chemical source term

$$\frac{\partial}{\partial t}\phi = \underbrace{M(\phi)}_{\text{convection/diffusion}} + \underbrace{N(\phi)}_{\text{chemical}} . \tag{9.94}$$

By inserting an intermediate step, the time integration can be split

$$\tilde{\phi}(x_i, t+\Delta t) - \phi(x_i, t) = \int_{t}^{t+\Delta t} M(\phi) dt ,$$

$$\phi(x_i, t+\Delta t) - \tilde{\phi}(x_i, t+\Delta t) = \int_{t}^{t+\Delta t} N(\phi) dt . \tag{9.95}$$

Each of these two steps can then be solved as shown above. The error is of the order $(\Delta t)^2$. In this context, we should point out that the CFD code KIVA calculates for convection and diffusion separately. Because of the different character of the fundamental differential operators (hyperbolic vs. elliptic), this method is not unreasonable.

- **Discretization and numerical solution of the momentum equation**

Finally, the momentum equation for the calculation of velocity and pressure by use of the continuity equation should be considered. For numerical reasons, it is recommendable to resort to so-called *staggered grids*, i.e. pressure and velocity are calculated on computational grids shifted to each other, the pressure for example in the cells (i.e. effectively their centers) and the velocity on the nodes.

The calculation of velocity commonly takes place iteratively, for which several algorithms are known (e.g. SIMPLE, PISO, SIMPISO, ...). In the final analysis, all have the fact in common that in the first step the momentum equation is solved for the velocities of the momentums kept constant. In the second step, pressure corrections are then calculated with the help of a Poisson equation for pressure. With these pressure corrections, new velocities are then calculated again, and thus again, until a pre-given break off threshold for the convergence is reached. In the incompressible case, the Poisson equation for the pressure shell be given which follows from the velocity equation by divergence formation

$$\Delta p = -\rho \frac{\partial v_i}{\partial x_j} \frac{\partial v_j}{\partial x_i} + 2 \frac{\partial^2}{\partial x_i \partial x_j}(\mu_t S_{ij}) . \tag{9.96}$$

9.1.4 Computational meshes

Mesh generation is often the most decisive and most strongly limiting factor in CFD calculations today. It cannot be repeated often enough that a good computation mesh is *the key* to success, much more essential than, for example, the introduction of modified turbulence models. A good computation mesh should consist of hexahedrons, be wall-adapted (i.e. also keep to the y_+ rule (9.45)) and be sufficiently fine/problem-adapted, so that all flow structures (free jets, flames, ...) can be well resolved (see Fig. 9.4 c and d). In praxis, these requirements can often not be fulfilled completely satisfactorily.

Another problem is mesh movement, which is solved very CFD code-specifically, so that one has to rely closely in generating moving meshes on the respective code philosophy. Only a few CFD codes exist (primarily FIRE, STAR CD, and KIVA) which provide for the functionality of a mesh movement, as is important in the treatment of in-cylinder problems (moving valves and piston). In stationary meshes, the situation is less critical, and the meshes can usually be exchanged. One can therefore also resort to independent mesh generation programs. As before, the construction of a qualitatively superior computational mesh still requires high effort – and above all experience.

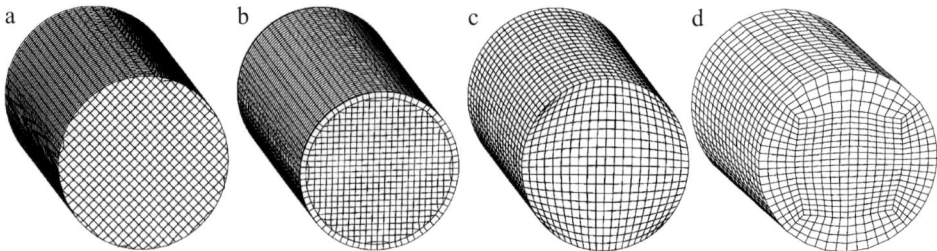

Fig. 9.4: Principle representation of various meshing concepts. To the far left a truncated cartesian mesh which has a very poor mesh structure. A cartesian mesh with a surface layer truncated only on the inside (second to the left) is already clearly better. Second to the right is a representation of a mesh with wall-adapted, hexahedral mesh structure, whereby there are, however, still deformations at the corners of the base meshes. The mesh on the far right displays an optimal, wall-adapted mesh structure

As opposed to this, in recent times a unfortunate trend is making itself known, also in the case of engine CFD applications "fast" mesh generation with automatic mesh generators, which simply cut off Cartesian meshes on the surface of the object to be modeled (see Fig. 9.4 a). What is still acceptable in the case of highly complex geometries like coolant flows, is unsuitable for the calculation of internal engine processes at least, due to the complex, wall-dominated flow structures and the accumulating errors during the transient simulation. In every case, one should pay attention in the case of such truncated meshes that at least one adaptive mesh wall layer exists (y^+ rule) and that the Cartesian mesh is only inscribed into the remaining internal space (see Fig. 9.4 b), or else the boundary layer will not be correctly represented. Furthermore, one must check visually whether bad mesh structures have emerged in the automatically produced mesh (an extremely narrow point, e.g. only resolved by a very small amount of cell layers). In addition, the time advantage in automatic mesh generation is usually smaller than one may at first estimate. For on the one hand, an essential part of the mesh generation work consists in the preparation of the surface meshes, and this effort cannot be spared in automatic mesh generation. On the other hand, adaptive, "hand-generated" meshes are often more versatile, because variants can be easily worked in; in the case of automatically generated meshes, a new mesh must be generated for each variant.

Another variant of automatic mesh generation consists in the production of tetrahedral meshes. But as we already mentioned in the numeric section, these are not suitable for codes that work in the finite volume method.

9.1.5 Examples

The CFD applications in the engine sector are, as already mentioned, diverse. In the following, the simulation of internal cylindrical flow structures in SI engines and diesel engines and the simulation of internal nozzle flow in diesel injectors will be looked at as examples.

- **Simulation of flow structures in the cylinder: the SI engine**

We will now consider the intake and compression flow in a SI engine. The geometry of the calculation mesh is represented in Fig. 9.5 (left).

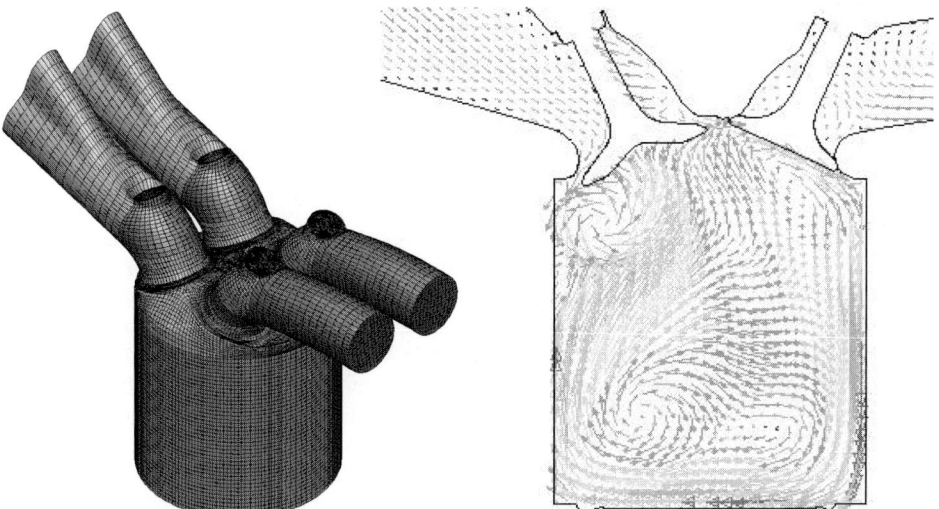

Fig. 9.5: Simulation of the cylinder flow of a SI engine. Basic geometry for the computational mesh (left) and the flow image in a section through valves at BDC (right)

Because of the mirror-symmetry, only a half-mesh must be taken into consideration, which saves calculation time. Full cylinder meshes of a typical passenger car engine in the BD including intake and exhaust tracts should consist of at least 1,000,000 mesh cells, in order to resolve the relevant flow structures incl. turbulence sufficiently. In Fig. 9.5 (right), the flow is shown in a section through the valves in the charge changing BD. One sees the typical tumble flow (large swirl in the clockwise direction). Above on the left, under the intake valves, another small "counter-tumble" exists (counter-clockwise swirl). Usually this swirl is first formed when the intake valves open, the actual tumble develops only later, with a larger valve lift. In the compression phase at the latest the counter-tumble is "eaten up" by the main tumble. In three dimensions, a tumble of course does not simply represent a cylinder. We are concerned in the case of the swirl axis (or better: swirl axes) with a complex, unsteady, 3-dimensional formation (see Fig. 9.6 left), which is frequently somewhat reminiscent of a ω.

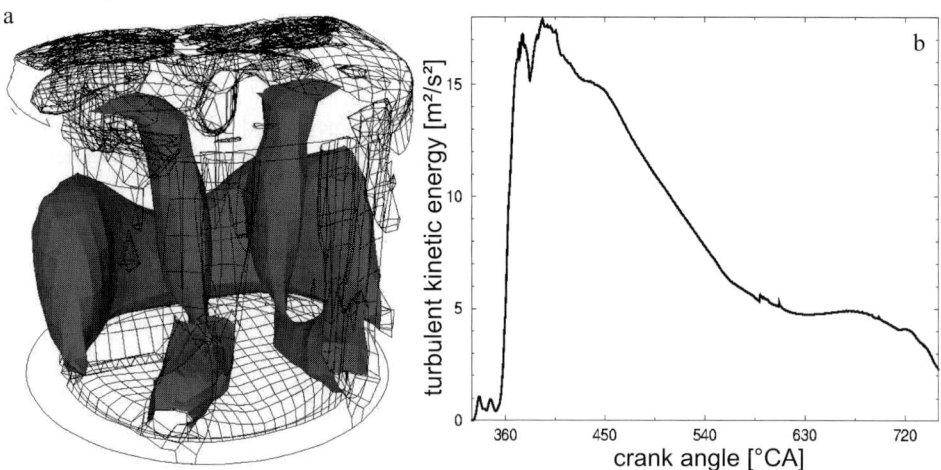

Fig. 9.6: Representation of the tumble axles in a SI engine at charge changing BDC (left) and of the time development of turbulence energy in the cylinder (right)

Towards the ignition TDC, the combustion space becomes increasingly flat, the tumble is "squashed" by the piston and disintegrates into turbulence. This can be clearly recognized in the turbulence evolution of Fig. 9.6 (right). At first, during the intake phase, the flow is quite directionless. Consequently, many areas of high shear exist, which creates a lot of turbulence. This corresponds to the very high turbulence values between 350 °CA and 450 °CA. At the BDC (540 °CA), the flow has however already become quite calm: only the tumble still exists (and possibly the counter-tumble), i.e. the large swirl structures. All small structures are already dissipated. Before the ITDC, from approx. 630 °CA on, the tumble itself disintegrates into turbulence, and in the temporal evolution a local maximum develops, or at least a plateau. Since the tumble scales with the speed, this is valid approximately for the plateau as well. However, the turbulence values at the ITDC (i.e. the "plateau") considerably influence the burning velocity (see chap. 4.1.3). This is essentially the reason why a SI engine to some extent functions independently of engine speed, i.e. the combustion velocity scales with the engine speed.

One also understands how the combustion can be influenced with the intake flow – one has to raise the tumble. This is the only form, in which flow energy can be conserved long enough, i.e. until the ITDC, in order to disintegrate then into turbulence. The turbulence, which is directly generated during the intake process, is dissipated at the ITDC. This process is also supported by the combustion chamber form ("pentroof"): in the case of a flat combustion space, the tumble would be "squashed" earlier, i.e. disintegrate, and could thus less incite combustion.

However, an excessively high turbulence value at the spark plug is less favorable, because it can lead to flame extinguishment. Another important factor for SI engine combustion is the local EGR distribution, which can also be determined by means of CFD.

- **Simulation of flow structures in the cylinder: diesel engines**

For the passenger car engine, a high swirl value is very important for a good mixture. Again, swirl generation and -evolution are quantities, which can be investigated very effectively by means of CFD.

A typical example is represented in Fig. 9.7 (left). On the left are the intake ducts, on the right the exhausts. For the sake of swirl production, the intake ducts have a helical form. Moreover, a valve-seat generated swirl, a so-called *valve seat cover* was introduced in addition, i.e. the front intake port has a cover in the valve seat. The swirl number development for the variant from Fig. 9.7 (left) with valve seat cover ("variant B") and an equivalent variant without valve seat cover ("variant A") can be seen in Fig. 9.7 (middle). The variant with valve seat cover provides a significant increase in the swirl number.

If we observe the special structure of the flow Fig. 9.7 (right), we can also easily see the reason: the duct without valve seat cover (in this case the right one) functions as a tangential duct. If the front intake port were to function as a tangential duct as well, both flows would more or less compensate for each other, and no resulting swirl would remain. The task of the valve seat cover is thus to block the tangential component of the flow, thus supporting the swirl port functionality.

Since the swirl, geometrically speaking, is much better adapted to the cylinder geometry than a tumble, it disintegrates to a much smaller extent during compression; instead, the flow rotating around the cylinder axis is compressed into the piston bowl, by means of which the velocity can increase because of the conservation of angular momentum (pirouette effect), see Fig. 9.7 (middle). The swirl even increases towards the end of compression.

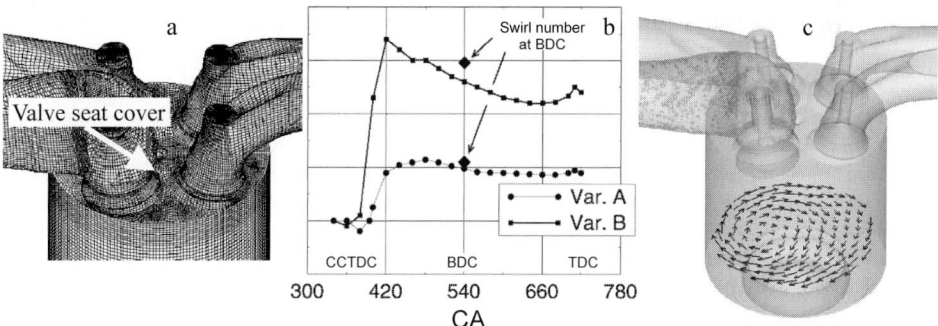

Fig. 9.7: Flow in the intake port of a diesel engine, port geometry (left), time development of the swirl number for two variants (middle), and the flow field at BDC (right)

The swirl number ω is defined as the ratio of the angular momentum L and the momentum of inertia θ related to the engine speed n

$$\omega = \frac{L}{2\pi n \theta} \text{ with } L = \int r v_{\tan} \mathrm{d}m \text{ and } \theta = \int r^2 \mathrm{d}m \;. \tag{9.97}$$

In this case, r designates the distance from the cylinder respectively the center of rotation, $v_{\tan}$ the tangential component of the flow velocity. With the calculation at every time step, we obtain transient swirl number evolution across the crank angle. In the flow bench test, a swirl value is determined as a function of the valve lift. Correlating the valve lift graph with the respective crank angle and converting the swirl value into an angular momentum flux, the latter can be integrated over the intake stroke and a "swirl number BDC" be determined. Because of the small losses in the case of a swirl flow around the cylinder axis, this approximation is usually sufficient, i.e. the real transient swirl number at BDC and the "swirl number BDC" do not deviate much from each other, see also Fig. 9.7 (middle). This is no longer valid for SI engines because of the much higher flow losses for tumble flow (in this case, the "tumble number BDC" is normally much larger than the real transient tumble number at BDC).

During compression, in addition to the swirl in the bowl, another secondary swirl is produced, which can be important for the mixture formation processes in the bowl (according to the respective injection). The orientation of this secondary swirl depends however on the swirl strength (Fig. 9.8). However, one must always keep in mind that this describes the flow before injection – the injection naturally changes the flow structures dramatically.

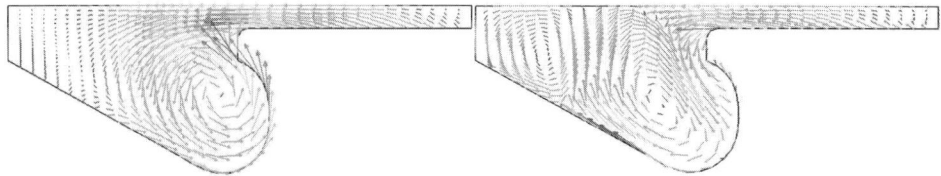

Fig. 9.8: Secondary swirl in the bowl of a diesel engine at 5 °CA before ignition TDC with BDC-speed at exactly 0 (left) and exactly 2.5 (right)

To sum, one can say that, similarly to the SI engine, all small-scale flow structures disintegrate during compression at the latest in the diesel engine as well. Only the large-scale swirl structures (incl. secondary swirls) survive and create, together with the injection, flow structures during the mixture formation phase, whereby the latter is the most dominant by far. For these reasons, it is quite common in the case of diesel engines to start calculation of the in-cylinder processes (mixture formation and combustion) at the closing of the intake valves assuming an ideal "cylinder-shaped" swirl structure. The only unknown, the swirl number, must then be taken from measurements (e.g. from the swirl number BDC), see above). This method is particularly recommendable in the case of a combustion process low in swirl (typically e.g. for commercial vehicle engines), because we spare ourselves the generation of a mesh with moving valves on the one hand (always costly) and can restrict the computation costs considerably in the case of symmetrical combustion chambers because of the high prob-

9.1 Three-dimensional flow fields

lem symmetry. For a (symmetrical) 6-hole nozzle for example, only a 60° combustion chamber sector with cyclical boundary conditions has to be examined.

- **Internal nozzle flow**

Another important CFD application is the simulation of internal nozzle flow, since it provides information about the initial conditions of injection spray simulation, which will be discussed in the next section.

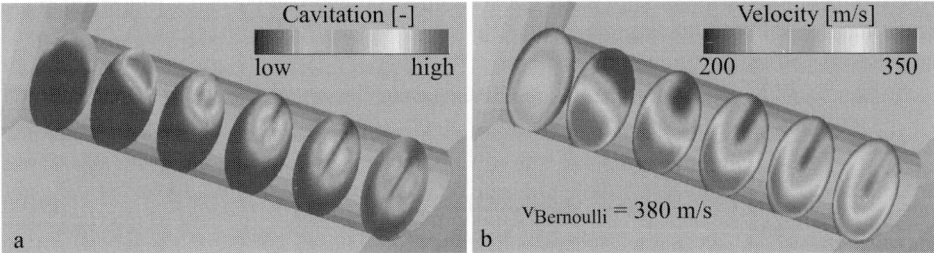

Fig. 9.9: Simulation of internal nozzle flow. a) Cavitation distribution of the sac hole nozzle of a commercial vehicle (left) b) Inhomogeneous velocity distribution in the spray hole (right)

Fig. 9.9 shows the computed cavitation and velocity distributions in different sections through a spray hole of a commercial vehicle-sac hole nozzle. We clearly recognize the appearance of cavitation in the spray hole. This is a typical event, especially for diesel injectors. It means that, at least at the location of the cavitation's origination, the vapor pressure of the fluid drops locally, and thus develop cavitation bubbles. These bubbles are transported with the fluid, and can grow, shrink, or implode, depending on the environmental conditions. Such a flow can no longer be calculated meaningfully with an incompressible, single-phase fluid model, because negative pressure would inevitably appear at the cavitation origin (the density would however remain constantly high, at the level of the fluid).

Today there are already CFD codes, which offer cavitation modeling on the basis of a turbulent dual-phase flow. In this case, a complete set of transport equations are solved for both phases (fluid on the one hand and gas in vesicular form as a disperse phase on the other), i.e. both phases can possess different velocity fields. However, both phases are coupled by diverse processes; for example bubble growth leads to mass exchange, and the bubble drag creates momentum exchange. The results in Fig. 9.9 are calculated with such a dual-phased code (FIRE).

In the borderline case of highly dominant exchange processes, both phases are firmly coupled, and a single-phase flow with two components arises. For this borderline case, many CFD codes offer cavitation modelings, and for most practical applications, this should be sufficient.

Finally, we will consider the initial conditions for injection spray simulation as the result of internal nozzle flow calculation. If A_{eff} is the effective cross section of the spray and v_{eff} its effective velocity, for the mass flow $\dot{m}$ and the momentum flux $\dot{I}$ at the nozzle orifice, we get:

$$\dot{m} = \rho_{fl} v_{eff} A_{eff} \quad \text{and} \quad \dot{I} = \rho_{fl} v_{eff}^2 A_{eff},\qquad(9.98)$$

with A_{eff} as the effective cross section limited towards the top by the geometrical orifice cross section A_{geo}; v_{eff} is limited by the Bernoulli velocity

$$v_{Bern} = \sqrt{\frac{2\Delta p}{\rho_{fl}}}$$

whereby Δp designates the difference between injection and combustion chamber pressure. The following loss coefficients

$$C_A = \frac{A_{eff}}{A_{geo}} \leq 1 \quad \text{(contraction coefficient) and} \qquad(9.99)$$

$$C_v = \frac{v_{eff}}{v_{Bern}} \leq 1 \quad \text{(velocity coefficient)} \qquad(9.100)$$

can be defined. With that, the discharge coefficient of the nozzle comes to

$$C_d = \frac{\dot{m}}{\rho_{fl} A_{geo} v_{Bern}} = C_v C_A. \qquad(9.101)$$

The discharge coefficient is now usually known or estimable from the injection rate measurements (often it is about 0.7), not however how it is divided into the coefficients C_A and C_v. These can be acquired from the nozzle flow simulation. In our case (Fig. 9.9), cavitation zones exist until the orifice exit, which lead to an reduction of the effective flow cross section, on the other hand the velocity in the cavitation-free zones is approximately equal to the Bernoulli speed. Numerical evaluation leads to a value of $C_v \approx 0.9$. Additional quantities from the internal nozzle flow simulation are the turbulent scales. These lead to the so-called primary breakup of the spray. Conversely, imploding cavitation bubbles lead to an increase in turbulence. Jets with increased spray breakup (because of strong atomization of droplets for example) look "bushier" and have other mixture formation properties, see König et al. (2002).

9.2 Simulation of injection processes

This section is exclusively devoted to the simulation of injection processes. In view of the fact that there are easily operable "spray modules" in all current CFD codes, this might be somewhat surprising. But if we look at the results gained from such modules critically, the problematic nature of this topic becomes evident. It must unfortunately be said that none of the codes for engine applications available on the market has a useful spray model. This problem can unfortunately not be solved finally in the context of this section.

In the following, we will first develop the standard spray model, the modeling of the single-droplet processes required for this, and finally the statistical-stochastical modeling of a droplet ensemble. In a second step, the numerical reasons for the failure of this model will be explained, and finally, various approaches and solution possibilities discussed.

9.2.1 Single-droplet processes

Single-droplet processes include the exchange processes of mass, momentum, and heat between a single droplet and the surrounding gaseous phase. The momentum exchange is described in a purely kinetic way in terms of drag, while mass and heat exchange with the environment is generated by diffusion and convection processes in the droplet surroundings.

Droplets are usually described with eight variables: location, velocity (each of them three variables), radius, and temperature. With a modeling of single droplet processes we find the equations of motion of these variables. Sometimes droplet vibration states are introduced as well. Their relevance however has not yet been convincingly demonstrated; we therefore leave them out.

- **Momentum exchange**

If a droplet of radius R, density ρ_{fl} and velocity $\vec{v}_{dr}$ moves in a gas of density ρ_g and velocity $\vec{v}_g$, a decelerating (i.e. directed against the velocity difference to the gaseous phase) force has an effect on the droplet

$$\vec{F} = \rho_{fl} \frac{4\pi}{3} R^3 \dot{\vec{v}}_{dr} = \frac{1}{2} \rho_g C_W \pi R^2 |\vec{v}_g - \vec{v}_{dr}| (\vec{v}_g - \vec{v}_{dr}) \ . \tag{9.102}$$

Together with the equation

$$\dot{\vec{x}}_{dr} = \vec{v}_{dr} \tag{9.103}$$

the droplet kinetics is determined. The drag coefficient C_W value is usually calculated as follows

$$C_w = \begin{cases} \dfrac{24}{\text{Re}_{dr}} \left(1 + \dfrac{\text{Re}_{dr}^{2/3}}{6}\right) & \text{for} \quad \text{Re}_{dr} \leq 1{,}000 \\ 0.424 & \text{for} \quad \text{Re}_{dr} > 1{,}000 \end{cases} \tag{9.104}$$

whereby

$$\text{Re}_{dr} = \frac{2r\rho_{fl} |\vec{v}_{dr} - \vec{v}_g|}{\mu_g} \tag{9.105}$$

designates the droplet-based Reynolds number, i.e. for large Reynolds numbers, the drag is quadratically depending on the velocity difference.

- **Mass and heat exchange**

The continuity equation and the vapor transport equation for the stationary, laminar case read

$$\frac{\partial}{\partial x_i}(\rho v_i) = 0 \quad \text{or} \quad \frac{\partial}{\partial x_i}(\rho v_i c) - \frac{\partial}{\partial x_i}\left(D\rho \frac{\partial}{\partial x_i} c\right) = 0 \ . \tag{9.106}$$

For reasons of analytical solvability, density, diffusion coefficient and temperature are set as constant. The equations (9.106) are now considered in the environment of a droplet under rotational symmetry; the droplet is thus at rest. The goal is the description of a stationary flow equilibrium for the vapor flux $\tilde{m}$ and the heat flux $\tilde{q}$ between the droplet surface and infinity. The integration of these equations from the droplet surface to a spherical shell of radius r with the help of the Gaussian law leads to

$$4\pi \rho r^2 v(r) = \text{const.(1)} , \qquad (9.107)$$

$$4\pi \rho r^2 v(r) c(r) - 4\pi D \rho r^2 \frac{dc(r)}{dr} = \text{const.(2)} , \qquad (9.108)$$

whereby const.(1) designates the total mass flux and const.(2) the vapor mass flux. Since effectively only vapor is flowing, both should be equal to $\tilde{m}$. After solving of (9.107) for v, inserting into (9.108), and prescribing of boundary conditions $c(R)$ and $c(\infty)$, integration leads to

$$v(r) = \frac{\tilde{m}}{4\pi \rho r^2} \quad \text{and} \qquad (9.109)$$

$$\tilde{m} = 4\pi D \rho R \ln\left(\frac{1 - c(\infty)}{1 - c(R)}\right) , \qquad (9.110)$$

whereby $c(R)$ can be calculated from the droplet temperature by means of the vapor pressure relation.

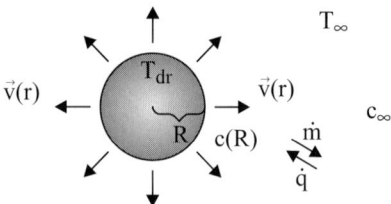

Fig. 9.10: Principle sketch for the representation of vapor and heat fluxes in the droplet environment

The equivalent treatment of the heat conduction equation assuming constant specific heat

$$\frac{\partial}{\partial x_i}(\rho v_i c_p T) - \frac{\partial}{\partial x_i}\left(\lambda \frac{\partial}{\partial x_i} T\right) = 0 \qquad (9.111)$$

leads to the following equation after inserting (9.109) using the boundary conditions at $T(R) = T_{dr}$ and T_∞

$$T(r) = T_\infty + \frac{1 - \exp\left(-\dfrac{\tilde{m} c_p}{4\pi r \lambda}\right)}{1 - \exp\left(-\dfrac{\tilde{m} c_p}{4\pi R \lambda}\right)} (T_{dr} - T_\infty) . \qquad (9.112)$$

The actual heat flow away from the droplet is only the diffusion flow from (9.111), the convection flow "steals" no specific energy from the droplet. We thus obtain

$$\tilde{q}(R) = -4\pi R^2 \lambda \frac{\partial T}{\partial r}(R) = \frac{\tilde{\dot{m}} c_p (T_{dr} - T_\infty)}{\exp\left(\frac{\tilde{\dot{m}} c_p}{4\pi R \lambda}\right) - 1} \qquad (9.113)$$

The (actually non-constant) physical quantities for density, diffusion constant, heat conductivity, and heat capacity are customarily calculated according to the 1/3-2/3 rule, as linear combinations of values at the surface and in infinity, whereby X is representative for the above quantities

$$X = \frac{X_{dr}}{3} + \frac{2 X_\infty}{3} \ . \qquad (9.114)$$

Since however the assumption of a droplet at rest is generally incorrect, in-flow effects are usually considered by means of the following correction according to Ranz-Marschall

$$\dot{m} = \tilde{\dot{m}} \frac{2 + 0.6 \text{Re}^{1/2} \text{Pr}^{1/3}}{2} \quad \text{and} \quad \dot{q} = \tilde{\dot{q}} \frac{2 + 0.6 \text{Re}^{1/2} \text{Sc}^{1/3}}{2} \ , \qquad (9.115)$$

whereby Pr and Sc designate the (laminar) Prandtl and Schmidt number

$$\text{Pr} = \frac{\mu c_p}{\lambda} \quad \text{and} \quad Sc = \frac{\mu}{\rho D} \ . \qquad (9.116)$$

For the correlation of $\dot{m}$ and $\dot{q}$ to the rates of change of the droplet variables T_{dr} and R, the following mass and heat balances can be formulated with the heat capacity of the fluid c_{fl} and the specific evaporation enthalpy $h_{evap}(T)$

$$4\pi R^2 \dot{q} = \underbrace{\dot{m} h_{evap}(T_{dr})}_{\text{evaporation}} + \underbrace{\rho_{fl} \frac{4\pi}{3} R^3 c_{fl} \dot{T}_{dr,heat}}_{\text{heating}} \ , \qquad (9.117)$$

$$4\pi R^2 \rho_{fl} \dot{R}_{evap} = -\dot{m} \ . \qquad (9.118)$$

At a given $\dot{m}$ and $\dot{q}$, $\dot{T}_{dr,heat}$ and $\dot{R}_{evap}$ can be determined.

In the literature, one can find several modifications of this modeling. Approaches exist, for example, which allow temperature gradients in the droplet. However, these effects appear to be rather unimportant in engine applications. It is still common to use single-component fuel models, although in the meantime interesting multi-component models already exist, which attempt to describe a component spectrum with a small number of form factors (see Lippert et al. (2000) and Pagel (2003)) by means of so-called "continuous" thermodynamics. Such models are especially interesting at high rest/evaporation times, for example in SI engine early direct injection or a HCCI process. Most CFD codes offer special "synthetic" single-component models for gasoline and diesel fuel. In case they should not be available, n-heptane for gasoline and dodecane for diesel are reasonable choices. One should however always keep in mind, that the properties of a mixture can never be exactly reproduced with a single-component fuel. Moreover, we should point out that we are only concerned with the physical properties, not the chemical properties of the fuels (in this case, for example, n-heptane would be a poor representative for gasoline – to consider only the knocking properties).

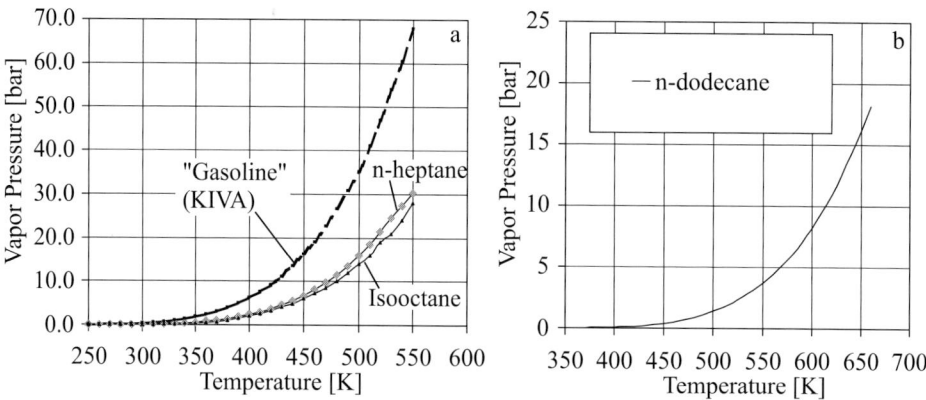

Fig. 9.11: a) Vapor pressure curve for gasoline-class fuels b) Vapor pressure curve of a component typical of diesel

What happens now to the fuel in the engine? For this, first consider a GDI engine with late injection. Still relatively cold droplets arrive in a combustion chamber with hot, compressed air. The surface of the droplets must accept on a state on the vapor pressure curve (see Fig. 9.11 a), it thus takes on the state attributed to the current droplet temperature, such that it "pushes out" a vapor covering with the partial pressure attributed by the vapor pressure curve. The droplets now heat up, i.e. they advance on the vapor pressure curve towards the top right. The vapor covering thickens thereby increasingly. Finally, the point is reached at which the partial pressure is equal to the vapor pressure[11] – the temperature now remains in this state, it boils.

Typically, the droplets are found very close to each other in the combustion chamber (especially in the spray area), all evaporate into a common vapor cloud, the states of which are, consequently, not far away from the vapor pressure curve. This is a confirmation for the reliability of the model of a uniform temperature in the inside of the droplet; because the temperature of the droplet surface cannot move far from the internal temperature, not only because the droplets are small, but also because the temperatures in the environment are limited.

In the diesel engine, the same basic process occurs, the droplets proceed on the vapor pressure curve again towards the top right. Especially in the case of diesel injection sprays, the droplets are even more tightly packed than in typical GDI sprays. However, because of the high combustion chamber pressures and the different vapor pressure curves of diesel-typical components, the critical point is now reached, as the end of the vapor pressure curve. After run-

---

[11] strictly speaking, the droplet temperature never reaches the boiling point exactly. It remains at a somewhat lower level at the wet bulb temperature.

ning through the critical point, there is no longer a phase limit. The CFD codes solve this usually by simply eliminating the droplet and attributing it to the vapor phase.

So-called "flashboiling" is a special case, which sets in when warm fuel is injected into an intake pipe or combustion chamber with a surrounding pressure lying below the vapor pressure corresponding to the droplet temperature. In this case, the spray immediately "explodes", so to speak. This phenomenon can not be calculated with the models described above, which describes the situation in the most CFD codes, even if in the meantime the first approaches towards a CFD modeling of flashboiling have been made, see for example Ra and Reitz (2002).

9.2.2 Spray statistics

An injection spray typically consists of hundreds of millions of droplets. Such an ensemble is (similar to turbulent gas flow) no longer deterministically calculable (i.e. on the basis of each single droplet). For the spray as droplet ensemble, it is therefore appropriate to introduce a statistical description, a droplet distribution density $p(\vec{x}, \vec{v}, R, T)$. It describes the probability of finding at location $\vec{x}$ a droplet with velocity $\vec{v}$, radius R, and temperature T.

Already 120 years ago, Ludwig Boltzmann was confronted with a similar problem as he sought to derive thermodynamics on the basis of statistics from the mechanics of atomic and molecular processes. For this he developed the *Boltzmann equation* which takes his name and with the help of that he could ground the 2nd law of thermodynamics atomically for the first time in history. This equation is the basis of the kinetic gas theory today. It can be employed for droplets with their particular dynamics, naming itself then simply the "*spray equation*" and forming the basis for all spray models.

However, this equation can no longer be integrated in a closed way. Instead, stochastic methods are used in the standard model – the solution is decided by "throwing dice" (the "Monte Carlo simulation").

Finally, the description of multi-particle processes like collision or breakup has to be discussed in order to close the spray modeling.

- **The spray-adapted Boltzmann equation**

For every single point of the eight-dimensional space of the variables $\vec{x}, \vec{v}, R, T$, the dynamics $\dot{\vec{x}}, \dot{\vec{v}}, \dot{R}, \dot{T}$ is described by (9.102), (9.103), (9.117) and (9.118). One can now summarize the eight variables in the 8-tupel $\alpha = (\vec{x}, \vec{v}, R, T)$ and write the equations of motion for it as

$$\dot{\alpha}_i = A_i(\alpha) \text{ with } i = 1,...,8 \ . \tag{9.119}$$

The goal is now to find an equation of motion for the distribution function $p(\alpha)$, of a "point cloud" in this phase space, so to speak. The solution of this problem is the so-called *Liouville equation*[12]

---

[12] In classical mechanics, the Liouville equation is normally formulated for conservative or Hamiltonian systems. The second term of the second line of (9.120) then vanishes because of the canonical Hamiltonian equations, i.e. the phase space volume is incompressible in this case.

$$\frac{\partial}{\partial t} p(\alpha,t) = -\sum_{i=1}^{8} \frac{\partial}{\partial \alpha_i} \left(A_i(\alpha) p(\alpha,t) \right)$$

$$= -\sum_{i=1}^{8} A_i(\alpha) \frac{\partial}{\partial \alpha_i} p(\alpha,t) - \left(\sum_{i=1}^{8} \frac{\partial A_i(\alpha)}{\partial \alpha_i} \right) p(\alpha,t). \quad (9.120)$$

The Liouville equation is a hyperbolic equation. Its characteristics are the equations of motion (9.119), as one see, with the help of the formulation in the lower line of (9.120). The formulation in the upper line of (9.120), on the other hand, shows that the probability is maintained, i.e.

$$\frac{\partial}{\partial t} \int_{\text{phase space}} d\alpha \, p(\alpha,t) = 0 \quad (9.121)$$

is valid. The Liouville equation as a equations of motion for the distribution function can already be sufficient for simple applications. In the general case however, all (discontinuous) multi-droplet processes like collision and breakup are not taken into consideration. This was also, applied to atoms and molecules, the problem of Boltzmann. He extended the Liouville equation by one source term, the so-called collision integral. First, one extracts the location $\vec{x}$ from the 8-tupel α.

$$\alpha = (\vec{x}, \beta) \text{ and } \beta = (\vec{v}, R, T). \quad (9.122)$$

For a general dual-particle collision term

$$I_{collision} = \int p(\vec{x},\beta_1) p(\vec{x},\beta_2) \sigma(\beta_1 \beta_2 \to \beta + ...; \vec{x}) d\beta_1 d\beta_2 \\ - \int p(\vec{x},\beta_1) p(\vec{x},\beta) \sigma(\beta_1 \beta \to ...; \vec{x}) d\beta_1 \quad (9.123)$$

can be formulated. The term $\sigma(\beta_1 \beta_2 \to \beta + ...)$ describes the conditional probability that with the presence of droplets β_1 and β_2 at location $\vec{x}$, a droplet of property β emerges through collision; $\sigma(\beta_1 \beta \to ...)$ stands for the conditional probability that droplet β collides with droplet β_1, the consequence of which is the disappearance of droplet β_1, hence the minus sign of the complete term. Three-particle collision processes (and processes of still higher orders of particles) are usually ignored, "single-particle processes" are nothing but disintegrations. With the above logic, one can write for them

$$I_{disintegration} = \int p(\vec{x},\beta_1) \sigma(\beta_1 \to \beta + ...; \vec{x}) d\beta_1 - p(\vec{x},\beta) \sigma(\beta \to ...; \vec{x}). \quad (9.124)$$

In summa, we obtain the spray-adapted Boltzmann equation

9.2 Simulation of injection processes

$$\frac{\partial}{\partial t} p(\alpha,t) = -\sum_{i=1}^{8} A_i(\alpha) \frac{\partial}{\partial \alpha_i} p(\alpha,t) - \left(\sum_{i=1}^{8} \frac{\partial A_i(\alpha)}{\partial \alpha_i} \right) p(\alpha,t)$$

$$+ \int p(\vec{x},\beta_1) p(\vec{x},\beta_2) \sigma(\beta_1 \beta_2 \to \beta + ...; \vec{x}) d\beta_1 d\beta_2 \quad (9.125)$$

$$- \int p(\vec{x},\beta_1) p(\vec{x},\beta) \sigma(\beta_1 \beta \to ...; \vec{x}) d\beta_1$$

$$+ \int p(\vec{x},\beta_1) \sigma(\beta_1 \to \beta + ...; \vec{x}) d\beta_1 - p(\vec{x},\beta) \sigma(\beta \to ...; \vec{x}).$$

This equation completely describes the spray dynamics to be calculated. A small restriction should however be made with reference to turbulent dispersion, i.e. the interaction of droplets with the flow turbulence. This is contained quite explicitly in (9.125), namely, in the first term on the right side, insofar as the 8-tupel α contains the instantaneous and not the averaged gas velocities. However, modeling approaches also exist which are based on the average gas velocity and correspond to an averaged form of (9.125), see O'Rourke (1989).

- **The numerical solution of the spray equation – the standard model**

Equation (9.125) is a high-dimensional, partial integral-differential equation. With direct methods, it is not solvable in the foreseeable future in eight-dimensional space. One only has to think of every dimension discretized into ten levels (this is surely still much too rough for many of the eight dimensions, e.g. for the spatial coordinates). In that case, a computational mesh of 10^8 cells has already been produced! For the solution of the equation therefore, another way must be taken. Considering the fact that the underlying Liouville equation is of hyperbolic character, it can be solved along its characteristics (which correspond to the droplet trajectories), i.e. by means of ordinary differential equations. For the case that the relevant area, in which the final droplet density is clearly other than zero, describes only a "low-dimensional surface" in eight-dimensional space, one possibly only needs a few trajectories.

With that, the following strategy is chosen: we introduce sufficiently many representative particles, so-called "parcels", which "follow" the characteristics of the Liouville equation (9.120). All continuous processes are thus already handled. The treatment of the collision processes, i.e. the terms on the right side of (9.125) is, however, equally obvious. The representative parcels experience these as a stochastic process, and they are formulated in the collision integrals exactly in this way. For all parcels found at a particular time in a volume element (practically speaking, "in a computational cell"), collision probabilities are calculated in accordance with the formulae (9.123) and (9.124), and, in accordance with these probabilities, the concrete actions are "diced out" (hence the notion "Monte Carlo simulation").

The actual continuous statistical description is thus modified into a discontinuous, equivalent stochastic description, the dynamics of the droplet distribution function is described by means of the dynamics of an ensemble of representative stochastic parcels. But now these parcels look very similar to the original droplets, all continuous processes progress similarly (momentum exchange with flow, evaporation, heating-up); one must only work stochastically for discontinuous processes (formation, breakup, collision). From this originates the name "discrete droplet model" (DDM). This is precisely the charm and intuitive comprehensibility of the stochastic model: one has the "feeling" of working with the individual drops, although one is ultimately only operating with stochastic parcels. All visualizations of injection spray is ultimately only operating with stochastic parcels. All visualizations of injection spray simulations show these stochastic parcels, never droplets.

The individual parcels represent elements of droplet flow, i.e. a description by means of parcels has a Lagrangian character, because it describes the flow by means of moving reference points. The number of parcels is completely independent of the number of droplets and should only be determined from considerations of statistical convergence.

In order to complete the modeling, a model-technical description of multi-particle processes must still be derived. This will take place in the following, after a mathematical excursion about determining random numbers.

- **Determination of random numbers**

In the stochastic modeling for solving the spray equation, the determination of random numbers following a pre-given probability distribution is an important building element. According to the standard, computers only make random number generators available for the generation of equally distributed random numbers between 0 and 1. The basic task now reads as follows

$$x \in X \subset \mathfrak{R}^n, \; f : \mathfrak{R}^n \to \mathfrak{R}, \; f(x) > 0, \; \int_X f(x)dx = 1 \; . \tag{9.126}$$

Drawing randomly an element $x \in X$ corresponding to the distribution function f means that if we repeat multiply the process, the elements $x_1, x_2, x_3,...$ should be distributed in accordance with the function f. To fulfill this task, two varying methods will be discussed:

Method I -- "Integrate and invert": this approach only works for one-dimensional distributions, i.e. $X = [a,b] \subset \mathfrak{R}$. We first calculate the distribution function $F(x)$

$$F(x) = \int_a^x f(\xi)d\xi \quad F:[a,b] \to [0,1] \; . \tag{9.127}$$

Because of the positivity of f, F is strictly monotonous and thus invertible.

Secondly, we determine the inverse function $F^{-1} : [0,1] \to [a,b]$ (if necessary, we have to numerically integrate and tabularize) and draw a random number $z \in [0,1]$. The value $x = F^{-1}(z)$ is then our desired random variable. To found this: with the probability dF, the random number lands in the interval $[x, x + dx]$, whereby

$$dx = \frac{dF}{f(x)} \; .$$

The density of probability p is given as the ratio of probability to interval length

$$p = \frac{dF}{dx} = f$$

Method II – "Draw and evaluate": this method is suitable for multi-dimensional spaces X as well.

First step: we determine an element $x \in X$ on the basis of equal distribution. With complicated quantities (e.g. the inside of a calculation area with complicated margins), one can proceed as follows: one inscribes $X \subset \Re^n$ into a "n-dimensional rectangle (quader)"

$$X \subset \tilde{X} = [a_1, b_1] \times [a_2, b_2] \times \ldots \times [a_n, b_n] \tag{9.128}$$

and, in an equally distributed manner, draws an element from $x \in \tilde{X}$ with the help of n random numbers $z_1, z_2, \ldots, z_n \in [0,1]$

$$x = (a_1 + z_1(b_1 - a_1), a_2 + z_2(b_2 - a_2), \ldots, a_n + z_n(b_n - a_n)) \ .$$

There are two possibilities: x lies in X, in which case it is our chosen element. Or x does not lie in X, then it is rejected and a new selection process started. In this way we are sure that all elements from X are chosen with equal probability.

Second step: the variable x chosen in the first step is evaluated. For this, $f_{max} = \max(f(\xi), \xi \in X)$. We draw a further random number $\tilde{z} \in [0,1]$ and compare it with the ratio

$$\zeta = \frac{f(x)}{f_{max}} \ .$$

If $\tilde{z} \leq \zeta$, x is accepted, otherwise it is rejected and, the process is commenced again from step one until an element is found and accepted. The reason for this: the probability that $\tilde{z} \leq \zeta$ is proportional to $f(x)$. In this way, every element x is chosen with a correct relative probability. Through the process of rejection and repetition in the case of a non-acceptable element x, we safeguard that the normalization of the probability density is fulfilled, and finally an element is selected with probability 1.

One typical task is to determine equally distributed spatial points within the calculation area. The calculation volume is however discretized into mesh cells, and thus we are actually dealing with the determination of mesh cells, under the boundary condition of equal distribution with reference to the volume. One should never in this situation think of simply drawing cell numbers at equal distribution! For the mesh cells possess, in general, (very) different volumes, and one must take this into consideration in the selection process. In this case, method II can ideally be used, such that we first draw random numbers equally distributed among the cell numbers and then evaluate the selected cell Z according to its volume V_Z, i.e. $f(Z) = V_Z$.

- **Modeling the nozzle**

The parcels have to be generated at the nozzle, and logically this occurs stochastically. Typically, one determines the injection direction per parcel in a pre-given solid angle or spray cone area, and possibly an initial droplet size in accordance with a drop size distribution as well. With the toolbox we developed just now, more complex initial conditions like correlations between injection direction and droplet size would also be realizable. Practically speaking however, the experimental data are usually lacking for the derivation of such complex constraints.

At this point, simulation results for internal nozzle flow can be helpful. However, sometimes it is quite pleasant to be somewhat free in the choice of initial conditions, as one thus has degrees of tuning freedom to adjust the simulation result (in this case e.g. spray shape and spreading) to experimental findings. One should however not take such a tuning too far! For example, it is obviously senseless to select the injection velocity higher than is permissible according to Bernoulli, even if this should be quite beneficial to the adjustment to experimental data (which is unfortunately often the case due to the mathematical/numerical deficiencies of the spray model).

It is recommendable, to give every parcel at the nozzle (independent of the drop size) the same mass. This corresponds to the approach, according to which the fuel mass is the quantity that is of actual interest to be discretized by parcels.

As an example, a method for the simulation of uniformly distributed injection into a spray cone area of angle φ will be derived. Two angles should be randomly chosen, the azimuth angle $\theta \in [0, 2\pi]$ and the polar angle $\gamma \in [0, \varphi]$. The azimuth angle θ may be selected in an equally distributed fashion, but γ may not. We remind the reader that the spatial angle measure has the form $\sin \gamma \, d\gamma d\theta$ in integration, and exactly this distribution must be selected. Because of the equal distribution of θ, we restrict ourselves to the choice of γ. This is solely a one-dimensional problem, and one can therefore use method I. The distribution function of γ has the form

$$F(\gamma) = \frac{\int_0^\gamma \sin \tilde{\gamma} \, d\tilde{\gamma}}{\int_0^\varphi \sin \tilde{\gamma} \, d\tilde{\gamma}} = \frac{1 - \cos \gamma}{1 - \cos \varphi} \, . \tag{9.129}$$

From a random number $z \in [0,1]$, we then obtain the following γ value

$$\gamma = \arccos(1 - z + z \cos \varphi) \, .$$

- **Modeling breakup processes**

Breakup processes influence the spray especially in its early phase close to the nozzle, and thus, practically speaking, form a unit together with the nozzle model. For example, a breakup model, which already provides small drops very fast, can be replaced by a nozzle model with small drops.

Because of the different mechanism, two kinds of spray breakup can be distinguished, primary breakup and secondary breakup. Primary breakup results from properties given to the spray already by the internal nozzle flow, like turbulence and cavitation (which creates turbulence again through cavitation bubble implosion). For secondary breakup, aerodynamic processes are relevant that are not a result of internal nozzle flow.

9.2 Simulation of injection processes

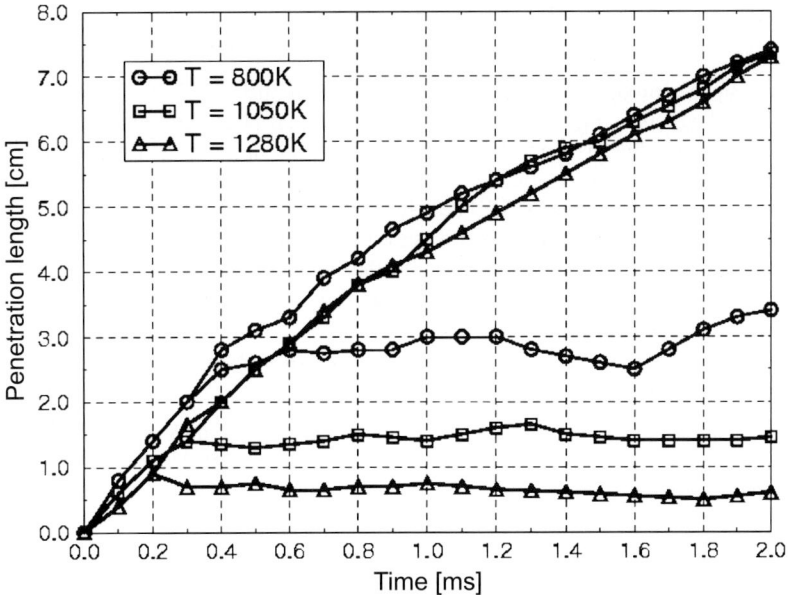

Fig. 9.12: Measurement of the penetration length of a diesel spray in a hot chamber at varying chamber temperatures

For primary breakup modeling, we need information about the internal nozzle flow, about its turbulence and cavitation distribution. Then, breakup time and length can be derived from the turbulent scales and cavitation bubble densities, see e.g. Tatschl et al. (2000). This information is however not often available. According to current insights, particularly in the case of modern diesel injection systems, a very strong primary breakup occurs; the spray leaves the nozzle in a "foamy" state so to speak. This offers the chance to start already with small droplets at the nozzle, i.e. no longer considering primary breakup explicitly. A typical drop size would be about 5 μm diameter. One sensible method of obtaining information through measurements about initial droplet sizes is described by Krüger (2001). It consists in injecting the spray into a hot chamber at varying chamber temperatures and, e.g. with sliding and schlieren and Mie scattering techniques, to determine the penetration depth of the liquid and gaseous phases (direct droplet size measurements are hopeless because of the density of diesel injection sprays, at least highly unreliable). Such an experimental result is shown in Fig. 9.12. At equal chamber densities, the gaseous phase penetrates approximately to the same extent, but the penetration depth of the liquid phase is much different, it reduces with increasing temperature. Our task is now to synchronize the spray breakup parameter and initial droplet size such that all of these penetration graphs can be reproduced with one set of parameters. This method led in the previous case to an initial Sauter diameter[13] of 5 µm at small further breakup rates. The prerequisite of this procedure is however a functioning spray model.

If we diminish the initial droplet size of a spray continuously, then the local homogeneous flow is initiated as a limiting case. In it, gaseous and liquid phases are in kinetic and thermodynamic equilibrium, because, on the one hand, very small drops have a relatively large flow resistance related to their mass, and as a result, no speed difference between the two phases is possible anymore; on the other hand, because of the high surface rate per volume unit, the gaseous phase must persist in a vapor pressure curve state. In fact, we now have a single-phase flow. Experimental investigations show that at least typical diesel injections can be relatively well described as a local homogeneous flow, see e.g. Siebers (1998). The experimental expression for "locally homogeneous flow" is "mixture-controlled". For the spray model, this means that the droplet size (and thus the spray breakup) is no longer a decisive factor if the droplet size is selected sufficiently low!

Secondary breakup processes have aerodynamic causes and thus show longer breakup lengths. They occur in competition with primary breakup. In dense diesel injections with strong primary breakup, they thus play a rather small role. In the case of gasoline direct injection on the other hand, there is hardly any or no primary breakup because of the minimal turbulence and cavitation rates of the internal nozzle flow (varying according to the injector type), and therefore secondary breakup can even play the dominant role. Especially in the case of the complex flow swirl structures of cone sprays, droplet sizes appear to have a real influence on the spray structure, and thus we are farther removed from the limiting case of local homogeneous flow.

In describing secondary breakup, we usually use an analysis of instability; the main effect is the so-called Kelvin-Helmholtz instability. Probably the best-known modeling approach is the WAVE model, see Reitz (1987). The wave length Λ growing the most and its rate of growth Ω read

$$\frac{\Lambda}{R} = 9.02 \frac{(1+0.45\, Oh^{0.5})(1+0.4\, T)^{0.7}}{(1+0.865\, We_{dr}^{1.67})^{0.6}} \quad \text{and} \tag{9.130}$$

$$\Omega \left(\frac{\rho_{dr}\, R^3}{\sigma_{dr}}\right)^{0.5} = \frac{(0.34 + 0.38\, We_g^{1.5})}{(1+Oh)(1+1.4\, T^{0.6})}, \tag{9.131}$$

whereby for the Weber numbers for liquid and gaseous phases We_{dr} and We_g and the Ohnesorge number Oh as well as the Taylor number T is valid:

$$We_{g/dr} = \frac{\rho_{g/dr}\, R\, v_{rel}^2}{\sigma_{fl}}, \quad Oh = \frac{\sqrt{We_{dr}}}{Re_{dr}}, \quad T = Oh\sqrt{We_g}, \tag{9.132}$$

---

[13] (The Sauter diameter d_s of a droplet distribution is defined as the mean value of d^3 divided by the mean value of d^2,

$$d_S = \langle d^3 \rangle / \langle d^2 \rangle$$

9.2 Simulation of injection processes

whereby σ_{fl} designates the surface tension. With these quantities, a breakup time τ_B and a stable radius R_s can be defined

$$\tau_B = \frac{3.788\, B_1\, R}{\Lambda\, \Omega}, \tag{9.133}$$

$$R_s = \begin{cases} B_0 \Lambda & \text{for} \quad B_0 \Lambda \leq R \\ \min\left(\sqrt[3]{\dfrac{3\pi R^2 v_{rel}}{2\Omega}},\, \sqrt[3]{\dfrac{3R^2 \Lambda}{4}}\right) & \text{for} \quad B_0 \Lambda > R \end{cases}, \tag{9.134}$$

whereby B_0 and B_1 are model constants.

There are still two options for the process of spray breakup: according to the first option, the droplets disintegrate more or less continuously until the stable radius – the radius change is described by the differential equation

$$\frac{dR}{dt} = -\frac{R - R_s}{\tau_B} = \dot{R}_{dis}. \tag{9.135}$$

According to the second idea, small, stable droplets are split off from a main drop. The implementation of both variants into the simulation model is obvious: in the first case, the spray breakup is treated as a continuous process, so to speak, the radius of each parcel decreases in accordance with (9.135), while the mass does not, i.e. the number of droplets corresponding to a parcel increases; a "repacking" occurs in the parcel. In the second case, subsidiary parcels consisting of small, stable droplets are created and split off from time to time.

In addition to the Kelvin-Helmholtz instabilities, one can also consider the Rayleigh-Taylor instabilities, see e.g. Patterson (1997) and Patterson and Reitz (1998); however, with realistic droplet sizes, these should not play such a large role. There are also completely different modeling ideas concerning secondary breakup, e.g. vibration-based models like the TAB (Taylor analogy breakup) model, which assume, that drop vibrations lead to breakup. It is however experimentally known, that these breakup types are no longer dominant at high Weber numbers. However, these methods lead to quantitatively comparable breakup time scales for reasons of dimensional analysis. In Fig. 9.13 a summary of the various aerodynamic breakup models is provided.

Quite often in the literature, the all-dominating role is attributed to spray breakup even in the propagation of the gaseous phase, which is physically simply wrong. Unfortunately however, this corresponds to the function that spray breakup models perform in CFD simulations, or with other words, spray breakup models are adjusted to compensate for mathematical/numerical model deficiencies.

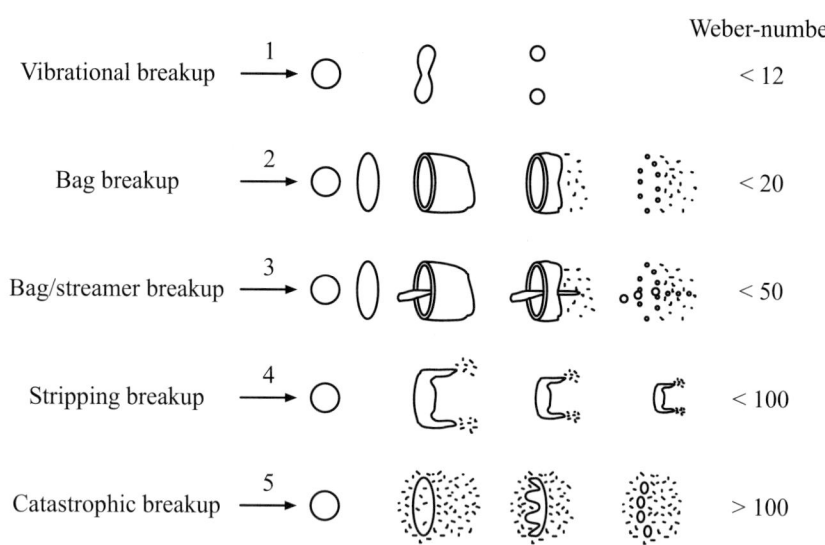

Fig. 9.13: Aerodynamic breakup mechanisms, acc. to Pilch et al. (1987)

- **Modeling collision processes**

Collision processes between two drops are divided into the two subprocesses "Collision with subsequent separation of droplets", or "grazing collision", and "coalescence". To distinguish between the two processes, the rotation energy of the agglomerate in the center-of-mass system is compared with the need in surface energy necessary for renewed droplet separation. The rotation energy E_{rot} in the center-of-mass system reads

$$E_{rot} = \frac{L^2}{2\theta}, \text{ with } L = \frac{m_1 m_2}{m_1 + m_2} b v_{rel} \text{ and } \theta = \frac{2(m_1 + m_2) R_{agg}^2}{5}. \qquad (9.136)$$

The difference of surface energies ΔE_{surf} amounts to

$$\Delta E_{surf} = 4\pi \sigma_{tr} \left(R_1^2 + R_2^2 - R_{agg}^2 \right). \qquad (9.137)$$

In this case $R_{1/2}$ and $m_{1/2}$ designate the radii and masses of droplets 1 and 2, R_{agg} the radius of the agglomerate of droplets 1 and 2. The quantity b designates the collision parameter. If the rotation energy predominates, a separation occurs, otherwise the coalescence remains, see Amsden et al. (1989).

For modeling in the stochastic model, collision probability is evaluated for the droplets of every parcel pair in a cell. The goal is a modeling in which the parcel number does not increase. Should coalescence occur, we allow all droplets in the parcel with the larger radius to react identically.

The drops of the parcel with the smaller radius taking part in the coalescence are eliminated from this parcel. In the statistic average, if enough parcels exist, a correct result is produced in this way. If a "grazing collision" occurs, the droplets taking part in the process are "fed back"

9.2 Simulation of injection processes

into their respective parcel with momentum and energy conservation. Effectively, both parcels have exchanged momentum and heat.

Further collision processes take place with the wall. In principle, we have to formulate reflection and atomization laws for the individual droplets. Furthermore, a formation of wall film occurs in this case. Wall film dynamics demand however a separate set of equations and solvers. Many CFD codes already offer approaches for this as well.

- **Modeling turbulent dispersion**

We designate as turbulent dispersion the interaction of droplets with the turbulence of the gas flow. Most CFD codes simply utilize the formulation for the drag (9.102), whereby the gas velocity appearing herein is composed of an averaged component plus a turbulent fluctuation

$$\vec{F} = \rho_{fl} \frac{4\pi}{3} R^3 \dot{\vec{v}}_{dr} = \frac{1}{2} \rho_g C_W \pi R^2 \left| \langle \vec{v}_g \rangle + \vec{v}_g'' - \vec{v}_{dr} \right| \left(\langle \vec{v}_g \rangle + \vec{v}_g'' - \vec{v}_{dr} \right). \tag{9.138}$$

That is, this formulation already contains the turbulent dispersion. The velocity fluctuations $\vec{v}_g''$, which a parcel "experiences", are drawn randomly according to the distribution function

$$G(\vec{v}_g'') = \frac{1}{\sqrt{2\pi}\sigma} \exp\left(-\frac{\vec{v}_g''^2}{2\sigma^2}\right) \quad \text{with } \sigma = \sqrt{\frac{2}{3}k}. \tag{9.139}$$

In this case, the lifespan of a turbulent fluctuation, τ_{corr}, as the minimum of the turbulent time scale and the time, which a drop needs to traverse a turbulent swirl, is calculated.

The CFD code KIVA also offers still another model, for the case that the fluctuation time duration τ_{corr} is smaller that the calculation step, see O'Rourke (1989). We will only briefly sketch the derivation here. For this, he uses an analytical solution of the linearized equation of motion with turbulent fluctuation as stochastic force, a so-called *Langevin equation*

$$\dot{\vec{x}}_{dr} = \vec{v}_{dr}$$

$$\dot{\vec{v}}_{dr} = D_R \left(\langle \vec{v}_g \rangle + \vec{v}_g'' - \vec{v}_{dr} \right) \quad \text{with } D_R = \frac{3}{8} \frac{\rho_g}{\rho_{fl}} c_w \frac{\left| \langle \vec{v}_g \rangle + \vec{v}_g'' - \vec{v}_{dr} \right|}{R}. \tag{9.140}$$

During the time τ_{corr} remains $\vec{v}_g''$ unchanged, after that a new value is randomly chosen according to distribution (9.139) (i.e. individually for each component). Fluctuations in different τ_{corr} intervals are thus statistically independent of each other. The result of the analytical integration and subsequent averaging is a distribution function $p(\vec{x}_{tr}, \vec{v}_{tr}, t; \vec{x}_0, \vec{v}_0, t_0)$, which reproduces the distribution of $\vec{x}_{dr}$ and $\vec{v}_{dr}$ at time t, under the initial conditions: $\vec{x}(t_0) = \vec{x}_0$ and $\vec{v}(t_0) = \vec{v}_0$. This distribution function is of the Gaussian type with the following variances and covariances

$$\sigma_{vv,ij} = \left[\langle v_i(t) v_j(t) \rangle - \langle v_i(t) \rangle \langle v_j(t) \rangle \right] =$$

$$\sigma^2 \frac{1 - \exp(-D_R \tau_{corr})}{1 + \exp(-D_R \tau_{corr})} \left(1 - \exp[-2 D_R \Delta t] \right) \delta_{ij} \tag{9.141}$$

$$\sigma_{xx,ij} = \langle x_i(t)x_j(t)\rangle - \langle x_i(t)\rangle\langle x_j(t)\rangle =$$
$$\sigma^2 \tau_{corr}\left(-\frac{2}{D_R}(1-\exp[-D_R\Delta t])+\Delta t\right)\delta_{ij} + \frac{1}{D_R^2}\sigma_{vv,ij} \quad (9.142)$$

$$\sigma_{xv,ij} = \langle x_i(t)v_j(t)\rangle - \langle x_i(t)\rangle\langle v_j(t)\rangle =$$
$$\sigma^2\left(\frac{\tau_{corr}}{2}(1-2\exp[-D_R\Delta t])+\frac{1-\exp(-D_R\tau_{corr})}{1+\exp(-D_R\tau_{corr})}\frac{1}{D_R}\exp[-2D_R\Delta t]\right)\delta_{ij} \quad (9.143)$$

whereby is valid: $\Delta t = t - t_0$. The distribution function explicitly reads

$$p(\vec{x}_{tr},\vec{v}_{tr},t;\vec{x}_0,\vec{v}_0,t_0) =$$
$$N^{-1}\exp\left[-q_1(\vec{v}_{tr}-\langle\vec{v}_{tr}\rangle)^2 - q_2(\vec{x}_{tr}-\langle\vec{x}_{tr}\rangle)^2 - q_3(\vec{v}_{tr}-\langle\vec{v}_{tr}\rangle)\cdot(\vec{x}_{tr}-\langle\vec{x}_{tr}\rangle)\right] \quad (9.144)$$

with

$$\langle\vec{v}_{tr}\rangle = \langle\vec{v}_g\rangle + (\vec{v}_0-\langle\vec{v}_g\rangle)\exp(-D_R\Delta t),$$

$$\langle\vec{x}_{tr}\rangle = \vec{x}_0 + \langle\vec{v}_g\rangle\Delta t + \frac{(\vec{v}_0-\langle\vec{v}_g\rangle)}{D_R}[1-\exp(-D_R\Delta t)],$$

$$q_1 = \frac{2\sigma_{xx}^2}{4\sigma_{xx}^2\sigma_{vv}^2 - \sigma_{xv}^4}, \quad q_2 = \frac{2\sigma_{vv}^2}{4\sigma_{xx}^2\sigma_{vv}^2 - \sigma_{xv}^4}, \quad q_3 = \frac{2\sigma_{vx}^2}{4\sigma_{xx}^2\sigma_{vv}^2 - \sigma_{xv}^4}, \quad (9.145)$$

$$N = \frac{2\pi}{\sqrt{4q_1q_2 - q_3^2}},$$

$$\sigma_{vv,ij} = \sigma_{vv}\delta_{ij}$$

and other corresponding (co)variances. Via square completion, we obtain for p

$$p(\vec{x}_{dr},\vec{v}_{dr},t;\vec{x}_0,\vec{v}_0,t_0) =$$
$$N^{-1}\exp\left[-q_1\left[\underbrace{(\vec{v}_{dr}-\langle\vec{v}_{dr}\rangle)}_{\delta\vec{v}} + \frac{q_3}{2q_1}\underbrace{(\vec{x}_{dr}-\langle\vec{x}_{dr}\rangle)}_{\delta\vec{x}}\right]^2 - \left(q_2 - \frac{q_3^2}{4q_1}\right)\underbrace{(\vec{x}_{dr}-\langle\vec{x}_{dr}\rangle)^2}_{\delta\vec{x}}\right], \quad (9.146)$$

i.e. every component of quantities $\delta\vec{v} + q_3/2q_1\,\delta\vec{x}$ and $\delta\vec{x}$ is Gauss-distributed with the variances

$$\frac{1}{2q_1} \text{ or } \frac{1}{2(q_2 - q_3^2/4q_1)},$$

moreover, they are all statistically independent. They can thus be "diced for" on a componential basis and $\delta\vec{v}$ determined then from that. The changes of $\vec{x}_{dr}$ and $\vec{v}_{dr}$ during time frame Δt are then calculated in the following way

9.2 Simulation of injection processes

$$\vec{x}_{dr}(t+\Delta t) - \vec{x}_{dr}(t) = \langle \vec{x}_{dr}(t+\Delta t) \rangle - \vec{x}_{dr}(t) + \delta\vec{x} \quad \text{with} \quad \langle \vec{x}_{dr}(t) \rangle = \vec{x}_{dr}(t)$$
$$\vec{v}_{dr}(t+\Delta t) - \vec{v}_{dr}(t) = \langle \vec{v}_{dr}(t+\Delta t) \rangle - \vec{v}_{dr}(t) + \delta\vec{v} \quad \text{with} \quad \langle \vec{v}_{dr}(t) \rangle = \vec{v}_{dr}(t).$$
(9.147)

In the limiting case $D_R \tau_{corr} \ll 1$, the distribution function p becomes the Greens function of a *Fokker Planck equation*

$$\frac{\partial}{\partial t} p(\vec{x}_{dr}, \vec{v}_{dr}, t) = -\frac{\partial}{\partial x_{dr,i}} \left[v_{dr,i} p(\vec{x}_{dr}, \vec{v}_{dr}, t) \right]$$
$$-\frac{\partial}{\partial v_{dr,i}} \left[D_R \left(\langle v_{g,i} \rangle - v_{dr,i} \right) p(\vec{x}_{dr}, \vec{v}_{tr}, t) \right]$$
$$+\frac{\sigma^2 D_R^2 \tau_{corr}}{2} \frac{\partial}{\partial v_{dr,i}} \frac{\partial}{\partial v_{dr,i}} p(\vec{x}_{dr}, \vec{v}_{dr}, t).$$
(9.148)

The original Fokker-Planck equation was derived in order to describe the *Brownian motion* (i.e. the movement of a particle in a fluid under the influence of thermal fluctuations). Equation (9.148) describes the movement of a droplet in a gas under the influence of turbulent fluctuations! In principle, this means that we put the spray-adapted Boltzmann equation through a turbulent averaging process, thereby creating a diffusion term in the velocity, i.e. our spray equation now reads

$$\frac{\partial}{\partial t} p(\alpha,t) = -\sum_{i=1}^{8} \frac{\partial}{\partial \alpha_i} \left[A_i(\alpha) p(\alpha,t) \right]$$
$$+ \frac{\partial}{\partial v_{dr,i}} \left[\frac{\sigma^2 D_R^2 \tau_{corr}}{2} \frac{\partial}{\partial v_{dr,i}} p(\alpha,t) \right] + I_{momentum},$$
(9.149)

whereby the drag term now only contains the averaged gas velocity. In this model, the turbulent fluctuations of the gaseous phase must no longer be explicitly given. However in this method as well, parcels must reproduce the local statistical distributions of the droplet phase. This is, of course, a considerable demand.

9.2.3 Problems in the standard spray model

As already mentioned, the standard spray model suffers under considerable deficiencies in practical applications. These will first be represented with a few examples in order then to explain the causes in more detail. If we compare the relative simplicity of modeling with stochastic parcels to the high complexity of the problem at hand, namely, the solution of an integro-differential equation in eight-dimensional space, then the failure of the spray model is actually not surprising at all. Indeed, the problems are obvious. On the one hand, there is the problem of the small length scales to be resolved at the nozzle hole, and on the other that of lacking statistical convergence.

- **Examples**

The spray penetration graphs shown in Fig. 9.14 (d) portray an experiment, in which a diesel injection spray is injected into a hot chamber. There is one penetration graph for the gaseous phase and one for the liquid phase. The calculations were carried out with different computational meshes; all of these provide different results, but none is congruent with the experiment. These meshes are all reasonable a priori, indeed, in comparison with typical engine meshes, they are even markedly high-quality. This failure is all the more drastic if one keeps in mind the simplicity of the underlying physical principle, the conservation of momentum.

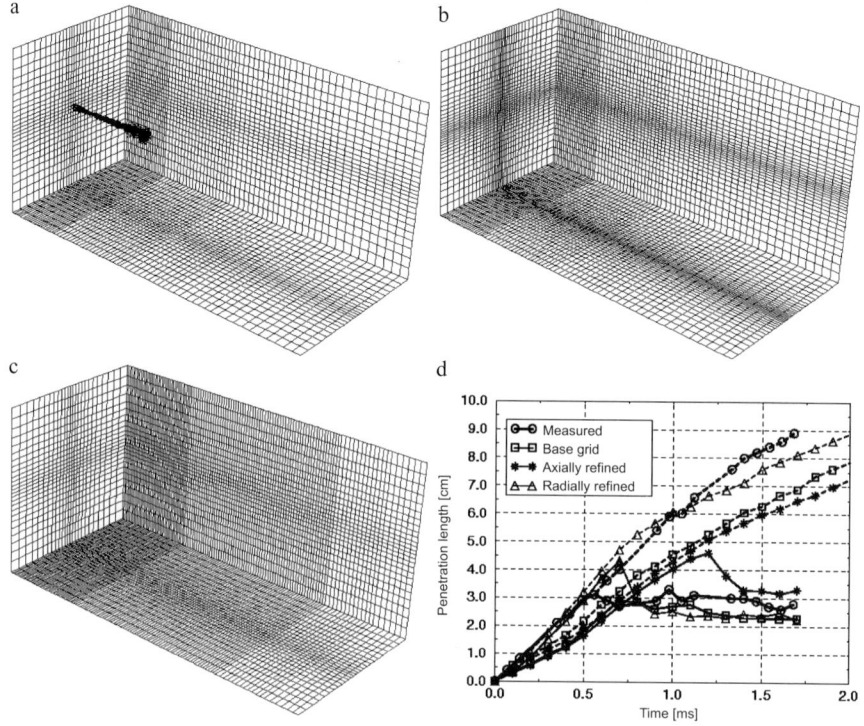

Fig. 9.14: Calculation of the dispersion of a diesel spray. a) Base mesh b) Radially refined mesh c) Axially refined mesh d) Penetration curves of the liquid and gas phases

Fig. 9.15 shows a strong dependence of the spray form on the injection direction relative to the computational grid. Also shown is the dependence of the spray shape on the turbulence in the chamber into which it is injected. This is also unphysical; experimentally, a spray in a turbulent engine differs only slightly from one in a high temperature chamber with little turbulence. The list of a priori absurd results can be almost arbitrarily lengthened, and anyone who has intensively worked with this modeling should be able to contribute his or her own examples.

9.2 Simulation of injection processes

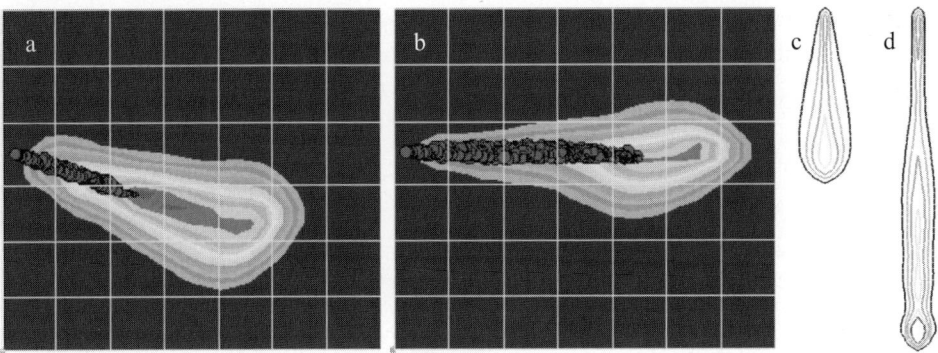

Fig. 9.15: Dependence of the spray shape on the orientation relative to the computational mesh (a and b). On the turbulence in the injection chamber (simulation of the injection into an environment with high (c) and low levels of turbulence (d))

In order nonetheless to achieve "validated" results, a number of modifications are introduced into the spray modeling, e.g. to describe the breakup (a very efficient tuning!). On the basis of the previous discussion, it should be clear what the most distinguished task of these "new models" is: the compensation of erroneous, unphysical dependences. We have thus actually already described the essential "physical content" of these models[14].

This "tuning" is very problematic, because in this way it is only guaranteed that obviously nonsensical results are avoided. The uncertainties and errors exist however also in smaller orders of magnitude and are inseparable from the physical effects to be investigated.

- **Problem: resolution of the spray**

One problem in spray modeling is immediately obvious and has also been quickly perceived: typical nozzle holes are so fine that they normally are not and cannot be numerically resolved. A nozzle hole of a passenger car diesel engine now ranges within the order of magnitude of 100 μm. One should still resolve it with 10 mesh cells, which would lead to a mesh cell edge size of 10 μm. Applied to a compression volume of 20 cubic centimeters, this would imply 20 billion mesh cells! Surely, one could still reduce the mesh cell number by meshes refined adaptively to the nozzle opening, and perhaps one could get by with less than 100 mesh cells per nozzle orifice (10x10). But an extremely high number of mesh cells cannot be avoided.

The consequence of poor mesh resolution is an erroneous calculation of the exchange processes between gaseous and liquid phase, see Fig. 9.16. Since a computational cell only knows one gas velocity, we obtain too small gas phase velocity locally at the spray location; as a result, the relative velocity between both phases responsible for mass and momentum exchange is overestimated. The (unphysical) effect of breakup mechanisms in the simulation now can also be understood: with large droplets, the zone with erroneous momentum exchange is skipped over, in some approaches it is even explicitly prevented in the area of the

---

[14] For example, breakup models should, physically speaking, hardly have an influence on the penetration depth of the spray tip, this depending as a first approximation only on the momentum flow. In praxis, almost every desired penetration depth can be "adjusted" by means of breakup tuning.

nozzle orifice ("intact core length"). Finally however, small droplets must quickly be created to take care of the necessary mixture formation quality.

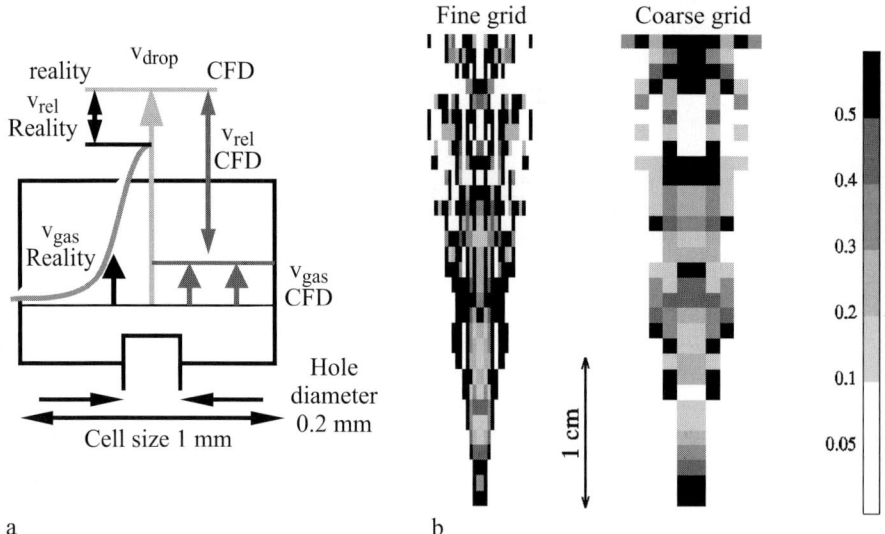

Fig. 9.16: Problems in spray modeling. a) Lacking nozzle hole resolution b) Poor statistical resolution

- **Problem: statistical convergence**

One could attempt to carry out individual calculations with a high mesh resolution, in order at least to prove the principal functionality of the spray model. If we use rotation-symmetrical calculations for example (with the nozzle hole axis as the symmetry axis), which are actually only two-dimensional, this should be possible. But in this case as well, we come across difficulties. There is even existing the (false) opinion that the spray model does not work for fine meshes at all.

But the actual problem is in the lacking statistical convergence per computational cell. The number of parcels per cell is relevant to this; if the number of mesh cells now increases because of mesh refinement, this ratio is naturally poorer. Consider Fig. 9.16 b), in which the relative standard deviation of liquid mass per mesh cell is shown in a stationary rotation-symmetrical spray simulation. For statistical convergence, it has to approach zero. Scaled up to three dimensions, 720,000 parcels were used, thus quite a high number. Nonetheless, the value of the relative standard deviation is already high for the coarse mesh, yet for the fine mesh still higher, in wide spray ranges over 50 %. To reach convergence, one should not only strive for the limit "edge length of the mesh cells approaches zero" but also the limit "parcel number per mesh cell approaches infinity". At the least, this ratio must be held at a high level. This implies however an immensely high number of parcels.

Moreover, the number of parcels required is extremely dependent of the applied models for reasons of statistical convergence: particularly critical are two-point functions like collisions, as these require (unrealistically) high numbers of parcels per cell. Also extremely critical is

9.2 Simulation of injection processes

the modeling of turbulent dispersion: the parcel distribution per cell must represent the turbulent fluctuations. This means higher demands with perspective of resolution such as one makes to the gaseous phase. For good reasons, a turbulence model is utilized there.

Finally, we should still discuss the fact that the failure of the spray model cannot be attributed to the non-existence of a continuum limit. Specifically, the assertion is often made that, in mesh refinement, eventually a drop is larger than a mesh cell, a case which is undefined.

Firstly, we should remark that parcels have no volume, i.e. no air displacement. They do possess a radius, but this has the character of an "internal" degree of freedom. However, one can measure the "size" of the parcel by the size of the source terms caused by it. But, into this size definition enters not only the drop size, but also the number of drops in the parcel! The more small cells exist, the more parcels are needed, and a parcel then contains less droplets, i.e. the source terms are "subdivided". In principle, more parcels can exist as droplets. We must always keep in mind that no droplets are "flying", but rather that a solution method for the calculation of the spray-adapted Boltzmann equation should be applied, which has however a well-defined (continuous) solution. Certainly, the "paradox" under consideration is typical: practically speaking, many mesh refinements fail due to the problem of "large droplets", since they do not go along with an increase in parcels. Yet this is not a problem of the stochastic approach in principle, but rather one of incorrect application.

9.2.4 Solution approaches

Concepts that avoid the problems described above are unfortunately not yet available in commercial engine CFD codes. However, approaches exist on the basis of self-implementations, see e.g. Otto et al. (1999). But extensive solutions are being developed in the meantime by the CFD code providers as well, like the Eulerian multi-phase methods. In the following, ways will be discussed which can be realized in the short or moderately short term by oneself (with a modeling effort that is not to be underestimated).

- **Formulation of a feasible Lagrangian model**

Firstly, we will show how the standard Lagrangian method can be transferred into a mathematically meaningful and feasible model. For this, the following measures must be taken:

   - The nozzle orifice must be resolved in the gaseous phase: this is indispensable.
   - The Pope correction should be used for the gas-phase turbulence (see Pope, 1978), i.e. an additional term is introduced into the ε-equation

$$\Delta Q = 0.2 \frac{k^2}{\varepsilon} \left[S_{ij} (\nabla \times \bar{v})_i (\nabla \times \bar{v})_j - (\nabla \times \bar{v})^2 Tr(S_{ij}) \right]. \quad (9.150)$$

Otherwise the spray turns out too "thick" and the penetration depth too small. One simple modification of constants, which also sees to a correct free-spray penetration depth, is a choice of $\varepsilon_1 \approx 1.55$.

In spray models, the following must be considered:

   - Models based on knowledge of two-point correlations (collisions), are to be eliminated, since two-point correlations are very poorly resolvable. Droplet coalescence is a process which runs against spray breakup and can therefore be considered simply by an effective breakup model. In gauging the breakup model constants, this takes place more or less automatically. Momentum diffusion by collisions can be described by an effective turbulent dispersion.

- Sufficiently high numbers of parcels must be secured, 30 parcels per cell is already a quit good value. If N cells of edge length Δl (in the spray direction) are found at the nozzle orifice, and if the injection velocity amounts to v_E, then about $30 \cdot N$ parcels in the time $\Delta t = \Delta l / v_E$ should be injected. This is of course valid under the assumption that along the spray direction the number of cells, which resolve the spray cross-section, does not increase.

- The modeling of turbulent dispersion is to be reformulated, because the standard formulation implies a "direct" resolution of gaseous phase and droplet turbulence, which can hardly be performed!

In analogy to the O'Rourke model, we introduce a diffusion method into the modeling of turbulent dispersion of the liquid phase, similar to the Businessq or eddy diffusivity approach for describing turbulent diffusion in the gaseous phase. This corresponds to the O'Rourke model in the limiting case of small droplets, i.e. for $D_R \tau_{corr} \gg 1$ and $D_R \Delta t \gg 1$.

If we average now across droplet velocity as well and introduce a special diffusion term into the Boltzmann equation, it then reads

$$\frac{\partial}{\partial t} p(\tilde{\alpha},t) = -\sum_{i=1}^{8} \left\langle \frac{\partial}{\partial \alpha_i} [A_i(\tilde{\alpha}) p(\tilde{\alpha},t)] \right\rangle + \frac{\partial}{\partial x_{dr,i}} \left[D_{dr} \frac{\partial}{\partial x_{dr,i}} p(\tilde{\alpha},t) \right] + \left\langle I_{breakup} \right\rangle. \quad (9.151)$$

The variable set $\tilde{\alpha}$, in contrast to a, does not contain the momentary, but rather the mean droplet velocity. In a small time interval, the mean droplet velocity and the diffusion constant are approximately constant. We then have to look at the following differential equation in the spatial coordinates

$$\frac{\partial}{\partial t} p(\vec{x}_{dr},t) + \frac{\partial}{\partial x_{dr,i}} \left[\left\langle v_{dr,i} \right\rangle p(\vec{x}_{dr},t) \right] - D_{dr} \Delta p(\vec{x}_{dr},t) = 0 \quad (9.152)$$

(in the sense of an "operator split", the remaining terms can be left out of (9.151), since the effects of diffusion are only calculated for a small time interval). The corresponding Greens function reads

$$p(\vec{x}_{dr},t;\vec{x}_0,t_0) = \frac{1}{\left(\sqrt{4\pi D_{dr} \Delta t}\right)^3} \exp\left[-\frac{(\vec{x}_{dr} - \langle \vec{x}_{dr} \rangle)^2}{4 D_{dr} \Delta t}\right], \quad (9.153)$$

with

$$\langle \vec{x}_{dr} \rangle = \vec{x}_0 + \vec{v}_{dr} \Delta t \text{ and } \Delta t = t - t_0.$$

For the variance of $\vec{x}_{dr}$ is valid

$$\sigma_{xx} = \sigma_{yy} = \sigma_{zz} = 2 D_{dr} \Delta t, \quad (9.154)$$

i.e., no more $\delta \vec{v}$ exists, only $\delta \vec{x}$, whose three components are normally-distributed with the variance $2 D_{dr} \Delta t$. The effect of droplet phase turbulence is now described by the variable

9.2 Simulation of injection processes

D_{dr}, turbulent fluctuations are directly converted into a spatial shift. How is this quantity D_{dr} to be selected? In the above limiting case of small droplets, (9.142) provides the relation for the variance

$$\sigma_{xx} = \sigma_{yy} = \sigma_{zz} = \sigma^2 \tau_{corr} \Delta t \; . \tag{9.155}$$

A comparison of coefficients with (9.154) supplies

$$D_{dr} = \frac{\sigma^2 \tau_{corr}}{2} \; . \tag{9.156}$$

We will now briefly discuss the velocity averaging of other terms in (9.151). The collision terms need, as shown above, no longer be considered. In the case of the breakup terms, the secondary breakup is dependent on velocity, but is actually only relevant at high velocity differences in the area of the nozzle, so that the turbulent fluctuations play a minor role. Moreover, model constants must be calibrated anyway, the secondary breakup models have a similarity-theoretical character. Therefore, one can simple replace the relative velocity of the droplet and the gas by the difference of the mean velocity of both phases.

In the terms describing heat-up and evaporation of droplets, velocity influences are considered in the pre-factors according to Ranz-Marschall, since the amount of velocity difference between droplet and gas appears in the droplet-related Reynolds number. These should indeed be corrected, i.e. averaged. For this, a simple method can be used. At first, one splits the velocity difference into a mean and a fluctuating term

$$\left\langle (\vec{v}_{dr} - \vec{v}_g)^2 \right\rangle = \left(\langle \vec{v}_{dr} \rangle - \langle \vec{v}_g \rangle \right)^2 + \left\langle (\vec{v}_{rel}'')^2 \right\rangle \; . \tag{9.157}$$

Assuming a droplet velocity fluctuation of $\vec{v}_0''$ at $t = 0$, a velocity fluctuation of the gaseous phase $\vec{v}_g'' = $ const. between $t = 0$ and $t = \tau_{corr}$, the velocity difference of the fluctuations can be calculated from (9.140)

$$\vec{v}_{rel}'' = (\vec{v}_0'' - \vec{v}_g'') \exp(-D_R t) \; . \tag{9.158}$$

After squaring and averaging from $t = 0$ to $t = \tau_{corr}$, we obtain

$$\left\langle (\vec{v}_{rel}'')^2 \right\rangle = \left(\left\langle (\vec{v}_0'')^2 \right\rangle + \left\langle (\vec{v}_g'')^2 \right\rangle \right) \frac{1 - \exp(-2D_R \tau_{corr})}{2 D_R \tau_{corr}} \approx \frac{1}{D_R \tau_{corr}} \sigma^2 \; , \tag{9.159}$$

whereby the equilibrium values from (9.139) and (9.141) are inserted

$$\left\langle (\vec{v}_g'')^2 \right\rangle = \sigma^2 \text{ and } \left\langle (\vec{v}_0'')^2 \right\rangle = \frac{1 - \exp(-D_R \tau_{corr})}{1 + \exp(-D_R \tau_{corr})} \sigma^2 \approx \sigma^2 \; . \tag{9.160}$$

This finally leads to an averaged Reynolds number

$$\left\langle \text{Re}_{dr}(|\vec{v}_{dr} - \vec{v}_g|) \right\rangle = \text{Re}_{dr}(|\langle \vec{v}_{dr} \rangle - \langle \vec{v}_g \rangle|) \cdot \sqrt{1 + \frac{1}{D_R \tau_{corr}} \left(\frac{\sigma}{|\langle \vec{v}_{dr} \rangle - \langle \vec{v}_g \rangle|} \right)^2} \; . \tag{9.161}$$

As an example for a calculation in such a modified Lagrangian model, a simulation of spray formation of a DISI nozzle will be presented. We are dealing in this case with a rotation-symmetrical calculation, so that a resolution of the nozzle hole is possible. This effective 2D-mesh (Fig. 9.17 a) still possesses however 55,000 cells, and we are working with 20,000 parcels. Transferred to the 3D-problem, this would correspond to 14 million parcels! In the case of the conical sprays of a gasoline direct injection, typical swirl structures form at the spray tip (outside and inside). If the conical angle is not large enough, the internal swirl is suppressed and a spray collapse takes place (see Fig. 9.17).

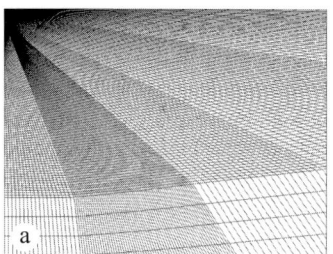
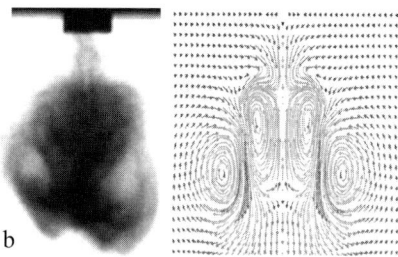

Fig. 9.17: 3D-simulation of direct fuel injection. a) Spray adapted computational mesh b) Collapse of the spray cone as a result of an insufficiently large spray cone angle (diagnostics and simulation)

- **Locally homogeneous flow**

In the limiting case of strong coupling between fluid and gas phases, which, for example, is the case with sufficiently small droplets, the special case of locally homogeneous flow arises; the liquid phase can now be described as a species in the context of a single-phase treatment, while the turbulent dispersion changes into turbulent diffusion of a species. In addition to the species transport equation of the liquid phase, we still require transport equations for the quantities "droplet radius" and "droplet temperature". In total, the equations for the liquid phase (mass fraction c_{fl}) read

$$\frac{\partial}{\partial t}(\rho c_{fl}) + \frac{\partial}{\partial x_i}(\rho v_i c_{fl}) - \frac{\partial}{\partial x_i}\left(\rho D_t \frac{\partial}{\partial x_i} c_{fl}\right) = \frac{3\rho c_{fl} \dot{R}_{evap}}{R}, \quad (9.162)$$

$$\frac{\partial}{\partial t}(\rho c_{fl} R) + \frac{\partial}{\partial x_i}(\rho v_i c_{fl} R) - \frac{\partial}{\partial x_i}\left(\rho D_t \frac{\partial}{\partial x_i}(c_{fl} R)\right) = \rho c_{fl}\left(4\dot{R}_{evap} + \dot{R}_{dis}\right), \quad (9.163)$$

$$\frac{\partial}{\partial t}(\rho c_{fl} T_{dr}) + \frac{\partial}{\partial x_i}(\rho v_i c_{fl} T_{dr}) - \frac{\partial}{\partial x_i}\left(\rho D_t \frac{\partial}{\partial x_i}(c_{fl} T_{dr})\right)$$
$$= \rho c_{fl} \dot{T}_{heat} + \frac{3\rho c_{fl} \dot{R}_{evap}}{R} T_{dr}, \quad (9.164)$$

whereby the terms $\dot{T}_{heat}$, $\dot{R}_{dis}$ and $\dot{R}_{evap}$ are given by (9.117), (9.118) and (9.135). The liquid phase should contribute to the local density, but not to the specific heat or to the pres-

9.2 Simulation of injection processes

sure. This can be realized within a commercial code, in that we choose for the species "liquid" an extremely high molecular weight. The scalars "droplet radius" and "droplet temperature" are passive. Of course, in the transport equation for the species "vapor" and in that for the internal energy of the gas phase, corresponding source terms must be inserted into the equations (9.162) - (9.164).

The source terms on the right side depend on the relative velocity of the drops to the gas. Already in the previous section, a mean Reynolds number was introduced, which only needs the difference of the mean velocities. Since a single-phase flow is being considered, this velocity difference is essentially caused by turbulent fluctuations. According to (9.159) must be set

$$|\bar{v}_{rel}| \approx \frac{\sigma}{\sqrt{D_R \tau_{corr}}} \ . \tag{9.165}$$

Because of the condition of small droplets, this model can be recommended especially for the simulation of diesel sprays. It can be employed with only a small amount of modeling effort in most CFD codes. However, nozzle-resolved meshes must be used as before; the Pope correction should be employed.

In order to resolve a more complex radius-temperature-velocity spectrum, the introduction of so-called "droplet classes" makes sense. Each of these droplet classes is its own species and describes droplets that are to be ascribed to a narrowly circumscribed radius, temperature, and velocity interval (these intervals are dependent on the space coordinates). Every droplet class is represented by its own set of equations (9.162) - (9.164).

- **The embedding of 1D-Euler methods and other approaches**

All previously described methods require the numerical resolution of the nozzle orifice. As obvious as this requirement may be, it is quite difficult to fulfill in praxis. Thus, so-called embedding methods will be discussed, with the help of which the demands on mesh resolution in the engine CFD code can be minimized. In such a method, the spray, i.e. the liquid and gaseous phases, are calculated in a nozzle-proximate area (ideally in the zone, in which the liquid phase appears) with an independent spray code on a special computational mesh (typically one- or two-dimensional). The exchange terms of both phases (with respect to impulse, mass, and energy) are calculated then in the engine CFD code coupled in at the corresponding location. In this engine CFD code, only the gaseous phase is calculated. For the thermodynamic boundary conditions, a re-coupling of the engine to the spray code is sensible.

The method is restricted to the calculation of linear injection sprays (from hole nozzles). It should only be applied as near to the nozzle as possible, where effects like cross-flow still play an insignificant role. Farther down the spray, one can, for example, switch over to the standard Lagrangian model at a defined location, see Fig. 9.18 b).

The effect of the embedding approach is based on the fact that a sufficiently high resolution can be represented in the spray code and that the spray propagation including all exchange processes between the phases are correctly calculated there. In the engine code, the resolution requirements are now reduced, since an interference of the resolution errors between both phases is avoided. The correct source terms are coupled into the engine calculation, an erroneous calculation of the gaseous phase induces no resulting errors in the source term. However, the resolution requirements on the engine CFD code is still high, and it cannot be recommended enough that one works with spray-adapted meshes (see Fig. 9.18 a).

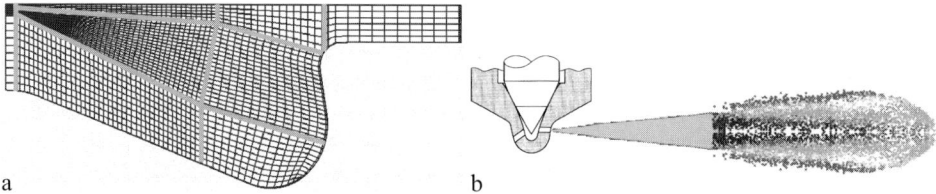

a   b

Fig. 9.18: Spray simulation for direct fuel injection in the embedding method. a) Spray adapted mesh, b) ICAS model concept: in the gray cone 1D-calculation, outside of the cone use of the standard model

Another quite useful measure is the introduction of a limitation to the spray cross-section radius l_{spray} of the turbulent length scale in the spray area. From this prescription, a constraint condition for ε is derived

$$\varepsilon \geq c_\mu^{3/4} \frac{k^{3/2}}{l_{spray}} \ . \tag{9.166}$$

In the case of a violation of this relation, ε is defined corresponding to the equal sign. In such a process, we are certainly approaching the limits of what is permissible in the fluid-mechanical sense; but we are after all primarily interested in a pragmatic approach. On the other hand, the method is both physically sensible and mathematically well-defined. In addition to the turbulent length scale, turbulent diffusion and viscosity ($\propto k^2/\varepsilon$) are also limited, a false momentum flow-off is therefore prevented, which would lead to a too small penetration depth. An essential aspect of the length scale limiter: it becomes automatically ineffective on fine, spray-resolving meshes (where the relation (9.166) is automatically fulfilled).

A comparison of the propagation of the gaseous phase in the engine and in the spray code (it is calculated in both codes!) provides additional certainty about the correctness of the calculation. The models discussed until now (the modified Lagrange method as well as the locally homogeneous flow) are suitable as spray codes. However, one-dimensional approaches are also being used, e.g. the ICAS model (integrated cross-averaged spray model). In this case, genuine 2-phase Euler equations for droplet classes are averaged across the spray cross-section. With this averaging, diffusion terms are also neglected, the essential effect of diffusion is included in the spray cone angle, which represents an entry parameter. The equations for the liquid phase are

$$\frac{\partial}{\partial t}\left(r^2 \rho c_{fl}\right) + \frac{\partial}{\partial r}\left(r^2 \rho v_{fl} c_{fl}\right) = r^2 \frac{3\rho c_{fl} \dot{R}_{evap}}{R} \ , \tag{9.167}$$

$$\frac{\partial}{\partial t}\left(r^2 \rho c_{fl} v_{fl}\right) + \frac{\partial}{\partial r}\left(r^2 \rho v_{fl}^2 c_{fl}\right) = r^2 \frac{3\rho c_{fl} \dot{R}_{evap}}{R} v_{fl} - r^2 \rho c_{fl} D_R (v_{fl} - v_g) \ , \tag{9.168}$$

9.2 Simulation of injection processes

$$\frac{\partial}{\partial t}\left(r^2\rho c_{fl}R_{dr}\right)+\frac{\partial}{\partial r}\left(r^2\rho v_{fl}c_{fl}R_{dr}\right)=r^2\rho c_{fl}\left(4\dot{R}_{evap}+\dot{R}_{dis}\right), \tag{9.169}$$

$$\frac{\partial}{\partial t}\left(r^2\rho c_{fl}T_{dr}\right)+\frac{\partial}{\partial r}\left(r^2\rho v_{fl}c_{fl}T_{dr}\right)=r^2\rho c_{fl}\dot{T}_{heat}+r^2\frac{3\rho c_{fl}\dot{R}_{evap}}{R}T_{dr}. \tag{9.170}$$

Equations for the gaseous phase can be formulated analogously

$$\frac{\partial}{\partial t}\left(r^2\rho c_g\right)+\frac{\partial}{\partial r}\left(r^2\rho v_g c_g\right)=-r^2\frac{3\rho c_{fl}\dot{R}_{evap}}{R}+E, \tag{9.171}$$

$$\frac{\partial}{\partial t}\left(r^2\rho c_g v_g\right)+\frac{\partial}{\partial r}\left(r^2\rho v_g^2 c_g\right)=-r^2\frac{3\rho c_{fl}\dot{R}_{evap}}{R}v_{fl}-r^2\rho c_{fl}D_R(v_g-v_{fl}). \tag{9.172}$$

In this case, E designates the entrainment, i.e. the intake of air as a source term of the spray gas mass. Equation (9.171) needs not be a solved together with the other equations, since the remaining equations (9.167) - (9.170) and (9.172) form a 5-dimensional equation system for five variables ($c_{fl}, v_g, v_{fl}, R, T_{dr}$); they can be used to calculate the entrainment source term instead. This model has a hyperbolic character, is well solvable and already contains, despite its extreme simplicity, many effects of spray dynamics. Unfortunately, a one-dimensional Euler model has not been represented as yet in any engine CFD codes. As in the three-dimensional case, many droplet classes can again be introduced, whereby each can now be described by a set of equations (9.167) - (9.170).

The detailed description of the practical employment of such a method (ICAS) in engine CFD codes can be found in Otto et al. (1999) or Krüger (2001). In Fig. 9.19 a) an application of the ICAS model on a HCCI mixture formation with a 30-hole nozzle is represented. The good prediction with reference to spray propagation and shape can be seen in Fig. 9.19 b).

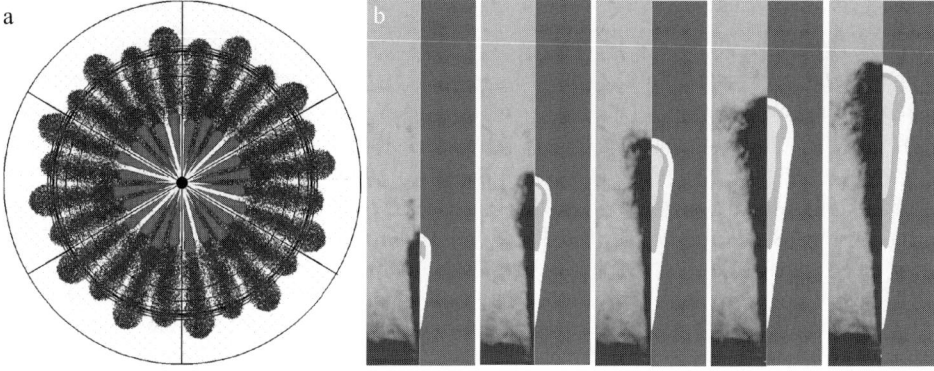

Fig. 9.19: Spray simulation with the ICAS approach (commercial diesel vehicle). a) Simulation of mixture formation in a 30-hole nozzle b) comparison measurement/simulation (without tuning)

- **The 3D-Euler method**

The final option is a complete, three-dimensional, genuinely multi-phased description of the injection spray (every class is now a phase, i.e. possesses its own velocity). This approach can indeed also be used in an embedding method; in this case a rotation-symmetrical calculation is the most sensible. Such methods are presently being developed in several engine CFD codes.

9.3 Simulation of combustion

This section is devoted to the fluid-mechanical simulation of turbulent combustion, for diesel and SI engines. In essence, we are "only" dealing with turbulent averaging of the source term of the species transport equations (9.18); however, it is immediately apparent that this is a difficult endeavor, as reaction kinetics are typically exponentially depending on the temperature. The necessary modeling effort for this is not insignificant – with the pure application of commercially standardized modelings, we (unfortunately) do not get very far.

We should point out that we are only concerned with engine combustion, i.e. with unsteady, turbulent combustion processes in complex, moving geometries, as a consequence of or in accompaniment with complex mixture processes. It thus becomes quickly clear that many combustion modelings developed for much simpler boundary conditions are adaptable to engines.

9.3.1 General procedure

First, the general procedure will be presented. The thermodynamic accurateness of the model and of the boundary conditions must first be secured. This takes place logically by using standardized 1D- and 0D-codes. Especially in diesel engine applications, one must take into consideration that the real gas behavior already plays a considerable role. However, component elasticity (piston, cylinder head screws) are also relevant, and both effects have (at least in their effect upon the pressure path) varying signs; ideally, they may compensate for each other. Real gas behavior is unfortunately unavailable in engine CFD codes at this time.

In the second step, one should turn one's attention to the pressure path in the expansion phase (i.e. after the end of combustion). Here exist again two types of sources of error, on the one hand a wall heat transfer calculated too low, which results in pressure values in the simulation which are too high, on the other hand a poor mixture formation (typically in diesel engine full load calculations), which is responsible for pressure values which are too low. However, even if the pressure graph should "fit" exactly, the does not yet necessarily indicate a correct description of combustion. It is not at all uncommon that both errors compensate for each other (of course only in the pressure graph). For the correct calculation of wall heat transfer, we refer to chap. 9.1.2: meshes with correct y^+ values must be utilized, and one should apply the Han-Reitz formulation (9.53). Nevertheless, one will still calculate wall heat transfer rates which are too low. An important reason for this is the lacking heat transfer because of soot radiation. As long as nothing better is available, the simplest way of "correcting" this is a scaling of the wall heat transfer, so that the desired value results globally. This target value can be obtained from a heat release rate analysis or estimated from a heat transfer formula like that of Woschni (1970), see chap. 7.1.1.

9.3 Simulation of combustion

The other case is more critical, namely, when in diesel or SI engines with stratified charge the pressure path after the end of combustion is too low (i.e. lower than the experimental graph or one calculated with a 0D-program). In that case, there is presumably a calculational mixture-formation deficiency. It must be stressed that here, typically, the 0D-simulation is more reliable in the comparison between 0D- and 3D-simulation. A 0D-program might possess an unsuitable (or poorly adjusted) combustion model for the concrete application case at hand, but after the end of combustion, all pressure graphs should lie close to each other, independently of the concrete heat release rate, provided only that an equal amount of fuel has been converted. 0D-programs work however mostly with experimentally well adjusted conversion rates. Not so the 3D-simulation. In this case the global conversion rate is not an input parameter, but results from the CFD calculation of flow, mixture formation, and combustion. If we now have, as the consequence of an erroneous mixture-formation calculation, a locally rich mixture zone with $\lambda < 1$, unburned fuel or combustible intermediary species (like H_2, CO, see below) must in this case continue to exist. And no 3D-combustion model in the world (which as such is always locally formulated) can solve this problem. It is therefore not worthwhile to look out for better combustion models, as the spray model is the responsible. The problem with the latter has already been discussed enough in the last section. An ad hoc measure, which indeed "helps", is increasing the injection velocity, usually to completely unphysical values beyond the Bernoulli velocity. This procedure is not recommended.

If the thermodynamics and mixture formation simulation are largely under control, i.e. the pressure path in compression and expansion is correct/plausible (according to the available occasions for comparison or validation), we can concern ourselves with the analysis of the actual combustion. Typically, the pressure path is not used for this, but rather the gross heat release rate. Experimentally, this one results from the pressure path by indication analysis; the heat release rate follows from this by means of a heat transfer model. In the 3D-simulation it seems to be obvious to obtain the gross heat release rate by summation of the heat release rates. This procedure is, however, incompatible with the experimental procedure! In order to see this, we proceed from a gas mixture in a chamber in pressure equilibrium but which has different temperatures locally. For the pressure p is then valid

$$p = \frac{\rho(x)\tilde{R}T(x)}{M}, \tag{9.173}$$

whereby $\rho(x)$ and $T(x)$ designate the local distributions of mass density and temperature, M is the molar mass. If we mix the system adiabatically, the total mass m and the internal energy U remain

$$m = \int_V dx \rho(x),$$

$$U = \int_V dx \rho(x) \int_{T_0}^{T(x)} d\vartheta\, c_V(\vartheta), \tag{9.174}$$

whereby $c_V(T)$ designates the specific heat capacity in dependence on temperature T. After mixing, a homogeneous density $\bar{\rho}$ and a homogeneous temperature $\bar{T}$ are initiated, which can be calculated from the conserved quantities m and U

$$\bar{\rho} = \frac{m}{V},$$

$$\int_{T_0}^{\bar{T}} d\vartheta\, c_V(\vartheta) = \frac{U}{m}. \tag{9.175}$$

How does the pressure now change? For the difference between the pressure after mixing p_M and the pressure p before mixing is valid

$$\Delta p = p_M - p = \frac{\bar{\rho}\tilde{R}\bar{T}}{M} - \frac{\overline{\rho(x)\tilde{R}T(x)}}{M} = \frac{\tilde{R}}{M}\left[\bar{\rho}\bar{T} - \frac{1}{V}\int_V dx\, \rho(x)T(x)\right]. \tag{9.176}$$

Furthermore

$$\bar{T} - T(x) = \frac{1}{c_V(\bar{T})}\left[\int_{T_0}^{\bar{T}} d\vartheta\, c_V(\vartheta) - \int_{T_0}^{T(x)} d\vartheta\, c_V(\vartheta) - \int_{T(x)}^{\bar{T}} d\vartheta\, (c_V(\vartheta) - c_V(\bar{T}))\right] \tag{9.177}$$

from which after multiplication with $\rho(x)$ and subsequent spatial averaging for Δp follows

$$\Delta p = \frac{\tilde{R}}{M\, c_V(\bar{T})}\left[\bar{\rho}\frac{U}{m} - \frac{U}{V} - \int_V dx\, \rho(x) \int_{T(x)}^{\bar{T}} d\vartheta\, (c_V(\vartheta) - c_V(\bar{T}))\right]$$

$$= \frac{\tilde{R}}{M\, c_V(\bar{T})} \int_V dx\, \rho(x) \int_{T(x)}^{\bar{T}} d\vartheta\, (c_V(\bar{T}) - c_V(\vartheta)). \tag{9.178}$$

We see that Δp at a constant, temperature-independent heat capacity is zero! Assuming that the specific heat increases under increasing temperature, i.e.

$$\frac{dc_V(T)}{dT} \geq 0$$

the expression

$$\int_{T(x)}^{\bar{T}} d\vartheta\, (c_V(\bar{T}) - c_V(\vartheta)) \tag{9.179}$$

is always positive, for $T(x) \geq \bar{T}$ as well as for $T(x) \leq \bar{T}$. Yet this means that $\Delta p \geq 0$! Mixture leads thus to an increase in pressure, it "looks like" a combustion!

Two cases will be considered: in the first, the heat release occurs homogeneously in space, in the second non-homogeneously, only after mixing does take place. The total system should be closed. Since pressure and internal energy are quantities of state, the final states must be identical in both cases, but because in the second case a pressure increase is associated with the mixture process, the pressure increase in the preceding process of non-homogeneous combustion must turn out less than in the homogeneous combustion of the first case. Applied to the engine problem, this means however that the gross heat release rate obtained from the 3D-

9.3 Simulation of combustion

simulation as a space integral, in contrast to the one gained from pressure path analysis, is shifted earlier (see Fig. 9.20). For a comparison of heat release rates from measurement and calculation, one should thus best put both pressure paths through a pressure indication analysis.

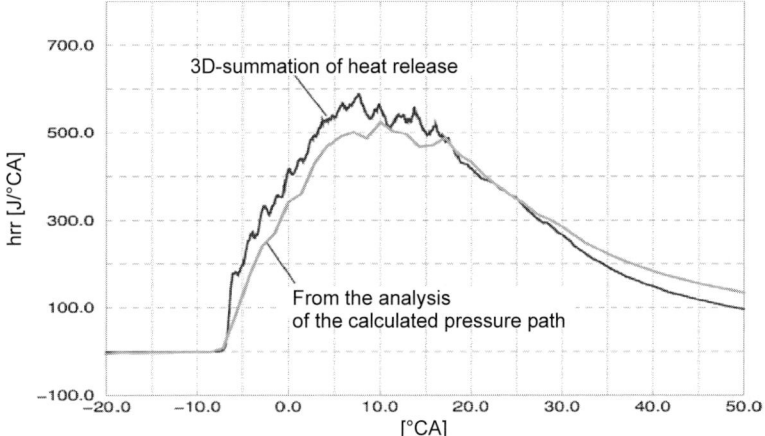

Fig. 9.20: Comparison of two gross heat release rates from the 3D-simulation, one produced via spatial integration and the other one via pressure path analysis

9.3.2 Diesel combustion

In the main phase of diesel combustion, it is a case of diffusion combustion. Autoignition and "premixture combustion" are considerably more influenced reaction-kinetically. The same is valid for pollutant formation.

The question of course arises of whether combustion simulation is necessary at all. Isn't the dominant process in the diesel engine mixture formation, is thus a simulation of mixture formation not sufficient to analyze engine behavior? In principle, this approach is, depending on the formulation of the question, not altogether wrong, but combustion helps to find errors in the mixture-formation simulation in purely "mixture-formation-dominated" cases, such as when larger amounts of unburned fuel, CO, or H_2 remain at the end of the calculation.

In the following, the simulation of heat release will first be treated and then the more complex phenomena of ignition and pollutant formation.

- **Simulation of heat release**

The main phase of diesel combustion can be seen as a turbulent diffusion flame, i.e. it progresses mixture-controlled, according to the formula "mixed = burned". With diffusion, turbulent diffusion is meant here. The simplest method of modeling a turbulent diffusion flame is the so-called "eddy breakup model". In such a model, into the species transport equations of type (9.32) source terms Q are introduced that scale with the species concentrations and the inverse turbulent length scale, i.e. they describe decay and formation processes that run with the turbulent breakup time

$$\tau_t \propto \frac{k}{\varepsilon},$$

e.g.

$$Q \propto \frac{c_A c_B}{\tau_t}. \tag{9.180}$$

Corresponding to the conversion rate, a heat release is calculated.

To describe diesel combustion, the combustion progress must still be modeled. The most well-known and used method for this is the mixing time-scale model, see Patterson and Reitz (1998). In it, from the turbulent and a chemical time scale (τ_t and τ_{chem}), an effective time scale τ_{eff} is formed, with which the combustion processes proceed

$$\tau_{eff} = \tau_{chem} + f\,\tau_t, \quad f = \frac{1-\exp(-r)}{1-\exp(-1)}, \quad \tau_{chem} \ll \tau_t = c_\mu \frac{k}{\varepsilon}, \tag{9.181}$$

whereby r designates the mass fraction of all reaction products. At the start of the reaction, $f = 0$, i.e. the effective time scale corresponds to the (small) chemical time scale, the reaction progresses very quickly, and a "premixed peak" occurs. With the increase in reaction products, f increases (to 1 maximally), and now is $\tau_{eff} \approx f \cdot \tau_t$ valid, i.e. we now have a diffusion combustion. The mixing time-scale model uses seven species, N_2, O_2, fuel, H_2O, CO_2, CO and H_2. From a given distribution of concentration ($c_{(k)}, k = 1...7$), the corresponding equilibrium distribution ($c_{(k)}^*, k = 1...7$) is calculated. We assume that every species tends towards local equilibrium with the time scale τ_{eff}

$$\rho\left(\frac{\partial}{\partial t} + v_i \frac{\partial}{\partial x_i}\right)c_k - \frac{\partial}{\partial x_i}\left(D_t \rho \frac{\partial}{\partial x_i} c_k\right) = \rho \frac{c_{(k)}^* - c_k}{\tau_{eff}}. \tag{9.182}$$

The chemical equilibrium code for the mixing time-scale model works with two lambda regimes. The limit between both of these regimes is given by the air-fuel ratio at which the fuel can be completely converted with the existing oxygen to CO and H_2. In the "rich" regime, the lambda values of which are lower than this limiting value, the equilibrium is determined such that the total available oxygen is utilized to create CO and H_2 from the fuel. In addition however, unburned fuel remains. In the "lean" regime, the lambda values of which are higher than the limiting value, we assume that no fuel remains. Besides the non-reactive N_2, five reactive species thus still remain, H_2O, H_2, O_2, CO and CO_2. Their equilibrium concentration is calculated by means of three elemental mass fraction conservation equations (for C, O, and H) as well as two relations, which follow from the law of mass action

$$\frac{[CO_2]}{[CO][O_2]^{0.5}} = K_C(p,T),$$

$$\frac{[H_2O]}{[H_2][O_2]^{0.5}} = K_H(p,T). \tag{9.183}$$

From these five equations, we obtain a polynomial of fourth order, which can be solved analytically with the corresponding solution formula. The results of this equilibrium solver are

9.3 Simulation of combustion

shown in Fig. 9.21. Corresponding to the reaction rates of the species, a source term for the enthalpy equation is determined by means of their specific reaction enthalpies $h_{(k)}$

$$q = \rho \sum_k h_{(k)} \dot{c}_{(k)} \; . \tag{9.184}$$

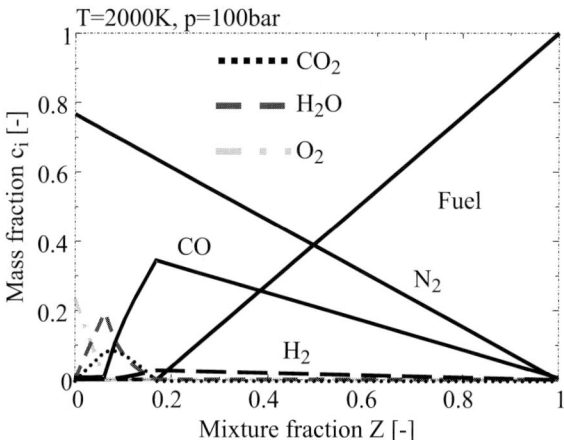

Fig. 9.21: Distributions of the seven species over the fuel mass fraction or the mixture fraction

With this seven-species method, local conditions can be very well prescribed, even in the rich area. Both in the modeling of reaction-kinetic influences and in turbulence interaction, the mixing time-scale model is however still very simple. Nevertheless, one can already analyze basic dependencies and mechanisms of diesel engine combustion with this modeling approach.

The modeling of the diffusion part can be improved with an approach corresponding to the mixing time-scale model, the so-called *pdf time-scale model*, see Rao and Rutland (2002). In this model, in addition to the species transport equations, transport equations for the mixture fraction and the mixture fraction variance are solved, from which the local mean values of the relevant species $c_{(k)}^{(pdf)}$ are determined. The species transport equations now read

$$\rho \left(\frac{\partial}{\partial t} + v_i \frac{\partial}{\partial x_i} \right) c_{(k)} - \frac{\partial}{\partial x_i} \left(D_t \rho \frac{\partial}{\partial x_i} c_{(k)} \right) = \rho \frac{c_{(k)}^{(pdf)} - c_{(k)}}{\tau_{chem}} \; . \tag{9.185}$$

The turbulent mixing process, which progresses with time scale τ_t, is described in the term $c_{(k)}^{(pdf)}$, since it tends in a time of the order of magnitude τ_t against the local equilibrium value in homogeneous mixing $c_{(k)}^*$

$$\dot{c}_{(k)}^{(pdf)} \approx \frac{c_{(k)}^* - c_{(k)}^{(pdf)}}{\tau_t} \; . \tag{9.186}$$

The premixed part is indeed still illustrated with the same "phenomenological" approach, but the turbulence interaction is now better described. This is helpful especially for full-load combustions, in which the premixed part is of only minor importance.

How is $c_{(k)}^{(pdf)}$ calculated? In the original formulation of the model, an averaging for this is carried out over the mixture fraction Z (with beta distribution) and the scalar dissipation rate χ (with Gaussian distribution)

$$c_{(k)}^{(pdf)} = \int dZ \int d\chi \, p_\beta(Z) \, pdf_{Gauss}(\chi) c_{(k)}(Z, \chi) \; . \tag{9.187}$$

In order to calculate this integral however, knowledge of the functions $c_{(k)}(Z,\chi)$ is necessary. The determination of these functions, so-called *flamelets*, demands costly reaction-kinetic calculations on laminar counter-flow flames. The scalar dissipation rate has the significance of an effective diffusion in the mixture fraction space and is responsible for so-called flame stretching effects, i.e. laminar flow equilibriums between diffusion and reaction. In principle, we thus leave the diffusion flame approach "mixed = burned", and the chemistry is no longer considered as "infinitely fast". Yet the reaction kinetics is very fast especially in the diesel engine diffusion flame phase because of high temperatures, and effects of finitely fast reaction kinetics are indeed already considered in the chemical time scale τ_{chem} in a very phenomenological manner. It is thus to be recommended that one neglects the flame stretching effects and applies the seven-species equilibrium kinetics as a reaction-kinetic basis as described above, see Steiner et al. (2004). The calculation and modeling cost is enormously reduced in this way. $c_{(k)}^{(pdf)}$ now amounts to

$$c_{(k)}^{(pdf)} = \int_0^1 dZ \, p_\beta(Z; \langle Z \rangle, \langle Z''^2 \rangle) c_{(k)}^*(Z) \; . \tag{9.188}$$

The distribution function p_β is the β function with the mean value $\langle Z \rangle$ and the variance Z''^2. The functions $c_{(k)}^*(Z)$ are the ones shown in Fig. 9.21. They are quite linear, piece by piece. With this, a very efficient integration scheme is possible for the integral (9.188), if one discretizes the Z-axis into intervals, in which the functions $c_{(k)}^*(Z)$ can be linearly (or quadratically) approximated. The product of a β-distribution function with a function linear in Z produces however exactly the linear combination of two β-distribution functions

$$N(a,b)^{-1}(1-Z)^{a-1} Z^{b-1} \cdot (A + B \cdot Z) = \\ AN(a,b)^{-1}(1-Z)^{a-1} Z^{b-1} + BN(a,b)^{-1}(1-Z)^{a-1} Z^b \; . \tag{9.189}$$

Thus, only integrals of the form

$$B(a,b;x) = N(a,b)^{-1} \int_0^x dZ \, Z^{a-1}(1-Z)^{b-1} \qquad 0 \leq x \leq 1 \tag{9.190}$$

must be solved, which are designated in the literature as incomplete beta-functions. To calculate them, very efficient algorithms are available. However, an efficient method of solving integrals (9.188) is decisive, as these have to be calculated in every time interval in every computational cell. And β-distribution functions may assume very unpleasant forms (at every Z-value, an arbitrarily sharp peak is possible).

9.3 Simulation of combustion

As we have mentioned, with the pdf time-scale model (together with an accurate spray model), very good full-load results can be achieved largely without tuning. A good example of this is depicted in Fig. 9.22.

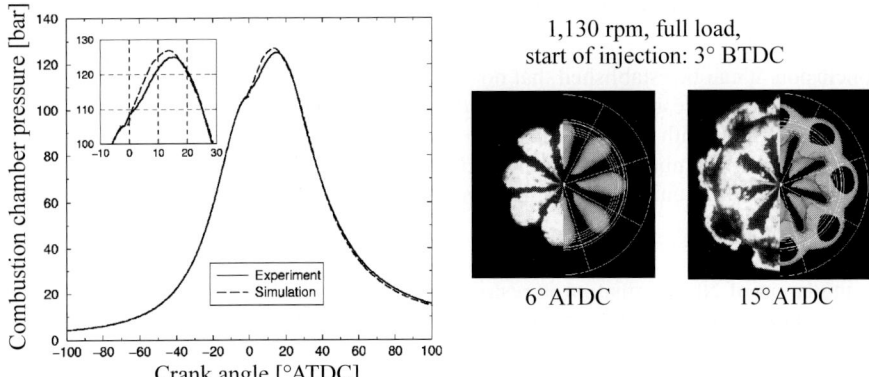

Fig. 9.22: Simulation of the combustion of a commercial vehicle diesel engine with the pdf time scale model. a) Comparison of experimental and calculated pressure paths b) Comparison of temperature-isosurface (simulation) with film shot from transparent engine

- **Ignition**

The simulation of ignition is a more difficult problem, since here the time scales of flow and (especially in this case very complex) reaction kinetics interfere. Well functioning reaction-kinetic mechanisms do indeed already exist for many hydrocarbons like heptane, which is quite well suited to model the autoignition behavior of diesel engines because of its comparable cetane number (about 50). But finally, this information does not help very much when the turbulence interaction is lacking.

One solution consists in ignoring turbulence interaction and introducing a source term based on laminar, detailed reaction kinetics. However, this approach is so imperfect that it is not worth the high calculation cost. Another often used alternative, which is at least less costly, consists in the use of phenomenological, reduced reaction kinetics; for example, the use of the adapted Shell model of Halstead et al. (1977) or even simpler phenomenological methods on the basis of a Wolfer equation, see also chap. 4.2.3. A typical modeling could look like this: one defines an indicator species c_I; if a predetermined threshold value $c_I^{(0)}$ is reached at a certain time, a local ignition occurs (i.e. the heat release model is activated). A transport equation is solved for c_I, for example with a source term according to Wolfer (1938), see also equation (5.41),

$$\rho\left(\frac{\partial}{\partial t}+v_i\frac{\partial}{\partial x_i}\right)c_I - \frac{\partial}{\partial x_i}\left(D_{v_t}\rho\frac{\partial}{\partial x_i}c_I\right) = A_{id}\,\rho\frac{p}{p_0}f(\lambda)\exp\left(-\frac{E_{id}}{T}\right). \qquad (9.191)$$

More costly methods with turbulence interaction work according to the flamelet concept. Unfortunately, the reaction kinetics is to slow to expect equilibrium. For this reason, we may proceed to describe only the source terms of a transport equation (e.g. for an ignition indicator) with a flamelet model (i.e. by means of mixture fraction averaging). As an indicator, CO has proven itself to be practical, among others, since the increase in CO-concentrations depicts the ignition process relatively monotonously. One difficulty consists however in the fact that now the CO source term is again dependent on the reaction progress (in our case, of CO concentration itself). The distribution of progress variables over the mixture fraction is unknown however.

In conclusion, it can be established that no patent formula exists; we are directed rather to the utilization of more or less inadequate methods and model tunings. In the case of typical diesel engine combustions with variations in ignition which are not all-too large, this is also not very critical; for HCCI combustions, this assertion is however no longer valid. Of the greatest value here would be a universally valid formulation.

- **NOx formation**

The simulation of NOx formation is restricted usually to thermal NOx and thus works with the Zeldovich mechanism, i.e. a transport equation is solved for the NOx concentration

$$\rho\left(\frac{\partial}{\partial t}+v_i\frac{\partial}{\partial x_i}\right)c_{\text{NOx}} - \frac{\partial}{\partial x_i}\left(D_t\rho\frac{\partial}{\partial x_i}c_{\text{NOx}}\right) = \\ Q_{Zeldovich}(c_{\text{NOx}},c_{\text{O}},c_{\text{OH}},c_{\text{H}},\lambda,p,T),$$ (9.192)

whereby the source term is calculated in accordance with the Zeldovich mechanism from the radical concentrations O, OH, and H; the radical concentration N is seen as being in partial equilibrium (see chap. 6.5.1).

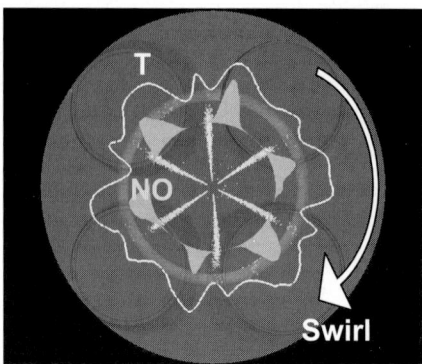

Fig. 9.23: Temperature and NO distribution in a passenger car diesel engine. On the lee-side (averted from the swirl) of the jets, the highest temperatures and thus the highest NO concentrations are reached

Since the time scale of NOx formation is very large (larger than the turbulent time scales – this precisely is the other limiting case of chemical turbulence interaction), most of the NOx is formed in the burned matter, and for this reason, turbulent temperature fluctuations are less relevant (the latter play the largest role in the burning zone). For this reason, NOx formation is usually calculated in a laminar fashion (i.e. purely reaction-kinetically on the basis of en-

9.3 Simulation of combustion

semble-mean values), as in (9.192). Whether or not this approach is sufficient cannot as of yet be conclusively determined, since the T-dependence of NOx formation is very high and thus the precision limits of statements from 3D-simulation are being approached.

Fig. 9.23 contains a calculated representation of a NOx distribution in a passenger car diesel engine along a ring curve on the spray level.

- **Soot formation**

To simulate soot formation and oxidation, different approaches exist, see also chap. 6.4.4. Firstly, there are the phenomenological models like Hiroyasu (formation) and Nagle-Strickland (oxidation). In this case, a transport equation for the soot mass fraction is solved

$$\rho\left(\frac{\partial}{\partial t}+v_i\frac{\partial}{\partial x_i}\right)c_{soot}-\frac{\partial}{\partial x_i}\left(D_t\rho\frac{\partial}{\partial x_i}c_{soot}\right)= $$
$$Q_{Hiroyasu}(\lambda,p,T)-Q_{Nagle-Strickland}(\lambda,p,T) \quad (9.193)$$

The prediction capability of this model is not very high however. Typically, the very high intermediary soot concentrations before the start of oxidation are only insufficiently reproduced.

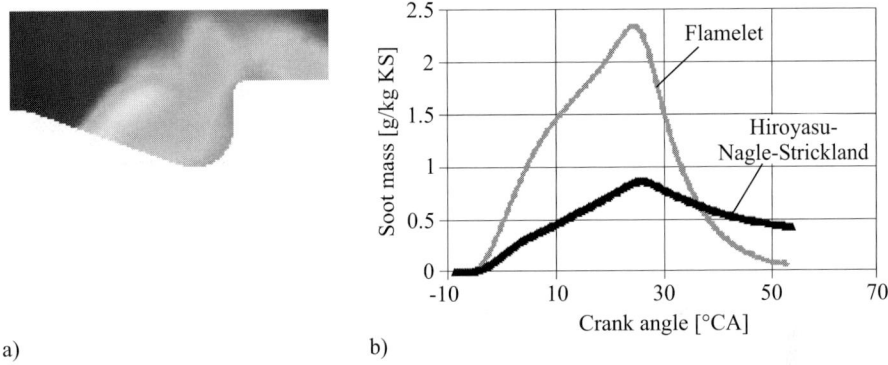

a)   b)

Fig. 9.24: 3D-simulation of the soot distribution in the diesel engine of a commercial vehicle. a) Soot distribution at 50° ATDC. The soot (bright) is found mainly in the bowl b) Amount of soot in the combustion chamber. Comparison of the Flamelet concept and the Hiroyasu-Nagle-Strickland model

Recently, a new flamelet-based method was suggested by Dederichs et al. (1999), which appears to be more predictive. In this case, the source terms of the soot transport equation are calculated in the flamelet method by means of an integral of type

$$Q_{soot}(\langle\chi\rangle;p,T)=\int dZ\, p_\beta(Z)Q_{soot}(Z,\langle\chi\rangle;p,T) \quad (9.194)$$

Since the expressions $Q_{soot}(Z,\langle\chi\rangle;p,T)$, as functions of Z, approximately have the form of β-functions, they can be approximated by the multiple of such a function, the coefficients and pre-factors of which are stored in a library. In this form, the Z-integration (9.194) takes on a simple form, since the product of two β-functions is again a β-function, and the integration can be performed analytically.

The integral soot mass development in the combustion chamber during combustion is depicted in Fig. 9.24 b), calculated with both Hiroyasu/Nagle-Strickland and flamelet. The flamelet results provide better results, as they predict high intermediary soot concentrations such as are also suggested in diagnostic investigations.

- **HC and CO emissions**

Further emissions playing an increasingly important role in modern diesel combustion processes, are unburned hydrocarbons (HC) and CO. Both are components of the seven-species modeling (HC only as fuel), i.e. the result of a combustion calculation with the seven-species model in principle provides HC and CO emissions. However, the CO and HC remaining in the calculation at end of combustion are typically mixture formation artifacts, i.e. the consequence of insufficient injection and mixture formation models. This is valid especially in the use of the "classical" Lagrangian spray model. Even in the application of improved spray models, as discussed in chap. 9.2.4, the prediction quality of the (turbulent) mixture conditions at the end of combustion will not be very high, deviations from the real evolution add up. Furthermore, detailed reaction kinetics comes into the foreground, and a "freezing" of the combustion occurs (i.e. we are confronted again with "large" chemical time scales). This is not well-described in present combustion models however.

In summary, it must be established that HC and CO emission are presently not calculable.

- **Outlook**

The previous has shown that flamelet models represent an essential progress with reference to chemistry-turbulence coupling. The main problem of the flamelet formulation has been pointed out: the consideration of progress variables at finitely fast reaction kinetics. It is not known how their distribution function is to be designed. Beyond the β-function for the mixture fraction, no a priori distribution function has proven successful.

Unsteady flamelets offer another way to deal with this problem. In this case, the combustion is modeled by means of several transient flamelets (these are diffusion reactors, i.e. one-dimensional objects in the mixture fraction space, with diffusion and reaction) with finite reaction kinetics (RIF: representative interactive flamelet, see Peters, 2000). However, one should not forget that this approach is not yet perfected; there are no closed application strategies, indeed, it is not even clear how many (and which) flamelets are to be used. If however not enough unsteady flamelets are applied, one indeed obtains an unsteady, turbulent reaction kinetics, but these instationarities have quite "strange" spatial coherence structures (because, simply speaking, large areas of space are described by a single diffusion reactor).

As opposed to this, the so-called CMC models ("conditional moment closure") are better founded theoretically, but are extremely costly in calculation time, and until now, they have not been applied to engine problems, see also Bilger (1993) as well as Klimenko and Bilger (1999).

A further option is the calculation of a posteriori distribution functions by means of so-called transported pdf models, see also Pope (1985). In this case, a local mixture condition (from the initial and end products of the reaction, thus incl. progress variables) is represented by means

9.3 Simulation of combustion

of an ensemble of homogeneous reactors. Every homogeneous reactor is represented by a parcel, which embodies a possible local species composition. Such a parcel moves with gas velocity in the combustion chamber, mixes with other homogeneous reactors in a computational cell (diffusion!) and its components react with each other (laminarly). This model thus corresponds to the parcel concept of the Lagrangian spray model. As a result, it suffers from the same weaknesses: it requires high calculation costs, statistical convergence is difficult to achieve, and small structures ("flame fronts") can hardly be resolved (for the latter, flamelet models with their stringent air-fuel correlations (mixture fraction distribution!) are best suited). A further problem consists in the fact that diffusion (i.e. the local mixing of the individual reactors) has to be modeled, i.e. cannot be derived directly from first principles. Since the available calculation capacities are rapidly increasing, the employment of transported pdf models in the near future is however conceivable. Suitable areas of use are problems with spatial gradients that are not too sharp, but still of high chemical complexity. The calculation of the (delayed) burn-out of HC and CO could be a good application, as well as the computation of autoignition (especially under HCCI conditions).

9.3.3 The homogeneous SI engine (premixed combustion)

One might believe that the simulation of the flame front combustion of a SI engine with homogeneous mixture should be relatively unproblematic, since the underlying physical processes are well-known and should be well describable. Unfortunately, exactly the opposite is the case. No engine CFD code exists that has a truly acceptable model for describing SI engine combustion at the ready. This also has to do – similarly to the spray models – with the high numerical requirements of flame front combustion models. In the following, various current methods will be discussed. At first however, we will devote ourselves to the leading insufficiency of contemporary combustion models, the lack of consideration of two-phases.

- **Two-phase behavior**

As has already been discussed in chap. 4.1.3, the wrinkled laminar flame front is very thin, usually only a few micrometers. We must now carry out an ensemble averaging. In it, the thin wrinkled laminar flame front (flame surface A_l) "blurs" and a thicker, "turbulent" flame front arises (the flame thickness is of the order of magnitude of the turbulent length scale), which is no longer correspondingly wrinkled (flame surface A_t). The propagation speeds s_l and s_t into the unburned mixture are different; their ratio corresponds to the reciprocal ratio of the flame surfaces

$$\frac{s_t}{s_l} = \frac{A_l}{A_t} \tag{9.195}$$

so that in the averaged as well as the non-averaged image (turbulent and laminar), the same burning rate is calculated ($A_l s_l = A_t s_t$). Because of the finite thickness of the turbulent flame front, an exact definition of the flame front position is not obvious, we can for example utilize the position of the 50 % conversion point. In Fig. 9.25, the relation between laminar and turbulent flame fronts is illustrated.

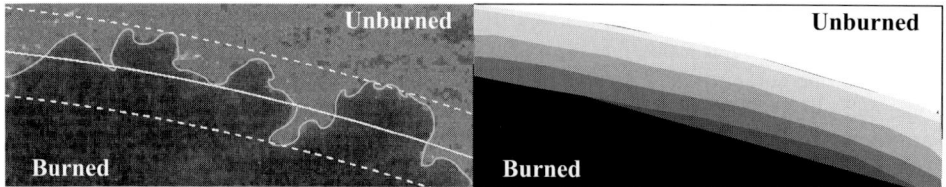

Fig. 9.25: Connection between laminar and turbulent flame fronts. a) Experimental laminar flame front, position and state of the turbulent flame front are shown b) The turbulent flame front in the averaged simulation

In principle, a step in density occurs in the laminar flame front (corresponding to the step in temperature), which corresponds to a step in the speeds. This situation is represented in Fig. 9.26 a) in the reference system of the flame front, i.e. the latter is at rest. With the laminar flame speed s_l, the unburned mixture enters into the stationary flame front, the burned mixture leaves it with another velocity v_b. Let ρ_b and ρ_{ub} be the densities in the burned and unburned mixtures. Due to mass conservation

$$\rho_b v_b = \rho_{ub} s_l \tag{9.196}$$

is then valid.

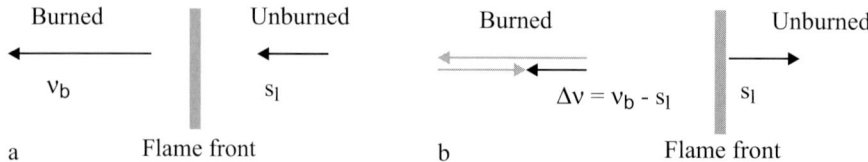

Fig. 9.26: Situation in the reference system a) of the flame front and b) of the unburned gasses

In Fig. 9.26 b), the situation is reproduced in the reference system of the unburned gas, i.e. a velocity shift transformation was simply carried out. We see that the speed makes a step between the unburned and burned mixture which is of quantity

$$\Delta v = v_b - s_l = \frac{\rho_{ub} - \rho_b}{\rho_b} s_l \;! \tag{9.197}$$

In the turbulent case, an overlapping of burned and unburned states occurs through the ensemble average, with different speeds, densities, temperatures, and turbulence levels. This is a two-phase flow! Unfortunately, premixed combustion is not handled this way in any of the engine CFD codes, which is the main inadequacy in their treatment of premixed combustion. Essential modeling progress is only to be expected by application of two-phase treatment (due to the obvious "jump relations" like (9.197)), an equivalent single-phase treatment of an averaged state is conceivable, from which both phases can be reconstructed.

The lacking consideration of two-phase behavior leads to various problems, one of which is the production of artificial turbulence in the flame zone. This is easy to see. The most important term of turbulence production reads (see (9.36))

9.3 Simulation of combustion

$$P = \tau_{R,ij} \cdot S_{ij} = F\left(\frac{\partial v_k}{\partial x_l}\right) \quad \text{with} \quad F(0) = 0 \ . \tag{9.198}$$

In the two-phase approach, this term would have to be interpreted correctly as the average of the term in the burned and unburned phases (v_b and v_{ub} describe the velocity of the burned and unburned phases)

$$P_{2\,phases} = cF\left(\frac{\partial v_{b,k}}{\partial x_l}\right) + (1-c)F\left(\frac{\partial v_{ub,k}}{\partial x_l}\right) \ . \tag{9.199}$$

In the standard, single-phase approach, one works only with a mean velocity v, which represents an overlapping of the velocities of the burned and unburned phases

$$v_k = cv_{b,k} + (1-c)v_{ub,k} \ . \tag{9.200}$$

With this, the term P is calculated. Upon the disappearance of the velocity gradients

$$\frac{\partial v_{b,k}}{\partial x_l} = \frac{\partial v_{ub,k}}{\partial x_l} = 0 \tag{9.201}$$

no turbulence is produced in the (correct) two-phase model. In the single-phase standard model however, already a c-gradient in the case of a simultaneous phase step in velocity produces (artificial) turbulence (at vanishing velocity gradients)

$$P = F\left(\frac{\partial v_k}{\partial x_l}\right) = F\left((v_{b,k} - v_{ub,k})\frac{\partial c}{\partial x_l}\right) \ . \tag{9.202}$$

Since F is essentially a quadratic function in the velocity gradients, the artificial turbulence is all the stronger, the thinner the flame front, because

$$P \propto \left[\Delta v \frac{\partial c}{\partial x}\right]^2 \approx \left[\Delta v \frac{1-0}{l_F}\right]^2 = \left[\frac{\Delta v}{l_F}\right]^2 \tag{9.203}$$

is roughly valid (l_F denote the flame thickness). The total turbulence production as an integral across the flame front then results in

$$P_{tot} \approx P l_F \propto \frac{(\Delta v)^2}{l_F} \ , \tag{9.204}$$

i.e. it diverges for $l_F \to 0$.

The error can assume dramatic proportions, when no model is used to stabilize the flame thickness. In this case, the flame produces turbulence, which thus assumes the highest values on the reverse side. This turbulence accelerates the flame (see for example the Damköhler relation (4.5)), the reverse side more than the forefront. The flame thus becomes ever faster and thinner. A thinner flame front creates however even more turbulence, with which the circle closes.

For the solution of this problem, the opportunity presents itself of suppressing the turbulence production term in the flame front. Especially for thin flame fronts, the errors caused by this are tolerable. However, no path can avoid a two-phase formulation in the long run.

- **The Magnussen model**

The simplest combustion model for premixed flames is the Magnussen model, consisting of a transport equation for the progress variable c ($c=0$: no conversion, $c=1$: conversion complete). In analogy to the breakup model for diffusion flames, the reaction rate is proportional to the inverse turbulent time scale ε/k, whereby it becomes clear that we are concerned with a turbulent premixed combustion. Moreover, the reaction rate must be zero for $c=0$ and $c=1$, in the burned and unburned mixture. The Magnussen model thus reads

$$\rho\left(\frac{\partial}{\partial t}+v_i\frac{\partial}{\partial x_i}\right)c-\frac{\partial}{\partial x_i}\left(D_t\rho\frac{\partial}{\partial x_i}c\right)=\alpha\rho_{ub}\frac{\varepsilon}{k}c(1-c)\ , \tag{9.205}$$

whereby α describes a model parameter. This model is scarcely used any longer, since is has serious insufficiencies; because of its simplicity, it is however well-suited to study the basic properties of an entire class of combustion models.

Equation (9.205) forms, after a starting time, a stable flame front profile that is independent of the exact initial conditions and runs through the combustion chamber like a dispersion-free wave with a defined propagation speed, see Fig. 9.27. This is a consequence of the non-linear source term. Such non-linear waves are known in various areas of physics and are called *solitary waves* or *solitons*. In contrast to this, the profile in the case of linear waves is not determined by the wave equation before, but is rather given by the initial conditions. Moreover, linear waves are usually subject to dispersion.

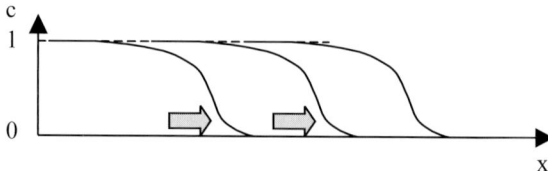

Fig. 9.27: Stable turbulent flame front profile

Yet problems result from this soliton property at the same time: profile determination takes place at all flame front locations numerically, i.e. by means of a solutions of a non-linear differential equation sometimes, with a very poor mesh resolution. Turbulent flame thicknesses lie in the order of magnitude of the turbulent length scale, and these often amount to a mere 1-2 mm under engine conditions. With a mesh, all size of 0.5 mm, this means four mesh cells per flame front, and that this is not necessarily sufficient for the discretization of a non-linear differential equation should be clear!

The question of propagation speed is answered in the so-called KPP theorem (Kolmogorov, Pichunov, Petrovski), see also Kolmogorov et al. (1937). The basic idea is that the propaga-

9.3 Simulation of combustion

tion speed of the flame front can be analyzed with the help of the spreading speed of its "bow wave", i.e. its forefront (see Fig. 9.28). Valid in this area is namely $c \approx 0$, and from this, (9.205) can be linearly approximated in c. $\rho \approx \rho_{ub}$ is additionally valid.

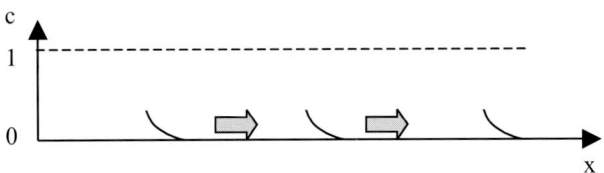

Fig. 9.28: Propagation of the face of the flame front

If we work in a single spatial dimension and assume constant turbulence values, we then obtain the equation

$$\rho_{ub}\frac{\partial c}{\partial t} + \rho_{ub} v \frac{\partial c}{\partial x} - \rho_{ub} D_t \frac{\partial^2 c}{\partial x^2} = \alpha \rho_{ub} \frac{\varepsilon}{k} c \ . \tag{9.206}$$

A stationary wave solution reads

$$c(x,t) = \gamma(x - (v + s_t)t) = \gamma(\xi) \ . \tag{9.207}$$

Insertion in (9.206) provides

$$-s_t \frac{d\gamma}{d\xi} - D_t \frac{d^2\gamma}{d\xi^2} = \alpha \frac{\varepsilon}{k} \gamma \ . \tag{9.208}$$

This equation has exponential functions as a solution. In the sense of Fig. 9.28, a solution of type

$$\gamma = \exp(-\omega \xi) \qquad \omega \geq 0 \tag{9.209}$$

should be sought. With it, we obtain

$$\omega = \frac{s_t \pm \sqrt{s_t^2 - 4\alpha \frac{\varepsilon}{k} D_t}}{2 D_t} \ . \tag{9.210}$$

This equation only has a real solution, if

$$s_t \geq 2\sqrt{\alpha \frac{\varepsilon}{k} D_t} \ .$$

The KPP theorem now asserts that the minimal speed value

$$s_{t,\min} = 2\sqrt{\alpha \frac{\varepsilon}{k} D_t} \tag{9.211}$$

is precisely that which arises when the flame front propagates from a limited location into a region with $c = 0$. In order to understand this, let us look again at the linearized equation (this time without the v-term for the sake of simplicity)

$$\frac{\partial c}{\partial t} - D_t \frac{\partial^2 c}{\partial x^2} = \alpha \frac{\varepsilon}{k} c \qquad (9.212)$$

the Greens function of which reads

$$c_G(x;t) = \frac{N}{\sqrt{t}} \exp\left(-\frac{x^2}{4D_t t} + \alpha \frac{\varepsilon}{k} t\right). \qquad (9.213)$$

This result is not difficult to derive from the case $\alpha = 0$ of the pure diffusion equation. The Greens function describes asymptotic propagation behavior (only of the frontage of the flame front!), which proceeds from a point source – the constant pre-factor N is irrelevant. In order to find the propagation speed, we have to calculate the function $x(t)$, for which c_G is stationary, i.e.

$$-\frac{x^2}{4D_t t} + \alpha \frac{\varepsilon}{k} t - \frac{1}{2}\ln(D_t t) = \text{const.} \,. \qquad (9.214)$$

For large t, the logarithmic term is negligible, and we obtain approximately

$$\left(\frac{x}{t}\right)^2 = 4\alpha \frac{\varepsilon}{k} D_t = s_{t,\min}^2 , \qquad (9.215)$$

i.e. we have again found our minimal speed (see (9.211)) as the speed of a flame proceeding from a point source! If we set

$$D_t = c_\mu \frac{k^2}{\varepsilon}$$

we then obtain

$$s_t = 2\sqrt{c_\mu \alpha}\, u'. \qquad (9.216)$$

This corresponds to the Damköhler relation (4.5) in the limiting case $s_t \gg s_l$.

With the help of the Greens function (9.213), we can also see that at suitable spatial pre-initialization (i.e. no point source) higher burning velocities can also be reached. For example, if we choose the initiation

$$c(x;t=0) = \exp(-\beta|x|) \text{ with } \beta < \frac{s_{t,\min}}{2D_t}. \qquad (9.217)$$

This leads for $t \geq 0$ to

$$c(x;t) = \int_{-\infty}^{\infty} dy \exp(-\beta|y|) \frac{N}{\sqrt{t}} \exp\left(-\frac{(x-y)^2}{4D_t t} + \alpha \frac{\varepsilon}{k} t\right). \qquad (9.218)$$

For the calculation of flame propagation for $x > 0$, this expression can be substituted by

9.3 Simulation of combustion

$$c(x;t) \approx \int_{-\infty}^{\infty} dy \frac{N}{\sqrt{t}} \exp\left(-\frac{(x-y)^2}{4D_t t} - \beta y + \alpha \frac{\varepsilon}{k} t\right)$$

$$= 2\sqrt{\pi D_t} \, N \exp\left(-\beta x + \left(D_t \beta^2 + \alpha \frac{\varepsilon}{k}\right) t\right),$$

(9.219)

because the integrand of (9.219) is a Gaussian function, the maximum of which lies at

$$y_{max} = x - 2D_t t \beta ,$$

the width at half maximum scales with $\sqrt{t}$ for $t \to \infty$.

The flame front propagation should now be described for large time periods, i.e. $x \cong s_t t$. From this follows for y_{max}

$$y_{max} = s_t t - 2D_t t \beta > (s_t - s_{t,min}) t \geq 0 ,$$

i.e. the range of the integrand of (9.219) that is considerably different than zero has, at least for large t, positive y-values. The equations (9.219) and (9.218) provide an identical propagation behavior.

The need for a stationary exponent leads for (9.219), lower line, to the relation

$$s_t(\beta) = \frac{x}{t} = D_t \beta + \frac{\alpha}{\beta} \frac{\varepsilon}{k} .$$

(9.220)

The minimum of this function (given by $ds_t/d\beta = 0$) reads again

$$s_t = 2\sqrt{\alpha \frac{\varepsilon}{k} D_t} \quad \text{for } \beta = \sqrt{\frac{\alpha \varepsilon}{D_t k}} .$$

(9.221)

However, this also means that for smaller β values there are higher flame propagation speeds. From this follows a further problem of the Magnussen model: it is very unstable at incorrect initializations, in general at c-values differing slightly from zero in front of the flame front (an initialization such as in (9.217) deviates indeed only slightly from zero). Graphically speaking, this is due to the fact that the propagation speed is determined by the forefront of the flame (i.e. small c-values), while the larger c-values "provide" for right profile within the flame front.

This numerical sensitivity comes to light especially near the wall. In principle, the KPP analysis should also be valid for the Magnussen model at the wall, i.e. the burning velocity (equation (9.216) should drop, since the turbulence is reduced at the wall (dissipation increases considerably). Typically, 3D-simulations supply the exact opposite behavior: extreme, completely unphysical flame accelerations in the area of the wall occur. To understand this phenomenon, we need to analyze the flame speed according to (9.211). At the wall

$$\varepsilon \propto \frac{1}{y} \xrightarrow{y \to 0} \infty$$

(9.222)

is valid, and as a result, the source term of equation (9.206) approaches infinity, while the diffusion term should tend towards zero, with a finite product. Because of a lack of numerical precision, diffusion does not approach zero however, and a certain numerical diffusion D_{num} remains, which is attached to the mesh resolution and numerical scheme. Consequently, the term (9.211) can exceed all limits in the area of the wall

$$s_{t,wall} = 2\sqrt{\alpha \frac{\varepsilon}{k} D_{num}} \propto \frac{1}{\sqrt{y}} \xrightarrow{y \to 0} \infty \ . \qquad (9.223)$$

To sum up, one can say that the Magnussen model produces a solitary wave, the profile and propagation speed of which are generated by means of a complex interaction of the source term and diffusion term. In the case of thin, turbulent flame fronts or in the area close to the wall, it suffers from serious numerical problems.

- **Flame surface density model**

The flame surface density model (also called coherent flame model) provides an improved physical description. In it, turbulence does not directly accelerate the flame (as in (9.205)), but wrinkles it more strongly. However, a more strongly wrinkled flame burns faster.

For this purpose, an additional transport equation is solved for the flame surface density Σ (flame surface per volume unit), or instead, for the specific flame surface $\sigma = \Sigma/\rho$. This equation exists in the various versions, see Poinsot and Veynante (2001). One typical variant reads

$$\rho\left(\frac{\partial}{\partial t} + v_i \frac{\partial}{\partial x_i}\right)\sigma - \frac{\partial}{\partial x_i}\left[\rho D_t \frac{\partial \sigma}{\partial x}\right] = \alpha_F \frac{\varepsilon}{k}\rho\sigma - \beta_F \frac{s_l}{c(1-c)}(\rho\sigma)^2 \ , \qquad (9.224)$$

whereby α_F and β_F (may) contain, besides model constants, functional dependencies on the turbulent and chemical time and length scales. The first term on the right side describes the flame surface production from turbulence, the second term represents a sink caused by burn-out. In addition, another transport equation must be solved for the progress variable

$$\rho\left(\frac{\partial}{\partial t} + v_i \frac{\partial}{\partial x_i}\right)c - \frac{\partial}{\partial x_i}\left(\rho D_t \frac{\partial}{\partial x_i}c\right) = \rho_{ub} s_l \rho\sigma \ . \qquad (9.225)$$

If an equilibrium between flame surface production and destruction exists in (9.224), then

$$(\rho\sigma)_{eq} s_l = \frac{\alpha_F}{\beta_F}\frac{\varepsilon}{k}c(1-c) \qquad (9.226)$$

is valid. With that, we obtain the Magnussen model from (9.225)! This approximation is however only fulfilled for large values of α_F and β_F.

9.3 Simulation of combustion

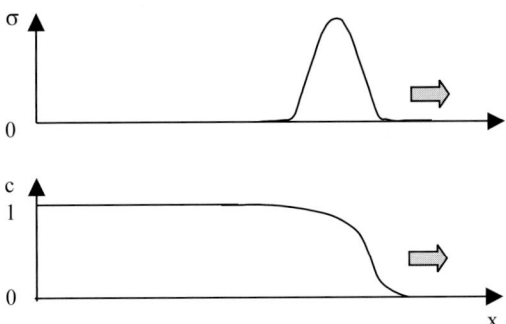

Fig. 9.29: Flame front profiles in c and σ

In the case of flame surface density models, the flame front is again well-described by a solitary wave (this time in the σ and c-field), with a well-defined propagation speed and well-defined c and σ profiles. The c field goes thereby from 0 (in front of the flame) to 1 (behind the flame), while σ starts at zero before the flame, increases to a maximum value in the flame, in order then to fall again to zero towards the rear flame front (see Fig. 9.29). To calculate the propagation speed of a planar flame front, we proceed similarly as in the Magnussen model, determining the propagation speed of the flame forefront; for this, (9.224) and (9.225) have to be linearized in σ and c, whereby $\rho = \rho_{ub}$. The equations thus obtained read

$$\rho_{ub}\left(\frac{\partial}{\partial t}+v\frac{\partial}{\partial x}\right)\sigma - \rho_{ub}D_t\frac{\partial^2\sigma}{\partial x^2} = \alpha_F\frac{\varepsilon}{k}\rho_{ub}\sigma - \beta_F\,s_l(\rho_{ub})^2\left(\frac{\sigma}{c}\right)\sigma\,, \tag{9.227}$$

$$\rho_{ub}\left(\frac{\partial}{\partial t}+v\frac{\partial}{\partial x}\right)c - \rho_{ub}D_t\frac{\partial^2 c}{\partial x} = (\rho_{ub})^2\,s_l\left(\frac{\sigma}{c}\right)c\,.$$

One recognizes that a stationary solution with $\sigma/c = $ const. exists

$$\begin{aligned}c(x,t) &= c_0\exp(-\omega(x-(v+s_t)t))\,,\\ \sigma(x,t) &= \sigma_0\exp(-\omega(x-(v+s_t)t))\,.\end{aligned} \tag{9.228}$$

Insertion into (9.227) provides

$$\begin{aligned}\omega s_t - D_t\omega^2 &= \alpha_F\frac{\varepsilon}{k} - \beta_F\,s_l\,\rho_{ub}\frac{\sigma_0}{c_0}\,,\\ \omega s_t - D_t\omega^2 &= s_l\,\rho_{ub}\frac{\sigma_0}{c_0}\,,\end{aligned} \tag{9.229}$$

which finally leads to

$$\omega = \frac{s_t \pm \sqrt{s_t^2 - 4D_t\,\dfrac{\alpha_F}{\beta_F+1}\dfrac{\varepsilon}{k}}}{2D_t}\,. \tag{9.230}$$

According to the KPP theorem, the propagation speed follows

$$s_t = 2\sqrt{D_t \frac{\alpha_F}{\beta_F + 1} \frac{\varepsilon}{k}} \qquad (9.231)$$

(Depending on the concrete formulation of the initial equation (9.224), this calculation may have a somewhat different outcome, one should only understand the result (9.231) examplarily.)

It is thus clear that the flame front models have a mathematical behavior comparable to the Magnussen model, propagation speed and flame front profile are again the result of the interaction of diffusion and the source terms in the relevant transport equations. And a good numerical resolution of the flame front is indispensable. But it is therefore also obvious that the flame surface density front models do not repair what is precisely the worst inadequacy of the Magnussen model, its poor numerical behavior!

Also, behavior near the wall continues to be critical. However, modeling methods exist, in which the wall problem no longer exists or is at least markedly reduced. In this way, for Poinsot and Menevaux, the function α_F contains the so-called ITNFS function Γ (Γ = intermittent turbulent net flame stretch, see Poinsot and Veynante, 2001),

$$\alpha_F = \alpha_0 \frac{\varepsilon}{k} \Gamma\left(\frac{l_t}{\delta_0}, \frac{\sqrt{k}}{s_l}\right), \qquad (9.232)$$

whereby δ_0 designates the laminar flame thickness. The function Γ, however, tends with a decreasing turbulent length scale towards zero, i.e. at the wall ($l_t \to 0$), the product $\varepsilon/k \, \Gamma$ also becomes equal to zero, no numerically caused divergence appears any longer! This formulation would also clearly be of help in the case of the Magnussen model[15].

Because of the resolution problems, flame surface density models are poorly adaptable to engine calculations; the available computational meshes are typically too coarse, especially in the case of late direct injection (high turbulence). Coherent flame models could be useful as a supplement to high resolution 2D-spray models as in chap. 9.2.4, however such calculations are rotation-invariant, and the ignition event is not.

Another option that can be seen sometimes is flame front adaptive mesh refinement. This is however not at all "up-to-date", the trend is unfortunately going in the opposite direction (see chap. 9.1.4). Nowadays, we can count ourselves as lucky if we have a separate mesh (which for premixed combustion is as homogeneous as possible) available for combustion simulation, one which no longer contains the valve structures.

It is often asserted that turbulent flame fronts are not at all that thin, but rather show a considerable flame thickness, which is experimentally proven, if one superimposes the flames of several cycles. In this case, ensemble averaging and cycle averaging are being mixed up again. As previously shown (see chap. 9.1.2), ensemble averaging only contains the averaging over turbulent ("coherent") fluctuations, which are produced at identical boundary conditions in the flow by the chaotic behavior of the underlying fluid dynamics. Cyclic fluctuations

---

[15] We should point out here that the ITNFS function has a physical meaning. But it has an extremely advantageous effect in the numerical sense as well.

contain additional fluctuations however, produced by fluctuations of the boundary or initial conditions (throttle, injection, residual gas, ignition, etc.). A turbulent combustion model includes only the turbulent, coherent fluctuations (thus the name "coherent flame model"); only these contribute to the formation of turbulent model flame front.

It is easy to see that one is not free in the choice of the ensemble under consideration. Typical incoherent fluctuations are, for example, fluctuations in the (effective) ignition time. Let

$$\langle \sigma \rangle_\varphi \text{ and } \langle c \rangle_\varphi$$

be the ensemble-averaged quantities σ and c with reference to the coherent fluctuations, at a fixed ignition time (or angle) φ. We then obtain the total mean values with reference to the coherent and incoherent fluctuations

$$\langle \sigma \rangle_{tot} = \int d\varphi \, f(\varphi) \langle \sigma \rangle_\varphi$$
$$\langle c \rangle_{tot} = \int d\varphi \, f(\varphi) \langle c \rangle_\varphi \tag{9.233}$$

at an ignition angle distribution function $f(\varphi)$. Because of their non-linearity however, the transport equations (9.224) and (9.225) is not invariant under such a transformation! I.e. if $\langle \sigma \rangle_\varphi$ and $\langle c \rangle_\varphi$ fulfill the equations (9.224) and (9.225), then this is not valid for $\langle \sigma \rangle_{tot}$ and $\langle c \rangle_{tot}$.

Thus we are not at liberty to choose the ensemble for averaging "suitably", the flame surface density models (like the Magnussen model) already correspond to a fixed choice, namely, that of the "minimal" ensemble, which contains only the coherent, intrinsic, fluid mechanical fluctuations.

- **The G-equation**

In order to escape the numerical problems of the flame front and Magnussen models, a combustion model is needed with a formulation in which the turbulent flame speed appears explicitly. Moreover, the sensitivity with reference to the resolution of the turbulent flame front should be as small as possible. The G-equation is such a model:

$$\frac{\partial G}{\partial t} + v_i \frac{\partial G}{\partial x_i} = s_t |\nabla G| \tag{9.234}$$

or

$$\frac{\partial G}{\partial t} + (v_i + s_t \hat{n}_i) \frac{\partial G}{\partial x_i} = 0 \text{ with } \hat{n} = -\frac{\nabla G}{|\nabla G|}. \tag{9.235}$$

This equation describes the propagation of a surface such that every surface element propagates with a propagation speed normal to it of amount s_t relative to the fluid. The surface is characterized by the amount of points, for which is valid $G(x) = 0$. Outside the flame surface, the variable G can be chosen arbitrarily, only that it be other than zero. The flame is thus first described in this image as an infinitely thin surface. Naturally, it cannot stay this way: aside from the physical realities, we should not permit a "step" in density and temperature into a CFD code. Thus, a finite flame thickness l_F has to be introduced. As a determination equation for l_F, the relation

$$l_F = b l_t \qquad (9.236)$$

for example may be used for a stationary flame, whereby $b \approx 2$. This is not valid near the wall however, since the flow is laminarized there. Model variants also exist with their own transport equations for the flame thickness, see Peters (2000).

$G = 0$ thus describes the central position of the flame front. As a profile for ∇c, a Gaussian function can be used; c is then defined as[16]

$$c(x) = \mathrm{erf}\left(\frac{2\,d(x)}{l_F}\right), \qquad (9.237)$$

if $d(x)$ describes the (positive or negative according to the position) distance of a given point to the flame front. However, we must still determine how operationally the distance of a spatial point from the flame front is to be calculated.

If s_t designates, as is usual, the turbulent flame speed relative to the unburned mixture, then (9.234) is valid only for the flame forefront, corresponding to the equations (9.227), i.e. for $\rho = \rho_{ub}$. In the case of a finite reaction progress with assigned density $\rho < \rho_{ub}$, an additional reverse flow (see (9.197)) must still be overcome, the difference velocity to the flame front (the speed of which relative to the unburned mixture amounts to s_t) is

$$v_\rho = s_t \frac{\rho_{ub} - \rho}{\rho}. \qquad (9.238)$$

In total therefore, one can write more generally

$$\rho \frac{\partial G}{\partial t} + \left(\rho v_i + \rho_{ub} s_t \hat{n}_i\right)\frac{\partial G}{\partial x_i} = 0. \qquad (9.239)$$

The G-equation is a hyperbolic equation with a transport velocity different from $\vec{v}$

$$\vec{v} + \frac{\rho_{ub}}{\rho} s_t \hat{n} \,;$$

it thus needs in principle its own solution algorithm, which is not provided in the standard CFD codes. Even in the case that a solution method were available, (9.239) can behave problematically outside the flame front, since no special behavior is prescribed there. A recommendable approach is thus to make the demand

$$|\nabla G| = 1, \qquad (9.240)$$

i.e. with this constraint ("grange" of the G field) the G values outside the flame front correspond to the distances to the flame front (negative before and positive behind the flame front).

---

[16] $\mathrm{erf}(x) = \dfrac{1}{\sqrt{\pi}} \int\limits_{-\infty}^{x} \exp(-x^2)\,dx$

9.3 Simulation of combustion

With this choice, the local definition of the progress variables can be given, according to (9.237) is valid

$$c(x) = erf\left(\frac{2G(x)}{l_F}\right). \qquad (9.241)$$

However, the property (9.40) is not preserved in time, the equation (9.239) has to be reinitialized! This means that after every time interval (or after a number of time intervals), the differential equation

$$\frac{\partial G(x,t,\tau)}{\partial \tau} = sign(G(x,t))(1-|\nabla G(x,t,\tau)|), \qquad G(x,t,0) = G(x,t), \qquad (9.242)$$

has to be solved for $\tau \to \infty$, it converges then towards $|\nabla G|=1$. This implies a considerable additional calculation effort indeed.

However, the great advantage of the G-equation is that we are concerned with a linear wave, at least for the planar case; the flame profile and propagation speed are uncorrelated. A low numerical resolution of the profile is thus not so critical.

Various formulae can be used for s_t, the flame speeds of the flame surface density model (e.g. (9.231)) as well. Frequently, phenomenological relations similar to the Damköhler relation (4.5) are utilized, like

$$s_t = s_l \left(1 + A \cdot \left(\frac{u'}{s_l}\right)^n\right). \qquad (9.243)$$

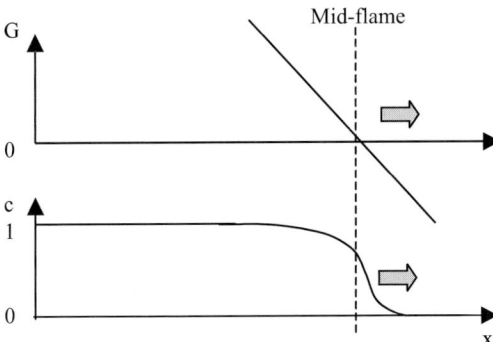

Fig. 9.30: Schematic representation of the simulation of flame front propagation with the G-equation with reinitialisation

A further problem is the G-equation's handling of the wall. The use of (9.236) at the wall entails problems, since the flame front would be very thin (laminarized). We can simply put a numerical lower limit on the flame front thickness or introduce an additional transport equation for the turbulent flame thickness l_F (as then l_F does not normally reach equilibrium).

Sometimes, besides the l_F equation, further transport equations are formulated for flame properties, for example for the flame surface density Σ (see also Peters (2000)). In the formulation of these transport equations, one should be careful that no violation of causality takes place; the propagation of flame properties must be compatible to the flame propagation itself. Only the flame variables at the flame location ($G = 0$) are physical and may have an influence on (physical) flame variables at a later point of time. One admissible transport equation for a flame property l_F reads[17]

$$\rho \frac{\partial l_F}{\partial t} + (\rho v_i + \rho_{ub} s_t \hat{n}_i) \frac{\partial l_F}{\partial x_i} - \frac{\partial}{\partial x_\parallel}\left(\rho D_t \frac{\partial}{\partial x_\parallel} l_F\right) = 2\rho D_t - c_s \rho \frac{\varepsilon}{k} l_F^2 , \qquad (9.244)$$

whereby

$$\hat{n} = -\frac{\nabla G}{|\nabla G|},$$

$$\frac{\partial}{\partial x_\parallel}\left(\rho D_t \frac{\partial}{\partial x_\parallel} l_F\right) := \frac{\partial}{\partial x_i}\left(\rho D_t \frac{\partial}{\partial x_i} l_F\right) - \hat{n}_i \frac{\partial}{\partial x_i}\left(\rho D_t \hat{n}_j \frac{\partial}{\partial x_j} l_F\right). \qquad (9.245)$$

- **The diffusive G-equation**

Unfortunately, no complete G-equation implementation with its own convection and re-initialization exists until now in engine CFD codes, STAR CD does however contain a diffusive G-equation for the progress variable (under the designation single-equation Weller model[18])

$$\rho\left(\frac{\partial c}{\partial t} + v_i \frac{\partial G}{\partial x_i}\right) - \frac{\partial}{\partial x_i}\left(\rho D_t \frac{\partial G}{\partial x_i}\right) = \rho_{ub} s_t |\nabla c| . \qquad (9.246)$$

In contrast to (9.239), this equation contains a turbulent diffusion term; moreover, it is used directly to calculate the progress variable c (i.e. a relation like (9.241) can now be dropped, it could however not even be applied any more, since a distance variable to the flame front is now lacking). The term proportional to s_t is treated as a source term.

The advantages of the formulation (9.246) are obvious: we are dealing here with a "conventional" scalar transport equation, which can be dealt with in the standard approach. Special effort for a particular convection or a re-initialization need not be made.

On the other hand, the solution quality reduces clearly with the source term treatment of the s_t term. And the re-initialization was necessary for the generation of a distance variable, which served itself for the calculation of flame thickness. It is here that the largest deficiency of the diffusive G-equation becomes apparent: it computes flames that are obviously too

---

[17] The physical information of a configuration should be invariable under a gauge transformation
$l_F(x) \to l_F(x) + G(x)\Phi(x)$ for any $\Phi(x)$. This demand can serve to evaluate suitable forms of the transport equations (9.244).

[18] The actual Weller model is however a two-equation model that resolves the flame front in detail and resembles the coherent flame model (see Weller (1993)).

thick, which deliquesce under the influence of diffusion ($l_{F,Diff-G} \propto 2\sqrt{D_t t}$). This does not at all characterize the behavior of a flame front however, which of course forms a solitary wave with a stationary profile, as we learned above. The propagation speed remains untouched by the diffusion term however, and the same is true of the global conversion rate. Thus, excessively thick flame fronts are effectively less harmful as one might at first assume. At least in the planar, one-dimensional case, the G-equation is form-invariance with respect to the transformation (9.233), i.e. in using a G-equation formulation, one indeed has a certain right to permit incoherent, flame-thickening fluctuations as well.

In conclusion, a great practical advantage of thick flames should not remain unmentioned: the artificial turbulence production in accordance with p. As long as the turbulence production problem is not solved, one can actually not at all avoid calculating with non-physically thick flames.

However, this means that there is a spatially erroneous flame distribution in the CFD calculation, especially towards the end of combustion. In particular when other physical phenomena running parallel to combustion are also investigated (e.g. knocking or mixture formation in stratified charge combustion), a false flame distribution causes problems.

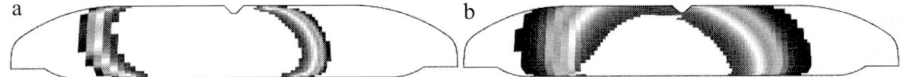

Fig. 9.31: Simulation of premixed combustion a) with the G-equation and b) with the diffusive G-equation

- **Ignition**

In the description of spark ignition and early flame kernel formation, one is still largely working with models that are quite phenomenological. One should at least care for a numerically correct procedure and not create any incontinuities in the boundary conditions. For example, the progress variable should not be set suddenly from 0 to 1 in the "ignition cell" or cells, but rather be continuously raised.

- **Knocking**

First approaches to knock calculation with CFD exist already in the literature; because of the limitations of the available combustion and wall heat transfer models, these attempts must however be approached with caution.

To describe autoignition reaction, a knock kinetics (for example the Shell model of Halstead et al. (1977), see chap. (4.2.3)) has to be solved in the CFD code. One must take care that this kinetics is only applied to the unburned phase. Every small increase in temperature because of an approaching main combustion (especially when we are dealing with a "numerical precurser") leads inevitably to knocking rates that are much too high. This is a serious problem especially in the case of thick, diffusive flames.

The consideration of the influence of turbulence is another, as yet unsolved problem. As a first approximation, fluctuations in temperature or in the mixture composition can be modeled by use of variance transport equations (9.55).

- **Evaluation and outlook**

Reliable combustion models in engine CFD codes should by all means be made available soon. The leading candidate for this seems to be the G-equation, in association with a two-phase approach for the improved separation of the burned and unburned phases.

Transported pdf methods are less suitable for the description of sharp flame fronts. But for the treatment of turbulence influences on knocking they could be quite suitable, so that in the medium or long term future, one of its first application areas could be developed.

9.3.4 The SI engine with stratified charge (partially premixed flames)

In the case of a stratified charge combustion, at the moment of flame front propagation, rich ($\lambda < 1$) and lean ($\lambda > 1$) mixtures are found in the combustion chamber at the same time, behind the flame front furthermore rich zones exists with reductant agents (probably CO in essence) and lean zones with oxygen. Thus a diffusion flame remains at the ($\lambda = 1$) isoline. This coupled structure of premixed and diffusion flame is also called a triple flame (see also Fig. 9.32).

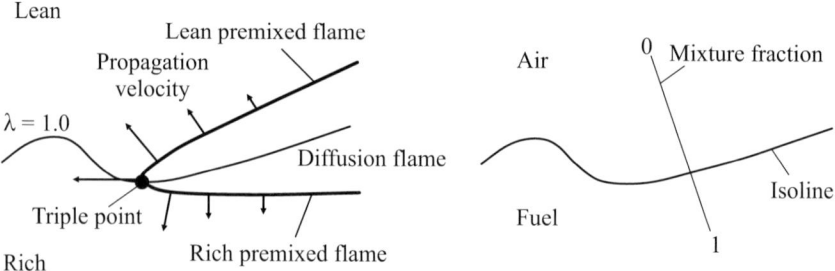

Fig. 9.32: Structure of a triple flame

In calculating stratified charge combustion, one often works with very insufficient means. Either pure diesel combustion models are applied, or the modeling comes from a premixed SI engine – diffusion flame effects then remain unconsidered. This is actually unnecessary; it is not hard to develop a suitable method for stratified charge simulation from already existing modeling elements of turbulent premixed and diffusion flames. Astonishingly, stratified charge combustion models behave more "good-natured" in many cases than pure premixed combustion models, since no flame fronts with such sharp temperature and density gradients appear.

As in the diesel model, one should first set the base species to be used for the description of the local gas conditions, the seven species from chap. 9.2.3 (fuel, N_2, O_2, H_2O, CO_2, CO, H_2) are practical here as well (as a first step, one can also work with three species, air, fuel, and product). If we use a G-equation to describe the premixed combustion, and a simple flamelet approach to describe the diffusion combustion, i.e. we transport the mixture fraction (also calculable from the species) and the mixture fraction variance, then the instantaneous species mass fractions can be determined as follows

9.3 Simulation of combustion

$$c_{(i)} = c_{(i),um} + c_{progr}\left(c_{(i),m} - c_{(i),um}\right) . \tag{9.247}$$

whereby

$$c_{(i),m} = \int_0^1 c_{(i)}(Z) p_\beta\left(Z;\langle Z\rangle, Z''^2\right) dZ \tag{9.248}$$

designates the equilibrium value for the current mixture state and

$$c_{(i),um} = \int_0^1 c_{(i)}(Z) p_\beta\left(Z;\langle Z\rangle, Z''^2 = \langle Z\rangle(1-\langle Z\rangle)\right) dZ \tag{9.249}$$

describes the corresponding unmixed state with maximum variance. A reaction is described by

$$dc_{(i)} = \underbrace{(c_{(i),m} - c_{(i),um}) \, dc_{progr}}_{\text{premixed combustion}} + \underbrace{c_{progr} \, dc_{i,m}}_{\text{diffusion comustion}} , \tag{9.250}$$

i.e. it splits up naturally into premixed and diffusion combustion. Correspondingly, the source term in the transport equation of the internal energy reads

$$q dt = \sum_k h_{(k)} \rho \, dc_{(k)} , \tag{9.251}$$

with the specific species formation enthalpies $h_{(i)}$. As alterative variants to (9.250), one can calculate the instantaneous species content from Z, Z''^2, and c and then compare it with the one that is only transported (with convection and diffusion). The difference of both quantities provides the reaction in the current time interval. Finally, the merely transported species are overwritten with the new total target values. The turbulent flame speed of the premixed flame can be obtained by flamelet-based averaging in the mixture fraction space

$$\langle s_t \rangle = \int_0^1 s_t(s_l(Z), Z) p_\beta\left(Z;\langle Z\rangle, Z''^2\right) dZ . \tag{9.252}$$

On formulations for $s_l(Z)$, see e.g. equation (4.1). Pollutant formation models can be adopted from the modeling of diesel combustion. Depicted in Fig. 9.33 is an example for such a 3D-simulation of a stratified charge combustion.

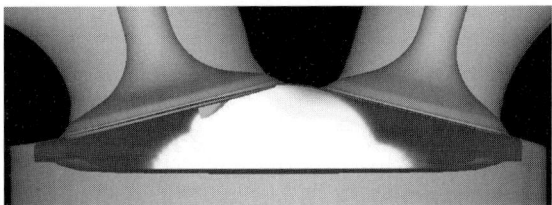

Fig. 9.33: Simulation of stratified charge combustion in a DISI engine

Literature

Abdel Hamid, S., Hagelstein, D., Rautenberg, H., und Seume, J. (2002): TC-Gesamtkennfeldbestimmung. FVV Forschungsvorhaben

Abramovich, G. N. (1963): The Theory of Turbulent Jets. MIT Press

Amsden, A. A., O'Rourke, P. O., Buttler, and T. D. (1989): KIVA-II: A Computer Program for Chemically Reactive Flows. Los Alamos National Laboratory Report, LA-11560-MS

Baehr, H. D. (2000): Thermodynamik, 10. überarb. Aufl., Springer-Verlag, Berlin, Heidelberg, New York

Baehr, H. D., und Stephan, K. (2004): Wärme- und Stoffübertragung. 4. Aufl., Springer-Verlag, Berlin, Heidelberg, New York

Baulch, D. L., Drysdale, D. D., Horne, D. G., und Lloyd, A. C. (1969): High Temperature Reaction Rate Data. Report No.4, University of Leeds

Baulch, D. L., Cobos, C. J., Cox, A. M., Frank, P., Haymann, G., Just, T., Kerr, J.A., Murrels, T., Pilling, M. J., Twe, J., Walker, R. W., and Warnatz, J. (1991): Compilation of rate data for combustion modelling. Supplement I. J. Phys. Chem. Ref. Data 22, 847

Barba, C., Burckhardt, C., Boulouchos, K., und Bargende, M. (1999): Empirisches Modell zur Vorausberechnung des Brennverlaufes bei Common-Rail-Dieselmotoren. MTZ 4, 262-270

Bargende, M. (1990): Ein Gleichungsansatz zur Berechnung der instationären Wandwärmeverluste im Hochdruckteil von Ottomotoren. Dissertation, TH Darmstadt

Bargende, M., Weining, H.-K., Lautenschütz, P., und Altenschmidt, F. (1997): Thermodynamik der neuen Mercedes-Benz 3 Ventil-Doppelzünder V-Motoren. In: Essers, U. (Hrsg.), Kraftfahrwesen und Verbrennungsmotoren, 2. Stuttgarter Symposium, Expert-Verlag

Baumgarten, C. (2003): Modellierung des Kavitationseinflusses auf den primären Strahlzerfall bei der Hochdruck-Dieseleinspritzung. Dissertation, Universität Hannover

Beitz, W., und Grote, K.-H. (1997): Dubbel Taschenbuch für den Maschinenbau. 19. Auflage SpringerVerlag

Belardini, P., Bertori, C., Cameretti, M. C., and Del Giacomo, N. (1994): A Coupled Diesel Combustion and Soot Formation Model for KIVA II Code: Characteristics and Experimental Validation. Int. Symp. COMODIA 94, 315-323

Bilger, R. W. (1993): Conditional Moment Closure for Turbulent Reacting Flow. Phys Fluids A 5 (2), 436-444

Blizard, N. S., and Keck, J. C. (1976): Experimental and Theoretical Investigation of a Turbulent Burning Model for Internal Combustion Engines, SAE-Paper 740191

Bockhorn, H. (1994): A Short Introduction to the Problem - Structure of the Following Parts. In: Bockhorn (Ed.) Soot Formation in Combustion, Springer-Verlag Berlin Heidelberg New York

Bockhorn, H. (1997): Private Mitteilung

Bockhorn, H., Hentschel, J., Peters, N., Weber, J., und Pittermann, R. (2003): Simulation der Partikelemission aus Dieselmotoren. 5. Dresdner Motorenkolloquium, 152-160

Borrmeister, J., und Hübner, W. (1997): Einfluß der Brennraumform auf die HC-Emission und den Verbrennungsablauf. Motortechnische Zeitschrift MTZ 58, 408-414

Bronstein, I. N., und Semendjagew, K. A. (1997): Taschenbuch der Mathematik, 3. überarb. und erweiterte Auflage, Verlag Harri Deutsch, Thun und Frankfurt am Main

Bühler, U. (1995): Prüfstandsuntersuchungen zur Dioxin-Emission von Verbrennungsmotoren. Dissertation, Universität Stuttgart

Bühler, U., Essers, U., und Greiner, R. (1997): Dioxin-Emission des Straßenverkehrs. Motortechnische Zeitschrift MTZ 58, 422-425

Cebeci, T. (2002): Convective Heat Transfer. Second Revised Ed. Springer-Verlag, Berlin, Heidelber, New York

Chaimowitsch, J. M. (1965): Ölhydraulik. VEB Verlag Technik, 6. Aufl., Berlin

Chmela, F., Orthaber, G., und Schuster, W. (1998): Die Vorausberechnung des Brennverlaufs von direkteinspritzenden Dieselmotoren auf der Basis des Einspritzverlaufs. MTZ 59, Heft 7/8

Constien, M. (1991): Bestimmung von Einspritz- und Brennverlauf eines direkteinspritzenden Dieselmotors. Dissertation, TU München

Csallner, P. (1981): Eine Methode zur Vorausberechnung der Änderung des Brennverlaufes von Ottomotoren bei geänderten Betriebsbedingungen. Dissertation, TU München

Damköhler, G. (1940): Der Einfluss der Turbulenz auf die Flammgeschwindigkeit in Gasgemischen. Z. Elektrochem., 46, 601-626

Dederichs, A. S., Balthasar, M., and Mauß, F (1999): Modeling of NOx and Soot Formation in Diesel Combustion, Oil & Science and Technology 54, 246-249

de Neef, A. T. (1987): Untersuchung der Voreinspritzung am schnellaufenden direkteinspritzenden Dieselmotor. Dissertation, ETH Zürich

Dent, J. C. (1971): Basis for the Comparison of Various Experimental Methods for Studying Spray Penetration. SAE Paper 710571

Doll, M. (1989): Beitrag zur Berechnung des stationären und transienten Betriebsverhaltens kleiner, schnellaufender Dieselmotoren mit unterteilten Brennräumen. Dissertation, TU München

Dorsch, H. (1982): Ladeluftkühlung bei aufgeladenen Personenwagen-Ottomotoren. MTZ 43, 201-205

Eilts, P. (1993): Modell zur Vorausberechnung des Brenngesetzes mittelschnellaufender Dieselmotoren. MTZ 54

Ferziger, J. H., and Perić, M. (1996): Computational Methods for Fluid Dynamics. Springer-Verlag, Berlin, Heidelberg, New York

Fenimore, C. P. (1979): Studies of fuel-nitrogen in rich-flame gases. 17th symp. Comb., 661, The Combustion Institute, Pittsburgh

Fieweger, K., und Ciezki, H. (1991): Untersuchung der Selbstzündungs- und Rußbildungsvorgänge von Kraftstoff/Luft-Gemischen im Hochdruckstoßwellenrohr. SFB 224 – Forschungsbericht

Findeisen, D., und Findeisen, F. (1994): Ölhydraulik. Springer-Verlag, 4. Aufl., Berlin, Heidelberg

Franzke, D. E. (1981): Beitrag zur Ermittlung eines Klopfkriteriums der ottomotorischen Verbrennung und zur Vorausberechnung der Klopfgrenze. Dissertation, Technische Universität München

Frenklach, M., and Wang, H. (1994): Detailed Mechanism and Modeling of Soot Particle Formation. In: Bockhorn, H. (ed) Soot Formation in Combustion. 165-192, Springer-Verlag, Berlin

Fröhlich, K., Borgmann, K., und Liebl, J. (2003): Potenziale zukünftiger Verbrauchstechnologien. 24. Internationales Wiener Motorensymposium, Fortschritts-Berichte VDI, Reihe 12, 539, 220-235

Fuchs, H., Pitcher, G., Tatschl, R., und Winklhofer, E. (1996): Verbrennungsmodell - Simulationsmodell ottomotorische Verbrennung, FVV, Heft 614, Vorhaben Nr. 516

Fusco, A., Knox-Kelecy, A. L., and Foster, D. E. (1994): Application of a Phenomenological Soot Model to Diesel Engine Combustion. Int. Symp. COMODIA 94, 571-576

Gerstle, M. (1999): Simulation des instationären Betriebsverhaltens hochaufgeladener Vier- und Zweitakt-Dieselmotoren. Dissertation, Universität Hannover

Görg, K. A. (1982): Berechnung instationärer Strömungsvorgänge in Rohrleitungen an Verbrennungsmotoren unter besonderer Berücksichtigung der Mehrfachverzweigung. Dissertation, Ruhr Universität Bochum

GRI-MECH 3.0 (2000): www.me.berkeley.edu/gri_mech

Große, L. (1962): Arbeitsmappe für Mineralölingenieure. VDI Verlag, 2. Aufl., Düsseldorf

Hahne, E. W. P. (2000): Technische Thermodynamik, 3., überarb. Auflage, Oldenbourg Wissenschaftsverlag GmbH; München, Wien

Halstead, M. P., Kirsch, L. J., Prothero, A., and Quinn, C. P. (1975): A Mathematical Model for Hydrocarbon Autoignition at High Pressures. Proceedings of the Royal Society, A346, 515-538, London

Halstead, M. P., Kirsch, L. J., and Quinn, C. P. (1977): The Autoignition of Hydrocarbon Fuels at High Temperatures and Pressures – Fitting of a Mathematical Model. Combustion and Flame 30, 45-60

Han, Z., and Reitz, R. D. (1995): A Temperature Wall Function Formulation for Variable-Density Turbulent Flows with Application to Engine Convective Heat Transfer Modeling. Int. J. Heat Mass Transfer Vol.40, 613-625

Hausen, H. (1976): Wärmeübergang im Gegenstrom, Gleichstrom und Kreuzstrom. Springer Verlag, 2. Aufl., Berlin, Heidelberg, New York

Heider, G. (1996): Rechenmodell zur Vorausberechnung der NO-Emission von Dieselmotoren. Dissertation, TU München

Heywood, J. B. (1988): Internal Combustion Engine Fundamentals. McGraw-Hill Book Company, New York

Hiroyasu, H., Kadota, T., and Arai, M. (1983a): Development and Use of a Spray Combustion Modeling to Predict Diesel Engine Efficiency and Pollutant Emission. Part 1: Combustion Modeling. Bulletin of the JSME, Vol. 26, 569-575

Hiroyasu, H., Kadota, T., and Arai, M. (1983b): Development and Use of a Spray Combustion Modeling to Predict Diesel Engine Efficiency and Pollutant Emission. Part 2: Computational Procedure and Parametric Study. Bulletin of the JSME, Vol. 26, 576-583

Hohenberg, G. (1980): Experimentelle Erfassung der Wandwärme von Kolbenmotoren. Habilitation, TU Graz

Hohlbaum, B. (1992): Beitrag zur rechnerischen Untersuchung der Stickstoffoxid-Bildung schnellaufender Hochleistungsmotoren, Dissertation, Universität Fridericiana, Karlsruhe

Huber, K. (1990): Der Wärmeübergang schnellaufender, direkteinspritzender Dieselmotoren. Dissertation, TU München

Jenni, E. (1993): Der BBC-Turbolader, Geschichte eines Schweizer Erfolgs. ABB Turbo Systems AG, Baden

Jischa, M. (1993): Herausforderung Zukunft: Technischer Fortschritt und ökologische Perspektiven, Spektrum Akad. Verlag, Heidelberg

Justi, E. (1938): Spezifische Wärme, Enthalpie, Entropie und Dissoziation technischer Gase. Springer Verlag, Berlin

Kaufmann, W. J., und Smarr, L. L. (1994): Simulierte Welten, Spektrum-Verlag, Heidelberg

Klaiß, Th. (2003): Selbstzündung und Wärmeübergang an der Klopfgrenze von Ottomotoren. Dissertation, Universität Hannover

Kleinschmidt, W. (1993): Der Wärmeübergang in aufgeladenen Dieselmotoren aus neuerer Sicht. 5. Aufladetechnische Konferenz, Augsburg

Kleinschmidt, W. (2000): Zur Simulation des Betriebes von Ottomotoren an der Klopfgrenze. Fortschritt-Berichte VDI, Reihe 12, Nr. 422. VDI-Verlag, Düsseldorf

Klimenko, A. Y., and Bilger, R. W. (1999): Conditional Moment Closure for Turbulent Combustion. Prog. Energy Comb. Sci. 25, 595-687

König, G., Blessing, M., Krüger, C., Michels, U., and Schwarz, V. (2002): Analysis of Flow and Cavitation Phenomena in Diesel Injection Nozzles and its Effects on Spray and Mixture Formation, 5th Internationales Symposium für Verbrennungsdiagnostik der AVL Deutschland, Baden-Baden

Kolesa, K. (1987): Einfluß hoher Wandtemperaturen auf das Betriebsverhalten und insbesondere auf den Wärmeübergang direkteinspritzender Dieselmotoren. Dissertation, TU München

Kolmogorov, A. N., Petrovsky, I. G., and Piskunov, N. S. (1937): Study of the diffusion equation with growth of the quantity of matter and its application to a biology problem. Bull. Univ. Moscou, Ser. Int., Sec. A 1, 1-25, translated in: P. Pelcé, Dynamics of Curved Fronts, Perspectives in Physics, Academic Press, New York, 1988

Krüger, Ch. (2001): Validierung eines 1D-Spraymodells zur Simulation der Gemischbildung in direkteinspritzenden Dieselmotoren. Dissertation, RWTH Aachen

Küntscher, V. (1987): Kraftfahrzeugmotoren, Auslegung und Konstruktion. VEB Verlag Technik, 1. Aufl., Berlin

Kuchling, H. (1988): Taschenbuch der Physik. Verlag H. Deutsch, 11. Aufl., Frankfurt a. Main

Kuder, J., und Kruse, Th. (2000): Parameteroptimierung an Ottomotoren mit Direkteinspritzung. MTZ 61, 378-384

Lange, J. (1996): Bestimmung der Carbonylverbindungen im Abgas von schwerölbetriebenen Dieselmotoren. Fortschritt-Berichte VDI, Reihe 15, Nr. 161, VDI Verlag, Düsseldorf

Li, H., Miller, D. L., and Cernansky, N. P. (1992): A Study on the Application of a Reduced Chemical Reaction Model to Motored Engines for Heat Release Prediction. SAE Technical Paper 922328, San Francisco

Li, H., Prabhu, S. K., Miller, D. L., and Cernansky, N. P. (1994): Autoignition Chemistry Studies on Primary Reference Fuels in a Motored Engine. SAE Technical Paper 942062, Baltimore

Li, H., Miller, D. L., and Cernansky, N. P. (1996): Development of a Reduced Chemical Kinetic Model for Prediction of Preignition Reactivity and Autoignition of Primary Reference Fuels. SAE Technical Paper 960498, Detroit

Lippert, A. M., Stanton, D. W., Reitz, R. D., Rutland, Ch. J., and Hallet W. L. H. (2000): Investigating the Effect of Spray Targeting and Impingement on Diesel Engine Cold Start, SAE Paper 2000-01-0269

Lucas, K. (2001): Thermodynamik. Die Grundgesetze der Energie- und Stoffumwandlungen. 3., korr. u. aktualisierte Aufl., Springer-Verlag, Berlin, Heidelberg

Maas, H. (1979): Gestaltung und Hauptabmessungen der Verbrennungskraftmaschine. Die Verbrennungskraftmaschine, Neue Folge, Band 1, Springer Verlag, Heidelberg

Maas, H., und Klier, H. (1981): Kräfte, Momente und deren Ausgleich in der Verbrennungskraftmaschine. Die Verbrennungskraftmaschine, Band 2, Springer Verlag, Heidelberg

Merker, G. P., und Baumgarten, C. (2000): Wärme- und Fluidtransport – Strömungslehre. B. G. Teubner-Verlag, Stuttgart

Merker, G. P., Hohlbaum, B., and Rauscher, M. (1993): Two-Zone Model for Calculation of Nitrogen-oxide Formation in Direct-Injection Diesel Engines, SAE-Paper 932454

Merker, G. P., und Gerstle, M. (1997): Evaluation on Two Stroke Engines Scavenging Models, SAE-Paper 970358

Merker, G. P., und Eiglmeier, C. (1999): Wärme- und Fluidtransport – Wärmeübertragung. Teubner-Verlag, Stuttgart

Merker, G. P., und Kessen, U. (1999): Technische Verbrennung: Verbrennungsmotoren. B. G. Teubner-Verlag, Stuttgart, Leipzig

Merker, G. P., und Schwarz, C. (2001): Technische Verbrennung – Simulation verbrennungsmotorischer Prozesse. Teubner-Verlag, Stuttgart

Merker, G. P., und Stiesch, G. (1999): Technische Verbrennung – Motorische Verbrennung. B. G. Teubner-Verlag, Stuttgart, Leipzig

Metghalchi, M., and Keck, J. C. (1980): Laminar Burning Velocity of Propane-Air Mixtures at High Temperature and Pressure. Combustion and Flame, 38, 143-154

Metghalchi, M., and Keck, J. C. (1982): Burning Velocities of Mixtures of Air with Methanol, Isooctane, and Indolene at High Pressure and Temperature. Combustion & Flame, 48, 191-210

Meyer, S., Krause, A., Krome, D., und Merker, G. P. (2002): Piezo-Common-Rail-System mit direkt gesteuerter Düsennadel. MTZ 63, 86-93

Miersch, J. (2003): Transiente Simulation zur Bewertung von ottomotorischen Konzepten. Dissertation, Universität Hannover

Moran, M. J., and Shapiro, H. N. (1992): Fundamentals of Engineering Thermodynamics. 2nd Ed., Wiley, New York

Münzberg, H. G., und Kurzke, J. (1977): Gasturbinen - Betriebsverhalten und Optimierung. Berlin, Heidelberg, New York

Nagle, J., and Strickland-Constable, R. F. (1962): Oxidation of Carbon between 1000-2000°C. Proc 5th Conf on Carbon, 1, 154-164, Pergamon Press, London, UK

Niemann, G. (1981): Maschinenelemente. Springer Verlag, 2. Aufl., Berlin, Heidelberg, New York

Nishida, K., and Hiroyasu, H. (1989): Simplified Three-Dimensional Modeling of Mixture Formation and Combustion in a D.I. Diesel Engine. SAE Technical Paper 890269

NIST JANAF (1993): Thermochemical Tables Database, Version 1.0

Oberg, H. J. (1976): Die Darstellung des Brennverlaufes eines schnelllaufenden Dieselmotors durch zwei überlagerte Vibe-Funktionen. Dissertation, Braunschweig

Östreicher, W. F. (1995): Neue Regelungsstrategien für Antriebsanlagen mit hochaufgeladenen, schnelllaufenden Viertakt-Dieselmotoren. VDI-Fortschrittsberichte, Reihe 12, Nr. 252, VDI-Verlag, Düsseldorf

O'Rourke, P. J. (1989): Statistical Properties and Numerical Implementation of a Model for Droplet Dispersion in a Turbulent Gas. Journal of Computational Physics, 83(2):345 - 360

Otto, F., Krüger, Ch., Wirbeleit, F., und Willand, J. (1999): Probleme und Lösungsansätze bei der Simulation der dieselmotorischen Einspritzung. Mess- und Versuchstechnik für die Entwicklung von Verbrennungsmotoren, Haus der Technik, Essen

Pagel, S. (2003): Verdampfungs- und Selbstzündungsmodelle für Mehrkomponentengemische. Dissertation, Universität Hannover

Patankar, S. V.(1980) : Numerical Heat Transfer and Fluid Flow, Hemisphere Publishing Corp., Mc-Graw Hill Comp.

Pattas, K. (1973): Stickoxidbildung bei der ottomotorischen Verbrennung. MTZ 34

Patterson, M. A. (1997): Modeling the Effects of Fuel Injection characteristics on Diesel Combustion and Emission, PH D Thesis, University of Wisconsin-Madison

Patterson, M. A., and Reitz, R. D. (1998): Modelling the Effects of Fuel Spray Characteristics on Diesel Engine Combustion and Emissions, SAE-Paper 980131

Peters, N. (2000): Turbulent Combustion. Cambridge University Press

Pflaum, W., und Mollenhauer, K. (1977): Wärmeübergang in der Verbrennungskraftmaschine. Springer Verlag, Berlin, Heidelberg, New York

Pilch, M., and Erdman, C. A. (1987): Use of Breakup Time Data and Velocity History Data to Predict the Maximum Size of Stable Fragments for Acceleration-Induced Breakup of a Liquid Drop. International Journal of Multiphase Flow, 13, 741-757

Pischinger, R., Kraßnig, G., Taućar, G., and Sams, Th. (1989): Thermodynamik der Verbrennungskraftmaschine. Neue Folge, Band 5, Springer-Verlag, Wien

Pischinger, R., Klell, und Sams, Th. (2002): Thermodynamik der Verbrennungskraftmaschine. 2te, überarbeitete Auflage, Springer-Verlag, Wien, New York

Poinsot, Th., and Veynante, D. (2001): Theoretical and Numerical Combustion. R.T. Edwards, Inc.

Pope, S. B. (1978): An explanation of the turbulent round-jet/plane-jet anomaly. American Institute of Aeronautics and Astronautics Journal, 16, 279-281

Pope, S. B. (1985): PDF methods for turbulent reactive flows. Prog. in Energy Comb. Sci. 19, 119-197

Ramos, J. I. (1989): Internal Combustion Engine Modelling, Hemisphere Publishing Corporation, New York

Ra, Y., and Reitz, R. D. (2002): A Model for Droplet Vaporization for Use in Gasoline and HCCI Engine Applications, ILASS Americas, 15 th Annual Conference on Liquid Atomization and Spray Systems, Madison, WI

Rao, S., Rutland, and C. J. (2002): "A Flamelet Timescale Combustion Model for Turbulent Combustion in KIVA", 12th Int. Multidim. Engine Modeling User`s Group Meeting at the SAE Congress

Reitz, R. D. (1987): Modeling Atomization Processes in High-Pressure Vaporizing Sprays. Atomization and Spray Technology, 3, 309-337

Reulein, C. (1998): Simulation des instationären Warmlaufverhaltens von Verbrennungsmotoren. Dissertation, TU München

Reulein, C., Schwarz, C., und Witt, A. (2000): Methodeneinsatz bei der Ermittlung des Potentials von Downsizing-Motoren. Tagung "Downsizing von Motoren", Haus der Technik, München

Scheele, M. (1999): Potentialabschätzung zur Verbesserung des indizierten Wirkungsgrades kleinvolumiger Ottomotoren, Diss., Universität Hannover

Schlösser, W. E. (1961): Ein mathematisches Modell für Verdrängerpumpen und -motore. Ölhydraulik und Pneumatik 5, 122 ff., Krausskopf, Mainz

Schmidt, Ch., Hohenberg, G., und Bargende, M. (1996): Arbeitsspiel bezogenes Luftverhältnis. MTZ 57, 572-578

Schreiber, M., Sasat Sakak, A., Lingens, A., and Griffiths, J. F. (1994): A Reduced Thermokinetic Model for the Autoignition of Fuels with Variable Octane Ratings. 25th Symposium (Int.) on Combustion, 933-940, Irvine California

Schreiner, K. (1993): Untersuchungen zum Ersatzbrennverlauf und Wärmeübergang bei Hochleistungsdieselmotoren. MTZ 54, 554-563

Schwarz, C., Woschni, G., und Zeilinger, K. (1992): Anfahrverhalten. FVV-Forschungsberichte, Heft 454, Forschungsvereinigung Verbrennungskraftmaschinen e. V., Frankfurt/Main

Schwarz, C. (1993): Simulation des transienten Betriebsverhaltens von aufgeladenen Dieselmotoren. Dissertation, TU München

Schwarzmeier, M. (1992): Der Einfluß des Arbeitsprozeßverlaufs auf den Reibmitteldruck von Dieselmotoren. Dissertation, TU München

Seifert, H. (1962): Instationäre Strömungsvorgänge in Rohrleitungen an Verbrennungskraftmaschinen. Springer Verlag, Berlin-Göttingen-Heidelberg

Semenov, N. N. (1935): Chemical Kinetics and Chain Reactions. Oxford University Press, London, UK

Siebers, D. L. (1998): Liquid-Phase Fuel Penetration in Diesel Sprays. SAE Paper 980809

Sitkei, G. (1963): Über den dieselmotorischen Zündverzug. MTZ 26, 190-194

Spalding, D. B. (1961): A single formula for the "law of the wall". ASME J. Appl. Mech., 28, 455-457

Spicher, U., und Worret, R. (2002): Entwicklung eines Kriteriums zur Vorausberechnung der Klopfgrenze. FVV Forschungsvorhaben Heft 741, Frankfurt/Main

Stegemann, J., Seebode, J., Baumgarten, C., and Merker, G. P. (2002): Influence of Throttle Effects at the Needle Seat on the Spray Characteristics of a Multihole Injection Nozzle. Proc. 18th ILASS-Europe Conf, 31-36, Zaragoza, Spain

Stegemann J. (2003): Dieselmotorische Einspritzverlaufsformung mit piezoaktuierten Einspritzsystemen. Dissertation, Universität Hannover

Stegemann, J., Meyer, S., Rölle, T., und Merker, G. P. (2004): Einspritzsystem für eine vollvariable Verlaufsformung. MTZ 65, 114-121

Steiner, R., Bauer, C., Krüger, Ch., Otto, F., and Maas, U. (2004): 3D-Simulation of DI-Diesel Combustion applying a Progress Variable Approach Accounting for Complex Chemistry , to be published at SAE 03/04

Stephan, K., und Mayinger, F. (1999): Thermodynamik, 14. Aufl., Bd. 2, Mehrstoffsysteme und chemische Reaktionen, Springer-Verlag, Berlin, Heidelberg, New York

Stephan, K., und Mayinger, F. (1998): Thermodynamik, 15. Aufl., Bd. 1: Einstoffsysteme. Grundlagen und technische Anwendungen. Springer-Verlag, Berlin, Heidelberg, New York

Stiesch, G. (1999): Phänomenologisches Multizonen-Modell der Verbrennung und Schadstoffbildung im Dieselmotor, Dissertation, Universität Hannover

Stiesch, G. (2003): Modeling Engine Spray and Combustion Processes. Springer-Verlag, Berlin, Heidelberg

Streit, E. E., and Borman, G. L. (1971): Mathematical simulation of large turbocharged two-stroke diesel engines. SAE-Paper 710176, International Congress & Exposition

Stromberg, H.-J. (1977): Ein Programmsystem zur Berechnung von Verbrennungsmotorkreisprozessen mit Berücksichtigung der instationären Strömungsvorgänge in den realen Rohrleitungssystemen von Mehrzylinder-Verbrennungsmotoren. Dissertation, Ruhr-Universität Bochum

Tabaczynski, R. J. (1980): Further Refinement and Validation of a Turbulent Flame Propagation Model for Spark Ignition Engines, Combustion and Flame 39, 111-121

Tatschl, R., v. Künsberg Sarre, C., Alajbegovic, A., and Winklhofer, E. (2000): Diesel Spray Break-up Modelling including Multidimensional Cavitation Nozzle Flow Effects. 16th ILASS-Europe 2000, Sep. 11-13, Valencia, Spain

Teigeler, M., Schmitt, F., Enderle, Ch., Wirbeleit, F. und Bockhorn, H. (1997): Mechanismen der NO_X-Bildung und –Reduktion. Ansätze zur innermotorischen NO_X-Absenkung unter dieselmotorischen Bedingungen. 2. Dresdener Motorenkolloquium, HTW-Dresden (FH), 15/16 Mai 1997

Thoma, M., Stiesch, G., und Merker, G. P. (2002): Phänomenologisches Gemischbildungs- und Verbrennungsmodell zur Berechnung von Dieselmotoren mit Voreinspritzung. 5. Int Symp für Verbrennungsdiagnostik, 91–101, Baden-Baden

Truckenbrodt, E. (1980): Fluidmechanik. 2. Aufl., Band 1, Springer Verlag, Berlin, Heidelberg, New York

Urlaub, A. (1994): Verbrennungsmotoren, 2. Auflage, Springer-Verlag, Berlin, Heidelberg, ISBN

van Basshuysen, R., und Schäfer, F., Hrsg., (2003): Handbuch Verbrennungsmotor – Grundlagen, Komponenten Systeme, Perspektiven, 2te Auflage, Verlag Friedrich Vieweg & Sohn, Braunschweig, Wiesbaden

Verein Deutscher Ingenieure, Hrsg., (1991): VDI-Wärmeatlas. 6. Aufl., VDI-Verlag, Düsseldorf

Vibe, I. I. (1970): Brennverlauf und Kreisprozeß von Verbrennungsmotoren. VEB Verlag Technik, Berlin

Vogel, C. (1995): Einfluß von Wandablagerungen auf den Wärmeübergang im Verbrennungsmotor. Dissertation, TU München

Vogt, R. (1975): Beitrag zur rechnerischen Erfassung der Stickoxidbildung im Dieselmotor. Dissertation, Universität Stuttgart

Warnatz, J., Maas, U., und Dibble, R. W. (1997): Verbrennung, 2. Auflage, Springer-Verlag, Berlin, Heidelberg

Warnatz, J., Maas, U., und Dibble, R. W. (2001): Verbrennung: Physikalisch-Chemische Grundlagen, Modellierung und Simulation, Experimente, Schadstoffentstehung. 3. Aufl., Springer-Verlag, Berlin, Heidelberg

Weisser, G., und Boulouchos, K. (1995): NOEMI – Ein Werkzeug zur Vorabschätzung der Stickoxidemissionen direkteinspritzender Dieselmotoren. 5. Tagung „Der Arbeitsprozeß des Verbrennungsmotors, TU Graz, 23-50

Weller, H. G. (1993): The Development of a New Flame Area Combustion Model Using Conditional Averaging. Thermo-Fluids Section Report TF/9307, Imperial College of Science, Technology and Medicine, London

Westbrook, C. K., and Dryer, F. L. (1981): Simplified Reaction Mechanism for the Oxidation of Hydrocarbon Fuels in Flames. Combust Sci Tech, 27, 31-43

White, F. M. (1991): Viscous Fluid Flow. Second Edit., McGraw-Hill, Inc. New York

Winkelhofer, E., Wiesler, B., Bachler, G., und Fuchs, H. (1991): Detailanalyse der Gemischbildung und Verbrennung von Dieselstrahlen. 3. Tagung "Der Arbeitsprozeß des Verbrennungsmotors", Graz

Witt, A. (1999): Analyse der thermodynamischen Verluste eines Ottomotors unter den Randbedingungen variabler Steuerzeiten. Dissertation, TU Graz

Witt, A., Siersch, W., und Schwarz, C. (1999): Weiterentwicklung der Druckverlaufsanalyse für moderne Ottomotoren. 7. Tagung "Der Arbeitsprozeß des Verbrennungsmotors", Graz, 53-67

Wolfer, H. H. (1938): Der Zündverzug beim Dieselmotor. VDI Forschungsheft 392

Wolters, P., Salber, W., Krüger, M., Körfer, T., und Dilthey, J. (2003): Variable Ventilsteuerung – Schlüsseltechnologie für Homogene Selbstzündung. 5. Dresdner Motorenkolloquium, S 118-129, Dresden, 2003

Woschni, G. (1970): Die Berechnung der Wandwärmeverluste und der thermischen Belastung der Bauteile von Dieselmotoren, MTZ 31, 491-499

Woschni, G., und Anisits, F. (1973): Eine Methode zur Vorausberechnung der Änderung des Brennverlaufs mittelschnelllaufender Dieselmotoren bei geänderten Betriebsbedingungen. MTZ 34, 106 ff, Franckh-Kosmos Verlags-GmbH, Stuttgart

Zacharias, F. (1966): Analytische Darstellung der thermischen Eigenschaften von Verbrennungsgasen. Dissertation, TU Berlin

Zapf, H. (1969): Beitrag zur Untersuchung des Wärmeübergangs während des Ladungswechsels im Viertakt-Dieselmotor. MTZ 30, 461ff, Franck-Kosmos Verlags-GmbH, Stuttgart

Zeldovich, Y. B. (1946): The oxidation of nitrogen in combustion and explosions. Acta Physiochimica 21, 577-628, USSR

Zellbeck, H. (1997): Neue Methoden zur Vorausberechnung und Onlineoptimierung des Betriebsverhaltens aufgeladener Dieselmotoren. 133, in: Aufladetechnische Konferenz, Dresden

Zinner, K. (1985): Aufladung von Verbrennungsmotoren. Springer-Verlag, Berlin

Index

Numerics
1st fundamental equation of turbocharging 40
1st law of thermodynamics 12, 90, 143
2nd fundamental equation of turbocharging 42
2nd law of thermodynamics 14, 331
2-stroke engine 195ff
2-stroke process 11, 195ff
3-D Euler method 354
4-stroke process 11
50% mass fraction burned (50 mfb) 91

A
abnormal combustion 69ff
acetylene hypothesis 128
activation energy 47
actual performance 28
additional compressor, electrically driven 289
adiabatic 20
    relation 15
air
    entrainment 101
    expenditure 30
    -guided methods 62
    ratio 26
alcohols 52f
aldehydes 52f, 58, 121ff
alkanes 51f, 124
    -, cyklo- 51f
alkenes 51, 56ff, 124
angular-momentum conserv. law 236f, 326f
anti-knock index 55
aromatic compounds 51f, 124
Arrhenius equation 47, 82
assembly of the reciprocating engine 7
atomic balances 46f
atomization 105
    law 343
autoignition 31, 33, 69, 81ff, 174, 357, 361,
        365, 379
    -, controlled 70ff
    process 82
averaging
    -, ensemble- 305, 374f
    -, Favre- 306f
    -, Reynolds- 305

B
balance
    -, atomic 46f
    -, energy- 12, 14f, 90, 143, 198ff
    -, mass- 12, 88, 143, 188, 196ff
    -, performance 39
benzene ring 51f, 124, 128f
Bernoulli velocity 325f, 355
beta function 311
    -, incomplete 360
boiling temperature 53
Boltzmann equation 331, 348
    -, spray adapted 331ff, 347
boost pressure control 264f
bore/stroke ratio 31
boundary
    -, cyclic 325
    layer 299, 307ff, 320
    -, thermal 119
breakup
    models, aerodynamic 339
    -, primary 336
    process 336
    -, secondary 336, 349
    time 106f
Brownian motion 343
burning
    velocity 66, 371f
Businessq approach 348

C
CAI (controlled auto-ignition) 70f, 95f
calculation time 98
caloric state equation 13, 178, 300
cam-operated injection system 73
carbon monoxide 117f
carbonyl compounds 122f
Carnot cycle 17ff
catalytic converter, storage-reduction 63
cavitation 325ff, 336ff
center, combustion (50mfb) 64, 91, 177, 262,
        275
central difference scheme 314ff
cetane number 55ff, 361
CFD codes 313, 320ff, 325ff, 351ff, 378ff
characteristic
    qualities 28ff

-s 209, 299, 333
values 28ff
charge
    air cooling (CAC) 205, 239ff
    changing 195ff
        losses 94f
    -, stratified 60, 380
charger, positive displacement 224f
charging 35ff, 214ff
    -, 1st fundamental equation of turbo- 40
    -, 2nd fundamental equation of turbo- 40
    -, external 35
    -, internal 35
    -, mechanical super- 35, 37ff
    methods 35ff
        -, comparison of different 43
    -, pressure 36
        -, turbo- 38ff, 43
    -, pulse 37
        -, turbo- 41ff
    -, register 35
    -, resonance 35
    -, swing-pipe 35
    -, turbo- 35
chemical
    equilibrium 26f, 44ff, 92ff
    potential 45
chondiodal curves 32
closed
    cycles 17ff
    systems 12ff
closure approach 312
CMC models 364
CO emission 364
coalescence 72, 80, 128, 340
coefficient
    -, flow 144ff, 212f, 226ff
    -, heat transfer 148f, 151ff, 157, 191, 198, 255
    -, rate 47
    -, reaction 82
    -, stoichiometric 44
    -, substitute conduction 158
coherent flame model 372ff
collision 72, 80, 331ff, 340f
    integral 332
    probability 332
    process, modeling 340
    process, three-particle 332
    term, dual-particle 332
combustion 86ff

-, abnormal 69ff
center (50 mfb) 64, 91, 177, 262, 275
chamber recirculation 71, 263
-, complete 26f
delay 64
-, diesel engine 72, 83ff, 98ff
-, diffusion 84f, 111f, 189, 357f, 378f
-, end of 91
-, flame front 365
function, (substitute) 262
-, homogeneous 70, 86f
-, imperfect 26f, 93f
-, incomplete 26f, 93f
-, knocking 69ff, 88, 174ff
lean 71
losses 94
-, main (diffusion) 85
models 2
-, normal 69
-, perfect 26f
-, post- 85f
-, premixed 84f, 189
process 11, 60ff, 65ff, 70ff, 86, 95ff, 358
    -, low in swirl 224
rate 108f
-, start of 91
-, stratified charge 379f
temperature 116
common rail system 73f, 76ff
comparative process, open 25ff
comparison fuel 107
complete combustion 26f
component model 178f
computational meshes 320f
compression ratio 31
compressor suck line 380
concept of composite methods 37
constant
    -pressure process 21ff, 24f
    -volume process 20f, 23f, 92
    -, von Karman 308
continuity equation 297
contraction coefficient 326
control, boost pressure 264f
controlled autoignition 70ff
convective derivative 298
convergence, statistical 356, 365
cooling
    -, charge air 205, 239ff
    cycle 245ff, 248ff, 256
    -, internal 63

Index

Courant number 317
crankshaft drive 7ff, 144
    -, kinematics of the 8f
critical pressure ratio 17
curve, fish hook 67
cycle
    averaging 374
    -, closed 17ff
    -, ECE 290ff
    fluctuations 305, 312, 374
cyclic boundary 325
cyclical fluctuations 67f
cyklo-alkanes 51f
cylinder-shaped swirl 324
cylinder
    -, flow over of a 303
    volume 9

D

Dalton's law 47
Damköhler relation 66, 370, 377
degree of conversion 27
    -, total 27
derivative
    -, convective 298
    -, substantive 298
diesel
    -, direct injection 73ff
    engine 33ff, 72ff
        combustion 72, 83ff, 98ff
        mixture formation 72
    injection 73ff, 328
difference scheme, central 314ff
diffusion
    combustion 84f, 111f, 189, 357f, 378f
    flame 357, 360, 380
    term 300ff, 305, 307, 313
    -, turbulent 305, 348, 350
dioxin 124ff
direct injection 61ff, 73ff, 89, 95, 186
    diesel engine 73ff
    SI engine 61ff, 89, 95, 186
discrete droplet model (DDM) 333
dispersion 368
    -, turbulent 333, 341ff, 348, 350
displacement 31
dissipation
    rate, scalar 310, 360
    -, turbulent 305f, 310
dissociation 58, 182f
distribution function

a posteriori 364
a priori 364
distributor injection pump (DIP) 74
division of losses 28
double Vibe
    function 162ff
    heat release rate 162ff
drag 327
drop
    size distribution 335
droplet
    classes 351ff
    distribution density 331
    ensemble 331
    evaporation 107f
    kinetics 327
    model, discrete (DDM) 333
    -related Reynolds number 349
    vibrations 339
    states 327
dual phase flow 325

E

eccentric rod relation 9
ECE cycle 290ff
eddy
    breakup model 357
    diffusivity approach 305, 348
    viscosity approach 305
effective pressure
    -, mean (mep) 29
        -, friction (fmep) 29, 257ff
        -, indicated (imep) 28
efficiency 29
    -, mechanical 29
electrically assisted turbocharger 289
electrically driven additional compressor 289
electromechanical valve train 146f, 156, 277f
element mass friction 302
emission
    -, CO- 364
    -, HC- 122, 364
    -, particulate matter 127ff
energy
    -, activation 47
    balance 12, 14f, 90, 143, 198ff
    conversation 5f
        -, general 5
        -, law of 141, 204f
        -, thermal 5
    -, internal 89, 177ff, 183ff, 300

-, specific 300
-, kinetic 301
-, thermal 26
-, total 301
-, turbulent, kinetic 99, 305
engine
   -, 2-stroke 195ff
   bypass 283ff
   -, diesel 33ff, 72ff
      -, direct injection 73ff
   friction 257ff
   -, hydrogen 95, 97
   knocking 60
   map 31ff, 43
   -, perfect 25ff, 92
   -, process, real 25ff
   -, reciprocating 6ff
      -, assembly of 7
   -, SI (spark ignition) 60ff
      -, direct injection 61ff
   suck line 281, 284
enthalpy
   -, evaporation 143
   -, free 44
   -, reaction 46
   -, thermal 301
   -, total 301
entrainment 354
   model 113
equation
   -, Arrhenius- 47, 82
   -, Boltzmann 334, 348
   -, core, for turbocharger 236
   -, elliptic 299
   -, flow 15, 198f, 209f, 213, 215
   -, Fokker Planck 346
   -, Fourier heat conduction 157
   -, gas
      -, ideal 154
      -, real 301
   -, heat conduction, Fourier 157
   -, heat transfer 151ff
   -, Helmholtz 299
   -, hyperbolic 299
   -, impulse 322
   -, Langevin- 341
   -, Liouville- 331ff
   -, Navier-Stokes- 201, 297ff
   -, Newtonian 158
   -, parabolic 299
   -, Poisson 299, 309

   -, potential (see Poisson)
   -, spray 331
equilibrium
   -, chemical 26f, 44ff, 92ff
   -, constant 46
   -, partial 48ff, 137
   -, water-gas 182
equivalent cross section, isentropic 40
ether 54
Euler
   coordinates 298
   equation 298
   -ian multi-phase methods 347
evaporation
   enthalpy 143
   rate 108
exhaust
   gas composition 116ff
   gas recirculation 71f, 166, 261ff, 279ff, 290f
   spread 146, 263, 279
external charging 35
extrapolation 218ff, 228, 229ff

F

Favre averaging 306f
Favre variance 310
Fenimore mechanism 133, 140
filling and emptying method 141f, 197, 199ff
finite volume method 313f, 321
fish hook curve 67
flame
   -, counter-flow 360
   -, diffusion 357, 360, 380
   extinguishments 119
   front 65f, 113f, 187ff, 365ff, 371ff, 378ff
      combustion 365
      density 372
   model, coherent 372ff
   speed 65f, 119, 371, 377
      -, laminar 65f, 114f, 366
      -, turbulent 65f, 375f, 381
   stretching effects 360
   surface 365
      density model 372
   thickness 365, 367f, 377ff
   wrinkling 66
flamelet 360ff, 363ff
   model 362ff
   -, representative interactive (RIF) 364
   -, unsteady 364

flashboiling 331
flexibility 33
flow
   bench 145, 324
   coefficient 144ff, 212f, 226ff
   -, compression 321
   compressor 214
   -, dual phase 325
   equation 15, 198f, 209f, 213, 215
   function 144, 215, 209
   -, intake 322
   -, internal nozzle 80, 321, 325f, 335ff
   -, local homogeneous 338, 350f
   over of a cylinder 303
   shear stress 307ff
   structure, internal cylindrical 321
   -, tumble 321
   turbine 225ff
fluctuations, cyclical 67f
Fokker Planck equation 346
formulation, conservative 303
Fourier heat conduction equation 157
free
   enthalpy 44
   jets 312, 320
   -wheel condition 40
friction
   mean effective pressure 29, 257ff
   torque 236, 237f
fuel consumption 67f
   -, specific 30
fully variable
   mechanical valve train 95, 146
   valve train 95, 146, 170, 277f
function
   -, beta- 311
      -, incomplete 360
   -, double vibe 162ff
   -, flow 15, 198f, 209f, 213, 215
   -, Greens 343, 348, 370
   -, substitute combustion 262
furane 124

G

G-
   equation 375ff
      -, diffusive 378ff
   field 376
gas
   dynamic, one-dimensional 142, 201ff, 205
   equation

   -, ideal 154
   -, real 301
force 9
   -, ideal 13f, 178
jet theory 99
path 197ff
theory, kinetic 331
Greens function 343, 348, 370
gross reaction equation 56

H

HC-emission 122, 364
HCCI 70, 86f, 353, 362, 365
heat
   conduction equation, Fourier 157
   conductivity 198
   exchangers 251
   flux 308
   path 88
   transfer 107, 147ff, 156f, 255
      coeff. 148f, 151ff, 157, 191, 198, 255
      equation 151ff
heat release 58, 108f, 113, 357ff
heat release rate 84ff, 88ff, 95, 98ff, 160ff, 169ff, 354ff
   analysis 354
   -, double VIBE 162ff
      -, substitute 164f
   -, polygon-hyperbola 98, 163, 164f
      -,substitute 164f
   precalculatin 165ff
      -, Vibe 173ff, 270
   -, substitute 100, 142, 165ff
   -, total 91, 104, 160ff
   -, Vibe 98, 173ff, 270
      -, substitute 160ff, 274
heating value
   -, lower (lhv) 30, 54, 143
   -, mixture 30
   -, upper 30
high pressure process 25
Hiroyasu-Nagle-Strickland-model 363f
homogeneous
   charge compression ignition (HCCI) 70
   combustion 70, 86f
hybrid method 316
hydrocarbons
   -, polycyclic aromatic 121, 128f
   -, unburned 118ff, 364

I

ICAS model 352f
ideal gas 13f, 178
   equation 154
ignition 63f, 379
   angle 64
   -, auto 69, 81ff
   -, controlled auto- (CAI) 70f, 95f
   delay 88, 103, 108, 111f, 166f
   duration 64
   -, homogeneous charge compression 70
   model 114f
   performance 54
   -, space 71
   -, surface 69f
   timing 64
imperfect combustion 25f, 93f
impulse 203f
   equation 322
   -, law of conversation 144, 203f
incomplete combustion 25f, 93f
indicated
   mean effective pressure (imep) 28
   performance 28
   work 28
inertia force
   -, oscillating 10
   -, rotating 10
injection
   -, direct 61ff, 73ff, 89, 95, 186
   -, multi point 31, 60f, 63, 95ff
   nozzle 73ff, 78ff
   -, pre- 110f
   pump, distributor (DIP) 74
   speed 112
   systems 73ff
instability
   -, analysis of 338
   -, Kelvin-Helmholtz 338f
   -, Rayleigh-Taylor 339
intact core length 346
intake
   flow 322
   spread 146, 263, 279
   swirl 150, 191
integration methods 200ff
internal
   charging 35
   cooling 63
   energy 89, 177ff, 183ff, 300
      -, specific 300

   performance 28
ion hypotheses 128
isentrope 15
isentropic
   equivalent cross section 40
   outflow speed 17
isobar 15
isochore 15
isotherm 15
ITNFS-Function 374

K

k, ε-model 154, 307
Karman vortex street 303f
Kelvin-Helmholtz instability 338f
ketones 53
kinetic energy 301
knock limit 64
knocking 69ff, 88, 174ff, 241, 37ff
   combustion 69ff, 88, 174ff
   criterion 175ff
   -, engine 60
Kolmogorov scale 304
KPP theorem 368f, 371, 374

L

Lagrang
   -ian coordinates 298
   -ian model 347
   -e method 352
laminar flame speed 65f, 114f, 366
Langevin equation 341
Laplace operator 298
large eddy simulation (LES) 312
law
   -, 1st, of thermodynamics 12, 90, 143
   -, 2nd, of thermodynamics 14, 331
   -, angular-momentum conserv. 236f, 326f
   -, atomization 341
   -, Daltons 47
   of conversation 141
      of energy 141, 204f
      of impulse 144, 203f
      of mass 141
         action 46, 358
   of the wall, logarithmic 308
   of the wall, turbulent 307
   -, reflection 341
Lax-Wendroff method 206ff
lean combustion 71
length scale, turbulent 304f, 374

Index   397

limitation of the turbulent length scale 352
Liouville equation 331ff
load regulation 60ff
loss distribution 88, 92ff, 96f
losses
    -, charge changing 94f
    -, combustion 94
    -, wall heat 94
lower heating value (lhv) 30, 54, 143

M

Magnussen model 368ff
main (diffusion) combustion 85
mass
    balance 12, 88, 143, 188, 196ff
    -, fraction, element 302
mean effective pressure (mep) 29
    -, friction (fmep) 29, 257ff
    -, indicated (imep) 28
mechanical
    (super-) charging 35, 37ff
    efficiency 29
    valve train, fully variable 95, 146
medium pressure 20f, 29, 68f
mesh
    generation 320ff
    generator, automatic 320
    movement 320
    refinement 347
    resolution 312, 372
    structure, wall-adapted 320
    -, surface 321
    -, tetrahedral 321
    -, truncated cartesian 320
method
    -, filling and emptying 141f, 197, 299ff
    -, hybrid 316
    -, Lax-Wendroff 206ff
    -, operator split 319
    -, Runge-Kutta 200
    -, zero-dimensional 141
methods
    -, air-guided 62
    -, concept of composite 37
    -, integration 200ff
    -, numeric solution 205ff
    -controlled 338
    -, diesel engine 72ff
    formation 60ff, 80ff, 105f
    fraction 301f, 309, 359, 362, 380f
    heating value 30

-, inhomogeneous 301
model 1
    -building 1ff, 199f
    categories 98
    -, coherent flame 372ff
    -, component 178f
    -depth 1
    -, discrete droplet (DDM) 333
    -, eddy breakup 357
    -, entrainment 353
    -, flame surface density 372
    -, flamelet 362ff
    -, Hiroyasu-Nagle-Strickland 363f
    -, ICAS 352f
    -, ignition 114f
    -, k,ε- 154, 307
    -, Lagrangian 347
    -, Magnussen 368ff
    -, mixing time-scale 358
    -, pdf time-scale 359
    -, Shell 82, 361, 379
    -, single-equation Weller 378
    -, single-zone cylinder 142ff
    -, spray 326ff, 343
    -, WAVE 338
modeling the nozzle 335
models
    -, CMC 364
    -, combustion 2
    -, fuel 329
    -, multi-component 329
    -, multi-zone 142
    -, non-parametric 2
    -, parametric 2
    -, phenomenological 98ff, 160, 174f
    -, quasidimensional 98
    -, Reynolds stress 312
    -, total-process 2
    -, transported-pdf- 364f
    -, two-zone 176f, 187ff, 193ff, 196
    -, wall temperature 157ff
molecular
    size 50
    structure 50
mono-scroll turbine 293, 296
mixture
multi
    particle processes 331
    point injection 31, 60f, 63, 95ff
    -zone models 142

N

Navier-Stokes equation 201, 297ff
neural network 160, 173
Newtonian
    equation 158
    method 147, 205
nitrogen oxides 116f, 132ff
non-parametric models 2
normal combustion
NOx
    building 362
    calculation 193ff
nozzle
    flow, internal 80, 321, 325f, 335ff
    hole resolution 346
numeric solution methods 205ff
numerical fluid mechanics 313
Nußelt number 148, 153, 205

O

octane number 54f
one-dimensional gas dynamic 142, 201ff, 205
open comparative process 25ff
open, stationary flooded system 12ff
OHC system 136f
Ohnesorge number 338
oil cycle 245ff, 251ff, 256
oil pump 252f
operator split method 319
oscillating inertia force 10
outflow function 16
outflow speed
    -, isentropic 17

P

p, v diagram 32
packet
    approach 104
    speed 106
PAH 121, 128f
parametric models 2
parcel 333ff, 346ff, 365
    number 346
partial equilibrium 48ff, 137
particulate matter emission 127ff
pdf time-scale model 359
perfect combustion 26f
performance
    balance 39
    -, effective 29
    -, indicated 28

-, internal 28
phase space 331f
piezo elements 88
pipe branching 212ff
pirouette effect 323
piston
    acceleration 10
    path 8
    speed 8, 10
Poisson equation 299, 319
pollutant formation 98, 116ff, 381
polycyclic aromatic hydrocarbons 121, 128f
polygon-hyperbola hrr 98, 163, 164f
    -, substitute 100, 142, 165ff
Pope correction 311, 347, 351
port fuel injecting SI engine 186
positive displacement charger 224f
post-combustion 85
potential
    -, chemical 45
    equation (see Poisson)
Prandtl number 149, 203, 306
premix
    -ed combustion 84f, 189
    peak 103, 358
    -ture combustion 357, 365ff
pressure
    charging 36
    correction 90, 319
    -, mean effective (mep) 29
        -, friction (fmep) 29, 257ff
        -, indicated (imep) 28
    -, medium 18f, 28ff, 67
    -, ratio, critical 17, 199
    trace analysis 88ff
    turbocharging 38ff, 43
probability density function (pdf) 310
procedure 11f
process
    -, aerodynamic 336
    -, autoignition 82
    -, breakup 336
    -, combustion 11, 60ff, 65ff, 70ff, 86, 95ff
        -, low in swirl 324
    -, constant-pressure 21ff, 24f
    -, constant-volume 20f, 23f, 92
    -, high pressure 25
    -, multi particle 331
    -, open comparative 25ff
    -, propagation 81
    -, real engine 25ff

Index

simulation 3
   variable 301, 362, 364, 368, 378f
prompt NO 138ff
propagation process 81
pulse charging 37
pulse turbocharging 41ff
pump
   -, oil 252f
   -, water 249f

Q

quasi-steady-state 48ff, 136

R

random numbers 334f
rate coefficient 47
Rayleigh-Taylor instability 339
reaction
   coefficient 82
   enthalpy 46
   -, gross 56
   rate 47
real
   engine process 25ff
   gas equation 301
reciprocating engines 6ff
   -, assembly of 7
recirculation
   -, combustion chamber 71, 263
   -, exhaust gas 71f, 166, 261ff, 279ff, 290f
reconstruction of the torque band 269ff
register charging 35
regulation
   -, load 60ff
reinitialization 370, 377, 378f
residual gas, amounts of 89
resolution
   -, nozzle hole 346
   of the spray 345f
   -, statistical 346
resonance charging 35
Reynolds
   averaging 305
   number 149, 205, 303ff
      -, droplet-related 349
   stress 305
rotating inertia force 10
Runge-Kutta method 200

S

Sauter diameter 107, 338
scalars
   active 302
   formal 301
   passive 301
scattering techniques, Mie 337
scattering techniques, schlieren 337
scavenging pressure 279
scheme
   -, central difference 314ff
   -, explicit 317
   -, implicit 316f
   -, monotone numerical 315
   -, upwind 315f
Schmidt number, turbulent 305
shear
   stress speed 307
   tensor 306
Shell model 82, 361, 379
Seiliger cycle 22f
SI (spark ignition) engine 60ff
   -, direct injection 61ff
   -, port fuel injecting 186
simulation 2ff
   -, direct numerical 313
   -, Monte-Carlo 331
   process 3
   -, soot formation 363
   -, soot oxidation 363
   -, spray 325, 334, 352f
   -, transient 320
single
   droplet processes 327ff
   -equation Weller model 378
   -zone cylinder models 142ff
solitons 373, 385
solution methods, numeric 205ff
sonic speed 17
soot
   development 129f, 131f, 363f
   oxidation 132
   radiation 354
space ignition 71
specific
   fuel consumption 30
   internal energy 300
speed 105
   -, flame 65f, 119, 371, 377
      -, laminar 65f, 114f, 366
      -, turbulent 65f, 375f, 381

-, shear stress 307
spray
   angle 101
   collapse 350
   dispersion 80f
   dynamic 353
   equation 331
   front 83, 100
   -guided methods 62
   model 326ff, 343
   penetration 82, 103, 114, 337, 347
      graphs 343ff
   -, primary breakup of the 326
   -, resolution of the 345f
   simulation 325, 334, 352f
      -, rotation symmetrical 346
spread 146
   -, exhaust 146, 263, 279
   -, intake 146, 263, 279
staggered grids 322
state equation
   -, caloric 13, 178, 300
statistical convergence 356, 365
steady-state, quasi 48ff, 136
stoichiometric coefficient 44
storage-reduction catalytic converter 63
stratified charge 61, 380
   combustion 379
stress tensor 297
substantive derivative 298
substitute
   combustion function 262
   conduction coefficient 158
suck line
   -, compressor 380
   -, engine 281, 284
sum convection 297
surface
   ignition 69f
   tension 339
surge
   line 216f, 223, 265, 288
   protection 265f
swing-pipe charging 35
swirl 101, 119, 323f, 362
   -, cylinder-shaped 324
   -, intake 150, 191
      -, valve seat generated 323
   number 323f
      -, time development of the 323
   -, secondary 324

system
   -, closed 12ff
   -, open, stationary flooded 12ff

T

tangential duct 323
Taylor number 338
thermal
   boundary layer 299, 307ff, 320
   efficiency 17f, 20ff
   energy 26
   enthalpy 301
   NO 133
   state equation 13, 88f, 300
   time scale, turbulent 304f, 368
thermodynamics
   -, 1st law of 12, 90, 143
   -, 2nd law of 14, 331
   of the internal combustion engine 12ff
thermostat 250f
torque band
   -, reconstruction of the 269ff
total
   degree of conversion 27
   energy 301
   heat release rate 91, 104, 160ff
   process analysis 245ff
   process models 2
transparent engine 361
transported pdf methods 380
triple flame 380
tumble, counter 322
tumble flow 321
turbine
   -, flow 225ff
   geometry, variable (VTG) 231, 289
   -, mono-scroll 293, 296
   -, twin-scroll 232ff
turbocharger
   -, core equation for 236
   -, electrically assisted 289
turbocharging 35
   -, constant-pressure 38ff
turbulence 303, 320ff
   -, artificial 367, 380
   level 66
   model modification 312
   -, spatially homogeneous 305
   structure, anisotropies in the 312
turbulent
   diffusion 305, 348, 350

Index

dispersion 333, 341ff, 348, 350
dissipation 305f, 310
flame speed 65f, 375f, 381
kinetic energy 99, 305
law of the wall 307
length scale 304f, 374
Prandtl number 306
Schmidt number 305
time scale 304f, 368
velocity scale 304f
viscosity 305
twin-scroll turbine 232ff
two-phase
   behavior 365ff
   flow, locally homogeneous 352
two-point correlations 347
two-zone model 187ff, 196f
type number 227

U

unburned hydrocarbons 118ff, 364
unit injector system 76
unit pump system 75
upper heating value 30
upwind scheme 315f

V

valve lift function 324
valve seat cover 323
valve train
   -, electromechanical 146f, 156, 277f
   -, fully variable mechanical 95, 146
variable turbine geometry (VTG) 230, 289
velocity
   -, Bernoulli- 325f, 355
   -, burning 66, 371f
   coefficient 326
   scale 304f
Vibe
   function, double 162ff
   heat release rate 98, 172ff, 270
      -, double 162ff
      -, substitute 164f
      -, substitute 16ff, 274
   parameter 169ff
violation of causality 378
viscosity
   -, term of 299
   -, turbulent 305
volume 197
   work 14

volumetric efficiency 30
von-Karman constant 308
vortex street, Karman 303f
VTG (variable turbine geometry) 231, 289

W

wall
   distance 307
   film 341
      dynamics 341
   friction 299
   -guided methods 62
   heat
      flow 247
      losses 94
      transfer 354
   -, law of the, logarithmic 308
   -, law of the, turbulent 307
waste gate 289
water
   gas equilibrium 182
   pump 249f
WAVE model 338
Weber number 338
white noise 305
work, indicated 28

Z

Zeldovich mechanism 72, 132, 140, 193f, 362
zero-dimensional method 141
zero-speed line 218